LONDON MATHEMATICAL SOCIETY LECTURE NOTE SERIES

Managing Editor: Professor I.M. James,
Mathematical Institute, 24-29 St Giles, Oxford

1. General cohomology theory and K-theory, P.HILTON
4. Algebraic topology, J.F.ADAMS
5. Commutative algebra, J.T.KNIGHT
8. Integration and harmonic analysis on compact groups, R.E.EDWARDS
9. Elliptic functions and elliptic curves, P.DU VAL
10. Numerical ranges II, F.F.BONSALL & J.DUNCAN
11. New developments in topology, G.SEGAL (ed.)
12. Symposium on complex analysis, Canterbury, 1973, J.CLUNIE & W.K.HAYMAN (eds.)
13. Combinatorics: Proceedings of the British Combinatorial Conference 1973, T.P.McDONOUGH & V.C.MAVRON (eds.)
15. An introduction to topological groups, P.J.HIGGINS
16. Topics in finite groups, T.M.GAGEN
17. Differential germs and catastrophes, Th.BROCKER & L.LANDER
18. A geometric approach to homology theory, S.BUONCRISTIANO, C.P. BOURKE & B.J.SANDERSON
20. Sheaf theory, B.R.TENNISON
21. Automatic continuity of linear operators, A.M.SINCLAIR
23. Parallelisms of complete designs, P.J.CAMERON
24. The topology of Stiefel manifolds, I.M.JAMES
25. Lie groups and compact groups, J.F.PRICE
26. Transformation groups: Proceedings of the conference in the University of Newcastle-upon-Tyne, August 1976, C.KOSNIOWSKI
27. Skew field constructions, P.M.COHN
28. Brownian motion, Hardy spaces and bounded mean oscillations, K.E.PETERSEN
29. Pontryagin duality and the structure of locally compact Abelian groups, S.A.MORRIS
30. Interaction models, N.L.BIGGS
31. Continuous crossed products and type III von Neumann algebras, A.VAN DAELE
32. Uniform algebras and Jensen measures, T.W.GAMELIN
33. Permutation groups and combinatorial structures, N.L.BIGGS & A.T.WHITE
34. Representation theory of Lie groups, M.F. ATIYAH et al.
35. Trace ideals and their applications, B.SIMON
36. Homological group theory, C.T.C.WALL (ed.)
37. Partially ordered rings and semi-algebraic geometry, G.W.BRUMFIEL
38. Surveys in combinatorics, B.BOLLOBAS (ed.)
39. Affine sets and affine groups, D.G.NORTHCOTT
40. Introduction to Hp spaces, P.J.KOOSIS
41. Theory and applications of Hopf bifurcation, B.D.HASSARD, N.D.KAZARINOFF & Y-H.WAN
42. Topics in the theory of group presentations, D.L.JOHNSON
43. Graphs, codes and designs, P.J.CAMERON & J.H.VAN LINT
44. Z/2-homotopy theory, M.C.CRABB
45. Recursion theory: its generalisations and applications, F.R.DRAKE & S.S.WAINER (eds.)
46. p-adic analysis: a short course on recent work, N.KOBLITZ
47. Coding the Universe, A.BELLER, R.JENSEN & P.WELCH
48. Low-dimensional topology, R.BROWN & T.L.THICKSTUN (eds.)

49. Finite geometries and designs, P.CAMERON, J.W.P.HIRSCHFELD & D.R.HUGHES (eds.)
50. Commutator calculus and groups of homotopy classes, H.J.BAUES
51. Synthetic differential geometry, A. KOCK
52. Combinatorics, H.N.V.TEMPERLEY (ed.)
53. Singularity theory, V.I.ARNOLD
54. Markov processes and related problems of analysis, E.B.DYNKIN
55. Ordered permutation groups, A.M.W.GLASS
56. Journees arithmetiques 1980, J.V.ARMITAGE (ed.)
57. Techniques of geometric topology, R.A.FENN
58. Singularities of smooth functions and maps, J.MARTINET
59. Applicable differential geometry, F.A.E.PIRANI & M.CRAMPIN
60. Integrable systems, S.P.NOVIKOV et al.
61. The core model, A.DODD
62. Economics for mathematicians, J.W.S.CASSELS
63. Continuous semigroups in Banach algebras, A.M.SINCLAIR
64. Basic concepts of enriched category theory, G.M.KELLY
65. Several complex variables and complex manifolds I, M.J.FIELD
66. Several complex variables and complex manifolds II, M.J.FIELD
67. Classification problems in ergodic theory, W.PARRY & S.TUNCEL
68. Complex algebraic surfaces, A.BEAUVILLE
69. Representation theory, I.M.GELFAND et. al.
70. Stochastic differential equations on manifolds, K.D.ELWORTHY

London Mathematical Society Lecture Note Series : 54

Markov Processes and Related Problems of Analysis

E.B.DYNKIN
Professor of Mathematics
Cornell University

CAMBRIDGE UNIVERSITY PRESS

Cambridge

London New York New Rochelle

Melbourne Sydney

CAMBRIDGE UNIVERSITY PRESS
Cambridge, New York, Melbourne, Madrid, Cape Town, Singapore, São Paulo

Cambridge University Press
The Edinburgh Building, Cambridge CB2 8RU, UK

Published in the United States of America by Cambridge University Press, New York

www.cambridge.org
Information on this title: www.cambridge.org/9780521285124

First published 1982
Re-issued in this digitally printed version 2008

A catalogue record for this publication is available from the British Library

Library of Congress Catalogue Card Number: 81–38438

ISBN 978-0-521-28512-4 paperback

CONTENTS

Preface

Most of the papers compiled in this volume have been published in Uspekhi Matematicheskikh Nauk and translated into English in the Russian Mathematical Surveys. The core consists of the series [IV], [V], [VI], [VII] presenting a new approach to Markov processes (especially to the Martin boundary theory and the theory of duality) with the following distinctive features:

1. The general non-homogeneous theory precedes the homogeneous one. This is natural because non-homogeneous Markov processes are invariant with respect to all monotone transformations of time scale – a property which is destroyed in the homogeneous case by the introduction of an additional structure: a one-parameter semi-group of shifts. In homogeneous theory, the probabilistic picture is often obscured by the technique of Laplace transforms.

2. All the theory is invariant with respect to time reversion. We consider processes with random birth and death times and we use on equal terms the forward and backward transition probabilities, i.e., the conditional probability distributions of the future after t and of the past before t given the state at time t. (This is an alternative to introducing a pair of processes in duality defined on different sample spaces.)

3. The regularity properties of a process are formulated not in topological terms but in terms of behaviour of certain real-valued functions along almost all paths. Specifying a countable family of the base functions, we introduce a topology in the state space such that almost all paths have certain continuity properties. However this can be done in many different ways with different exceptional sets of paths. It is reminiscent of the situation with coordinate systems: there exist many of them and we have no reason to prefer any special one.

Two recent papers [VII] and [VIII] are closely related to the main series.

An earlier article [I] (its title is used as the title of the volume) presents the state of the theory of Markov processes in 1959. At this time the theory was in the process of extensive development and Markov processes attracted researchers around the world. The article is a report on the work done by a group of young mathematicians at Moscow University (almost all of them were in their twenties). A number of open problems and prospective directions have been mentioned in the article. Two of them: additive functionals of Markov processes and applications of Ito's stochastic differential equations to partial differential equations – became a major area of research in subsequent years. Three years later the progress was reported in monograph [1].

The boundary theory of Markov processes is one of the principal subjects of the volume. In [II] a boundary value problem with a directional derivative for the Laplace equation is studied. At that time the general theory had not been sufficiently developed and the first sections of [II] are devoted to adjustment of Martin's method.

A general boundary theory is presented in [IV] and [V]. It is based on a theorem concerning the decomposition of certain classes of measures into extreme elements. An improved version of this theorem with applications to a number of other problems is contained in [VIII]. The key role is played by a special type of sufficient statistics. Under minimal assumptions on a transition function, the corresponding entrance and exit spaces are evaluated in [IX] using a combination of the boundary theory and the ergodic theory.

The relation of the general boundary theory to Hunt's boundary theory for Markov chains can be easily seen in an earlier paper [III]. The main difference is in the way a Markov chain (X_t, P) with given transition probabilities is associated with an excessive measure μ. In Hunt's theory $\mu(x)$ is the expected number of hittings x by X_t during the life interval $[\alpha, \beta]$. In our approach $\mu(x) = P\{\alpha \leqslant t, X_t = x, t \leqslant \beta\}$. This modification makes possible the generalization presented in [IV] and [V].

Papers [VI] and [VII] are devoted to the problem of constructing Markov processes whose paths have certain regularity properties. The class of regular processes investigated in [VI] is close to the class of right processes introduced by Meyer and studied by Getoor [4]. The theory of Markov representations of stochastic systems developed in [VII] presents an alternative to the classical theory of duality due to Hunt, Kunita, Watanabe, Getoor, Sharp and others. The relation between both theories is discussed in [2]. Additive functionals of stochastic systems have been studied in [3]. Interesting results in spirit of [VII] have been obtained by Kuznecov [5], [6], [7], [8], and [9] and Mitro [10], [11].

For this edition the author has revised the entire text of the English translations. A few slips in the originals (some of them noticed by Kuznecov) have also been corrected.

References

[1] E. B. Dynkin, Markovski protsessy, Fizmatgiz, Moscow 1963.
Translation: Markov processes, Springer–Verlag, Berlin–Göttingen–Heidelberg 1965.

[2] E. B. Dynkin, On duality for Markov processes, In Stochastic Analysis, Avner Friedman and Mark Pinsky (editors), Academic Press, New York–San Francisco–London, 1978. 63–77.

[3] E. B. Dynkin, Markov systems and their additive functionals, Ann. Probab. 5 (1977), 653–677.

[4] R. K. Getoor, Markov processes: Ray processes and right processes, Springer–Verlag, Berlin–Heidelberg–New York 1975.
[5] S. E. Kuznecov, Construction of a regular split process, Teor. Verojatnost. Primenen. 22 (1977), 791–812.
[6] S. E. Kuznecov, Behaviour of excessive functions along the trajectories of a Markov process, Teor. Verojatnost. Primenen. 24 (1979), 486–502.
[7] S. E. Kuznecov, Construction of regular split processes (random interval), In "Multidimensional Statistical Analysis", Central Institute for Mathematics and Economics, Moscow, 1979, 123–164.
[8] S. E. Kuznecov, Criteria of regularity for Markov processes, In "Multidimensional Statistical Analysis", Central Institute for Mathematics and Economics, Moscow, 1979, 165–176.
[9] S. E. Kuznecov, Transformation of a Markov process by means of excessive functions, In "Multidimensional Statistical Analysis", Central Institute for Mathematics and Economics, Moscow, 1979, 177–186.
[10] J. B. Mitro, Dual Markov processes: Construction of a useful auxiliary process, Z. Wahrscheinlichkeitsth. verw. Gebiete, 47 (1979), 139–156.
[11] J. B. Mitro, Dual Markov functionals: Applications for a useful auxiliary process, Z, Wahrscheinlichkeitsth. verw. Gebiete 48 (1979), 97–114.

Roman figures refer to the articles collected in this volume (see the table of contents).

MARKOV PROCESSES AND RELATED PROBLEMS OF ANALYSIS[1]

1. The intimate connection between Markov processes and problems in analysis has been apparent ever since the theory of the former began to develop. It is not without reason that A. N. Kolmogorov's paper [39] (Russian translation [38]) of 1931, which is of fundamental importance in this domain, was entitled "On analytical methods in probability theory". The investigation of these connections also forms, to a large extent, the subject matter of A. Ya. Khinchin's book of 1933 on "Asymptotic laws of the theory of probability" [52] (Russian translation [51]).

In the fifties, and more particularly during the last five years, the theory of Markov processes entered a new period of intense growth. If previously the connections between probability theory and analysis were somewhat one-sided, probability theory applying results and methods of analysis, now the opposite tendency increasingly asserts itself, and probabilistic methods are applied to the solution of problems of analysis. Methods belonging to the theory of probability not only suggest a heuristic approach, but also, in many cases, yield rigourous proofs of analytic results. Applications of the methods of the theory

---

1 This paper is an expanded version of a survey read by the author at a meeting of the Moscow Mathematical Society held on October 20th, 1959, and devoted to the activities of the seminar directed by E. B. Dynkin at the University of Moscow.

of semigroups of linear operators have led to far-reaching advances in the classification of wide classes of Markov processes. New and deep connections between the theory of Markov processes and potential theory have been discovered. The foundations of the theory have been critically re-examined; the new concept of a strongly Markovian process has acquired a crucial importance in the whole theory of Markov processes. Intensive research on these lines is being done in the whole world, attracting many distinguished mathematicians: Feller, Doob, Hunt, Ray, Chung, Kac, and many others in the U.S.A.; Ito, Yosida, Maruyama, and their disciples in Japan; Kendall, Reuter, and others in England; Fortet in France. Soviet mathematicians are also taking an active part in this creative competition. A group of mathematicians cultivating the new approach in the theory of Markov processes is gathered in the seminar led by the present author at the University of Moscow. This survey covers the most important results obtained by members of the seminar from the academic year 1955–56 onwards, particular attention being paid to the latest results obtained during the last couple of years. Naturally, the work done by foreign mathematicians on the subjects studied by the seminar will also be discussed; however, in this respect, the survey cannot claim to be complete.

A short account of the indispensable general concepts of the theory of Markov processes is given in §1. A survey of the main lines of investigation followed by the seminar is given in §2–7. The concluding §8 contains information about the membership of the seminar and its history.

§1. Introduction

2. What is a Markov process? We shall begin with an important class of Markov processes, called diffusion processes, which describe a physical phenomenon known as *Brownian motion*. It is well known that particles of dye-stuff immersed in a liquid move chaotically, changing the direction of their motion all the time. This movement is due to collisions of the particles with molecules of the liquid. The first mathematical theory of Brownian motion was created by Einstein and Smoluchowski. In a contemporary form, due to A. N. Kolmogorov, this theory is shaped as follows: the main mathematical entity is a function $P(t, x, \Gamma)$, which represents the probability of a particle being in the set Γ after a length of time t from a moment when it was at the point x. Concerning the properties of this function, some assumptions are made from which it follows that

$$P(t, x, \Gamma) = \int_{\Gamma} p(t, x, y)\, dy,$$

where $p(t, x, y)$ is the fundamental solution of the parabolic equation

$$\frac{\partial p}{\partial t} = \sum a_{ij}(x) \frac{\partial^2 p}{\partial x_i \partial x_j} + \sum b_i(x) \frac{\partial p}{\partial x_i}. \tag{1}$$

This result makes it possible to apply the theory of differential equations to the solution of a variety of important problems on Brownian motion. On the other hand, many other, not less important, questions do not fit into this theory. For instance, one may want to know how quickly particles of dyestuff will be deposited on an absorbing screen; but it is impossible to solve this problem by means of the function $P(t, x, \Gamma)$ without making additional assumptions.

3. A more complete mathematical model of Brownian motion should give an account not only of probabilities involving one moment of time, but also of those which involve the whole course of the process; the whole trajectory x_t should be the object of the theory. The random character of the motion is expressed mathematically by the assumption that $x_t = x_t(\omega)$, where ω is an element of the set Ω, which is "the space of elementary events" on which a system of probabilistic measures P_x is given. The sets A on which $P_x(A)$ is defined are described as events associated with the process, and the value of $P_x(A)$ is interpreted as the probability of the event A under the condition that the motion began at the point x. In particular, one of the events associated with the process is $\{x_t \in \Gamma\}$. The probability $P_x\{x_t \in \Gamma\} = P(t, x, \Gamma)$ is called the transition function of the process; whereas in the first model it was regarded as the only mathematical characteristic of the model, it now occupies a subordinate position.

The mathematical entity we have arrived at is precisely a Markov process in the modern sense. It is a pair (x_t, P_x), where $x_t = x_t(\omega)$ is a function of $t \geqslant 0$ and of $\omega \in \Omega$, and P_x is a system of probability measures in the space Ω. The phase space to which the values of x_t belong is, in the case of Brownian motion, a domain of the three-dimensional space. In general, however, it is an arbitrary set E for which a system of "measurable subsets" has been defined. One essential condition has to be satisfied by the function x_t and the measure P_x: it is the Markovian principle that the future should be independent of the past when the present is known. More precisely, given the value of x_t, prospects of the future motion of the particle should not depend on its movement before the instant t.

4. The general scheme connecting Markov processes with analysis is based on the concept of a shift of a function defined over the phase space. The value of t being arbitrarily fixed, let $f(x)$ be a measurable function over the phase space. Then $f(x_t)$ is well defined over Ω; the integral of this function with respect to the measure P_x is precisely the value of the shifted function at the point x. Thus

$$T_t f(x) = M_x f(x_t) = \int_E P(t, x, dy) f(y),$$

where $P(t, x, \Gamma)$ is the transition function. The shift of a function is a linear operator. The Markovian principle implies $T_s T_t = T_{s+t}$ $(s, t \geqslant 0)$, so that the operators T_t form a semigroup. Consider now the "operator of an infinitely small shift".

$$Af(x) = \lim_{t \to 0} \frac{T_t f(x) - f(x)}{t}. \quad (2)$$

This operator is called the *infinitesimal* operator of the Markov process. If Af is defined, $T_t f$ is the solution of the equation

$$\frac{\partial u}{\partial t} = Au \quad (3)$$

which satisfies the initial condition $u(0, x) = f(x)$.

For the diffusion process (which describes Brownian motion), the infinitesimal operator is given by[1]

$$Af = \sum a_{ij}(x) \frac{\partial^2 f}{\partial x_i \partial x_j} + \sum b_i(x) \frac{\partial f}{\partial x_i} \quad (4)$$

(for every x, the numbers $a_{ij}(x)$ form a positive semi-definite matrix). In this case, (3) is essentially equivalent to (1).

In general, the transition function of a process can be regarded as the fundamental solution of equation (3). But having at our disposal a system of measures P_x allows us to create, within the framework of our theory, a much greater variety of constructions and transformations than that which would be possible within the pure theory of differential equations. For instance, in the formula defining the shift operator T_t, one can replace the constant t by a random time τ. The operators T_τ can no longer be expressed in terms of the transition function. Furthermore, by means of such operators, one can express, for instance, the solution of Dirichlet's problem for an arbitrary elliptic equation in an arbitrary domain.

§2. General problems in the theory of Markov processes

5. These problems concern the foundations of the theory of Markov processes and have mostly a set-theoretical character. I shall briefly discuss three problems of this kind.

The first concerns the nature of the trajectories of the process. Consider, for instance, a diffusion process whose infinitesimal operator is defined by (4). Are all the trajectories of the process continuous? This question is of crucial importance not only in the study of Brownian motion, but also in the qualitative solution of any analytical problem connected with the differential operator (4). What is the answer to this question? In the first place, the question is not quite correctly stated. The point is that the set of trajectories is not uniquely defined by the infinitesimal operator or by the transition function. Therefore, a more

1 Strictly speaking, the operator A is given by (4) for all functions with continuous partial derivatives up the second order. However, its domain contains some less regular functions as well. Thus the infinitesimal operator is an extension of (4).

correct statement of the question would be this: is there a Markov process admitting an infinitesimal operator (4), and such that all its trajectories are continuous? The answer is "Yes". In the theory of Markov processes, a diffusion process is always understood to be one admitting an operator (4), and having the property that all its trajectories are continuous. Such a process with continuous trajectories is essentially defined by the operator A.

A general criterion allowing one to decide to which transition functions there correspond Markov processes with continuous trajectories was given by E. B. Dynkin in 1952 [35]. A little later it was found independently by Kinney [82]. The condition is very simple; however, it is only sufficient, and not necessary. In 1957, L. V. Seregin [43] deduced another, slightly more complicated, but stronger, criterion which, in a wide class of cases, is not only sufficient, but also necessary for the continuity of the trajectories.

On the basis of the Dynkin–Kinney criterion, a simple sufficient condition for the continuity of the trajectories can be given in terms of the infinitesimal operator. The essential part of this condition requires that the operator should have a local character, i.e. that $Af(x_0)$ should not vary when the function is modified outside a neighbourhood of x_0. Clearly, this condition is satisfied by all differential operators. Hence one can obtain the proposition previously mentioned about the continuity of the trajectories of diffusion processes. In papers by Dynkin, Kinney, and Seregin, beside conditions for the continuity of the trajectories of a process, conditions are also deduced for their continuity to the right, and for their having no discontinuities of the second kind. With respect to the trajectories of a special class of Markov processes, very subtle conditions for their continuity to the right were found by A. A. Yushkevich [57], [58].

6. A second problem arising in the theory of Markov processes concerns the domain of validity of the Markovian principle of the future being independent of the past when the present is known.

Decompose a trajectory of the process into two parts: up to the time τ when a set Γ is first reached, and after this time. Assume that x_τ is known. Is the knowledge of the trajectory before the time τ relevant to the prediction of the motion after the time τ? Physical intuition requires a negative answer. However, such an answer does not follow from the definition of a Markov process, since this definition involves a fixed time t, and not a random time τ. We describe as strongly Markovian those Markov processes for which the principle that the future should be independent of the past when the present is known applies not only to a fixed time, but also to a well-defined class of random times τ.

The first paper in which the strongly Markovian property of some processes was rigourously stated and proved was written by J. L. Doob [64]; it contained a discussion of a special class of Markov processes with denumerable phase spaces. More general processes with denumerable sets of states were investigated from this view point by A. A. Yushkevich [56] in 1953.

The study of strongly Markovian processes as a class in its own right was initiated in papers by E. B. Dynkin [22], [23], [26] and E. B. Dynkin and A. A. Yushkevich [36][1] in 1955–56. Dynkin showed that, starting from the strongly Markovian property, and imposing definite conditions of continuity on the trajectories of the processes, one can compute their infinitesimal operators. In their joint paper, Dynkin and Yushkevich were the first to give a general definition of a strongly Markovian process; they constructed examples of Markov processes which are not strongly Markovian, and deduced sufficient conditions for a Markov process to be strongly Markovian.

These conditions require that, in an appropriate topology, all the trajectories should be continuous to the right and that shift operators should transform continuous bounded functions into continuous functions. It is easily seen that diffusion processes satisfy both these conditions, and, therefore, are strongly Markovian.

The strongly Markovian property was further analyzed in a succession of papers (A. A. Yushkevich [57], [59], R. Blumenthal [60], E. B. Dynkin [27], G. Maruyama [84], P. Lévy [83], D. Ray [87]) which appeared during the last three years. In his very interesting paper, D. Ray showed that under quite general assumptions it is possible to extend the phase space of a Markov process in such a way as to make the process strongly Markovian.

The basic results concerning strongly Markovian processes, as well as the essential criteria for the continuity of trajectories, are discussed in E. B. Dynkin's monograph [33].

7. The third set-theoretical problem which I wish to mention concerns the introduction of an intrinsic topology in the phase space. Topology plays no part in the definition of a Markov process. The phase space E is an arbitrary abstract set in which a system of measurable subsets has been singled out. However, we have seen that the study of Markov processes requires the introduction of some kind of topology in the phase space. In 1959, E. B. Dynkin [31] proposed the following definition of an intrinsic topology: The set Γ is called open if, for every $x \in \Gamma$ a trajectory starting from x remains in Γ for a positive length of time with probability 1[2]. Interesting properties of the intrinsic topology have been proved for standard processes, a wide class of processes which includes all processes important for applications, in particular, all diffusions. For such processes it is shown that a point x belongs to the intrinsic closure of a set Γ if, and only if, a particle starting from x visits Γ with probability 1 during an arbitrarily short time interval. A function $f(x)$ is continuous in the intrinsic topology if, and only if, with probability 1, $f(x_t)$, regarded as a function of t, is continuous to the right; the last result is due to I. V. Girsanov [15].

1 The year 1956 also brought three American papers (Hunt [78], Chung [62], Ray [86]) investigating independently of Yushkevich and Dynkin, various forms of the strongly Markovian property for some special classes of Markov processes.

2 In the case of Brownian motion, which corresponds to Laplace's operator, this topology coincides with the topology that was previously investigated by H. Cartan [61] and J. L. Doob [65], [66].

I. V. Girsanov [15] and M. G. Shur [54] proved that the continuity of a function $f(x)$ in the intrinsic topology is invariant with respect to shifts. The condition that shifts should transform every continuous function into a continuous function plays a very important part in the theory of Markov processes; we have seen that it is one of a set of two conditions which are sufficient to ensure the strongly Markovian character of a Markov process. Roughly speaking, this condition means that trajectories starting from neighbouring points behave in a similar way. The part played by this condition was first pointed out by W. Feller, and this is why processes satisfying it are described as Fellerian. By proving that, in the intrinsic topology, every standard process is Fellerian, Shur and Girsanov obtained a result of fundamental interest.

The intrinsic topology is by no means the only interesting topology for the theory of Markov processes. In particular, I. V. Girsanov [15] proposed an interesting definition of a uniform structure connected with a process. For Brownian motion, this structure is induced by the usual Euclidean metric.

§3. The form of an infinitesimal operator. Generalized diffusion processes

8. Which operators are infinitesimal operators of Markov processes? This question is of crucial importance in the theory of Markov processes, because, under very general assumptions, the transition function can be built up from the infinitesimal operator in a unique way (see [25]), and the knowledge of the transition function allows one to obtain some insight into the whole class of processes corresponding to this function. On the other hand, the answer to the same question affects the analyst too, since it tells him which operators are susceptible of treatment by probabilistic methods.

An important tool for the investigation of the forms of differential operators is supplied by a general theorem due to E. B. Dynkin [22], [26]. Its statement is expressed by the formula

$$Af(x) = \lim_{U \downarrow x} \frac{T_{\tau_U} f(x) - f(x)}{M_x \tau_U}. \tag{5}$$

Here U is a neighbourhood of x, and τ_U the time of the first exit from U; the passage to the limit takes place when U is contracted into x.

One notices the analogy between this formula and formula (2), which defines the infinitesimal operator. However, despite the outward similarity of these two formulae, the passage from one of them to the other is far from trivial. In this passage, essential use is made of the fact that the process in question is strongly Markovian and continuous to the right. Strictly speaking, the basic theorem, as stated above, applies only to Fellerian processes, but in a slightly different form it can be extended to non-Fellerian processes as well.

The main term in the right-hand side of (5) can be expressed as follows:

$$T_{\tau_U} f(x) = M_x f[x(\tau_U)] = \int_E f(y) \Pi_x(dy),$$

where Π_x denotes the probability function of the point reached by the particle at the time of its exit from U. If the process is continuous, this probability is concentrated on the boundary of U. In this case, the recipe, given by (5), for obtaining the infinitesimal operator A is strongly reminiscent of the well-known recipe for obtaining the Laplace operator from the operator of averaging over a sphere; the only difference lies in the fact that in the general case the mean is taken with respect to a non-uniform measure, instead of the uniform measure used for the Laplace operator. It is natural to describe any operator obtainable by means of such a recipe as a *generalized second-order elliptic differential operator*. The adoption of this term is further justified by the fact that such operators have many of the properties of conventional elliptic operators, and also by the fact that if, in a domain, the limit in the right-hand side of (5) exists for functions giving coordinates and pairwise products of coordinates, then, for any twice differentiable function, this limit can be expressed by a conventional (possibly degenerate) elliptic differential operator.

Thus, if a Fellerian process is continuous, its infinitesimal operator is a generalized second-order elliptic differential operator. In this sense, *any continuous Fellerian process can be regarded as a generalized diffusion process*.

9. The contention that any Fellerian process with continuous trajectories is a generalized diffusion process is further strengthened in the case of processes on the straight line.

In this case the differential operator is found to be a generalized second derivative

$$Af(x) = D_v D_u f(x), \tag{6}$$

where $D_u f$ is the derivative with respect to the function u, i.e. the limit of the ratio of the increment of f to that of u. In this formula, u and v are arbitrary increasing functions; u must be continuous (v can be discontinuous). If u and v are twice differentiable, the operator (6) can be expressed in the form of

$$Af(x) = a(x)\frac{d^2 f}{dx^2} + b(x)\frac{df}{dx}, \tag{6'}$$

so that we are confronted with an ordinary diffusion process. It can be said that the general continuous one-dimensional process given by (6) is a diffusion process for which the coefficients $a(x)$ and $b(x)$ are (in a certain sense) generalized functions.

Formula (6) is an easy consequence of (5), but it was first obtained by W. Feller ([73], see also [77]) by an entirely different, purely analytical method. Feller's remarkable contribution provided one of the main stimuli for the development of the theory of Markov processes in the last few years.

10. Recently, a series of results was obtained concerning the properties of continuous processes corresponding to operators given by (6). In particular, A. D. Venttsel' [5] proved that the transition function of such a process can be expressed in the form

$$P(t, x, \Gamma) = \int_{\Gamma} p(t, x, y)\, dv(y),$$

where $p(t, x, y) = p(t, y, x)$ is the fundamental solution of the equation

$$\frac{\partial p}{\partial t} = D_v D_u p.$$

An important step forward was also made by A. D. Venttsel' [4] in connection with another question, which was investigated in an outstanding paper by I. G. Petrovskii [41] as early as the thirties. The probabilistic meaning of the problem is this: to find the order of magnitude of the greatest deviation, from the initial position x, of the moving particle during a time t with $t \to 0$. By means of a suitable transformation of the x axis, the general case can be reduced to that of $u(x) \equiv x$ (in which case $b(x) \equiv 0$). I. G. Petrovskii gave a complete solution of the problem for the process corresponding to the operator $\frac{d^2}{dx^2}$. It is easy to show that with any coefficient $a(x)$, behaving in a regular way, the overall picture will remain the same. A. D. Venttsel' investigated the general case of processes ruled by the operator (6), and showed that, here, various qualitative departures from Petrovskii's results were possible. Roughly speaking, in the case treated by Petrovskii, the greatest deviation of the particle from its initial position during a length of time t is of the order of $t^{1/2}$. Venttsel' showed that if the coefficient $a(x)$ has a singularity at the initial point of a motion, this deviation can be of any order t^{α}, where $0 < \alpha < 1$. For points of discontinuity of $v(x)$, this deviation can be of the order of t.

More general differential operators

$$a(x)\frac{d^2 f}{dx^2} + b(x)\frac{df}{dx} + c(x) f, \text{ where } c \leqslant 0.$$

can also be fitted into probabilistic schemes. To such operators there correspond Markov processes terminating at random instants. As recently shown by E. B. Dynkin [90], the general form of such terminating continuous processes on the straight line is obtained by replacing (6) by

$$Af = D_v D_u (qf), \tag{7}$$

where q is a function which is convex from above.

11. In the process of deducing (6) from the basic theorem (5), one obtains simple expressions for the functions u and v in terms of entities characterizing the process from a probabilistic view point, and having an intuitive meaning. If the motion takes place on a segment, these entities are: the probability $p(x)$ of attaining the right-hand end before the left-hand end when starting from x and the mean time $m(x)$ elapsing between the start from x and the arrival at the boundary.

Analogous entities with intuitive meanings can also be introduced for many-dimensional processes. In a survey read before the III All-Union Congress of Mathematicians, E. B. Dynkin [29] proposed the following problem: for which classes of many-dimensional processes do these two entities completely determine the process? In 1958, I. V. Girsanov [14] showed that one of these classes is formed by multi-dimensional diffusion processes with non-degenerate matrices $a_{ij}(x)$.

12. I have dwelt mostly on processes with continuous trajectories. However, quite a few substantial results have been obtained concerning processes with discontinuous trajectories. In particular, E.B. Dynkin [28] gave a full classification of jump processes. These are processes in which the whole motion proceeds in jumps, i.e. in which the particle remains in its initial position for some positive time, then jumps to a new point, then to another point, and so on. The main difficulty encountered in the study of such processes was due to the possibility of transfinite sequences of jumps.

§4. Harmonic, subharmonic, and superharmonic functions associated with a Markov process

13. I recall the general definition of a superharmonic function. A function $f(x)$ is *superharmonic* if

(a) the mean of f over any sphere centred at x is smaller than, or equal to, $f(x)$;

(b) $f(x)$ is continuous from above.

If the first condition is satisfied when the phase "smaller than, or equal to" is replaced by "equal to", f is described as *harmonic*; if it is satisfied when the same phrase is replaced by "bigger than, or equal to", f is called subharmonic. To fix ideas I shall confine myself to superharmonic functions.

To every Markov process one can attach a class of functions which are analogous to ordinary superharmonic functions. Conditions (a) and (b) are replaced by

(a′) $T_\tau f(x) \leqslant f(x)$ (τ being the first exit time from an arbitrary open set);

(b′) $T_{\tau_n} f(x) \to f(x)$ if $P_x\{\tau_n \to 0\} = 1$.

In the case of a diffusion process, corresponding to the Laplace operator, conditions (a′) and (b′) are equivalent to (a) and (b). The proof is far from simple, but it is easy to explain why (a′) implies (a). Let τ be the first exit time from the solid sphere bounded by S. In view of the invariance of the Laplace operator with respect to all rotations, a particle starting from the centre of the sphere will have a uniform probability distribution on the sphere S when it arrives there. Hence $T_\tau f(x) = M_x f(x_\tau)$ coincides with the mean of S.

The concept of a superharmonic function associated with a Markov process was introduced in a paper by E. B. Dynkin [31] [1]. In the same paper it was proved that the class of nonnegative superharmonic functions associated with a Markov process was identical with that of excessive functions, introduced by G. A. Hunt [79] in 1957. Hunt defines an excessive function as one which is nonnegative and is not increased by shifts. More precisely, it should satisfy the conditions:

(a″) $T_t f(x) \leqslant f(x)$ for any t;

(b″) $T_t f(x) \to f(x)$ as $t \to 0$.

Hunt proved a series of theorems about excessive functions. Taking into account the connection between excessive and superharmonic functions, one can draw the following conclusions from Hunt's results:

1° Call a function f *smooth* if Af is defined. A smooth function is superharmonic if, and only if, $af \leqslant 0$.

2° Any nonnegative superharmonic function is the limit of a non-decreasing sequence of smooth superharmonic functions.

3° All superharmonic functions are continuous in the intrinsic topology.

14. A concept intimately connected with that of an excessive function is the concept of an excessive random variable, which emerged quite recently (see E. B. Dynkin [34], [90]). Its definition requires a minor digression.

A mapping $t \to t + h$ of the time transforms the random variable x_t into x_{t+h}. This transformation can be naturally extended to all the random variables associated with a Markov process. The accepted notation for this operator is θ_h.

A nonnegative random variable ξ will be called *excessive* if:

(α) $\theta_h \xi \leqslant \xi$ for any h;

(β) $\theta_h \xi \to \xi$ as $h \to 0$.

As an example of an excessive random variable, we can take

$$\xi = \int_0^\infty f(x_t)\,dt, \text{ where } f \geqslant 0.$$

Indeed,

$$\theta_h \xi = \int_0^\infty f(x_{t+h})\,dt = \int_h^\infty f(x_t)\,dt,$$

which shows that the conditions (α) and (β) are satisfied.

It is easy to verify that if ξ is an excessive random variable, then

$$f(x) = M_x \xi \tag{8}$$

---

[1] Harmonic, subharmonic, and superharmonic functions associated with discrete-time Markov processes were considered by W. Feller [74] and J. L. Doob [69]. For some special diffusion processes, these functions were discussed in papers by J. L. Doob [64]–[66].

is an excessive function. It follows from a remarkable theorem proved by V. A. Volkonskii [12], [13] that any bounded excessive function admits a representation of the type of (8). L. V. Seregin constructed examples of unbounded excessive functions which cannot be represented by (8). It would be very interesting to know how wide a class of unbounded excessive functions can be represented by this formula.

15. As a special case of (8), superharmonic functions can be represented in the form of potentials. The Newtonian potential in three dimensions can be defined by

$$V(x) = \int_E \frac{f(y)\,dy}{\| x - y \|} \tag{9}$$

the integral being taken with respect to the Lebesgue measure, $\| x \|$ denoting the length of the vector x, and $f(y)$ being a nonnegative function which can be regarded as the density of a distribution.

One finds that (9) can be re-written in the form

$$V(x) = M_x \int_0^\infty f(x_t)\,dt, \tag{10}$$

where x_t is a diffusion process in three dimensions, which corresponds to the Laplace operator. Formula (10) is meaningful for any Markov process and makes it possible to construct potentials for any such process. Note that (10) is a special case of the representation of a superharmonic function by means of an excessive random variable, ξ being chosen to be $\int_0^\infty f(x_t)\,dt$.

One more result obtained at our seminar and concerning superharmonic functions must be mentioned: It is a theorem by M. G. Shur [55], which shows that Riesz's classical theorem about the decomposition of any superharmonic function into a sum of a harmonic function and of a potential can be extended to the class of superharmonic functions associated with any non-degenerate diffusion process (i.e. with any elliptic second-order differential operator).

§5. Additive functionals and associated transformations of Markov process

16. A system of random variables φ_t is called an *additive functional* of a Markov process if

(1) $\theta_h \varphi_t = \varphi_{t+h} - \varphi_t$,

(2) φ_t is defined by the course of the process up to the time t.

As an example of an additive functional of a Markov process, we can take

$$\varphi_t = \int_0^t f(x_t)\,dt.$$

For Brownian motion, another important example of an additive functional is known; it is what we call a stochastic integral,

$$\varphi_t = \int_0^t b(x_t)\,dx_t$$

(on this subject see [91]).

If φ_t is an additive functional, the function

$$\alpha_t = e^{\varphi_t}$$

is called a *multiplicative functional.* It still satisfies (2), while the first condition is replaced by a similar one in which subtraction is replaced by division.

17[1]. The importance of multiplicative functionals for the theory of Markov processes can be seen from what follows: Put

$$\widetilde{P}(t, x, \Gamma) = \int_{x_t \in \Gamma} \alpha_t P_x(d\omega).$$

One finds that if a_t is a multiplicative functional for which $M_x a_t \leqslant 1$, there exists a Markov process with $\widetilde{P}(t, x, \Gamma)$ as its transition function. In general, this process may be a terminating one.

The following natural question arises: when can the process corresponding to the transition function $\widetilde{P}(t, x, \Gamma)$ be so chosen that it admits the same trajectories as the initial process? One finds that it suffices that there be a random variable ξ with the following properties:

(1) $\qquad a_t\theta_t\xi \leqslant \xi$ for every t,

(2) $\qquad M_x\xi = 1$ for every x,

(3) $\qquad \theta_h\xi \to \xi$ as $h \to 0$.

If this condition is satisfied, a process admitting $\widetilde{P}(t, x, \Gamma)$ as a transition function can be constructed with the same set of trajectories, but, in general, the trajectories will have to terminate at random.

We shall consider three important classes of such transformations.

18. If, in the condition (1), equality takes place, one will not have to terminate the trajectories. In this case, the passage to the new process is reduced to a transformation of the measures according to the formula

$$\widetilde{P}_x(A) = \int_A \xi P_x(d\omega). \tag{11}$$

This case occurs, in particular, if the initial process is a terminating diffusion

---

[1] The results discussed in Nos. 17–20 are contained in the papers [34] and [92] by E. B. Dynkin (see also [90]).

process, and the functional a_t is given by

$$\alpha_t = \exp\left[-\int_0^t b(x_u)dx_u - \frac{1}{2}\int_0^t b^2(x_u)dx_u\right].$$

ξ can be taken to be $\lim_{t \to \zeta} a_t$, , where ζ is the time of the termination of the process. The transformation (11) leads to the addition of the term $b(x)\frac{d}{dx}$ to the infinitesimal operator of the process. In the many-dimensional case, by means of a similar construction, one can add an arbitrary first-order differential operator to the infinitesimal operator of any non-degenerate diffusion process (see [16], and also [42], where further references can be found).

19. A second important example of a transformation of a process arises if the functional a_t satisfies $a_t \leqslant 1$ for every t. Then the conditions (1)–(3) are satisfied with $\xi = 1$. This operation on processes was discussed at an earlier stage by E. B. Dynkin [33] (see also [32]). Processes thus obtained are described as *sub-processes* of the initial process.

To get a sub-process, it suffices to terminate the trajectory of the initial process with a certain probability distribution, a_t being the probability that this termination will take place after the time t.

20. The third important example of a transformation of a process is connected with excessive random variables. Let ξ be any excessive random variable, and $f(x) = M_x \xi$ the corresponding excessive function. Then

$$\alpha_t = \frac{f(x_t)}{f(x_0)}$$

is a multiplicative functional, and the couple $a_t, \frac{\xi}{f(x_0)}$ satisfies the conditions (1)–(3). We obtain an interesting transformation of the process, which transforms the infinitesimal operator according to the formula

$$\widetilde{A}g = \frac{1}{f}A(fg).$$

This transformation had not been discussed previously.

21. An important transformation based on additive functionals does not fit the scheme discussed above. This is the random time change.

Let $\tau(t)$ be a monotonically increasing random function. We transform the function x_t into $\widetilde{y}_t = x_{\tau(t)}$. The transformed process is Markovian if $\tau(t)$ is the inverse function to an additive functional. (The proof of this result requires some restrictions on the initial process (x_t, P_x)).

The random time change in Markov processes was first discussed in a paper by V. A. Volkonskii [9], in which it was proved, in particular, that all the one-dimensional continuous processes can be obtained from the simplest diffusion process associated with the operator $\frac{d^2}{dx^2}$ by means of a monotonic transformation of the x axis and of a random time change. It should be noted that the random time change which corresponds to the additive functional $\varphi_t = \int_0^t f(x_u)du$, leads to the multiplication of the infinitesimal operator of the process by $f(x)^{-1}$.

22. It seems to me that what has been said shows clearly the importance of additive and multiplicative functionals for the theory of Markov processes. Hence the great interest of the problem of finding all the additive functionals of any given process. V. A. Volkonskii [13] succeeded in finding all the non-negative additive functionals of Brownian motion in one dimension. However, if we either do away with the condition that the functional should be non-negative, or pass to several dimensions, we are confronted with a problem which is still unsolved. We feel that this is one of the most topical problems in the theory of Markov processes.

The complex of problems connected with transformations of Markov processes is of great interest, since, here, we face a new calculus which enriches our store of analytic tools.

§6. Stochastic integral equations

23. Let x_t be a process corresponding to the operator $\frac{d^2}{dx^2}$, and $\widetilde{x}_t$ a process corresponding to the operator

$$L = \frac{1}{2}\sigma^2(x)\frac{d}{dx} + m(x)\frac{d}{dx}.$$

As shown by K. Ito [80], [81], the processes x_t and $\widetilde{x}_t$ can be so chosen as to be connected by the relation

$$\widetilde{x}_t = \widetilde{x}_0 + \int_0^t m(\widetilde{x}_t)\, dt + \int_0^t \sigma(\widetilde{x}_t)\, dx_t. \tag{13}$$

The right-hand side contains a stochastic integral which already appeared in section 16. Equation (13) admits a simple physical interpretation. When transcribed in the form of

$$d\widetilde{x}_t = m(\widetilde{x}_t)\, dt + \sigma(\widetilde{x}_t)\, dx_t, \tag{14}$$

it means that when the position $\widetilde{x}_t = x$ of the particle is known, the motion during a short time interval will be composed essentially of a determinate motion with velocity $m(x)$, and of a diffusion (corresponding to the operator $\frac{d^2}{dx^2}$) with a coefficient equal to $\sigma(x)$.

Formula (13) can be regarded as an integral equation allowing us to express

$\tilde{x}_t$ in terms of x_t. This equation can be solved by applying the method of successive approximations. There is an analogous integral equation for multidimensional processes. Thus we have a most convenient analytic tool for the reduction of problems concerning the general elliptic second-order differential operator to problems on the Laplace operator. Heuristically speaking, one can say, for instance, that Ito's method allows us to replace the study of the partial differential equations for the transition probabilities of a process by that of the ordinary differential equations for its trajectories. It is true that these differential equations are stochastic, but they can be approached in the same way as conventional differential equations, viz. by reducing them to integral equations, which can be solved by successive approximations.

24. The method of Ito's stochastic differential equations has been successfully applied to the solution of a number of concrete analytic and probabilistic problems.

Using this method, I. V. Girsanov [18] showed that, under mild assumptions of non-degeneracy for the operator (4), the corresponding diffusion process has the property that shifts $T_t (t > 0)$ transform all bounded Borel-measurable functions into continuous functions. This is a strengthening of Feller's condition, discussed at the end of §2, and we describe processes having this property as *strongly Fellerian.* The concept of a strongly Fellerian process was introduced and discussed by I. V. Girsanov in [17].

A. V. Skorokhod [88], [89] has studied boundary problems for one-dimensional diffusion processes by means of the method of stochastic differential equations.

M. I. Freidlin [44] applied this method to the investigation of boundary problems for elliptic equations which degenerate in the interior of the domain. It should be pointed out that, in contrast with the various methods of the pure theory of differential equations, Ito's method is completely insensitive to the degeneration of the differential operator. This circumstance allowed M. I. Freidlin to make substantial progress in a problem which, until then, had been fairly intensively studied by methods of classical analysis.

Another remarkable advantage of Ito's stochastic equations resides in the fact that their application is not made appreciably more complicated by an increase in the number of dimensions. The passage from one to n dimensions is almost automatic, and K. Dambis [19] recently showed that the passage to infinitely many dimensions is almost equally simple. The stochastic differential equations which he constructed for diffusion processes in the Hilbert space make it possible to study elliptic and parabolic differential equations in functions of infinitely many arguments. This theory may reveal itself as very comprehensive, but so far only the first steps have been made.

§7. Boundary problems in the theory of differential equations and the asymptotic behaviour of trajectories

25. In general, the specification of an analytic formula of the type of (4) does not fully define the infinitesimal operator of the process. The domain of functions for which an infinitesimal operator is defined is an essential part of the concept of such an operator. Apart from some well-defined requirements of regularity, the domain in question is determined by boundary conditions. Generally speaking, boundary conditions can be imposed in various ways, and the investigation of all the possible types of boundary conditions is an important problem. One finds that every type of boundary condition is associated with a well-defined type of asymptotic behaviour of the trajectories of the process as they approach the boundary.

W. Feller [70], [71], [76] [1] described all the possible types of boundary conditions for one-dimensional diffusion and generalized diffusion processes, covering all the cases in which the differential operator degenerates in the neighbourhood of the boundary. The many-dimensional problem is substantially more difficult. Much work on it had been done by specialists in differential equations; now substantial progress has been made in this domain by means of probabilistic methods. A. D. Venttsel' [3], [6] studied the most general forms of boundary conditions for non-degenerating elliptic equations when the boundary is regular, and he found new types of such conditions, which had never been discussed previously in the theory of differential equations.

Of foreign work done on boundary problems during the last few years, one must mention a series of papers by J. L. Doob [67], [68] dealing with the first boundary problem in its most general setting.

R. Z. Khas'minskii [46] studied the first boundary problem for elliptic equations degenerating on the boundary of the domain, and was able to obtain results which were much more general and complete than those previously published by M. V. Keldysh, M. I. Vishik, and others.

As previously mentioned in §6, M. I. Freidlin [44] investigated the first boundary problem for equations which degenerate in the interior of the domain.

Beside boundary problems, other asymptotic problems are also of interest in the theory of differential equations; the question how a solution $u(t, x)$ of a parabolic differential equation behaves when $t \to \infty$ is an example of such a problem. In probability theory, propositions on this kind of behaviour are called *ergodic theorems*; they can be studied by investigating the asymptotic behaviour of the trajectories of the process for $t \to \infty$. The case of one-dimensional generalized Brownian motion was investigated by Maruyama and

[1] The first of Feller's papers contains a substantial gap, first filled by A. D. Venttsel' in [2].

Tanaka [85]. For the many-dimensional case, interesting results were obtained by R. Z. Khas'minskii [49].

§8. Concluding remarks

26. In this brief survey, I have been unable to discuss even briefly quite a number of papers written by members of the seminar. I shall enumerate some of these contributions: a paper by F. I. Karpelevich, V. H. Tutubalin, and M. G. Shur [37] on the connection between Brownian motion in the Lobachevskii plane and the physical theory of waveguides; a paper by M. G. Shur on ergodic properties of Markov chains [53]; a paper by R. Z. Khas'minskii on limiting distributions of additive functionals of a Markov process [50], and another one [47] by him about the positive solutions of the equation $Af(x) + c(x) f(x) = 0$ (in this paper it is shown, in particular, that the smallest eigenvalue of an elliptic operator has properties of stability with respect to some considerable changes in the domain); papers by C. S. Leung [40] and A. D. Venttsel' [7] on conditional Markov processes; etc.

27. Finally, I should like to say a few words about the history of our seminar and its composition. During the academic year 1957–58, the seminar on Markov processes was detached from the general seminar or probability theory which began its work under the direction of the present speaker at the Moscow State University during the academic year 1954–55. However, it should be noted that subjects connected with new parts of the theory of Markov processes determined the basic tendencies of the general seminar as early as 1955. A. D. Venttsel', V. A. Volkonskii, I. V. Girsanov, E. B. Dynkin, L. V. Seregin, V. Tutubalin, M. I. Freidlin, R. Z. Khas'minskii, M. G. Shur, and A. A. Yushkevich were permanent and active members of the seminar. Some mathematicians, without being permanent members, obtained a series of interesting results in connection with problems which arose at the seminar, and delivered there lectures which were followed by lively discussions. These mathematicians were: A. V. Skorokhod, who completed his graduate study at the University of Moscow in 1957, and who now works at the University of Kiev; and Yu. V. Blagoveshchenskii, who works in the United Institute for Nuclear Research in Dubno. From 1958 onwards, our Chinese colleague C. S. Leung has been taking an active part in the Seminar.

During the academic year 1958–1959, ten students wrote their dissertations within the framework of the Seminar. Many of these dissertations contain independent results which are being prepared for publication.

The following list of references shows all the work done at the Seminar on Markov processes at the University of Moscow, independently of its being, or not being, mentioned in the present survey. Other papers are included in the list only if they are quoted in this article.

Received by the editors on October 29, 1959.

References

[1] Yu. V. Blagoveshchenskii, On diffusion processes with a small dispersion (Preprint).
[2] A. D. Venttsel', Operator semigroups corresponding to the generalized second-order differential operator, Dokl. Akad, Nauk SSSR, 111, No. 2 (1956), 269–292.
[3] A. D. Venttsel', On boundary conditions for multi-dimensional diffusion processes, Teorya Veroyatnost. i Premenen., **4**:2 (1959), 172–185.
[4] A. D. Venttsel', Local behaviour of trajectories of diffusion processes, Proceedings of All-Union Conference on Probability and Math. Statistics, Erevan 1960, 236–238.
[5] A. D. Venttsel', On the density of transition probabilities in a one-dimensional diffusion process, Teorya Veroyatnost. i Primen. **6**:4 (1961), 439–446.
[6] A. D. Venttsel', General boundary problems associated with diffusion processes, Uspehi Mat. Nauk, **15**:2 (1960), 202–204.
[7] A. D. Venttsel', Conditional Markov processes (Preprint).
[8] V. A. Volkonskii, A multidimensional limit theorem for homogeneous Markov chains with a denumerable set of states, Teorya Veroyatnost. i Primenen., **2**:2 (1957), 230–255.
[9] V. A. Volkonskii, Random transformations of time in strongly Markovian processes, Teorya Veroyatnost. i Primenen. **3**:3 (1958), 332–350.
[10] V. A. Volkonskii, Continuous one-dimensional Markov processes and additive functionals defined on them, Teorya Veroyatnost. i Primenen. **4**:2 (1959), 208–211.
[11] V. A. Volkonskii, The construction of non-homogeneous Markov processes with the help of a random transformation of time, Teorya Veroyatnost. i Primen. **6**:1 (1961), 47–56.
[12] V. A. Volkonskii, Additive functionals of Markov processes, Dokl. Akad. Nauk SSSR, **127**, No. 4 (1959), 735–738.
[13] V. A. Volkonskii, Additive functionals of Markov processes, Trudy Moskov. Mat. Obshch., **9** (1960), 143–189.
[14] I. V. Girsanov, A property of non-degenerate diffusion processes, Teorya Veroyatnost. i Primen. **4**:3 (1959), 355–361.
[15] I. V. Girsanov, On some topologies associated with a Markov process Dokl. Akad. Nauk SSR, **129**, No. 3 (1959), 488–491.
[16] I. V. Girsanov, On the transformation of a class of stochastic processes by means of an absolutely continuous change of measure, Teorya Veroyatnost. i Primen. **5**:3 (1960), 314–330.
[17] I. V. Girsanov, Strongly Fellerian processes 1, Teorya Veroyatnost. i Primen., **5**:1 (1960), 7–28.
[18] I. V. Girsanov, Strongly Fellerian processes 2, (1960). (Preprint).
[19] K. E. Dambis, Stochastic integral equations in the Hilbert space, Degree dissertation, Moscow State University, 1959.
[20] E. B. Dynkin, New analytic methods in the theory of Markovian stochastic processes, Vestnik Leningrad. Univ. **11** (1955), 247–266.
[21] E. B. Dynkin, Functionals of trajectories of Markovian stochastic processes, Dokl. Akad. Nauk SSSR **104** (1955), 691–694.
[22] E. B. Dynkin, Infinitesimal operators of Markovian stochastic processes, Dokl. Akad. Nauk SSSR, **105** (1955), 206–209.

[23] E. B. Dynkin, Continuous one-dimensional Markov processes, Dokl. Akad. Nauk SSSR, **105** (1955), 405–408.
[24] E. B. Dynkin, Markov processes with continuous time. Proceedings of the III All-Soviet Mathematical Congress, vol. 2, Moscow (1956), 44–47.
[25] E. B. Dynkin, Markov processes and semigroups of operators, Teorya Veroyatnost. i Primen. **1**:1 (1956), 25–37.
[26] E. B. Dynkin, Infinitesimal operators of Markov processes, Teorya Veroyatnost. i Primen. **1**:1 (1956), 38–60.
[27] E. B. Dynkin, Inhomogeneous strongly Markovian processes, Dokl. Akad. Nauk SSSR, **113** (1957), 261–263.
[28] E. B. Dynkin, Markovian jump processes, Teorya Veroyatnost. i Primen. **3**:1 (1958), 41–60.
[29] E. B. Dynkin, New methods in the theory of Markov processes, Proceedings of the III All-Soviet Mathematical Congress, vol. 3, Moscow (1958), 334–342.
[30] E. B. Dynkin, One-dimensional continuous strongly Markovian processes, Teorya Veroyatnost. i Primen. **4**:1 (1959), 3–54.
[31] E. B. Dynkin, Intrinsic topology and excessive functions associated with a Markov process, Dokl. Akad. Nauk SSSR, **127**, No. 1 (1959), 17–19.
[32] E. B. Dynkin, Markov processes and their sub-processes, Izv. Akad. Nauk Armyan, SSR (1960).
[33] E. B. Dynkin, Osnovaniya teorii markovskikh protsesov (The foundations of the theory of Markov processes), Moscow, 1959.
[34] E. B. Dynkin, Transformations of Markov processes associated with additive functionals, Proceedings of the Fourth Berkeley Symposium on Mathematical Statistics and Probability. (1960).
[35] E. B. Dynkin, Criteria for continuity and for absence of discontinuities of the second kind in trajectories of a Markovian random process, Izv. Akad. Nauk SSSR Ser. Mat. **16** (1952), 563–572.
[36] E. B. Dynkin and A. A. Yushkevich, Strongly Markovian processes, Teorya Veroyatnost. i Primen. **1**:1 (1956), 149–155.
[37] F. I. Karpelevich, V. I. Tutubalin, and M. G. Shur, Limit theorems for the composition of distributions in the Lobachevskii plane and space, Teorya Veroyat. i Primen. **4**:4 (1959), 432–436.
[38] A. N. Kolmogorov, Analytic methods in probability theory, Uspehi Mat. Nauk, V (1938), 5–41.
[39] A. Kolmogorov, Über die analytischen Methoden in der Wahrscheinlichkeitsrechnung, Math. Ann., **104** (1931), 415–458.
[40] C. S. Leung, Conditional Markov processes, Teorya Veroyatnost. i Primen., **5**:2 (1960), 227–228.
[41] I. Petrovsky, Über das Irrfahrtproblem, Math. Ann. **109** (1934), 425–444.
[42] Yu. V. Prokhorov, Convergence of stochastic processes and limit theorems in probability theory, Teorya Veroyatnost. i Primen. **1**:2 (1956), 177–238.
[43] L. V. Seregin, Conditions for the continuity of stochastic processes, Teorya Veroyatnost. i Primen. **6**:1 (1961), 3–30.
[44] M. I. Freidlin, The first boundary problem for degenerate elliptic equations (Preprint)
[45] R. Z. Khas'mynskii, The probability distribution of functionals of trajectories of a random process of the diffusion type, Dokl. Akad. Nauk SSSR **104**, No. 1 (1955), 22–25.

[46] R. Z. Khas'mynskii, Diffusion processes and elliptic differential equations degenerating on the boundary of the domain, Teorya Veroyatnost. i Primen. **3**:4 (1958), 430–451.
[47] R. Z. Khas'mynskii, Positive solutions of the equation $\mathfrak{A}u + vu = 0$, Teorya Veroyatnost. i Primen. **4**:3 (1959), 332–341.
[48] R. Z. Khas'mynskii, Ergodic properties of diffusion processes and the stabilization of solutions of parabolic equations (Preprint).
[49] R. Z. Khas'mynskii, Ergodic properties of recurrent diffusion processes and the stabilization of solutions of the Cauchy problem, Teorya Veroyatnost. i Primen. **5**:2 (1960), 196–214.
[50] R. Z. Khas'mynskii, Limiting distributions of sums of conditionally independent random variables associated with a Markov chain, Teorya Veroyatnost. i Primen. **6**:1 (1961), 119–125.
[51] A. Ya. Khinchin, Asimptoticheskie zakony teorii veroyatnostei (Asymptotic laws of the theory of probability) Moscow–Leningrad, ONTI (1936).
[52] A. Khintchine, Asymptotische Gesetze der Wahrscheinlichkeitsrechnung, Berlin, 1933.
[53] M. G. Shur, Ergodic properties of Markov chains which are invariant on homogeneous spaces, Teorya Veroyatnost. i Primen. **3**:2 (1958), 137–152.
[54] M. G. Shur, On the Feller property of Markov processes, Dokl. Akad. Nauk SSSR, **129**, No. 6 (1959), 1250–1253.
[55] M. G. Shur, Harmonic and superharmonic functions associated with diffusion processes, Sibirskii Mat. Sb **1** (1960), 277–296.
[56] A. A. Yushkevich, The differentiability of transition probabilities of a homogeneous Markov process with a denumerable number of states, Degree thesis, Moscow State University, 1953; Uchenye Zapiski Moskovskogo Gosundarstvennogo Universiteta **9**, No. 186 (1959), 141–159.
[57] A. A. Yushkevich, Strongly Markovian processes, Teorya Veroyatnost. i Primen. **2**:2 (1957), 187–213.
[58] A. A. Yushkevich, Some properties of Markov processes with denumerable numbers of states, Proceed. All-Union Conference on Probability and Math. Statistics, Erevan 1960, 239–246.
[59] A. A. Yushkevich, On the definition of a strongly Markovian process, Teorya Veroyatnost. i Primen. **5**:2 (1960), 237–243.
[60] R. Blumenthal, An extended Markoff Property, Trans. Amer. Math. Soc. **85** (1957), 52–72.
[61] H. Cartan, Théorie générale du balayage en potentiel newtonien, Ann. Univ. Grenoble, Sect. Sci. Math. Phys. (N.S.) **22** (1946), 221–260.
[62] K. L. Chung, Foundations of the theory of continuous parameter Markov chains, Proc. Third Berkeley Symposium Math. Stat. and Prob., Vol. 2 (1956), 30–40.
[63] K. L. Chung, On a basic property of Markov chains, Ann. of Math. **68**:1 (1958), 126–149.
[64] J. L. Doob, Markoff chains – denumerable case, Trans. Amer. Math. Soc. **58**:3 (1945), 455–473.
[65] J. L. Doob, Semimartingales and subharmonic functions, Trans. Amer. Math. Soc. **77**:1 (1954), 86–121.
[66] J. L. Doob, A probability approach to the heat equation, Trans. Amer. Math. Soc. **80**:1 (1955), 216–280.

[67] J. L. Doob, Probability Methods applied to the first boundary value problem, Proc. Third Berkeley Symposium Math. Stat. and Prob., Vol. 2 (1956), 49–80.
[68] J. L. Doob, Probability theory and the first boundary value problem, Illinois J. Math., 2:1 (1958), 19–36.
[69] J. L. Doob, Discrete potential theory and boundaries, J. Math. Mech. 8 (1956), 433–458.
[70] W. Feller, The parabolic differential equations and the associated semi-groups of transformations, Ann. of Math. **55** (1952), 468–519 (Translation in Matematika **1**:4 (1957), 105–153).
[71] W. Feller, Diffusion processes in one dimension, Trans. Amer. Math. Soc. **77**:1 (1954), 1–31 (Translation in Matematika **2**:2 (1958), 119–146).
[72] W. Feller, The general diffusion operator and positivity preserving semi-groups in one dimension, Ann. of Math. **60**:3 (1954), 417–436.
[73] W. Feller, On second order differential operators, Ann. of Math. **61**:1 (1955), 90–105.
[74] W. Feller, Boundaries induced by nonnegative matrices, Trans. Amer. Math. Soc. **83**:1 (1956), 19–54.
[75] W. Feller, On boundaries and lateral conditions for the Kolmogorov differential equations, Ann. of Math. **65**:3 (1957), 527–570.
[76] W. Feller, Generalized second order differential operators and their lateral conditions, Illinois J. Math. **1**:4 (1957), 459–504.
[77] W. Feller, On the intrinsic form for second order differential operators, Illinois J. Math. **2**:1 (1958), 1–18.
[78] G. A. Hunt, Some theorems concerning Brownian motion, Trans. Amer. Math. Soc. **81**:2 (1956), 294–319.
[79] G. A. Hunt, Markov processes and potentials I–III, Illinois J. Math. **1** (1957), 44–93; **1** (1957), 316–369; **2** (1958), 151–213.
[80] K. Ito, On a stochastic integral equation, Proc. Japan. Acad. **22** (1946), 43–35.
[81] K. Ito, On stochastic differential equations, Mem. Amer. Math. Soc. **4** (1951), 1–51 (Translation in Matematika **1**:1 (1957), 78–116)
[82] J. R. Kinney, Continuity properties of sample functions of Markov processes, Trans. Amer. Math. Soc. **74** (1953), 280–302.
[83] P. Lévy, Processus markoviens et stationaires. Cas dénombrable, Ann. Inst. H. Poincaré **16**:1 (1958), 7–25.
[84] G. Maruyama, On the strong Markov property, Mem. Fac. Sci. Kyusyu Univ. Ser. A**13**:1 (1959), 17–29.
[85] G. Maruyama and H. Tanaka, Some properties of one dimensional diffusion processes, Mem. Fac. Sci. Kyusyu Univ. Ser. A**11**:2 (1957), 117–141.
[86] D. Ray, Stationary Markov processes with continuous paths, Trans. Amer. Math. Soc. 82:2 (1956), 452–493.
[87] D. Ray, Resolvents, transition functions and strongly Markovian processes, Ann. of Math. **70**:1 (1959), 43–78.
[88] A. V. Skorokhod, Stochastic equations for diffusion processes with boundaries, Teroya Veroyatnost. i Primen. **6** (1961), 267–298.
[89] A. V. Skorokhod, The diffusion process with retarded reflection on the boundary, (Preprint).
[90] E. B. Dynkin, Markovskie protsessy (Markov processes), Forthcoming book.

[91] E. B. Dynkin, Additive functionals of a Wiener process defined by stochastic integrals, Teorya Veroyatnost. i Primen. **5** (1960), 441–452.
[92] E. B. Dynkin, Some transformations of Markov processes, Dokl. Akad. Nauk SSSR, **133**:2 (1960), 269–272.

Translated by S. K. Zaremba

MARTIN BOUNDARIES AND NON-NEGATIVE SOLUTIONS OF A BOUNDARY VALUE PROBLEM WITH A DIRECTIONAL DERIVATIVE

Contents

Introduction

1. In 1941 R. S. Martin in the paper [24] proposed a method of characterizing all positive harmonic functions defined in an arbitrary region of Euclidean l-space. The author, a young mathematician at the Illinois University, died shortly after his paper appeared. The importance of his results was not immediately appreciated. Apparently M. Brelot ([16], [17]) was the first to turn his attention to Martin's ideas, Martin's results were further elucidated by papers of Choquet concerning convex cones in linear topological spaces, ([19], [20]; see also [21] where there is an extensive bibliography). After the appearance of Doob's paper [22], Martin's ideas attracted the attention of specialists in probability theory. In recent years the number of works dealing with various aspects and applications of Martin's theory has increased rapidly.

2. With a view to presenting Martin's basic ideas we shall first of all consider the unit disc D. It is well known that every non-negative harmonic function on D can be represented in the form

$$h(z) = \int_{\partial D} k_w(z)\, \mu(dw),$$

where

$$k_w(z) = \frac{1-|z|^2}{|z-w|^2},$$

∂D is the boundary of D, and μ is a finite measure on ∂D. The representation given by (1) holds in a more general case where D is any bounded l-dimensional region whose boundary is sufficiently smooth. (In this case, the function $k_w(z)$ can be obtained by differentiating the Green's function in the direction of the inward normal to ∂D.) Thus, each non-negative harmonic function on D is expanded in a family of functions, each member of the family being in one-to-one correspondence with the points of the boundary of D. Martin's main result consists of the following observation: that a similar expansion is possible for an arbitrary region D, no matter how unpleasant, provided that one considers in place of the ordinary boundary ∂D in Euclidean space a certain "intrinsic" boundary B (which we henceforth refer to as the *Martin Boundary*). Roughly speaking, in order to obtain the Martin boundary from the ordinary boundary one has to identify certain points and "split up" certain other points, i.e. replace each of such points by a whole collection of new points. Furthermore, Martin showed that in the general situation it is necessary to carry the integration, not over the whole boundary B, but rather over a certain subset B_e of it (the so-called *set of minimal points*): it is only by means of such a reduction that the integral representation becomes unique.

Martin's construction was later extended from harmonic functions to solutions of elliptic differential equations ([15]), as well as to certain other kinds of equation (integral and difference equations among others); such extensions are connected with Markov chains and Markov processes ([22], [23], [25]).

In this paper we shall require another extension of Martin's results enabling one to describe non-negative solutions of boundary value problems. An exact formulation of the problem and the required extension of Martin's theory will be given in §1.[1]

Questions in the theory of convex cones, which arise naturally at this point, are considered in §2.

The rest of this paper is devoted to a study of a special boundary value problem, the so-called boundary value problem with a directional derivative.

3. The homogeneous boundary value problem with directional derivatives, or in short "Problem $\mathcal{A}$", can be formulated as follows. Let D be a region in

[1] However that may be, we do not presuppose any familiarity with other accounts of Martin's theory.

Euclidean n-space with a sufficiently smooth boundary ∂D. Let $v(z)$ be a given vector field defined on ∂D and varying smoothly over ∂D. It is required to study all the harmonic functions h that satisfy the boundary condition

$$\frac{\partial h(z)}{\partial v} = 0 \qquad (z \in \partial D), \tag{2}$$

where $\frac{\partial}{\partial v}$ denotes the derivative in the direction of the vector v.

If the field v is nowhere tangent to the boundary, then the only solutions to Problem $\mathfrak{A}$ are the constants. But if the field is tangent at certain points, the problem may admit many non-trivial solutions.

The problem with directional derivatives has been studied by several authors, starting with Poincaré. Almost all these papers deal with the two-dimensional case.[1] Liénard has suggested a simple method for characterizing all solutions to Problem $\mathfrak{A}$ that are smooth right up to the boundary. The method consists in reducing the problem to a boundary value problem for analytic functions (see e.g. V. I. Smirnov [12], 118).

In the present paper we shall study all non-negative solutions to Problem $\mathfrak{A}$, where arbitrary singularities are allowed at points of the boundary where the field is tangent. We shall only consider the two-dimensional case. Nevertheless, the suggested approach can be useful for the analysis of the multi-dimensional problem as well. In our formulation the problem becomes essentially a local one. (The whole thing reduces to a study of the behaviour of the Green's function near the exceptional points of the boundary.) In the classical approach on the other hand, the problem is a "global" one, and some non-trivial topology is involved in its solution. (In the case of a two-dimensional region bounded by a simple closed contour C, the number of linearly independent, classical solutions is determined by the index of the vector field v. In the multi-dimensional case it is necessary, of course, to deal with more complicated topological invariants.)

4. An exact formulation of the problem for non-negative harmonic functions which satisfy condition (2) is given in §3. The main results are stated in §8. We now give a brief description of these results.

Let D be a two-dimensional region[2] bounded by a smooth closed contour C. Since the length of the vector $v(z)$ has no real significance, we may suppose it to be equal to one. Thus, the vector $v(z)$ is uniquely determined by the angle θ which it forms with the forward tangent to C. The field is tangent to C at those points where θ is 0 or π. We assume that there is only a finite number of such

---

[1] Certain results for the 3-dimensional case have recently been obtained by A. V. Bitsadze. In the two-dimensional case, the boundary value problem with directional derivatives has been studied for the more general elliptic differential equation $\Delta u + au_x + bu_y = f$ in a monograph by I. N. Vekua entitled "Generalized analytic functions" Moscow, 1959.

[2] In view of the local character of our problem, the topology of D has no essential significance, and the results can be extended to multiply-connected regions.

points. The points at which θ or $\pi - \theta$ changes sign are called *exceptional*. We denote the set of these points by Γ. (If at some point θ or $\pi - \theta$ vanishes without changing sign, then by an arbitrarily small deformation of the field it is possible to get rid of the tangency in a neighbourhood of this point. Therefore we shall not regard such points as exceptional and make no special provision for them in the solutions.)[1]

Multiplication of the field $v(z)$ by -1 does not change the boundary condition (2). Hence if we single out an arbitrary exceptional point, we can suppose that θ takes the value 0 (and not π) at that point. We shall say then that the point α is positive (resp. negative) if the sign of θ changes from plus to minus (resp. from minus to plus) as the point advances in the positive direction. The set of all positive points is denoted by Γ_+ and the set of negative ones by Γ_-.

We shall show that *the Martin boundary B decomposes into connected components B_α corresponding to the exceptional points α. If $\alpha \in \Gamma_-$, the component B_α consists of a single point; if $\alpha \in \Gamma_+$, B_α is a closed interval.*

To each point w of the Martin boundary there corresponds a non-negative solution $k_w(z)$ to Problem $\mathcal{A}$, and every non-negative solution of Problem $\mathcal{A}$ has an expansion by $k_w(z)$. However, as we have already mentioned, not all of the functions $k_w(z)$ are used in the expansion, only some of them (corresponding to the subset B_e of B). In our case B_e *turns out to be finite, and consists of three kinds of point*:

(a) *the component B_α, if $\alpha \in \Gamma_-$*;

(b) *an end-point of the interval B_α, if $\alpha \in \Gamma_+$*;

(c) *an interior point of the interval B_α corresponding to those $\alpha \in \Gamma_+$ at which both the function θ and its derivative with respect to arc length vanish.* (We denote the set of these points α by Γ_+^0).

We shall denote solutions corresponding to points of type (a) by u_α, of type (b) by p_α^+, p_α^-, and of type (c) by $\tilde{u}_\alpha$. Then *every non-negative solution to Problem $\mathcal{A}$ can be written uniquely in the form*

$$h(z) = \sum_{\alpha \in \Gamma_-} a_\alpha u_\alpha(z) + \sum_{\alpha \in \Gamma_+} \{c'_\alpha p_\alpha^-(z) + c''_\alpha p_\alpha^+(z)\} + \sum_{\alpha \in \Gamma_+^0} a'_\alpha \tilde{u}_\alpha, \tag{3}$$

where a_α, c'_α, c''_α, a'_α are non-negative numbers.

The function h is bounded if and only if $a_\alpha = a'_\alpha = 0$. (This follows from the fact that p_α^- and p_α^+ are bounded, while u_α, $\tilde{u}_\alpha$ are unbounded near α but bounded outside some neighbourhood of α.)

The classical solutions, smooth up to the boundary, have the form:

$$h(z) = \sum_{\alpha \in \Gamma_+} c_\alpha \{p_\alpha^-(z) + p_\alpha^+(z)\}. \tag{4}$$

---

[1] It is possible, of course, to regard such points also as exceptional. The corresponding modifications in the results and proofs do not present any great difficulty.

Furthermore, the constants c_α are interrelated – one relationship for each point of the set Γ_-. These relationships are independent, so that the minimum number of linearly independent classical solutions is equal to the difference between the number of positive and the number of negative exceptional points. (This is the well-known Argument Principle.)

Note that the function 1 (which is a bounded non-negative solution of Problem $\mathscr{A}$) can be written in the form (4) as follows:

$$1 = \sum_{\alpha \in \Gamma_+} \{p_\alpha^+(z) + p_\alpha^-(z)\}. \tag{5}$$

M. B. Malyutov [9] was the first to give a description, in general form, of the bounded solutions of Problem $\mathscr{A}$, and to isolate from among them the classical solutions. For this purpose he relies substantially on arguments of a probabilistic character.

In the present paper probabilistic methods are not used; even so, it is worth giving at least a brief indication of the intuitive probabilistic interpretation of some of the basic results.

The Laplace operator is an infinitesimal operator of an elementary Markov stochastic process with continuous trajectories, a so-called Wiener process (see e.g. [4]). Condition (2) can be interpreted as the condition that a wandering particle be reflected from the boundary in the direction of the vector $v(z)$ (or $-v(z)$ if $v(z)$ is directed outwards of D). Such an interpretation is impossible for the exceptional points, and we assume that the process terminates as soon as the particle strikes such a point. It turns out that the motion starting at the point z terminates at the point $\alpha \in \Gamma_+$ with probability $p_\alpha^+(z) + p_\alpha^-(z)$. The probability of reaching a negative point is zero. The trajectory can only enter tangentially to the contour C at α, either from the positive side (with probability $p_\alpha^+(z)$) or from the negative side (with probability $p_\alpha^-(z)$). From this point of view equation (5) assumes on a natural meaning.

The probabilistic interpretation for unbounded solutions is somewhat more complicated. In this connection we refer the reader to papers by Doob [22] and Hunt [23].

§1. Boundary value problems for Laplace's equation and the Martin boundary

1.1. Let D be an arbitrary region in Euclidean l-space. We shall be investigating the harmonic functions on D that satisfy some boundary condition $\mathscr{R}$.

The notion of a boundary condition is defined as follows. Let us agree to use the term *neighbourhoods of the boundary of D* for sets of the form $D \setminus K$, where K is any compact set contained in D. We consider all possible functions whose domain of definition is a neighbourhood of the boundary of D. We are given a set $\mathscr{R}$ of such functions which satisfies the conditions:

1.1.A. *If f_1 and f_2 coincide on a neighbourhood of the boundary and $f_1 \in \mathscr{R}$, then $f_2 \in \mathscr{R}$.*

1.1.B. *If* $f_1, f_2 \in \mathcal{R}$, *then* $c_1 f_1 + c_2 f_2 \in \mathcal{R}$ *for any real* c_1, c_2.

1.1.C. *If in some neighbourhood of the boundary a sequence of harmonic functions* f_n *converges to the function* f, *and if* $f_n \in \mathcal{R}$ $(n = 1, 2, \ldots)$, *then* $f \in \mathcal{R}$.

Thus, $\mathcal{R}$ determines a boundary condition. In what follows, the statement "f satisfies the boundary condition $\mathcal{R}$" will mean the same thing as the statement "$f \in \mathcal{R}$". We shall call harmonic functions that satisfy the boundary condition $\mathcal{R}$ *solutions of the boundary value problem* $\mathcal{R}$.

R. S. Martin in his paper [24] studies harmonic functions defined on a region D that are free from any boundary conditions whatever. In order to include Martin's case into our general scheme it suffices to choose for $\mathcal{R}$ the set of *all* functions defined near the boundary of D.

1.2. We shall study a restricted class of boundary value problems by supposing that it is possible to select from $\mathcal{R}$ a subset $\mathcal{R}_+$ satisfying the following requirements:

1.2.A. *If* f_1 *and* f_2 *coincide on some neighbourhood of the boundary and* $f_1 \in \mathcal{R}_+$, *then* $f_2 \in \mathcal{R}_+$.

1.2.B. *If* $f_1, f_2 \in \mathcal{R}_+$, *then* $c_1 f_1 + c_2 f_2 \in \mathcal{R}_+$ *for any non-negative numbers* c_1, c_2.

1.2.C. *If* $f_n \to f$ *uniformly on some neighbourhood of the boundary and if* $f_n \in \mathcal{R}_+$ $(n = 1, 2, \ldots)$, $f \in \mathcal{R}$, *then* $f \in \mathcal{R}_+$.

1.2.D. *If* $f \in \mathcal{R}$ *and* $f \geqslant 0$ *on some neighbourhood of the boundary, then* $f \in \mathcal{R}_+$.

1.2.E. (MINIMUM PRINCIPLE). *Let* K *be a compact set contained in* D *and let* $D_0 = D \setminus K$. *Let* $f \in \mathcal{R}_+$ *be continuous on*[1] $D_0 \cup \partial K$ *and harmonic throughout* D_0. *Then either* $f \geqslant 0$ *on the whole of* D_0, *or there exists a point* $z_0 \in \partial K$ *such that* $f(z_0) < f(z)$ *for all* $z \in D_0$.

It follows from conditions 1.2.D. and 1.2.E. that among the solutions of the boundary value problem $\mathcal{R}$, those that belong to $\mathcal{R}_+$ are precisely the solutions which are non-negative throughout D.

For Martin's case $\mathcal{R}_+$ may be defined by the condition: $f \in \mathcal{R}_+$ if f is bounded below and if

$$\lim_{z \to z_0} f(z) \geqslant 0 \tag{1.1}$$

for every regular point z_0 of the boundary of D;[2] in the event that the region D is unbounded and $l \geqslant 3$, there is the additional requirement that

$$\lim_{|z| \to \infty} f(z) \geqslant 0. \tag{1.2}$$

The validity of conditions 1.1.A.–1.1.C. and 1.2.A.–1.2.D. for this situation is obvious. The validity of condition 1.2.E. follows from Evans's well-

1 By ∂D we mean the boundary of K.

2 For the definition of regular point, see for instance [10], § 31 or [5].

known theorem which states that if a function f is harmonic on a region D_0, is bounded below in D_0, satisfies condition (1.1) for any regular point z_0 in ∂D_0 (and satisfies condition (1.2) in the event that D_0 is unbounded and $l \geqslant 3$), then $f \geqslant 0$ throughout D_0. A proof of Evans's theorem can be found, for example, in the article by M. V. Keldysh [5]. (See Chap. II, in particular Lemma II; Keldysh's formulation differs slightly from the one we have given, but by modifying his argument a little it is not difficult to arrive at our required result.)[1]

We denote by $\mathscr{R}_0$ the set of all functions f such that $f_n \in \mathscr{R}_+$ and $-f \in \mathscr{R}_+$. It is clear that $\mathscr{R}_0$ satisfies conditions 1.1.A. and 1.1.B. We shall show that it satisfies the following condition:

1.2.C′. *If a sequence of harmonic functions $f_n \in \mathscr{R}_0$ converges locally uniformly to f in some neighbourhood of the boundary, then $f \in \mathscr{R}_0$.*[2]

For suppose that the sequence f_n converges locally uniformly to f in a neighbourhood V of the boundary ∂D, and that U is a region subjected to the following conditions: a) $U \subset D$; b) $\partial U \subset V$; c) $U \cup \partial U$ is compact. The sequence f_n converges uniformly on compact sets, and from the Minimum Principle 1.2.E. it follows that it converges uniformly on V. According to 1.1.C. the limit function belongs to $\mathscr{R}$; by virtue of 1.2.C. it belongs to $\mathscr{R}_0$.

1.3. We wish to construct an integral representation for all non-negative solutions of the boundary value problem $\mathscr{R}$. As a starting point we make use of the Green's function, postulating its existence.

Thus, we suppose that the following condition is fulfilled:

1.3.A. *For each $w \in D^\delta$ there exists a harmonic function $h_w(z)$ on D such that the function $g_w(z) = h_w(z) + \gamma(w - z)$ belongs to $\mathscr{R}_0$. Here D^δ is a neighbourhood of the boundary of D, and*

$$\gamma(z) = \begin{cases} -|z| & \text{for } l = 1, \\ -\ln|z| & \text{for } l = 2, \\ |z|^{2-l} & \text{for } l > 2. \end{cases}$$

It follows from the Minimum Principle 1.2.E. that the function $g_w(z)$ is uniquely defined by condition 1.3.A. We shall call it the *Green's function*.

We shall suppose, in addition, that the following postulate is fulfilled:

1.3.B. *The partial derivatives of $g_w(z)$ with respect to the coordinates of the point w exist and are continuous in w and z for all $w, z \in D^\delta$, such that $w \neq z$.*

---

[1] Evans's Theorem is a sharpening of the well-known minimum principle for harmonic functions: if a function f is harmonic on the region D_0, satisfies condition (1.1) for all points $z_0 \in \partial D_0$ (and satisfies condition (1.2) in the case where D is unbounded and $l \geqslant 3$), then $f \geqslant 0$ throughout D_0 (see, for instance, I. G. Petrovskii [10], §28).

[2] A sequence $f_n(z)$ is said to converge to $f(z)$ locally uniformly in the region A if for each $z_0 \in A$ there is a neighbourhood U of the point z_0 such that for some N $f_n(z)$ is defined throughout U $(n > N)$ and $\sup_{z \in U} |f_n(z) - f(z)| \to 0$ as $n \to \infty$. If a sequence of non-negative harmonic functions converges at each point of some region, then the convergence is always locally uniform.

The partial derivatives of $g_w(z)$ with respect to the coordinates of w are locally uniform limits of harmonic functions belonging to $\mathscr{R}_0$. Consequently (see 1.2.C′) they too belong to $\mathscr{R}_0$.

In Martin's case the set $\mathscr{R}_0$ consists of the bounded functions tending to zero as z approaches a regular point of the boundary (and also as $z \to \infty$, if D is unbounded and $l \geqslant 3$). In order to construct the Green's function it suffices to find a harmonic function $h_w(z)$ that coincides with $-\gamma(w-z)$ at all regular points of the boundary, and converges at infinity to zero if D is unbounded. The existence of such a function is proved, for instance, in [10] (§§31–32). It is known (see, for example, Keldysh [5], Ch. V) that the function $h_w(z)$ can be written in the form

$$h_w(z) = -\int_{\partial D} h(z, dy)\, \gamma\,|w-y|,$$

where $h(z, A)$ is a harmonic function with respect to $z \in D$ and a finite measure with respect to $A \subseteq \partial D$ (the so-called harmonic measure). From this the validity of condition 1.3.B. follows easily.

1.4. Let U be any region with a smooth boundary ∂U such that $U \cup \partial U$ is compact and is contained in D and that $\partial U \subset D^\delta$. Our immediate object is to construct a continuous function h on $D \setminus U$ that is harmonic on $D_0 = D \setminus (U \cup \partial U)$, belongs to $\mathscr{R}_0$, and coincides on ∂U with a previously defined continuous function φ.

We denote by $q_w(z)$ ($w \in \partial U$, $z \in D_0$) the derivative of $g_w(z)$ with respect to w taken in the direction of the outward normal to ∂U. For any $w \in \partial U$, $q_w(z)$ is a harmonic function on D_0 belonging to $\mathscr{R}_0$. By 1.3.B, $q_w(z)$ is continuous with respect to w.

We shall look for the function h in the form

$$h(z) = \int_{\partial U} q_w(z)\, F(w)\, dw, \tag{1.3}$$

where dw is an element of volume on the smooth manifold ∂U, and F a continuous function. We divide the manifold ∂U into a finite number of cells $A_1^n, \ldots, A_{m_n}^n$ with diameter less than $1/n$. Choosing arbitrary points $w_k^n \in A_k^n$ and denoting the volume of A_k^n by C_k^n we put

$$h_n(z) = \sum_{k=1}^{m_n} q_{w_k^n}(z)\, F(w_k^n)\, C_k^n.$$

It is obvious that the h_n are harmonic functions and that the sequence h_n converges to h locally uniformly on D_0. Hence by 1.2.C′, $h \in \mathscr{R}_0$.

If in (1.3) we replace the function $q_w(z)$ by the corresponding normal derivative of the function $\gamma(w-z)$, we obtain the usual formula for the potential for a double layer. By 1.3.A, h differs from this potential by a function that is continuous on ∂U. Hence, by a well-known property of the potential for a double layer (see, for example, I. G. Petrovskii [10], §34) it follows that the limiting value of $h(z)$, when z approaches $z_0 \in \partial U$ from the outside of U, is equal to

$$cF(z_0) + \int_{\partial U} q_w(z_0) F(w)\, dw,$$

where c is a positive constant depending only on the dimension of the space (e.g. $c = \pi$, when $l = 2$). In order, therefore, to obtain a function F with the required properties it is sufficient to solve the integral equation

$$f(z) = cF(z) + \int_{\partial U} q_w(z) F(w)\, dw \qquad (z \in \partial U) \tag{1.4}$$

and substitute the solution F in (1.3).

The Fredholm Theory (see, for example, E. Goursat [3], 609) is applicable to the kernel $q(z, w) = q_w(z)$ and to prove that (1.4) has continuous solutions we merely have to satisfy ourselves that for $f = 0$ the equation has only the zero solution (the Fredholm Alternative).

Consider the potential for a single layer

$$Q(z) = \int_{\partial U} g_w(z) F(w)\, dw.$$

It is well known (I. G. Petrovskii [10], §34) that the derivatives Q^+ (respectively Q^-) of this potential along the outward (respectively inward) normal to ∂U are given by the formulae

$$Q^+ = -h - cF, \quad Q^- = -h + cF. \tag{1.5}$$

The equation (1.4) can be written in the form $f = cF + h$, so that for $f = 0$ we have $Q^+ = 0$.

It is also well known (I. G. Petrovskii [10] §28) that if a harmonic function Q is continuous on the closed ball S and if the derivative of this function in the direction of the inward normal to ∂S vanishes at some point $z_0 \in \partial S$, then the function Q cannot satisfy the inequality $Q(z_0) < Q(z)$ for all interior points of S. If the boundary ∂U is sufficiently smooth, then for any point $z_0 \in \partial U$ we can construct a ball containing z_0 such that all of its interior points belong to D_0. Hence it is impossible that $Q(z_0) < Q(z)$ for all $z \in D_0$. But $Q \in \mathscr{R}_0 \subseteq \mathscr{R}_+$, and according to the Minimum Principle 1.2.E, $Q \geqslant 0$ throughout D_0. Similarly $-Q > 0$ throughout D_0, and therefore $Q = 0$ throughout D_0. Now Q is continuous on D. Thus, $Q = 0$ on ∂U, and consequently $Q = 0$ on U (see footnote 1, page 31). It follows that $Q^- = 0$ on ∂U, and from (1.5) it is clear that $F = 0$.

1.5. Let us now consider any non-negative solution f to the boundary value problem $\mathscr{R}$. Making use of the construction of 1.4. we "touch up" the function f near the boundary ∂D and obtain an integral representation of the touched-up function f. The integral representation of f itself will then be obtained by a passage to the limit.

Thus, let U be a region with a smooth boundary such that $U \cup \partial U$ is a compact subset of D and $\partial U \subset D^\delta$. By 1.4. we can construct a function $h \in \mathscr{R}_0$ that is harmonic on $D_0 = D \setminus (U \cup \partial U)$, continuous on $D \setminus U$ and coincides with f on ∂U. By 1.2.D, $f \in \mathscr{R}_+$ and by 1.2.B, $f - h \in \mathscr{R}_+$. But $f - h = 0$ on ∂U, and by the Minimum Principle 1.2.E, $f - h \geqslant 0$ in $D \setminus U$. For similar reasons $h \geqslant 0$ in $D \setminus U$.

We put

$$\tilde{f}(z) = \begin{cases} f(z) & \text{for } z \in U, \\ h(z) & \text{for } z \in D \setminus U. \end{cases}$$

On the strength of the inequality $f \geqslant \tilde{f} \geqslant 0$ it is not difficult to establish that $\tilde{f}$ is superharmonic on D (see I. G. Petrovskii [10], §35, Theorem 5). Consider any region $\tilde{D}$ for which: a) $U \cup \partial U \subseteq \tilde{D}$, b) $\tilde{D} \cup \partial \tilde{D}$ is compact and contained in D. By Riesz' Theorem (see for example I. I. Privalov [11] (p.159) or E. B. Dynkin [4] (14.14.A)) a finite measure μ on D can be chosen such that the difference

$$\delta(z) = \tilde{f}(z) - \int_{\tilde{D}} \gamma(z - w)\,\mu(dw)$$

is a harmonic function on $\tilde{D}$. Since $\tilde{f}$ is harmonic on U and on $\tilde{D} \cap D_0$, the measure μ is zero on these sets.[1] Consequently the measure μ is concentrated on ∂U.

It follows from 1.3.A that the function

$$H(z) = \tilde{f}(z) - \int_{\partial U} g_w(z)\,\mu(dw) = \delta(z) - \int_{\partial U} h_w(z)\,\mu(dw)$$

is also harmonic on $\tilde{D}$. On the other hand, H is harmonic on D_0. Hence H is a harmonic function throughout D. Since h and g_w belong to $\mathscr{R}_0$, so does H and by the Minimum Principle 1.2.E, $H = 0$. Thus,

$$\tilde{f}(z) = \int_{\partial U} g_w(z)\,\mu(dw) \qquad (z \in D).$$

1.6. We now construct a sequence of regions U_n with smooth boundaries so that a) $U_n \cup \partial U_n$ are compact subsets of D, b) $\partial U_n \subset D^\delta$, c) $U_n \uparrow D$. By 1.5 to each of these regions there corresponds a superharmonic function $\tilde{f}_n(z)$ and a finite measure μ_n concentrated on ∂U_n, where

---

[1] This follows easily from the fact that the measure μ is uniquely determined by the superharmonic function f.

$$\tilde{f}_n(z) = \int_{\partial U_n} g_w(z)\, \mu_n(dw) \qquad (z \in D). \tag{1.6}$$

Clearly $\tilde{f}_n(z) \to f(z)$, and therefore

$$f(z) = \lim_{n \to \infty} \int_{\partial U_n} g_w(z)\, \mu_n(dw) \qquad (z \in D). \tag{1.7}$$

Proceeding from this formula we shall give an integral representation of the function f. Here the following result will play an important part.

HELLY'S THEOREM.[1] *If ν_n is a sequence of measures on a compact space E such that the values $\nu_n(E)$ are bounded, then it is possible to construct a measure ν on E and to select from ν_n a subsequence ν_{n_k} such that for any continuous function $F(w)$ $(w \in E)$,*

$$\lim_{k \to \infty} \int_E F(w)\, \nu_{n_k}(dw) = \int_E F(w)\, \nu(dw).$$

Note that if the measures ν_n, starting at a certain n, are all concentrated on a compact set $\tilde{E} \subset E$, then the measure ν is also concentrated on $\tilde{E}$. In effect it is clear from Helly's theorem that in this case

$$\int_E F(w)\, \nu(dw) = 0$$

for every continuous function F that vanishes on $\tilde{E}$.

In order to apply Helly's Theorem to the measures μ_n, one has first of all to construct a compact set on which all these measures are concentrated. We choose n_0 so that $U_{n_0} \supset D \setminus D^\delta$. (This is possible because the sets $U_1 \subset U_2 \subset \ldots$ form a covering of the compact set $D \setminus D^\delta$.) Put $D^b = D \setminus U_{n_0}$. Obviously $\partial U_n \subset D^b$ for $n \geqslant n_0$, and hence all the measures μ_n $(n \geqslant n_0)$ are concentrated on D^b. Since $D^b \subseteq D^\delta$, the functions $g_w(z)$ are defined for all $w \in D^b$, and we can write equation (1.7) in the form

$$f(z) = \lim_{n \to \infty} \int_{D^b} g_w(z)\, \mu_n(dw) \qquad (z \in D). \tag{1.8}$$

---

[1] Helly proved this theorem for the case where E is a closed interval. The proof can be found in any university text-book on probability theory. The more general proof is not difficult to obtain if one combines the following two facts: 1) On the Banach space C of all continuous functions on the compact set E, any non-negative linear functional l can be expressed as an integral with respect to a finite measure μ; furthermore $\| l \| = \mu(E)$ (see for example P. Halmos [14], §56); 2) from each sequence of linear functionals whose norms are bounded one can select a weakly convergent subsequence (see for example L. A. Lyusternik and V. I. Sobolev [8], Ch. III, §24).

However, the set D^b is not compact. In order to overcome this difficulty we construct a compact set E that satisfies the following property:

1.6.A. *The set D^b is homeomorphic to a subset of the compact set E.*

We shall identify this subset with D^b, still denoting it by D^b. It can be assumed without loss of generality that the following postulate is fulfilled:

1.6.B. *The closure of D^b in E coincides with E.*

Indeed, to satisfy 1.6.B. it suffices to remove from E all points that are not in the closure of D^b.

We extend the measures μ_n onto E by putting $\mu_n(E \setminus D^b) = 0$. A second obstacle to the application of Helly's Theorem is the fact that the sequence $\mu_n(E) = \mu_n(D^b)$ is not necessarily bounded. We can cope with this complication by introducing the new measures

$$\nu_n(dw) = g_w(z_0)\,\mu_n(dw),$$

where z_0 is a point of U_1. It follows from (1.6) that

$$\nu_n(E) = \nu_n(\partial U_n) = \tilde{f}_n(z_0) = f(z_0),$$

and as a result the sequence $\nu_n(E)$ is bounded. We can now write (1.8) in the form

$$f(z) = \lim_{n\to\infty} \int_E k_w(z)\,\nu_n(dw) \qquad (z \in D), \tag{1.9}$$

where

$$k_w(z) = \frac{g_w(z)}{g_w(z_0)} \qquad (w \in E,\ z \in D). \tag{1.10}$$

The function $F(w) = k_w(z)$ is continuous on D^b. We shall suppose that the compactification E of D^b can be chosen so that the following condition holds:

1.6.C. *For any $z \in D$, the function $F(w) = k_w(z)$ can be continuously extended onto E.*

We shall denote the extended function again by $k_w(z)$.

It follows from Helly's Theorem and (1.9) that

$$f(z) = \int_E k_w(z)\,\nu(dw) \qquad (z \in D), \tag{1.11}$$

where ν is a measure on E. We put $D_m^b = D^b \cap U_m$ $(m \geqslant n_0)$. Since $n > m$ the measures ν_n are concentrated on the compact set $E \setminus D_m^b$, the measure ν is also concentrated on $E \setminus D_m^b$. Consequently it is concentrated on $B = E \setminus D^b = \bigcap_{m \geqslant n_0} (E \setminus D_m^b)$ and we have the required expansion, namely

$$f(z) = \int_B k_w(z)\,\nu(dw) \qquad (z \in D). \tag{1.12}$$

We shall call this the *Martin expansion*.

1.7. We now have to construct the compact set E satisfying the properties 1.6.A.–1.6.C. Observe that with each point w of this compact set a non-negative function $k_w(z)$ is associated. If $w \in D^b$, this function can be expressed as in (1.10) and we see that it is a harmonic function on the region D_w obtained from D by removing the single point w. Note that if $w_n \to w \in D^b$, then $k_{w_n}(z)$ converges to $k_w(z)$ locally uniformly in D_w. On the other hand, if $w_n \to w \in B$, then $k_{w_n}(z)$ converges to $k_w(z)$ locally uniformly in D. Hence for $w \in B$, $k_w(z)$ is a harmonic function on D that is equal to 1 when $z = z_0$.

These arguments lead naturally to the following construction. We denote by $\mathscr{H}_0$ the set of all functions $k_w(z)$ $(w \in D^b)$ and by $\mathscr{H}_1$ the set of all non-negative harmonic functions on D that are equal to 1 when $z = z_0$. We introduce a topology on the set $\mathscr{H} = \mathscr{H}_0 \cup \mathscr{H}_1$ by putting $f_n \Rightarrow f$, if $f_n(z) \to f(z)$ locally uniformly in D $(f \in \mathscr{H}_1)$ or in D_w $(f = k_w(z))$. It is easy to see that with this topology the set $\mathscr{H}$ is compact,[1] $\mathscr{H}_1$ is a closed subset, and the mapping $w \to k_w$ is a homeomorphism of D^b onto $\mathscr{H}_0$. Denoting by E the closure of $\mathscr{H}_0$ in $\mathscr{H}$ we obtain a compact set satisfying all the conditions 1.6.A.–1.6.C.

1.8. It is clear from the reasoning at the beginning of 1.7. that the compact set E is determined uniquely to within homeomorphism by the conditions 1.6.A.–1.6.C. Moreover, the set $B = E \setminus D^b$ does not depend on the choice of $D^b \subset D$. Some of functions $k_w(z)$ can be outside $\mathscr{R}$. Put $w \in \widetilde{B}$ if $w \in B$ and $k_w \in \mathscr{R}$. Under certain conditions formula (1.12) holds with B replaced by $\widetilde{B}$. Namely, it is sufficient that a set of functions $\mathscr{R}_w$ can be associated with every w such that:

1.8.α. If $f_n \in \mathscr{R}_w$ converge to f on $D \cap V$ where V is a neighbourhood of w and if f_n are harmonic then $f \in \mathscr{R}_w$.

1.8.β. $\bigcap\limits_{w \in B} \mathscr{R}_w = \mathscr{R}$.

Indeed let w_0 belong to the support of a measure ν. By 1.8.α, $k_w \in \mathscr{R}_{w_0}$ for all $w \in B \setminus \{w_0\}$. Hence if $f \in \mathscr{R}$ then

$$f_U(z) = \int_U k_w(z)\,\nu(dw) = f(z) - \int_{B \setminus U} k_w(z)\,\nu(dw) \in \mathscr{R}_{w_0}$$

and by 1.8.α

$$k_{w_0}(z) = \lim_{U \downarrow w_0} \frac{1}{\nu(U)} \int_U k_w(z)\,\nu(dw) \in \mathscr{R}_{w_0}.$$

---

[1] It follows from Harnack's inequality (see for example M. Brelot [18], p.166–168), that the functions of $\mathscr{H}_1$ are uniformly bounded on any compact subset of D. Hence the compactness of $\mathscr{H}_1$ follows, for example, from Theorem 7, § 30 of I. G. Petrovskii's book [10]. The compactness of $\mathscr{H}$ is an obvious consequence of the compactness of $\mathscr{H}_1$.

We conclude from 1.8.α that $k_{w_0} \in \mathscr{R}$.

We call $\widetilde{B}$ the Martin boundary for the problem $\mathscr{R}$.

The representation (1.12) is not, as a rule, unique. In § §3–7 we shall deal with the case where one can distinguish in $\widetilde{B}$ a subset B_e subject to the following conditions:

1.8.A. The functions $k_u(z)$ $(u \in B_e)$ are linearly independent.

1.8.B. Each function $k_w(z)$ $(w \in \widetilde{B})$ can be represented in the form

$$k_w(z) = \sum_{u \in B_e} a_u(w)\, k_u(z). \tag{1.13}$$

It is obvious that the functions $a_u(w)$ $(u \in B_e)$ are uniquely determined and are continuous on $\widetilde{B}$. The expansion (1.12) can be rewritten in the form

$$f(z) = \sum_{u \in B_e} \widetilde{\nu}(u)\, k_u(z), \tag{1.14}$$

where

$$\widetilde{\nu}(u) = \int_{\widetilde{B}} a_u(w)\, \nu(dw). \tag{1.15}$$

One can interpret the sum on the right hand side of (1.14) as an integral of $k_w(z)$ with respect to a measure $\widetilde{\nu}$ concentrated on B_e. Thus, in the Martin expansion one need only look at those measures that are concentrated on B_e. It is easy to see that in this way the expansion becomes unique.

In the general case the set $\widetilde{B}$ in (1.12) can still be reduced to a subset B_e (not in general a finite one). Under certain specific conditions, after this reduction the expansion becomes unique. This result comes out of the general theory of cones in linear topological spaces. A brief outline of the theory of cones will be given in the next section.

§2. Cones in linear topological spaces[1]

2.1. Let $\mathscr{L}$ be a linear space over the field of real numbers, together with a topology in which the function $x + y$ $(x, y \in \mathscr{L})$ is continuous in x and y combined, and the function cx $(x \in \mathscr{L}$, c a scalar) is continuous in c and x combined. Then we say that a *linear topological space* is given. We shall suppose unless it is expressly stated otherwise, that the following two conditions are fulfilled:

2.1.A. (SEPARATION AXIOM.) *For each element $x \neq 0$ there is a neighbourhood of zero that does not contain x.*

---

[1] The contents of this section are not used in the sequel.

2.1.B. (LOCAL CONVEXITY.) *For each neighbourhood U of 0 there is a convex*[1] *open set V such that $0 \in V \subseteq U$.*

A subset $\mathscr{K}$ of a linear space $\mathscr{L}$ is called a *cone* if it contains with any two elements also their sum, and with any element also its product with an arbitrary positive number. Every cone is a convex set.

We call a *base of the cone* $\mathscr{K}$ the intersection of $\mathscr{K}$ with a linear variety $\{x: l(x) = 1\}$, where l is any linear functional taking positive values on $\mathscr{K}$.[2]

2.2. The most important problem in the theory of cones is that of expressing an arbitrary element of the cone in terms of extremal elements.

An element f of the cone $\mathscr{K}$ is called *extremal* if the equation $f = f_1 + f_2$ $(f_1, f_2 \in \mathscr{K})$ implies that $f_1 = c_1 f$, $f_2 = c_2 f$ (where c_1, c_2 are scalars). In 1940 M. G. Krein and D. P. Milman proved that if a cone in a linear topological space possesses a compact base, then it coincides with the closed convex hull of its extremal points. An essential further step was taken in 1956 by Choquet who proved the following theorem:

CHOQUET'S THEOREM.[3] *If the cone $\mathscr{K}$ in a linear topological space $\mathscr{L}$ possesses a compact metrizable base $\mathscr{B}$, then every element f of $\mathscr{K}$ can be represented in the form*

$$f = \int_{\mathscr{K}_e} k\mu(dk), \tag{2.1}$$

where $\mathscr{K}_e$ is the set of all extremal elements of $\mathscr{K}$ that belong to $\mathscr{B}$, and μ is a finite measure[4] *on $\mathscr{K}_e$.*

For the representation (2.1) to be unique it is necessary and sufficient that the following conditions are satisfied (where $f \prec g$ means that $g - f \in \mathscr{K}$):

2.2.A. *For any $f_1, f_2 \in \mathscr{K}$ there exists $g \in \mathscr{K}$ such that $f_1 \prec g$, $f_2 \prec g$, and if $f_1 \prec h$, $f_2 \prec h$, then either $g \prec h$ or $g = h$.*

2.2.B. *For any $f_1, f_2 \in \mathscr{K}$ there exists $g \in \mathscr{K}$ such that $g \prec f_1$, $g \prec f_2$ and if $h \prec f_1$, $h \prec f_2$, then either $h \prec g$ or $h = g$.*

Equation (2.1) implies that for any linear functional l on the space $\mathscr{L}$,

$$l(f) = \int_{\mathscr{K}_e} l(k)\mu(dk). \tag{2.2}$$

Conditions 2.2.A.–2.2.B. imply that $\mathscr{K}$ is a *lattice* with respect to the partial ordering $f \prec g$ (see for example A. G. Kurosh [7]).

---

[1] A set A is said to be *convex* if it contains with any two elements x and y also all elements of the form $px + qy$ $(p \geqslant 0, q \geqslant 0, p + q = 1)$.

[2] A real-valued continuous function $l(x)$ $(x \in \mathscr{L})$ is called a *linear functional on* $\mathscr{L}$ if $l(x + y) = l(x) + l(y)$ $(x, y \in \mathscr{L})$ and $l(cx) = cl(x)$ $(x \in \mathscr{L}$, c a scalar). Sets of the form $\{x: l(x) = c\}$ (l being a linear functional, c being a fixed scalar) are called *linear varieties*. Every linear variety that does not contain the origin can be written uniquely in the form $\{x: l(x) = 1\}$.

[3] See [19], [20], [21].

[4] It can be proved that $\mathscr{K}_e$ is a Borel set (in fact, a countable intersection of open sets).

2.3. As an example of a linear topological space (satisfying conditions 2.1.A.–2.1.B.) we can take the class $\mathscr{L}(D)$ of all functions on an l-dimensional region D that are expressible as the difference of two non-negative harmonic functions. The topology is given by means of locally uniform convergence. The set $\mathscr{K}(D)$ of all non-negative harmonics on D is a cone in this space. The set $\mathscr{K}(\mathscr{R})$ of all non-negative solutions of the boundary value problem $\mathscr{R}$ (which is clearly the same as $\mathscr{K}(D) \cap \mathscr{R}$) is also a cone.

The functions in $\mathscr{K}(D)$ that take the value 1 at a fixed point $z_0 \in D$ form a base $\mathscr{B}(D)$ of $\mathscr{K}(D)$. This base is compact (see footnote 1 on p.37). In order to put a metric on $\mathscr{B}(D)$ it suffices to write

$$\rho(f, g) = \max_{z \in D_0} |f(z) - g(z)|, \tag{2.3}$$

where D_0 is any closed ball contained in D. For a harmonic function vanishing on D_0 vanishes on the whole region D. Hence (2.3) defines a metric on $\mathscr{B}(D)$. Further, it is clear that if f_n converges locally uniformly to f, then $\rho(f_n, f) \to 0$. On the other hand, if $\rho(f_n, f) \to 0$ ($f_n \in \mathscr{B}(D)$), then the sequence f_n cannot have a cluster point different from f, and in view of the compactness of $\mathscr{B}(D)$, f_n converges locally uniformly to f.

The set $\mathscr{B}(\mathscr{R}) = \mathscr{B}(D) \cap \mathscr{R}$ is a base of $\mathscr{K}(\mathscr{R})$. This set is compact because of 1.1.C. It is also clear that it is metrizable. By Choquet's Theorem every element of $\mathscr{K}(\mathscr{R})$ admits the representation (2.1). If $z \in D$, then the formula $l_z(f) = f(z)$ defines a linear functional on $\mathscr{L}(D)$. Applying (2.2) to this functional we arrive at the expression

$$f(z) = \int_{\mathscr{K}_e} k(z)\, \mu(dk) \qquad (z \in D). \tag{2.4}$$

2.4. We shall show that any extremal element f of $\mathscr{K}(\mathscr{R})$ belonging to the base $\mathscr{B}(\mathscr{R})$ coincides with one of the functions k_w ($w \in \mathscr{B}$).

For by 1.6 f can be represented in the form (1.12). Putting $z = z_0$ in (1.12) and taking into account the fact that $f(z_0) = k_w(z_0) = 1$ we observe that $\nu(\widetilde{B}) = 1$. Hence there is a point $w_0 \in \widetilde{B}$ any neighbourhood of which has positive measure ν. For each neighbourhood U of the point w_0 (relative to $\widetilde{B}$) we put

$$f_U(z) = \int_U k_w(z)\, \nu(dw).$$

Since f_U and $f - f_U$ belong to $\mathscr{K}(\mathscr{R})$ and f is an extremal element of $\mathscr{K}(\mathscr{R})$, we have $f_U(z) = c_U f(z)$, where c_U is a constant. Putting $z = z_0$ we observe that $c_U = \gamma(U) > 0$. As a consequence

$$f(z) = \frac{1}{\nu(U)} \int_U k_w(z)\, \nu(dw).$$

If the diameter of the neighbourhood U tends to 0, then the integral on the right hand side tends to $k_{w_0}(z)$. Hence $f = k_{w_0}$.

We shall write $w \in B_e$ whenever w is an extremal element of $\mathcal{K}(\mathcal{R})$. Formula (2.4) can be rewritten in the form

$$f(z) = \int_{B_e} k_w(z)\, \mu(dw) \quad (z \in D). \tag{2.5}$$

(For the cone $\mathcal{K}(D)$, this formula was first obtained by Martin.)

According to Choquet's theorem, in all cases where the set $\mathcal{K}(\mathcal{R})$ satisfies the conditions 2.2.A–2.2.B the representation (2.5) is unique. We shall show that such is the situation in Martin's case, i.e. when $\mathcal{K}(\mathcal{R}) = \mathcal{K}(D)$.

We choose a sequence of regions D_n with smooth boundaries ∂D_n so that the sets $\hat{D}_n = D_n \cup \partial D_n$ are compact and satisfy the conditions $\hat{D}_n \subset D_{n+1}$, $\cup D_n = D$. Let $f_1, f_2 \in \mathcal{K}(D)$. Then the function $m = \min(f_1, f_2)$ is superharmonic, while the function $M = \max(f_1, f_2)$ is subharmonic. Denote by h_n (respectively H_n) the harmonic functions on D_n continuous on $\hat{D}_n$ and coinciding with m (respectively M) on ∂D_n. It is easy to see that

$$m \geqslant h_1 \geqslant h_2 \geqslant \ldots \geqslant h_n \geqslant \ldots \geqslant 0,$$
$$M \leqslant H_1 \leqslant H_2 \leqslant \ldots \leqslant H_n \leqslant \ldots \leqslant f_1 + f_2,$$

and that the functions

$$g(z) = \lim h_n(z), \quad G(z) = \lim H_n(z)$$

are, respectively, the greatest non-negative harmonic function not exceeding f_1 and f_2, and the smallest harmonic function less than or equal to f_1 and f_2.

For the cone $\mathcal{K}(\mathcal{R})$ it is generally more difficult to verify conditions 2.2.A–2.2.B than to prove directly the uniqueness of the expansion (2.5). We do not know of any examples of boundary value problems for which the expansion (2.5) is non-unique.

§3. The boundary value problem with a directional derivative (Problem $\mathcal{A}$)

3.1. Let D be a plane region bounded by a smooth closed contour C, and $v(z)$ a continuous vector field on C. Our aim is to study the harmonic functions on D that are bounded below and whose directional derivatives along v vanish on C.

We now give a more precise statement of the problem. A *closed contour* C in the plane R^2 is given by a function $c(t)$ $(-\infty < t < +\infty)$ taking values in R^2 and subject to the condition that for some $d > 0$, $c(t + d) = c(t)$ for all t, and $c(t_1) \neq c(t_2)$ for $|t_1 - t_2| < d$. The contour C is called *smooth* if $c(t)$ is differentiable and if its derivative $c'(t)$ is nowhere zero and is Hölder continuous.[1]

A *continuous vector field* on C we mean a function $v(z)$ $(z \in C)$ taking values in R^2, Hölder continuous, non-vanishing, and satisfying the following condition: the vector $v(z)$ is tangent to C at only a finite number of points z.

If C is smooth and if in some neighbourhood of the point t_0 the function $v[c(t)]$ has a Hölder continuous derivative with respect to t, then we shall say that the field v is *smooth* in a neighbourhood of $c(t_0)$.

We shall call the point $z_0 \in C$ *exceptional* if the projection of $v(z)$ along the outward normal to C changes sign[2] at z_0. We denote by Γ the set of all exceptional points and we write $C^* = C \setminus \Gamma$.

A harmonic function $h(z)$ in D will be called a *solution of Problem $\mathcal{A}$ for the contour C and the field v* if it satisfies the following boundary condition: at every non-exceptional point of C the directional derivative of h along v vanishes.

More precisely this means the following:

3.1.A. *The partial derivatives of h can be continuously extended to the set $D^* = D \cup C^*$.*

3.1.B. *If $z_0 \in C^*$ and the vector $v(z_0)$ has coordinates v_1 and v_2, then*[3]

$$v_1 h_{x_1}(z_0) + v_2 h_{x_2}(z_0) = 0.$$

3.2. Problem $\mathcal{A}$ may be considered as a special case of the boundary value problems described in §1. For let $\mathcal{A}$ denote the class of all functions h defined in a neighbourhood of the boundary of D and satisfying the conditions:

3.2.A. *h is continuously differentiable on a set $U = D \setminus K$ (where K is a compact subset of D) and its partial derivatives can be continuously extended to $U \cup C^*$.*

3.2.B. The postulate 3.1.B is satisfied.

It is clear that the set of solutions of Problem $\mathcal{A}$ can be characterized as the set of all functions that are harmonic on D and belong to $\mathcal{A}$.

The set $\mathcal{A}$ evidently satisfies conditions 1.1.A–1.1.B. In §4 it will be shown that it also satisfies condition 1.1.C. Later on we shall define a set $\mathcal{A}_+$ satisfying the conditions 1.2.A–1.2.E, we shall compute the Green's function for Problem $\mathcal{A}$, and using the results of §1 we shall describe all the non-negative solutions to this problem.

3.3. Let α_1 and α_2 be two points on the closed contour C. we can choose in a unique way numbers t_1 and t_2 such that $c(t_1) = \alpha_1$, $c(t_2) = \alpha_2$ and

---

[1] That is to say, there exist positive constants k and λ such that for all t_1, t_2, $|c'(t_1) - c'(t_2)| \leqslant k|t_1 - t_2|^\lambda$. (The elements of R^2 can be interpreted as vectors with $|z|$ denoting the length of z.)

[2] If this projection vanishes without change of sign, then by an arbitrary small deformation of v we can arrange matters so that the deformed field is not tangent to C near z_0. For this reason such points will not be considered exceptional.

[3] It follows from condition 3.1.A that the function h itself can be continuously extended to D^*. We shall talk about the values of the function h and its partial derivatives at the point $z_0 \in C^*$; these will be the appropriate limits of the values at $z \in D$ as $z \to z_0$.

$0 \leqslant t_1 < \infty$, $t_1 \leqslant t_2 < t_1 + d$. We denote the set of all points $c(t)$, where $t_1 \leqslant t \leqslant t_2$, by $[\alpha_1, \alpha_2]$ and we call it the closed arc with α_1 as the initial point and α_2 as the terminal point. The open arc (α_1, α_2) and the semi-open arcs $[\alpha_1, \alpha_2)$ and $(\alpha_1, \alpha_2]$ are defined similarly.

The arc $[\alpha_1, \alpha_2]$ is said to be smooth if $c'(t)$ does not vanish and is Hölder continuous for $t_1 \leqslant t \leqslant t_2$.

Let C be a smooth closed contour or a smooth arc and let $v(z)$ be a (Hölder) continuous vector field on C. We denote by $\Psi(z)$ the tangent of the angle between the vectors[1] $c'(t)$ and $v(z)$ $(z = c(t))$. The set Γ of exceptional points as defined in 3.1 coincides with the set of points at which $\Psi(z)$ changes sign as it passes through zero. We call an exceptional point *positive* (respectively, *negative*) if $\Psi(z) > 0$ (respectively, $\Psi(z) < 0$) in front of it. The set of all positive exceptional points will be denoted by Γ_+ and the set of all negative exceptional points by Γ_-.

3.4. We shall interpret 2-dimensional vectors as complex numbers. Let $v(z)$ be a vector field on a smooth closed contour C. We write

$$\frac{v[c(t)]}{c'(t)} = \rho(t)\, e^{i\theta(t)}, \tag{3.1}$$

assuming $\rho(t) \geqslant 0$ and the function $\theta(t)$ to be continuous. Under these conditions $\rho(t)$ is uniquely determined, and $\theta(t)$ is uniquely determined to within a constant $2k\pi$ (where k is an integer). Clearly $n = \frac{1}{2\pi}[\theta(t+d) - \theta(t)]$ is an integer, which is independent of t. The number $l = -n - 1$ (equal to the number of turns completed by the vector $v[c(t)]$ as t goes from 0 to $-d$) is called the *index of the field v*.

Note that $\Psi[c(t)] = \tan\theta(t)$. Making use of this it is not difficult to evaluate l in terms of the number l_+ of elements of Γ_+ and the number l_- of elements of Γ_-. The formula is

$$l + 1 = \frac{1}{2}(l_+ - l_-). \tag{3.2}$$

§4. Reduction of Problem $\mathcal{A}$ – some particular solutions

4.1. Let C and $\widetilde{C}$ be two smooth closed contours and let D and $\widetilde{D}$ be the regions enclosed by them. It is well known that:

4.1.A. *For any $z_0 \in D$, $\widetilde{z}_0 \in \widetilde{D}$ and for any $a \in [0, 2\pi]$ there exists a unique conformal mapping F from D onto $\widetilde{D}$ such that $F(z_0) = \widetilde{z}_0$ $\arg F'(z_0) = a$.*

4.1.B. *The mapping F can be extended to a one-to-one continuous mapping from the closed region $D \cup C$ onto a closed region $\widetilde{D} \cup \widetilde{C}$.*

---

[1] The direction of the vector $c'(t)$ will be called the positive direction of the tangent to C at the point $z = c(t)$.

4.1.C. *The derivative $F'(z)$ is Hölder continuous on $D \cup C$ (i.e. $|F'(z_1) - F'(z_2)| < k|z_1 - z_2|^\lambda$ for some positive constants k, λ for all $z_1, z_2 \in D \cup C$).*

Assertions 4.1.A–4.1.C follow from the well-known theorems of Riemann, Caratheodory and Kellog (see e.g. G. M. Goluzin [2], pp. 30, 50, and 468).

Let a continuous vector field $v(z)$ be given on the contour C. Then the formula

$$\tilde{v}[F(z)] = F'(z)\, v(z)$$

defines a continuous vector field $\tilde{v}$ on the contour $\tilde{C}$. It is easy to see that the function $\tilde{h}(z)$ is a solution of Problem $\mathcal{A}$ for the contour $\tilde{C}$ and the vector field $\tilde{v}$ if and only if the function $h(z) = \tilde{h}[F(z)]$ is a solution of Problem $\mathcal{A}$ for the contour C and the field v. Thus, it is sufficient to investigate Problem $\mathcal{A}$ for any single smooth closed contour. It is most convenient to take the unit circle for this contour.[1]

4.2. Let h be a harmonic function on the domain D, and let us write

$$\mathfrak{S}h = \frac{\partial h}{\partial x} - i\,\frac{\partial h}{\partial y}\,. \tag{4.1}$$

We consider the harmonic function $\hat{h}$ conjugate to h. The function $H = h + i\hat{h}$ is an analytic function on D and

$$\mathfrak{S}h\,(z) = H'(z).$$

It follows from this that $\mathfrak{S}h$ is an analytic function on D and, if $\mathfrak{S}h = f$, then

$$h(z) = \operatorname{Re} \int_{z_0}^{z} f(z)\, dz + \text{const} \tag{4.2}$$

(where z_0 is any point in D).

We shall denote by $\mathfrak{R}f$ the right hand side of (4.2). It is not difficult to see that if $h = \mathfrak{R}f$, then $\mathfrak{S}h = f$.

Thus, the operators $\mathfrak{S}$ and $\mathfrak{R}$ are mutually inverse and establish a one-to-one correspondence[2] between the harmonic functions on D considered to within a constant term and the analytic functions on D.

---

[1] Let $c(t)$ be an admissible parameterization of the contour C. Then $\tilde{c}(t) = F[c(t)]$ is an admissible parameterization of $\tilde{C}$. It is easy to see that

$$\frac{\tilde{v}[\tilde{c}(t)]}{\tilde{c}'(t)} = \frac{v[c(t)]}{c'(t)}\,.$$

In particular, the angle $\theta(t)$ for the contour $\tilde{C}$ and the field $\tilde{v}$ is the same as that for the contour C and the field v.

[2] This correspondence holds good even for a multiply connected region D. In this case, however, we have to consider not all the analytic functions on D, but only those for which (4.2) determines a single-valued function on D.

Note that the analytic function $\mathfrak{S}h$ is continuous in a neighbourhood of $z \in C$ if and only if the function h is continuously differentiable in a neighbourhood of this point. Furthermore, if $v(z)$ is a vector field on C and $f = \mathfrak{S}h$, then

$$\frac{\partial h}{\partial v} = \operatorname{Re} vf. \tag{4.3}$$

Hence the function h is a solution of Problem $\mathscr{A}$ for the contour C and the vector field v if and only if the function $f = \mathfrak{S}h$ satisfies the following conditions:

4.2.A. *f is an analytic function on D which is continuous on D^*.*

4.2.B. $\operatorname{Re} v(z) f(z) = 0$ *for* $z \in C \setminus \Gamma$.

We shall call a function f satisfying conditions 4.2.A–4.2.B a *solution of Problem* $\mathscr{B}$. Thus, the operators $\mathfrak{S}$ and $\mathfrak{R}$ set up a one-to-one correspondence between the solutions of Problem $\mathscr{A}$ (considered to within a constant term) and the solution of Problem $\mathscr{B}$.

4.3. We fix an arbitrary point z_0 of D and denote by D_0 the region obtained from D by removing z_0. We call a complex function $S(z)$ $(z \in D \cup C)$ a *characteristic function* of the vector field v if it satisfies the following conditions:

4.3.A. *The functions $S(z)$ and $T(z) = S(z)^{-1}$ are analytic on D_0 with possibly a pole at z_0.*

4.3.B. *$S(z)$ and $T(z)$ are Hölder continuous on the closed region obtained from $D \cup C$ by deleting a neighbourhood of z_0.*

4.3.C. *$S[c(t)]$ differs from $e^{i\theta(t)}$ by a positive factor (where $\theta(t)$ is the angle between the vector $v[c(t)]$ and the forward-pointing tangent to C at $c(t)$).*

By virtue of 4.3.C, $\arg S[c(t)] = \theta(t) + 2\pi k$, and hence the increase in this argument on going once around the contour C in the positive sense is equal to $-2\pi(l+1)$ (see 3.4). Consequently (see, for example, V. I. Smirnov [12], 94) the function $S(z)$ has a pole of order $l+1$ at z_0 (if $l+1>0$), or a zero of order $-l-1$ (if $l+1<0$). (If $l+1=0$, then $S(z)$ is regular and non-zero at z_0). If two functions satisfy conditions 4.3.A–4.3.C, then their ratio is an analytic function on D continuous on $D \cup C$ and positive on C. Such a function is necessarily a constant. Consequently the characteristic function $S(z)$ is determined uniquely to within a positive constant multiplier. The construction of $S(z)$ will be carried out in 4.4.

We put

$$r(z) = |S(z)|, \qquad \tau(z) = \operatorname{Re}\frac{S(z)}{r(z)}, \qquad \nu(z) = \operatorname{Im}\frac{S(z)}{r(z)}.$$

It follows from condition 4.3.C that

$$\tau[c(t)] = \cos\theta(t), \quad \nu[c(t)] = \sin\theta(t).$$

Note that

$$S(z) = r(z)\,[\tau(z) + i\nu(z)], \quad r(z) > 0 \quad (z \in D_0 \cup C). \tag{4.4}$$

In particular, when $z = c(t)$,

$$S(z) = r(z)\, e^{i\theta t}. \tag{4.5}$$

Next we observe that for $\gamma \in \Gamma$

$$S(\gamma) = \tau(\gamma)\, r(\gamma), \quad \nu(\gamma) = 0, \quad \tau(\gamma) = \pm 1. \tag{4.6}$$

Without altering the solutions to problems $\mathscr{A}$ and $\mathscr{B}$ we can replace the field v by $-v$. Moreover, $S(z)$ is only defined to within a positive constant factor. Therefore, if we are only interested in a single point $\gamma \in \Gamma$, we may assume without loss of generality that

$$S(\gamma) = \tau(\gamma) = r(\gamma) = 1. \tag{4.7}$$

It is clear from (4.4)–(4.5) that the function $\Psi(z)$ introduced in 3.3 admits the following representation:

$$\Psi(z) = \frac{v(z)}{\tau(z)}.$$

Hence, if z moves around the contour C in a positive direction, then $v(z)$ changes sign at each point $\gamma \in \Gamma$. Here, if $\gamma \in \Gamma_+$, then the sign changes from plus to minus, while if $\gamma \in \Gamma_-$, then the sign changes from minus to plus (on the assumption that $\tau(\gamma) = 1$).

From this point on we shall assume that $| v(z) | = 1$, that C is the unit circle ($c(t) = e^{it}$) and $z_0 = 0$. Because of (3.1), when $z = e^{it}$,

$$v(z) = ize^{i\theta t}. \tag{4.8}$$

From (4.5) and (4.8) we have

$$izS(z) = r(z)\, v(z). \tag{4.9}$$

4.4. We shall now carry out the construction of the function $S(z)$.

Let l be the index of the vector field v. Then the function $\chi(t) = \theta(t) + (l + 1)\, t$ has period 2π and can be regarded as a continuous function on C. This function is obviously Hölder continuous. Hence (see, for example, V. I. Smirnov [12], 117) Schwarz's formula

$$\sigma(z) = \frac{1}{2\pi} \int_{-\pi}^{\pi} \chi(t)\, \frac{e^{it} + z}{e^{it} - z}\, dt$$

defines on $C \cup D$ a Hölder continuous complex function $\sigma(z)$ which is regular on D and satisfies the condition $\operatorname{Re} \sigma(e^{it}) = \chi(t)$. Note that the function $\sigma(z) + i(l + 1)$ in z has real part $\theta(t)$ for $z = e^{it}$. Hence for $z = e^{it}$ the function

$$S(z) = e^{i[\sigma(z) + i(l+1) \ln z]} = z^{-l-1}\, e^{i\sigma(z)}$$

differs from $e^{i\theta(t)}$ by a positive factor. Consequently this function satisfies condition 4.3.C. It is easy to see that conditions 4.3.A–4.3.B are also satisfied.

If the vector field $v(z)$ is smooth in some neighbourhood of the point $\alpha = e^{it_0}$, then the function $\theta(t)$ possesses a Hölder continuous derivative $\theta'(t)$ in a neighbourhood of t_0. In this case, the sufficiently small $\epsilon > 0$ the function $\sigma'(t)$ is also Hölder continuous[1] on the set $A_\epsilon = \{z: |z| \leqslant 1, |z - e^{it_0}| < \epsilon\}$. Obviously, the functions $S'(z)$ and $T'(z)$ are also Hölder continuous as A_ϵ.

4.5. By using the characteristic function $S(z)$ we can reduce Problem $\mathscr{B}$ to a simpler Problem $\mathscr{C}$, which depends only on the set Γ and the index C, and on no other detail of construction of the vector field v.

We call a complex function $F(z)$ ($z \in D^*$) a *solution of Problem* $\mathscr{C}$ if:

4.5.A. $z^l F(z)$ *is regular on* D *and continuous on* D^*,

4.5.B. $\operatorname{Re} F(z) = 0$ *for* $z \in C \setminus \Gamma$.

Suppose that the functions f and F are connected by the relation

$$f(z) = \frac{T(z)}{iz} F(z) \quad (z \in D^*). \tag{4.10}$$

Then

$$z^l F(z) = iz^{l+1} S(z) f(z).$$

According to 4.3 the function $z^{l+1} S(z)$ is regular, non-vanishing on D, and continuous on D^*. Hence the function F satisfies condition 4.5.A if and only if f satisfies condition 4.2.A. Moreover, it is clear from (4.9) that F satisfies condition 4.5.B if and only if f satisfies condition 4.2.B.

Thus, as F ranges through all the solutions of Problem $\mathscr{C}$, (4.10) gives all the solutions of Problem $\mathscr{B}$. Consequently (see 4.2), the formula

$$h(z) = \Re f(z) = \operatorname{Re} \int_0^z \frac{T(z)}{iz} F(z)\, dz \tag{4.11}$$

gives all the solutions of Problem $\mathscr{A}$ to within an additive constant.

4.6. We put $z^* = \bar{z}^{-1}$. (The points z, z^* are obtained from each other by taking inverses with respect to the unit circle.) We shall write

$$F^*(z) = \overline{F(z^*)}.$$

Consider the regions

$$D_0 = \{z: 0 < |z| < 1\}, \quad D_1 = \{z: 1 < |z| < \infty\}, \quad \hat{D} = D_0 \cup C^* \cup D_1.$$

---

[1] To prove this assertion it is sufficient to express $\sigma(z)$ in the form

$$\sigma(z) = a + \frac{z}{\pi i} \int_C \frac{\chi(\ln w)}{w} \frac{dw}{w - z}$$

(where a is a constant), and to make use of the well-known properties of Cauchy-type integrals (see, e.g. F. D. Gakhov [1], 4.4).

We shall call a complex function $F(z)$ $(z \in \hat{D})$ a *solution to Problem* $\hat{\mathscr{C}}$ if:

4.6.A. *$F(z)$ is an analytic function on $\hat{D}$.*

4.6.B. $F(z^*) = -\overline{F(z)}$ $(z \in \hat{D})$.

4.6.C. *$z^l F(z)$ converges to a finite limit as $z \to 0$.*

It is clear that if the function F is a solution to Problem $\hat{\mathscr{C}}$, then its restriction to the set D^* is a solution to Problem $\mathscr{C}$. We shall prove that the converse statement is also true. For let $F(z)$ $(z \in D^*)$ be a solution of Problem $\mathscr{C}$. We extend this function onto $\hat{D}$ by putting

$$F(z) = -\overline{F(z^*)} \quad \text{for} \quad z \in D_1. \tag{4.12}$$

It is easy to see that the extended function is analytic on D_1 and continuous on $\hat{D}$. By the well-known principle of continuity (see, for example, R. Courant [6], Chap. 5, §2) the extended function is analytic on $\hat{D}$. It is clearly a solution of Problem $\hat{\mathscr{C}}$.

We observe that if $F(z)$ is an analytic function in $\hat{D}$ for which the following finite limits exist

$$\lim_{z \to 0} z^l F(z), \qquad \lim_{z \to \infty} z^{-l} F(z), \tag{4.13}$$

then the formula

$$\hat{F}(z) = F^*(z) - F(z) \tag{4.14}$$

defines a solution to Problem $\hat{\mathscr{C}}$, and in this way all the solutions of this problem are obtained.

4.7. We now show that the set $\mathscr{A}$ satisfies condition 1.1.C. Let h_n be a sequence of harmonic functions that are defined on a neighbourhood of the boundary of the disc D and satisfy conditions 3.2.A.–3.2.B. We suppose that $h_n \to h$ on U. Beginning with the functions h_n and h we define by (4.1) analytic functions f_n and f on U, and we then define, using (4.10), analytic functions F_n and F. Let U_1 be the region obtained from U by taking inverses with respect to the circle C. Using (4.12) we extend the functions F_n onto the region $\hat{U} = U \cup C^* \cup U_1$. Clearly the functions f_n satisfy conditions 4.2.A–4.2.B, the functions F_n satisfy conditions 4.5.A–4.5.B, while the extended functions F_n satisfy the postulates 4.6.A–4.6.B (replacing D by U, D^* by U^*, D_1 by U_1, and $\hat{D}$ by $\hat{U}$). It follows from the convergence of h_n to h in U that $f_n \to f$ and $F_n \to F$ in U. The analytic functions F_n converge, therefore, in U to some function coinciding with F on U. It is clear that this limit function satisfies conditions 4.6.A–4.6.B. Thus, the function F satisfies conditions 4.5.A–4.5.B, while the function h satisfies conditions 3.2.A-3.2.B.

4.8. We shall now consider some particular solutions of the problems described above. To begin with, let $l \geqslant 0$.

It is easy to see that the functions

$$\left.\begin{aligned} &\Phi_k(z) = i(z^{-k}+z^k) \qquad (k=0,\ 1,\ 2,\ \ldots,\ l),\\ &\Phi_{-k}(z) = z^{-k}-z^k \qquad (k=1,\ 2,\ \ldots,\ l),\\ &\Phi_\gamma(z) = \frac{1}{2}\frac{z+\gamma}{z-\gamma} \qquad (\gamma\in\Gamma) \end{aligned}\right\} \tag{4.15}$$

are solutions of Problem $\mathscr{C}$.

Using formula (4.11) we find the corresponding solutions to Problem $\mathscr{A}$

$$\left.\begin{aligned} &h_k(z) = \operatorname{Re}\int_0^z \frac{T(z)}{z}(z^{-k}+z^k)\,dz \qquad (k=0,\ 1,\ 2,\ \ldots,\ l),\\ &h_{-k}(z) = \operatorname{Re}\int_0^z \frac{T(z)}{iz}(z^{-k}-z^k)\,dz \qquad (k=1,\ 2,\ \ldots,\ l),\\ &h_\gamma(z) = \operatorname{Re}\int_0^z \frac{T(z)}{2iz}\frac{z+\gamma}{z-\gamma}\,dz \qquad (\gamma\in\Gamma). \end{aligned}\right\} \tag{4.16}$$

The functions h_k and h_{-k} are continuous on the closed disc $D \cup C$; the function h_γ is also continuous on $D \cup C$, except at the point γ where it has a singularity. We shall now investigate the nature of this singularity.

We note that

$$h_\gamma(z) = \varphi_\gamma(z) + A_\gamma(z),$$

where

$$\varphi_\gamma(z) = \operatorname{Re}\left[\frac{1}{i}T(\gamma)\int_0^z \frac{dz}{z-\gamma}\right],$$

$$A_\gamma(z) = \operatorname{Re}\left[\frac{1}{2i}\int_0^z \frac{(z+\gamma)z^{-1}T(z)-2T(\gamma)}{z-\gamma}\,dz\right].$$

According to 4.3, the function $T(z)$ has a zero at the point 0 of order $l+1 \geqslant 1$, and it follows from 4.3.B that the function $(z+\gamma)z^{-1}T(z)$ is Hölder continuous on $D \cup C$. Consequently, the absolute value of the integrand in the expression for $A_\gamma(z)$ is majorized by the function $k|z-\gamma|^{\lambda-1}$, where λ and k are positive constants. Therefore the integral converges absolutely and uniformly in the closed disc $D \cup C$, and A_γ is a continuous harmonic function on this disc. Furthermore, taking (4.6) into account we have

$$\varphi_\gamma(z)=\tau(\gamma)^{-1}r(\gamma)^{-1}\,\mathrm{Re}\,\frac{1}{i}\int_0^z\frac{dz}{z-\gamma}=\tau(\gamma)^{-1}r(\gamma)^{-1}\arg\frac{\gamma-z}{\gamma}, \tag{4.17}$$

where the quantity $\arg\left(\frac{\gamma-z}{\gamma}\right)$ above takes values in the interval $\left(-\frac{\pi}{2},\frac{\pi}{2}\right)$. The function $\varphi_\gamma(z)$ is continuous everywhere on $D\cup C$ except at the point γ. Its limit on approaching γ along the contour C and in the direction of the vector $v(\gamma)$ is equal to $-\frac{\pi}{2}r(\gamma)$; from the opposite direction the limit is $\frac{\pi}{2}r(\gamma)$ (we recall that $\tau(\gamma)=\pm 1$). If we are only interested in the one point $\gamma\in\Gamma$, then according to 4.3 we can assume without loss of generality that $\tau(\gamma)=r(\gamma)=1$, and

$$\varphi_\gamma(z)=\arg\frac{\gamma-z}{\gamma}. \tag{4.17'}$$

We shall now look at the case $l=-m<0$. This time the functions $\Phi_\gamma(z)$ no longer satisfy condition 4.5.A and are therefore not solutions of Problem $\mathscr{C}$ (although they satisfy condition 4.5.B). On the other hand, the functions

$$\frac{2z^m\gamma^{1-m}}{z-\gamma}=\frac{z+\gamma}{z-\gamma}+1+2z\gamma^{-1}+\ldots+2z^{m-1}\gamma^{1-m} \tag{4.18}$$

satisfy condition 4.5.A, but not condition 4.5.B.

Let $a(\gamma)$ be a real-valued function on the set Γ such that

$$\sum_{\gamma\in\Gamma}a_\gamma\gamma^{-k}=0\quad(k=0,1,2,\ldots,m-1). \tag{4.19}$$

Then the function

$$F(z)=\sum_{\gamma\in\Gamma}a_\gamma\Phi_\gamma(z)=\sum_{\gamma\in\Gamma}a_\gamma\frac{z^m\gamma^{1-m}}{z-\gamma} \tag{4.20}$$

satisfies conditions 4.5.A–4.5.B and so is a solution to Problem $\mathscr{C}$. The corresponding solution of Problem $\mathscr{A}$ is given by the formula

$$h(z)=\Re\left(\frac{T}{iz}F\right)=\mathrm{Re}\int_0^z\left[\frac{T(z)}{z}\sum_{\gamma\in\Gamma}\frac{a_\gamma}{z-\gamma}\left(\frac{\gamma}{z}\right)^{l+1}\right]dz. \tag{4.21}$$

We observe that

$$h(z)=\sum_{\gamma\in\Gamma}a_\gamma\varphi_\gamma(z)+B(z), \tag{4.22}$$

where φ_γ is given by (4.17), and

$$B(z) = \mathrm{Re} \int_0^z \sum_{\gamma \in \Gamma} \frac{a_\gamma \gamma^{l+1}}{i(z-\gamma)} [T(z) z^{-l-1} - T(\gamma) \gamma^{-l-1}] \, dz$$

is a continuous harmonic function on the closed disc $D \cup C$.

4.9. We now consider the case $l = -m < 0$ more closely.

In §5 it will be proved that if the set Γ_+ is empty, then the only solutions to Problem $\mathfrak{A}$ that are bounded below are the constants. We can therefore presuppose that $l_+ >$ and consequently (see 3.2)) $l_- = 2m - 2 + l_+ \geqslant 2m - 1$. Select any subset Γ_1 of Γ_- consisting of $2m - 1$ points, and put

$$P_\gamma(w) = \gamma^{m-1} w^{1-m} \prod_{\substack{\beta \in \Gamma_1 \\ \beta \neq \gamma}} \frac{w-\beta}{\gamma-\beta}, \tag{4.23}$$

$$\Lambda(w, z) = \frac{1}{2} \sum_{\gamma \in \Gamma_1} P_\gamma(w) \frac{z+\gamma}{z-\gamma} = \sum_{n=-m+1}^{m-1} w^{-n} \Lambda_n(z). \tag{4.24}$$

THEOREM 4.1. *Suppose that the function F is regular in a neighbourhood of the origin and that*

$$c_n = \frac{1}{n!} F^{(n)}(0).$$

Put

$$\mathfrak{C} F = F + \sum_{n=0}^{m-1} (c_n \Lambda_n + \bar{c}_n \Lambda_{-n}), \tag{4.25}$$

where the functions $\Lambda_n(z)$ are defined by (4.24). *Then either the functions F and $\mathfrak{C} F$ both satisfy or fail to satisfy condition* 4.5.B. *The function $z^l \mathfrak{C} F$ is regular in a neighbourhood of the origin. If F is regular on D and continuous on D^*, then the function $\mathfrak{C} F$ satisfies*[1] *condition* 4.5.A.

PROOF 1°. Note that $P_\gamma^* = P_\gamma$. It follows from this that

$$\Lambda(w^*, z) = -\overline{\Lambda(w, z^*)}, \qquad \Lambda_{-n} = -\Lambda_n^*. \tag{4.26}$$

Therefore the function $g = \mathfrak{C} F - F$ satisfies the relation $g^* = -g$. This means that $\mathrm{Re}\, g = 0$ on C, and that the functions F and $\mathfrak{C} F = F + g$ either both satisfy or both fail to satisfy condition 4.5.B.

2°. We shall write $f \equiv g$ if the function $(f - g) z^l$ is regular on a neighbourhood of the origin. We shall show that

[1] If $l \geqslant 0$ then it is natural to assume that the set Γ_1 is empty, $\Lambda(w, z) = 0$ and $\mathfrak{C} F = F$. Theorem 4.1 is trivial under these circumstances.

$$\left.\begin{aligned}\Lambda_0(z) &\equiv -\frac{1}{2},\\ \Lambda_n(z) &\equiv -z^n, \quad \Lambda_{-n}(z) \equiv 0 \quad (n=1, 2, \ldots, m-1).\end{aligned}\right\} \tag{4.27}$$

According to (4.18)

$$\frac{z+\gamma}{z-\gamma}+1+2\sum_{n=1}^{m-1} z^n\gamma^{-n}\equiv 0.$$

Hence

$$\Lambda(w, z) \equiv -\frac{1}{2}A_0(w)-\sum_{n=1}^{m-1} A_n(w)z^n, \tag{4.28}$$

where

$$A_n(w)=\sum_{\gamma} P_\gamma(w)\gamma^{-n}.$$

Note that the function $\widetilde{P}_\gamma(w) = w^{m-1}P_\beta(w)$ is a polynomial of degree not greater than $2m-2$, and $\widetilde{P}_\gamma(\gamma) = 1, \widetilde{P}_\gamma(\beta) = 0$ when $\beta \neq \gamma$ ($\beta \in \Gamma_1$). It follows that the function $w^{m-1}A_n(w) - w^{m-1-n}$ is a polynomial of degree not greater than $2m-2$ vanishing at all $2m-1$ points of the set Γ_1. Such a polynomial is identically zero. Hence $A_n(w) = w^{-n}$. Substituting these values in (4.28) we obtain

$$\Lambda(w, z) \equiv -\frac{1}{2}-\sum_{n=1}^{m-1} w^{-n}z^n. \tag{4.29}$$

A comparison of (4.24) with (4.29) now gives (4.27).

It is clear from (4.27) that

$$\mathfrak{C}F \equiv F-\sum_{n=0}^{m-1} c_n z^n \equiv 0.$$

Consequently the function $z^l\ \mathfrak{C}\ F$ is regular on a neighbourhood of the origin.

3°. The functions $\Lambda_n(z)$ can be expressed as linear combinations of the functions $\dfrac{z+\gamma}{z-\gamma}$ ($\gamma \in \Gamma_1$). Hence they are regular on D and continuous on D^* Clearly the function $\mathfrak{C}F - F$ also has these properties. Hence if F is regular on D and continuous on D^*, then these same properties will hold for the function $\mathfrak{C}$ and by 2° also for the function $z^l\ \mathfrak{C}\ F$.

REMARK 1. It follows from Theorem 4.1 and from 4.5 that if F is regular on D, continuous on D^*, and satisfies condition 4.5.B, then the function

$$h(z)=\operatorname{Re}\int_0^z \frac{T(z)}{iz}\,\mathfrak{C}F(z)\,dz \tag{4.30}$$

is a solution of Problem $\mathcal{A}$. A particularly important role is played by the solutions h_β that are obtained by means of this formula from the functions

$\Phi_\beta(z) = \frac{1}{2}\,\frac{z+\beta}{z-\beta}$. (For $l \geqslant 0$ formula (4.30) agrees with (4.16).)

REMARK 2. Note that for the function

$$\Phi_w(z) = \frac{1}{2}\,\frac{z+w}{z-w} \tag{4.31}$$

the following equation holds:

$$\mathfrak{C}\Phi_w(z) = \Phi_w(z) - \Lambda(w, z) + \sum_{n=1}^{m-1} (w^n - w^{*n})\,\Lambda_{-n}(z). \tag{4.32}$$

In particular, when $\alpha \in C$,

$$\mathfrak{C}\Phi_\alpha(z) = \Phi_\alpha(z) - \Lambda(\alpha, z) \tag{4.33}$$

and $\mathfrak{C}\Phi_\alpha(z) = 0$ when $\alpha \in \Gamma_1$ (and consequently $h_\alpha = 0$).

It is evident from (4.32) that the function $[\Phi_w(z) - \Lambda(w, z)]\,z^l$ is regular in a neighbourhood of the origin. A closer analysis shows that

$$z^l\,[\Phi_w(z) - \Lambda(w, z)] = \frac{w^{1-m}}{z-w} - \sum_{\gamma \in \Gamma_1} \frac{\gamma^{1-m} p_\gamma(w)}{z-\gamma}. \tag{4.34}$$

§5. The Minimum Principle and its consequences

5.1. Let $h \in \mathcal{A}$. (We recall that this means that h is subject to conditions 3.2.A–3.2.B.) We write $h \in \mathcal{A}_+$ if the function h is bounded below near C and

$$\varliminf_{z \to \gamma} h(z) \geqslant 0 \quad \text{for all} \quad \gamma \in \Gamma_+. \tag{5.1}$$

It is clear that the set $\mathcal{A}_+$ fulfills conditions 1.2.A–1.2.D. The basic problem of this section is to show that the Minimum Principle 1.2.E is also satisfied. More precisely we shall prove the following:

THEOREM 5.1. *Let D be a region bounded by a smooth contour C, K a compact subset of D, and $D_0 = D \setminus K$. It will be assumed that the function $h \in \mathcal{A}_+$ is continuous on $D_0 \cup \partial K$ and is harmonic on D_0. If the set Γ_+ is non-empty, then at least one of the following two statements is true*:

a) *there is a point $z_0 \in \partial K$ such that $h(z_0) < h(z)$ for all $z \in D_0$*;

b) *$h(z) \geqslant 0$ for all $z \in D_0$.*

If the set Γ_+ is empty, then either statement a) *is true or h is constant throughout D_0.*

By applying this theorem to the case where U is empty we obtain:

COROLLARY. *If the set Γ_+ is non-empty, then each solution to Problem $\mathcal{A}$ that is bounded below and belongs to $\mathcal{A}_+$ is non-negative, and each bounded solution to Problem $\mathcal{A}$ that belongs to $\mathcal{A}_0$ vanishes. If the set Γ_+ is empty, then every solution of Problem $\mathcal{A}$ that is bounded below is a constant.*

The proof of Theorem 5.1 will be given in 5.5 after we have proved a number of lemmas. In order to explain the role of these lemmas, let $h \in \mathcal{A}_+$ be non-constant and continuous on the compact set $C \cup D_0 \cup \partial K$. Then the minimum of h on $C \cup D_0 \cup \partial K$ is attained at some point z_0. Lemma 5.1 shows that z_0 cannot belong to D_0, and if $z_0 \in C$, then the vector $v(z_0)$ is tangent to C, or in other words, $\Psi(z_0)=0$. If $z_0 \in \partial K$, then condition a) is valid. If $z_0 \in C$, then by Lemma 5.4 the function $\Psi(z)$ must change sign at z_0, and hence $z_0 \in \Gamma$. It then follows from Lemma 5.2 that $z_0 \notin \Gamma_-$. This means that $z_0 \in \Gamma_+$. By virtue of (5.1) $h(z_0) \geqslant 0$, and consequently condition b) holds.

Lemma 5.3 is a subsidiary result, which is needed for the proof of Lemma 5.4. Further consequences of Theorem 5.1 will be deduced in 5.6.

5.2. LEMMA 5.1. *Let D, C, K, D_0 have the same meanings as in Theorem* 5.1. *Let $D_0 \subseteq Q \subseteq C \cup D_0 \cup \partial K$. Suppose that the function $h(z)$ $(z \in Q)$ is non-constant, continuous on Q, harmonic on D_0, and attains its minimum on Q at some point $z_0 \in Q$. Then $z_0 \in C \cup \partial K$.*

If z_0 belongs to a smooth arc (α, β) of the contour C, and if h at z_0 has a directional derivative along a vector v and v makes an acute angle with the inward normal, then

$$\frac{\partial h}{\partial v}(z_0) > 0. \tag{5.2}$$

PROOF. The first assertion is well-known (see. for example, I. G. Petrovskii [10], §28). Let us prove the second assertion. By a conformal transformation the general case can be reduced to the case where C is a circle. We construct a closed disc contained in $C + D_0$ and having a single point z_0 in common with C. Clearly a strict minimum of h on this disc is attained at z_0. According to a well-known lemma (see I. G. Petrovskii [10], §28, Lemma 1) the derivative of h along any direction that makes an acute angle with the inward normal to C is positive.

5.3. LEMMA 5.2. *Let $[\alpha, \beta]$ be a smooth arc of the closed contour C, and let $\gamma \in (\alpha, \beta)$. Let a vector field v be given on (α, β) so that*

$$\begin{aligned} \Psi(z) < 0 \quad &\textit{for} \quad z \in (\alpha, \gamma); \\ \Psi(z) > 0 \quad &\textit{for} \quad z \in (\gamma, \beta). \end{aligned} \tag{5.3}$$

We denote by D the region bounded by C, and by Q the set obtained from $D \cup C$ by removing the point γ.

Suppose that the function $h(z)$ $(z \in Q)$ is continuous and bounded below on Q, harmonic on D, and that it satisfies the condition

$$\frac{\partial h}{\partial v} = 0 \tag{5.4}$$

at all points of (α, β) (except γ).

Then

$$\inf_{Q} h = \inf_{C \setminus (\alpha, \beta)} h. \tag{5.5}$$

PROOF. Let $[\alpha_j, \beta_j]$ be a smooth arc of the contour C_j, where C_j bounds the region D_j, and let $\gamma_j \in (\alpha_j, \beta_j)$ ($j = 1, 2$). Consider a conformal mapping F_j of D_j onto the unit disc and construct a bilinear transformation G taking $F_2(\alpha_2)$, $F_2(\gamma_2)$, $F_2(\beta_2)$, respectively, into $F_1(\alpha_1)$, $F_1(\gamma_1)$, $F_1(\beta_1)$. The mapping $F = F_1^{-1} G F_2$ takes C_2, D_2, $[\alpha_2, \beta_2]$, γ_2, respectively, into C_1, D_1, $[\alpha_1, \beta_1]$, γ_1. It is conformal on D_2, continuous on $D_2 \cup C_2$, and has a non-zero continuous derivative $F'(z)$ at all points of the arc (α_2, β_2). Using this mapping we can see without too much difficulty that Lemma 2 is valid for any contour C, smooth arc $[\alpha, \beta]$, and point γ, provided that it is valid for some special choice of contour, arc $[\alpha, \beta]$, and point γ.

Let us choose for C the contour that forms the boundary of the semi-disc

$$D = \{z = x + iy: \quad y > 0, \quad x^2 + y^2 < 1\},$$

for the arc $[\alpha, \beta]$ its diameter $[-1, 1]$, and for γ the point 0. Consider the function

$$b(z) = \ln |z|^{-1}$$

and put

$$K_\varepsilon = \{z: \quad y \geqslant 0, \ b(z) = b(\varepsilon)\},$$
$$Q_\varepsilon = \{z: \quad y \geqslant 0, \ b(1) \leqslant b(z) \leqslant b(\varepsilon)\}.$$

The boundary of Q_ϵ consists of the two semicircles K_1 and K_2 and the portion L_ϵ of the diameter determined by the condition $b(1) < b(x) < b(\epsilon)$. Note that on L_ϵ

$$\frac{\partial b}{\partial x} = -x^{-1}, \quad \frac{\partial b}{\partial y} = 0. \tag{5.6}$$

Let $v_1(z)$, $v_2(z)$ be the coordinates of $v(z)$. We write

$$\hat{v}(z) = v(z) \operatorname{sign} v_2(z).$$

For each $z \in L_\epsilon$, $\hat{v}(z)$ forms an acute angle with the inward normal and, because of (5.6) and (5.3),

$$\frac{\partial b}{\partial \hat{v}} \leqslant 0. \tag{5.7}$$

Denote the greatest lower bounds of the function h over the sets Q and $K_1 = C \setminus (\alpha, \beta)$ by q and k, respectively. The function

$$\mathcal{H}_\varepsilon(z) = h(z) - k + (k - q)\frac{b(z) - b(1)}{b(\varepsilon) - b(1)}$$

is continuous on Q_ϵ and, by virtue of (5.4) and (5.7)

$$\frac{\partial \mathcal{H}_\varepsilon}{\partial \hat{v}} = \frac{\partial h}{\partial \hat{v}} + \frac{k-q}{b(\varepsilon)-b(1)} \frac{\partial b}{\partial \hat{v}} = \frac{k-q}{b(\varepsilon)-b(1)} \frac{\partial b}{\partial \hat{v}} \leqslant 0$$

for any $z \in L_\epsilon$. By Lemma 5.1, the minimum of $\mathcal{H}_\varepsilon$ over the set Q_ϵ is attained on $K_1 \cup K_\epsilon$. Since $\mathcal{H}_\varepsilon$ is non-negative on $K_1 \cup K_\epsilon$, it is non-negative on Q_ϵ. Making $\epsilon \downarrow 0$ we arrive at the inequality $h(z) \geqslant k$ throughout Q. Consequently (5.5) holds.

LEMMA 5.3. *Let $[\alpha, \beta]$ be a smooth arc of the closed contour C, and let a vector field v be given on (α, β) such that $\Psi(z) > 0$. Denote by D the region bounded by the contour C and by Q the set obtained from $D \cup C$ by removing the point α. Suppose that the function $h(z)$ ($z \in Q$) is continuous and bounded below on Q, harmonic on D, and that it satisfies the condition*

$$\frac{\partial h}{\partial v} = 0$$

at all points of the arc (α, β). Then

$$\inf_Q h = \inf_{C \setminus [\alpha, \beta)} h.$$

PROOF. Just as in Lemma 5.2 it suffices to prove our lemma for some particular contour C and particular arc $[\alpha, \beta]$.

In order to carry out the proof, it suffices to repeat the arguments used in the proof of Lemma 5.2, putting

$$D = \left\{ z:\ y > 0,\ \left| z - \tfrac{1}{2} \right| < \tfrac{1}{2} \right\},\ \alpha = 0,\ \beta = 1,\ b(z) = x\,|z|^{-2}.$$

The only difference is that the function H_ϵ is not defined at the point 0 belonging to Q_ϵ. Hence the fact that H_ϵ is non-negative does not follow directly from Lemma 5.1 and requires a separate proof.

Let $\delta > 0$, and put

$$H_\varepsilon^\delta = H_\varepsilon + \delta \ln |z|^{-1}.$$

Note that $\partial H_\epsilon^\delta / \partial \hat{v} \leqslant 0$ for all $z \neq 0$ in L_ϵ. For any $0 < r < 1$ the function H_ϵ^δ is continuous on the set

$$Q_\varepsilon^r = Q_\varepsilon \cap \{z:\ |z| \geqslant r\}.$$

By Lemma 5.1 the minimum of this function over Q_ϵ^r is attained at some point of the boundary that does not belong to L_ϵ. If r is sufficiently small, then at all such points the function H_ϵ^δ is non-negative. This means that it is non-negative throughout Q_ϵ^r. Hence we can conclude that $H_\epsilon^\delta(z) \geqslant 0$ for all $z \neq 0$ in Q_ϵ. Letting $\delta \downarrow 0$ we arrive at the conclusion that $H_\epsilon \geqslant 0$ for all $z \neq 0$ in Q_ϵ.

5.4. LEMMA 5.4. *Let* $[\alpha, \beta]$ *be a smooth arc of the closed contour C and* $\gamma \in (\alpha, \beta)$. *Let v be a given Hölder continuous vector field on the arc* $[\gamma, \beta)$ *such that*

$$\Psi(\gamma) = 0, \quad \Psi(z) > 0 \quad for \quad z \in (\gamma, \beta). \tag{5.8}$$

Let D be the region bounded by C. Suppose that $h(z)$ $(z \in D \cup C)$ *is continuous on a neighbourhood of* γ, *harmonic on D, and continuously differentiable at all points of* (γ, β), *with*

$$\frac{\partial h}{\partial v} = 0.$$

Then for sufficiently small $\epsilon > 0$ *the greatest lower bounds of h over the sets* $U_\epsilon (D \cup C) \cap \{z: |z-\gamma| \leqslant \varepsilon\}$ *and* $K_\epsilon = \{D \cup C\} \cap \{z: |z-\gamma| = \varepsilon\}$ *are equal.*

REMARK. It is clear that Lemma 5.4 remains valid if the arc $[\gamma, \beta)$ is replaced by $(\alpha, \gamma]$, and (5.8) by the condition

$$\Psi(\gamma) = 0, \quad \Psi(z) < 0 \quad \text{for} \quad z \in (\alpha, \gamma).$$

To verify this we need only use the transformation $F(z) = \bar{z}$.

PROOF OF LEMMA 5.4. Without loss of generality D can be considered to be the unit disc. We extend the field v onto the whole circle C in such a way that it is Hölder continuous, so that $\Psi(z)$ is non-negative on some arc $[\alpha', \gamma)$ and that the index l is zero. In relation to this field the point γ belongs to Γ_-, and by (4.16) to this point there corresponds a function h_γ continuous throughout $D \cup C$ except at γ; the restriction of this function to C has a discontinuity of the first kind at γ. Choose constants a and b so that the function $g = ah_\gamma + b$ has the limits -2 and $+2$ as the point z approaches γ along the arcs (α, γ) and (γ, β), respectively.

The boundary of U_ϵ consists of the arc K_ϵ and an arc (α_1, β_1) of the unit circle C. Choose $\epsilon > 0$ so small that h is continuous on U_ϵ and that

$$\left.\begin{aligned} g(z) &< -1 \quad \text{for} \quad z \in (\alpha_1, \gamma), \\ g(z) &> 1 \quad \text{for} \quad z \in (\gamma, \beta_1). \end{aligned}\right\} \tag{5.9}$$

Denote by u and k the greatest lower bounds of h over the sets U_ϵ, K_ϵ, respectively, and let $q = \sup_{U_\epsilon} |g|$. The function

$$H(z) = h(z) - u - \frac{g(z)+1}{q+1}(k-u)$$

satisfies the conditions of Lemma 5.3 with respect to the region U_ϵ and the arc (γ, β_1). Since $H(z)$ is non-negative on $K_\epsilon \cup (\alpha_1, \gamma)$, it is non-negative throughout U_ϵ, and we have

$$h(z) \geqslant u + \frac{g(z)+1}{q+1}(k-u) \qquad (z \in U_\epsilon).$$

By (5.9), $h(z) \geqslant u + \dfrac{2}{q+1}(k-u)$ for $z \in (\gamma\, \beta_1)$, and hence

$$u \geqslant h(\gamma) \geqslant u + \frac{2}{q+1}(k-u) \qquad (z \in (\gamma, \beta_1)).$$

Hence $k = u$, and the lemma is proved.

5.5. PROOF OF THEOREM 5.1. The theorem is obvious if the function h is constant; suppose then that it is not constant. Put $z \in \Gamma_0$ if $z \notin \Gamma$ and the vector $v(z)$ is tangent to C. Denote by D_ϵ the set of all points $z \in D_0^* = D^* \setminus K$ that are at a distance not less than ϵ from all points of $\Gamma \cup \Gamma_0$. Let $c = \inf h(z)$ for $z \in D_0^*$.

We suppose at first that Γ_+ is non-empty. Consider the set A_ϵ consisting of all points $z \in D_0^*$ that are at a distance not less than ϵ from Γ_+. If $c < 0$, then because of (5.1) $\inf\limits_{z \in A_\epsilon} h(z) = c$ for sufficiently small $\epsilon > 0$. It follows from Lemmas 5.2, 5.4 that if $\epsilon > 0$ is sufficiently small, then $\inf\limits_{D_\epsilon} h = c$. But the function h is continuous on the set D_ϵ, and therefore there exists $z_0 \in D_\epsilon$ such that $h(z_0) = c$. However, by Lemma 5.1 the point z_0 can belong neither to D_0 nor to $C \cap D_\epsilon$. Consequently $z_0 \in \partial K$, and assertion a) is true.

If the set Γ_+ is empty, then the same argument leads to the desired conclusion if A_ϵ is replaced by D_0^*.

5.6. Using Theorem 5.1 we shall now give a description of all the bounded solutions h of Problem $\mathfrak{A}$ for which the limit

$$\lim_{z \to \alpha} h(z) = b_\alpha$$

exists for any $\alpha \in \Gamma_+$. We put $h(\alpha) = b_\alpha$, so that the function h becomes defined and continuous on the set $D^* \cup \Gamma_+$.

THEOREM 5.2. *For any function $b(\alpha)$ $(\alpha \in \Gamma_+)$ there exists a unique bounded solution h to Problem $\mathfrak{A}$ satisfying the conditions*

$$h(\alpha) = b(\alpha). \tag{5.10}$$

If the index $l \geqslant 0$, then each bounded solution that is continuous on $D^* \cup \Gamma_+$ is uniquely expressible in the form

$$h(z) = a + \sum_{k=-l}^{l} a_k h_k(z) + \sum_{\gamma \in \Gamma_-} a_\gamma h_\gamma(z), \tag{5.11}$$

where h_k and h_γ are the functions defined by (4.16) *and a, a_k, a_γ are constants.*

If $l < 0$ and $l_+ > 0$, then every bounded solution continuous on $D^* \cup \Gamma_+$ can be uniquely written in the form

$$h(z) = a + \sum_{\gamma \in \Gamma_-} a_\gamma h_\gamma(z), \tag{5.12}$$

where a, a_γ are constants connected by the relations

$$\sum_{\gamma \in \Gamma_-} a_\gamma \gamma^k = 0 \qquad (|k| < |l|). \tag{5.13}$$

PROOF. 1°. Let $l \geqslant 0$. To begin with we show that the functions $1, h_k, h_\gamma$ are linearly independent and consequently the representation (5.11) is unique. For let

$$h = a + \sum a_k h_k + \sum a_\gamma h_\gamma, \quad F = \sum a_k \Phi_k + \sum a_\gamma \Phi_\gamma.$$

It is evident that $h = \Re\left(F \frac{T}{iz}\right)$. Hence if $h = 0$, then by 4.2, $F\frac{T}{iz} = \mathfrak{S}h = 0$ and consequently $\Sigma a_k \Phi_k + \Sigma a_\gamma \Phi_\gamma = 0$. It is easy to deduce from this that all the coefficients a_k, a_γ are zero. This implies that a is also zero.

2°. We now show that the coefficients in (5.11) can be selected so as to satisfy (5.10).

For the conditions (5.10) give rise to a system of l_+ linear equations in $2l + 2 + l_-$ unknowns. Now according to (3.2) $2l + 2 + l_- = l_+$, so that the number of equations is the same as the number of unknowns. By the corollary to Theorem 5.1 and 1° the corresponding homogeneous system only has the zero solution. Consequently the determinant of the system is non-zero and the system 5.10 has a solution.

3°. Now let $l \leqslant 0$, $l_+ > 0$. We suppose that the function h is defined by (5.12) and that the coefficients satisfy (5.13). Observe that $h = \Re\left(\frac{T}{iz}\mathfrak{C}F\right)$, where $F = \sum_{\gamma \in \Gamma_-} a_\gamma \Phi_\gamma$. By 4.9 it follows from (5.13) that $\mathfrak{C}F = F$. Hence $h = \Re\left(\frac{T}{iz}F\right)$. Repeating the arguments of 1° we conclude that if $h = 0$, then $a_\gamma = 0$ for all $\gamma \in \Gamma_-$, and that $a = 0$.

Next we show that the coefficients in (5.12) can be chosen so that the function h satisfies (5.10). For the conditions (5.10) together with the equations (5.13) lead to a system of $-2l - 1 + l_+$ linear equations in $l + 1$ unknowns. Again the number of equations is the same as the number of unknowns, and the corresponding homogeneous system has only the zero solution. This implies that the system (5.10), (5.12) has a solution. Since this system has along with each complex solution also the complex conjugate solution, it has a real solution.

4°. It remains to note that by the corollary to Theorem 5.1, conditions (5.10) uniquely determine the solution to Problem $\mathcal{A}$.

5.7. By Theorem 5.2 there exists for each point $\alpha \in \Gamma_+$ a unique solution to Problem $\mathcal{A}$ that is equal to one at α and to zero at the remaining points of Γ_+. Let us denote this solution by $p_a(z)$. By Theorem 5.1, all its values lie between zero and one.

Clearly a solution h that satisfies conditions (5.10) is given by the formula

$$h(z) = \sum_{\alpha \in \Gamma_+} b(\alpha)\, p_\alpha(z). \tag{5.14}$$

For $b(\alpha) = 1$ condition (5.10) is satisfied by the function $h(z) = 1$. Hence

$$1 = \sum_{\alpha \in \Gamma_+} p_\alpha(z). \tag{5.15}$$

§6. The Green's Function

6.1. If the set Γ_+ is empty, then the set of all non-negative solutions of the boundary value Problem $\mathcal{A}$ consists of the non-negative constants (see Corollary 1 to Theorem 5.1). The only case that has to be further studied is when the set Γ_+ is non-empty. In this section the Green's function for Problem $\mathcal{A}$ (see 1.3) will be constructed. That is, for each $w \in D_0 = D \setminus \{0\}$ a function $g_w(z)$ $(z \notin D)$ will be constructed satisfying the conditions:

6.1.A. $g_w(z) = h_w(z) - \ln|z - w|$, *where $h_w(z)$ is a harmonic function on the region D.*

6.1.B. *The partial derivatives of $g_w(z)$ can be continuously extended to D^*, and at each point $z \in C^*$ the directional derivative along $v(z)$ is zero.*

6.1.C. $\lim\limits_{z \to a} g_w(z) = 0$ *for all $\alpha \in \Gamma_+$.*

6.1.D. *The function $g_w(z)$ is bounded in a neighbourhood of any point $\beta \in \Gamma_-$.*

From Theorem 5.1 it follows easily that the Green's function is non-negative.[1]

6.2. In 4.2 the operators $\mathfrak{S}$ and $\mathfrak{R}$ were described; they establish a correspondence between the harmonic and the analytic functions on D. We now extend these operators to a somewhat wider class of functions, Namely if

$$\tilde{h}(z) = h(z) + a \ln|z - w|,$$

where h is a harmonic function on D, $w \in D$, and a is a real constant, then we put

$$\mathfrak{S}\tilde{h}(z) = \mathfrak{S}h(z) + \frac{a}{z - w},$$

and if

$$\tilde{f}(z) = f(z) + \frac{a}{z - w},$$

where f is regular analytic function on D, $w \in D$, and a is a real constant, then we put

$$\mathfrak{R}\tilde{f}(z) = \mathfrak{R}f(z) + a \ln|z - w|.$$

[1] If the set Γ_+ is empty, then the Green's function does not exist. For in this case the function $f - c$ belongs to the set $\mathcal{A}_0$ if f does, where c is any constant. Consequently the inequality $g_w(z) - c > 0$ must hold for any c, which is clearly impossible.

It is easy to see that for $z \neq w$ formula (4.1) holds, as before. Formula (4.2) also remains valid, provided one integrates along any curve joining z_0 to z and not passing through w. If the path of integration is smooth, then it can pass through w; however, in this case the integral has to be understood to be in the sense of the Cauchy principal value.

It is easy to see (see 4.2) that the function g_w satisfies conditions 6.1.A–6.1.B if and only if the function $G_w = \mathfrak{S} g_w$ satisfies the conditions:

6.2.A. $G_w(z) + \dfrac{1}{z-w}$ *is analytic on* D *and continuous on* D^*.

6.2.B. $\operatorname{Re} v(z)\, G_w(z) = 0$ *for* $z \in C \setminus \Gamma$.

Let $G_w(z)$ satisfy conditions 6.2.A–6.2.B. Consider the function

$$q_w(z) = \Re G_w(z) = \operatorname{Re} \int_0^z G_w(z)\, dz \tag{6.1}$$

and put

$$g_w(z) = g_w(z) - \sum_{a \in \Gamma_+} g_w(a)\, p_a(z), \tag{6.2}$$

where the $p_a(z)$ are defined in 5.7. Clearly g_w satisfies not only conditions 6.1.A–6.1.B but also conditions 6.1.C–6.1.D. Consequently it is the Green's function.

6.3. THEOREM 6.1. *Let v be a Hölder continuous vector field on the unit circle C with index l and characteristic function $T(z)$. Suppose that the set Γ_+ is non-empty. Put*

$$L_w(z) \doteq \Phi_w(z) - \Lambda(w, z), \tag{6.3}$$

where $\Lambda(w, z)$ is defined by (4.24) *when $l < 0$ and is zero when $l \geqslant 0$, and where*

$$\Phi_w(z) = \frac{1}{2}\frac{z+w}{z-w} \tag{6.4}$$

Furthermore let

$$G_w(z) = \frac{T(z)}{z}\, [\overline{S(w)}\, L_{w^*}(z) - S(w)\, L_w(z)]. \tag{6.5}$$

Then the function $g_w(z)$ ($w \in D$, $w \neq 0$; $z \in D^*$) defined by (6.1)–(6.2) *is the Green's function for Problem* $\mathfrak{A}$.

PROOF. We shall prove that the function

$$F_w(z) = i\overline{S(w)}\, L_{w^*}(z) - iS(w)\, L_w(z) \tag{6.6}$$

satisfies the conditions:

a) the function $A(z) = z^l \left[F_w(z) + \dfrac{iS(z)z}{z-w} \right]$ is regular on D and continuous on D^*;

b) $\operatorname{Re} F_w(z) = 0$ for $z \in C^*$.

Clearly the functions F_w and G_w are related by equation (4.10). By 4.5, 6.2.A follows from a) and 6.2.B from b). Hence (see 6.2) the function $g_w(z)$ given by (6.1)–(6.2) is the Green's function.

The function $A(z)$ is clearly regular on D and continuous throughout D^*, except possibly at the points 0 and w. Its regularity at the point 0 follows from remark 2 in 4.9. If $w \in D$, then $w^* \neq w$ and

$$\lim_{z \to w} A(z) = \left\{ i\overline{S(w)}\, L_{w^*}(w) + iS(w)\, \Lambda(w, w) + iS'(w)\, w + \frac{i}{2} S(w) \right\} w^l.$$

It follows that for $w \neq 0$ the function $A(z)$ is regular at w. Thus, condition a) is proved.

Furthermore, it is easy to see that $\Phi_{w^*} = -\Phi_w^*$. By 1° in the proof of Theorem 4.1, $\Lambda(w^*, z) = -\overline{\Lambda(w, z^*)}$. Hence $L_{w^*} = -L_w^*$, and

$$F_w(z) = [iS(w)\, L^*(z)]^* - iS(w)\, L_w(z). \tag{6.7}$$

The validity of b) follows immediately from this.

REMARK. It follows easily from Theorem 6.1 that the Green's function $g_w(z)$ satisfies condition 1.3.B on the region D^δ obtained from D by removing the point 0.

§7. Asymptotic behaviour of the Green's function

7.1. By (4.4)

$$S(w) = r(w)\, [\tau(w) + i\nu(w)]$$

for all $w \in D \cup C$ except $w = 0$. If we write for short

$$\tilde{T}(z) = \frac{T(z)}{z}, \quad L^\nu(w, z) = L_{w^*}(z) + L_w(z), \quad L^\tau(w, z) = L_{w^*}(z) - L_w(z), \tag{7.1}$$

we can write (6.5) in the form

$$G_w(z) = \tilde{T}(z)\, r(w)\, \{\tau(w)\, L^\tau(w, z) - i\nu(w)\, L^\nu(w, z)\}. \tag{7.2}$$

We shall study the behaviour of this function as $w \to \alpha \in \Gamma$. By (4.6), $\gamma(\alpha) = 0$. According to (4.7) we can suppose without loss of generality that $S(\alpha) = \tau(\alpha) = r(\alpha) = 1$. It follows from (7.1), therefore, that

$$G_w(z) = \tilde{T}(z)\, (L^\tau - i\nu L^\nu) + o(L^\tau) + o(\nu L^\nu), \tag{7.2'}$$

where the asymptotic estimate is uniform for all z in D^*.

7.2. Denote by D_ϵ^* the set obtained from D^* by deleting an ϵ-neighbourhood of the point α. We shall derive further estimates for $G_w(z)$ and $g_w(z)$ acting uniformly on each region D_ϵ^* ($\epsilon > 0$).

Let us write $B = o(A)$ whenever the ratio $\dfrac{B}{A}$ tends to zero uniformly in each region D_ϵ^* ($\epsilon > 0$) as $w \to \alpha$.

We put[1]

$$U_\alpha(z) = \frac{\partial L_w(z)}{\partial w}\Big|_{w=\alpha} = \frac{z}{(z-\alpha)^2} - \sum_{\gamma\in\Gamma_1} P'_\gamma(\alpha)\,\Phi_\gamma(z)\,{}^{1}). \tag{7.3}$$

Note that

$$L^\tau = (w^* - w)\left\{\frac{z}{(z-w^*)(z-w)} - \sum_{\gamma\in\Gamma_1} \Phi_\gamma(z)\,\frac{P_\gamma(w^*) - P_\gamma(w)}{w^* - w}\right\} =$$

$$= U_\alpha(z)(w^* - w) + o(w^* - w), \tag{7.4}$$

$$L_w(z) = L_\alpha(z) + U_\alpha(z)(w - \alpha) + o(w - \alpha), \tag{7.5}$$

$$L^\nu = 2L_\alpha(z) + U_\alpha(z)(w + w^* - 2\alpha) + o(w - \alpha) + o(w^* - \alpha). \tag{7.6}$$

Every point w sufficiently near to α can be written uniquely in the form

$$w = \alpha e^{is}(1 - t) \qquad (|s| < \pi). \tag{7.7}$$

Hence the scalars (s, t) can be considered as local coordinates in a neighbourhood of the point α. Clearly $t = 0$ when $w \in C$, and $t > 0$ when $w \in D$. The point α has the coordinates (0, 0). As $w \to \alpha$ we have

$$\left.\begin{aligned} w^* - w &= \alpha e^{is}\left(\frac{1}{1-t} - 1 + t\right) = 2t\alpha + o(t),\\ w - \alpha &= \alpha((1-t)e^{is} - 1) = \alpha(is - t) + o(s) + o(t),\\ w^* - \alpha &= \alpha\left(\frac{1}{1-t}e^{is} - 1\right) = \alpha(is + t) + o(s) + o(t). \end{aligned}\right\} \tag{7.8}$$

From (7.4), (7.6), (7.8) we obtain

$$\left.\begin{aligned} L^\tau &= 2t\alpha U_\alpha(z) + o(t),\\ L^\nu &= 2L_\alpha(z) + 2is\alpha U_\alpha(z) + o(s) + o(t). \end{aligned}\right\} \tag{7.9}$$

Combining (7.2′) and (7.9) we arrive at the formula

$$G_w(z) = 2\tilde{T}(z)\{-iL_\alpha(z)\nu + \alpha(s\nu + t)U_\alpha(z)\} + o(\nu s) + o(t). \tag{7.10}$$

Put

$$u_\alpha(z) = 2\,\mathrm{Re}\int_0^z \tilde{T}(z)\,\alpha\left[\frac{z}{(z-\alpha)^2} - \sum_{\gamma\in\Gamma_1}\Phi_\gamma(z)P'_\gamma(\alpha)\right]dz. \tag{7.11}$$

By (4.33), $L_\alpha(z) = \mathfrak{C}\Phi_\alpha(z)$. From (6.1), (7.10), (7.11) and Remark 1 in 4.9 it follows that

[1] If $l \geqslant 0$, then Γ_1 is empty and the second term vanishes.

$$q_w(z) = 2h_\alpha(z)\nu + (s\nu + t)u_\alpha(z) + o(\nu s) + o(t). \tag{7.12}$$

7.3. Now let $\alpha \in \Gamma_-$. Then we can choose the set Γ_1 so that $\alpha \in \Gamma_1$. According to Remark 2 in 4.9, $h_a(z) = 0$, and formula (7.12) takes on the form

$$q_w(z) = (s\nu + t)u_\alpha(z) + o(s\nu) + o(t). \tag{7.12'}$$

We shall suppose that in a neighbourhood of the point α the function S(w) satisfies a Lipschitz condition: $|S(w_2) - S(w_1)| \leqslant k\,|w_2 - w_1|$. (By 4.4 a sufficient condition for this to hold is, for instance, that the vector field $v(z)$ is smooth in a neighbourhood of α.)

We denote by $\theta(s)$ the angle between the vector field at the point αe^{is} and the positively directed tangent to C at this point; and we put $\theta(s, t) = \theta(s)$. By 4.3

$$\sin\theta(s) = \nu(\alpha e^{is}) = \nu(s, 0).$$

The function $\nu(w)$ as well as $S(w)$ satisfies the Lipschitz condition near α and consequently

$$\nu(s, t) = \nu(s, 0) + O(t) = \theta + o(\theta) + O(t),$$

where $\theta = \theta(s, t)$. It follows that (7.12′) can be rewritten in the form

$$q_w(z) = (s\theta + t)u_\alpha(z) + o(s\theta) + o(t).$$

Put $\rho = s\theta + t$. According to 3.3, $s\theta \geqslant 0$ in a neighbourhood of α. Therefore the numbers $s\theta\rho^{-1}$ and $t\rho^{-1}$ lie between zero and one, and

$$q_w(z) = \varrho u_\alpha(z) + o(\varrho).$$

By (6.2) this implies

$$\lim_{w\to\alpha} \varrho^{-1} g_w(z) = u_\alpha^0(z), \tag{7.13}$$

where

$$u_\alpha^0(z) = u_\alpha(z) - \sum_{\gamma\in\Gamma_1} u_\alpha(\gamma)\,p_\gamma(z). \tag{7.14}$$

7.4. We can even avoid the additional assumption made at the beginning of 7.3. For if we can show that

$$\varliminf_{w\to\alpha} \frac{\nu s}{t} \geqslant 0, \tag{7.15}$$

then from (7.12′) we can derive the formula

$$\lim_{w\to\alpha} (\nu s + t)^{-1} g_w(z) = u_\alpha^0(z), \tag{7.13'}$$

which is completely analogous to (7.13).

We now prove the inequality (7.15). Since $\nu \to 0$ as $w \to \alpha$, (7.15) holds with equality, provided $\frac{s}{t}$ remains bounded. Hence we may suppose that $\frac{t}{s} \to 0$. Let

(β, γ) be any arc containing a, and put

$$B_1(w)=\frac{1}{\pi}\arg\frac{\alpha-w}{\beta-w}\frac{\beta-\gamma}{\alpha-\gamma},\qquad B_2(w)=\frac{1}{\pi}\arg\frac{\gamma-w}{\alpha-w}\frac{\alpha-\beta}{\gamma-\beta},$$

$$B(w)=\frac{1}{\pi}\arg\frac{\alpha-w}{\gamma-w}\frac{\gamma-\alpha}{\beta-\alpha},$$

where the values of the arguments are taken in the interval $(-\pi, \pi]$. The functions B_1, B_2 and B_3 are harmonic on D and coincide on the contour C with the functions determining the arcs (β, α), (α, γ) and (γ, β), respectively. A simple computation shows that, as $w \to \alpha$ and $\frac{t}{s} \to 0$,

$$\left.\begin{aligned} &B_1(w)=b_1\frac{t}{s}+o\left(\frac{t}{s}\right) \quad \text{if} \quad s>0,\\ &B_2(w)=b_2\frac{t}{s}+o\left(\frac{t}{s}\right) \quad \text{if} \quad s<0\\ &B(w)=bt+o(t) \end{aligned}\right\} \tag{7.16}$$

(where b_1, b_2 and b are constants). Let $\epsilon > 0$. Choose the arc (β, γ) so that the function

$$f(w)=\operatorname{Im} S(w)=r(w)\,\nu(w)$$

satisfies the inequalities $0 < f(w) < \epsilon$ on (α, γ) and $-\epsilon < f(w) < 0$ on (β, α). (We recall that by 4.3, $r(w) > 0$ and the function $\nu(w)$ changes sign from $-$ to $+$ at α.) The function $f(w)$ is harmonic in the annulus $Q=\left\{w\colon \frac{1}{2}<|w|<1\right\}$. Hence for any constants a and c the functions

$$H_1(w)=f(w)+\varepsilon B_1(w)+aB(w)+c\ln|w|^{-1},$$
$$H_2(w)=-f(w)+\varepsilon B_2(w)+aB(w)+c\ln|w|^{-1}$$

are harmonic in Q. If a and c are sufficiently large, then the functions H_1 and H_2 are non-negative on the boundary of Q, and this implies that they are non-negative on Q. Hence in some neighbourhood of α,

$$r(w)\,\nu(w)\frac{s}{t} \geqslant \begin{cases} -\varepsilon B_1(w)\dfrac{s}{t}-aB(w)\dfrac{s}{t}+c\dfrac{s}{t}\ln|w| & \text{for } s>0,\\[2ex] \varepsilon B_2(w)\dfrac{s}{t}+aB(w)-c\dfrac{s}{t}\ln|w| & \text{for } s<0. \end{cases}$$

But $r(w) \to 1$ and $t^{-1}\ln|w| \to 1$ as $w \to \alpha$, and by (7.16) we have

$$\varliminf_{w\to\alpha}\frac{\nu(w)\,s}{t} \geqslant -\varepsilon \min(b_1, -b_2).$$

Hence (7.15) follows, because $\epsilon > 0$ was chosen arbitrarily.

7.5. Now let $\alpha \in \Gamma_+$. By (7.12)

$$q_w(z)=2h_\alpha(z)\,\nu+u_\alpha(z)\,t+o(\nu)+o(t). \tag{7.17}$$

Substituting this expression in (6.2) we arrive at the formula

$$g_w(z) = 2\widetilde{h}_\alpha(z)\,\nu + \widetilde{u}_\alpha(z)\,t - q_w(\alpha)\,p_\alpha(z) + o(\nu) + o(t), \tag{7.18}$$

where

$$\left.\begin{aligned} \widetilde{h}_\alpha(z) &= h_\alpha(z) - \sum_{\gamma\in\Sigma_+,\ \gamma\neq\alpha} h_\alpha(\gamma)\,p_\gamma(z), \\ \widetilde{u}_\alpha(z) &= u_\alpha(z) - \sum_{\gamma\in\Gamma_+,\ \gamma\neq\alpha} u_\alpha(\gamma)\,p_\gamma(z). \end{aligned}\right\} \tag{7.19}$$

For $z = \alpha$ (7.17) is inapplicable, and the behaviour of $q_w(\alpha)$ as $w \to \alpha$ must be investigated separately. This will be carried out in 7.6–7.8.

In the course of our calculations we shall come across a number of constants the exact values of which are unimportant for our purposes. We shall denote these constants by $a_1, a_2, \ldots$ if they are real, and by $A_1, A_2, \ldots$ if they are complex.

We shall suppose that w does not lie on the radius $[0, \alpha]$. With this assumption all integrals that occur converge absolutely.

Denote by Π the interval $[\alpha_0, \alpha]$, where α_0 is any fixed internal point of the radius $[0, \alpha]$. By (6.1) and (7.2),

$$q_w(\alpha) = q_w(\alpha_0) + r(w)\,\tau(w)\,\mathrm{Re}\int_\Pi \widetilde{T}(z)\,L^\tau(w, z)\,dz - r(w)\,\nu(w)\,\mathrm{Re}\int_\Pi \widetilde{T}(z)\,i\,L^\nu(w, z)\,dz. \tag{7.20}$$

It follows from (6.3), (6.4) and (7.1) that

$$\left.\begin{aligned} L^\nu(w, z) &= 1 + \frac{w^*}{z - w^*} + \frac{w}{z - w} - 2\Lambda(\alpha, z) + o(1), \\ L^\tau(w, z) &= \frac{w^*}{z - w^*} - \frac{w}{z - w} - [\Lambda_w(\alpha, z) + o(1)]\,(w^* - w), \end{aligned}\right\} \tag{7.21}$$

where $\Lambda_w = \dfrac{\partial\Lambda}{\partial w}$ and the estimate $o(1)$ is uniform for $z \in \Pi$.

Put

$$\mathscr{D}(w) = \int_\Pi \frac{T(z)}{z - w}\,dz. \tag{7.22}$$

The equation $w\widetilde{T}(z) = T(z) - \widetilde{T}(z)\,(z - w)$ yields

$$\int_\Pi \frac{w}{z - w}\,\widetilde{T}(z)\,dz = \mathscr{D}(w) + A_1,$$

and by (7.21)

$$\int_{\Pi} \tilde{T}(z)\, L^{\nu}(w, z)\, dz = \mathscr{D}(w^*) + \mathscr{D}(w) + A_2 + o(1),$$

$$\int_{\Pi} \tilde{T}(z)\, L^{\tau}(w, z)\, dz = \mathscr{D}(w^*) - \mathscr{D}(w) + A_3(w^* - w) + o(w^* - w).$$

Substituting these values in (7.20) and using (7.17) and (7.8) we obtain

$$q_w(\alpha) = -r(w)\,\nu(w)\, I(w) + r(w)\,\tau(w)\, J(w) + a_1 \nu(w) + a_2 t + o(\nu) + o(t), \quad (7.23)$$

where

$$I(w) = \operatorname{Re} i\,[\mathscr{D}(w^*) + \mathscr{D}(w)], \qquad J(w) = \operatorname{Re}\,[\mathscr{D}(w^*) - \mathscr{D}(w)]. \quad (7.24)$$

7.6. We shall have to deal with many-valued functions. Let us agree on each occasion to choose branches of such functions that are regular on the region E obtained by removing from the complex plane the ray emanating from the point α in the direction of α_0. In particular, we shall write

$$\ln \frac{w-\alpha}{\alpha} = \ln \left| \frac{w-\alpha}{\alpha} \right| + i \arg \frac{w-\alpha}{\alpha} \qquad (w \in E), \quad (7.25)$$

where the value of the argument is taken in the interval $(-\pi, \pi)$.

Let

$$\left.\begin{aligned} M_+(w) &= \ln \frac{w^*-\alpha}{\alpha} + \ln \frac{w-\alpha}{\alpha}, \\ M_-(w) &= \ln \frac{w^*-\alpha}{\alpha} - \ln \frac{w-\alpha}{\alpha}. \end{aligned}\right\} \quad (7.26)$$

It is not difficult to see from the relations

$$\frac{w^*-\alpha}{\alpha}\,\frac{w-\alpha}{\alpha} = -\frac{|w-\alpha|^2}{\alpha \bar{w}}, \qquad \frac{w^*-\alpha}{w-\alpha} = -\frac{\overline{w-\alpha}}{w-\alpha}\,\frac{1}{\bar{\alpha}\,\bar{w}}$$

that when $|w| \leqslant 1$, $s \neq 0$,

$$\left.\begin{aligned} M_+(w) &= 2 \ln |w-\alpha| + i\pi \operatorname{sign} s + o(1), \\ M_-(w) &= t + i \arg \frac{w^*-\alpha}{w-\alpha} + o(t). \end{aligned}\right\} \quad (7.27)$$

In order to study the behaviour of the functions $\mathscr{D}(w)$, $I(w)$, $J(w)$ as $w \to \alpha$ we shall make use of the following well-known proposition in the theory of functions of a complex variable (see, e.g., [1], 8.1):

7.6.A. *Let Π be a smooth arc beginning at α_0 and ending at α. Let $F(z)$ be a Hölder continuous complex function on Π. Then the formula*

$$f(w) = \int_{\Pi} \frac{F(z)}{z-w}\, dz$$

defines outside Π a regular analytic function f. As $w \to \alpha$,

$$f(w) = F(\alpha) \ln \frac{w-\alpha}{\alpha} + A + o(1),$$

where by $\ln \dfrac{w-\alpha}{\alpha}$ *we mean a branch that is regular outside* Π, *and where* A *is a complex constant.*

From this proposition it follows, in particular, that

$$\mathscr{D}(w) = \ln \frac{w-\alpha}{\alpha} + A_4 + o(1)$$

(because $T(\alpha) = 1$). Hence from (7.24)–(7.27) we have

$$I(w) = \mathrm{Re}\,[iM_+(w) + 2A_4 i] + o(1) = -\pi\, \mathrm{sign}\, s + a_3 + o(1). \qquad (7.28)$$

7.7. In order to give an estimate of J it is necessary to add the supplementary hypothesis that the vector field v is *smooth in a neighbourhood of the point* α (see 3.1). According to 4.4 and by virtue of this assumption, the derivative of the function $T(z)$ is Hölder continuous in a neighbourhood of α; therefore Proposition 7.6.A is applicable to the integral

$$\tilde{\mathscr{D}}(w) = \int_{\Pi} \frac{T'(z)}{z-w}\, dz\,.$$

Integrating by parts we easily derive the formula

$$\mathscr{D}'(w) = \tilde{\mathscr{D}}(w) + \frac{T(\alpha)}{w-\alpha} - \frac{T(\alpha_0)}{w-\alpha_0}\,. \qquad (7.29)$$

Clearly

$$\mathscr{D}(w^*) - \mathscr{D}(w) = \int_{K_w} \mathscr{D}'(x)\, dx \qquad (w \in E), \qquad (7.30)$$

where K_w is the line segment joining w and w^*. According to 7.6.A the function $\tilde{\mathscr{D}}(x)$ is regular outside Π, and as $x \to \alpha$

$$\tilde{\mathscr{D}}(x) = T'(\alpha) \ln \frac{x-\alpha}{\alpha} + A_5 + o(1). \qquad (7.31)$$

Combining (7.29), (7.30) and (7.31) we have

$$D(w^*) - D(w) = T'(\alpha)(w^* - \alpha) \ln \frac{w^*-\alpha}{\alpha} - T'(\alpha)(w-\alpha) \ln \frac{w-\alpha}{\alpha} + \ln \frac{w^*-\alpha}{\alpha} - \ln \frac{w-\alpha}{\alpha} + A_6(w^* - w) + o(w^* - w),$$

and by virtue of (7.26), (7.27) and (7.8),

$$D(w^*) - D(w) = \frac{1}{2} T'(\alpha) M_+(w)(w^* - w) +$$
$$+ \frac{1}{2} T'(\alpha) M_-(w)(w^* + w - 2\alpha) + M_-(w) + A_6(w^* - w) + o(w^* - w) =$$
$$= 2tT'(\alpha)\alpha \ln|w - \alpha| + it\pi T'(\alpha)\alpha \operatorname{sign} s +$$
$$+ i\{T'(\alpha)\alpha(e^{is} - 1) + 1\} \arg \frac{w^* - \alpha}{w - \alpha} + tA_7 + o(t). \quad (7.32)$$

Taking the real part we arrive at the expression

$$J(w) = -2\varkappa_1 t \ln|w - \alpha| - \varkappa_2 \pi t \operatorname{sign} s + j(w) + a_4 t + o(t), \quad (7.33)$$

where

$$\varkappa_1 = -\operatorname{Re} \alpha T'(\alpha), \qquad \varkappa_2 = \operatorname{Im} \alpha T'(\alpha), \quad (7.34)$$
$$j(w) = \{\varkappa_1 \sin s + \varkappa_2 (1 - \cos s)\} \arg \frac{w^* - \alpha}{w - \alpha}. \quad (7.35)$$

Let $w \to \alpha$. We shall say that w tends to α tangentially to C if $\frac{s}{t} \to \infty$, and we shall say that a strictly non-tangential approach takes place if the ratio $\frac{s}{t}$ remains bounded.

Note that if w tends to α tangentially to C, then by (7.8)

$$\frac{w^* - \alpha}{w - \alpha} = 1 - 2i\frac{t}{s} + o\left(\frac{t}{s}\right)$$

and consequently

$$j(w) = -2\varkappa_1 t - \varkappa_2 st + o(st). \quad (7.36)$$

We shall now show that

$$\varkappa_1 = \nu_s(\alpha), \qquad \varkappa_2 = \nu_t(\alpha) \quad (7.37)$$

(where ν_s and ν_t denote the partial derivatives of the function with respect to s and t). It follows from (4.4) and the relation $\nu^2 + \tau^2 = 1$ that $T(z) = R(z)\,[\tau(z) - i\nu(z)]$ $(R(z) = r(z)^{-1} > 0)$. Differentiating this equation with respect to s and t, and using the relations $\nu^2(w) + \tau^2(w) = 1$, $\nu(\alpha) = 0$, $\tau(\alpha) = R(\alpha) = 1$, we arrive at the formulae

$$T_s(\alpha) = R_s(\alpha) - i\nu_s(\alpha), \qquad T_t(\alpha) = R_t(\alpha) - i\nu_t(\alpha).$$

But $T_s(\alpha) = T'(\alpha)\, i\alpha$, $T_t(\alpha) = -T'(\alpha)\alpha$. Hence

$$-\alpha T'(\alpha) = iR_s(\alpha) + \nu_s(\alpha) = R_t(\alpha) - i\nu_t(\alpha),$$

and (7.37) follows from (7.34).

By the Mean Value Theorem $\nu(s, t) - \nu(s, 0) = \nu_t(s, \tilde{t})\, t$, where $0 < \tilde{t} < t$. Since $\nu_t(s, \tilde{t}) \to \kappa_2$ as $w \to \alpha$, we have

$$\nu(s, t) = \nu(s, 0) + \varkappa_2 t + o(t) = \sin\theta + \varkappa_2 t + o(t) = \theta + \varkappa_2 t + o(\theta) + o(t), \quad (7.38)$$

where $\theta = \theta(s) = \theta(s, t)$. Hence, in particular, it is clear that $\kappa_1 = \nu_s(0, 0) = \theta'(0)$, and by 4.3, $\kappa_1 \leqslant 0$.

7.8. Combining formulae (7.23), (7.28), (7.33) and (7.38) we have[1]

$$q_w(\alpha) = -\eta + \theta\pi \operatorname{sign} s + a_5\theta + a_6 t + o(\theta) + o(t)\,^1), \tag{7.39}$$

where

$$\eta = 2\varkappa_1 t \ln|w-\alpha| - j(w). \tag{7.40}$$

Hence by (7.18)

$$g_w(z) = \eta p_\alpha(z) + 2\pi\theta\left[\hat{h}_\alpha(z) - \frac{1}{2}\operatorname{sign} s p_\alpha(z)\right] + t\hat{u}_\alpha(z) + o(\theta) + o(t), \tag{7.41}$$

where

$$\hat{h}_\alpha(z) = \pi^{-1}\tilde{h}_\alpha(z) - a_7 p_\alpha(z), \qquad \hat{u}_\alpha(z) = \tilde{u}_\alpha(z) + 2\varkappa_2\tilde{h}_\alpha(z) - a_8 p_\alpha(z). \tag{7.42}$$

Consequently

$$\left.\begin{aligned} g_w(z) &= \eta p_\alpha(z) - 2\pi\theta p_\alpha^+(z) + t\hat{u}_\alpha(z) + o(\theta) + o(t) \qquad \text{for } s > 0,\\ g_w(z) &= \eta p_\alpha(z) + 2\pi\theta p_\alpha^-(z) + t\hat{u}_\alpha(z) + o(\theta) + o(t) \qquad \text{for } s < 0, \end{aligned}\right\} \tag{7.43}$$

where

$$\left.\begin{aligned} p_\alpha^+(z) &= \frac{1}{2} p_\alpha(z) - \hat{h}_\alpha(z),\\ p_\alpha^-(z) &= \frac{1}{2} p_\alpha(z) + \hat{h}_\alpha(z). \end{aligned}\right\} \tag{7.44}$$

To begin with, let $\kappa_1 \neq 0$, and put

$$\hat{t} = t \ln \frac{1}{|w-\alpha|}$$

We shall prove that $j(w) = o(t)$ and consequently $\eta = -2\kappa_1\hat{t} - j(w) \sim -2\kappa_1\hat{t}$. Suppose that this is not so. Then there exists a sequence $w_n \to \alpha$ and a number $\epsilon > 0$ for which $|j(w_n)| \geqslant \epsilon\,|\hat{t}_n|$ $(n = 1, 2, \ldots)$. By virtue of (7.35) $\left|\frac{j(w_n)}{s_n}\right|$ is bounded. Thus, the sequence $\epsilon\left|\frac{\hat{t}_n}{s_n}\right|$ is also bounded, the sequence $\left|\frac{t_n}{s_n}\right|$ tends to zero, and because of (7.36) $\left|\frac{j(w_n)}{\hat{t}_n}\right| \to 0$. The contradiction so obtained proves our assertion.

Suppose that $w \to \alpha$ so that $\frac{\pi}{\kappa_1}\frac{\theta}{\hat{t}} \to \lambda$. It is clear from (7.43) that

[1] Since the function $r(w)\,\tau(w)$ is Hölder continuous in a neighbourhood of α, $[r(w)\,\tau(w) - 1]\ \ln|w-\alpha| \to 0$ as $w \to \alpha$.

$$\lim \frac{g_w(z)}{|2\pi\theta|} = \begin{cases} p_\alpha^+(z) & \text{for } \lambda = +\infty, \\ p_\alpha^-(z) & \text{for } \lambda = -\infty \end{cases} \tag{7.45}$$

and

$$\lim \frac{g_w(z)}{|2\varkappa_1 \hat{t}|} = \begin{cases} p_\alpha(z) + \lambda p_\alpha^+(z) & \text{for } 0 \leqslant \lambda < +\infty, \\ p_\alpha(z) - \lambda p_\alpha^-(z) & \text{for } -\infty < \lambda \leqslant 0. \end{cases} \tag{7.46}$$

Taking into account the fact that $p_\alpha = p_\alpha^+ + p_\alpha^-$, and putting $\lambda^+ = \max(\lambda, 0)$, $\lambda^- = \max(-\lambda, 0)$ we can rewrite this last formula in the form

$$\lim |2\varkappa_1 \hat{t}|^{-1} g_w(z) = (1 + \lambda^+) p_\alpha^+(z) + (1 + \lambda^-) p_\alpha^-(z). \tag{7.47}$$

When the approach of w to α is strictly non-tangential, $\lambda = 0$ and

$$\lim |2\varkappa_1 \hat{t}|^{-1} g_w(z) = p_\alpha(z). \tag{7.48}$$

Note that $\theta \sim \kappa_1 s$, so that

$$\lambda = \lim \frac{\pi\theta}{\hat{t}}.$$

We have assumed so far that w does not lie on the radius $[0, \alpha]$ and hence $s \neq 0$. We shall now free ourselves from this restriction. Let $w_n \to \alpha$, $w_n \in [0, \alpha]$. In view of the continuity (with respect to w) and the positiveness of the functions $g_w(z)$ and $\hat{t}(w)$ we can construct a sequence w_n such that

$$|w_n - \tilde{w}_n| \to 0, \qquad \frac{\hat{t}(\tilde{w}_n)}{\hat{t}(w_n)} \to 1, \qquad \frac{g_{\tilde{w}_n}(z)}{g_{w_n}(z)} \to 1.$$

The validity of (7.48) has been proved for the sequence $\tilde{w}_n$. Hence this formula is clearly valid for the sequence w_n as well.

7.9. We suppose now that $\kappa_1 = 0$ and we shall prove that

$$j(w) = o(t). \tag{7.49}$$

If this were not so, then there would exist an $\epsilon > 0$ and a sequence $w_n \to \alpha$ such that $|j(w_n)| \geqslant \epsilon t_n$. By (7.35)

$$\varepsilon \overline{\lim} \frac{t_n}{s_n^2} \leqslant \overline{\lim} \frac{j(w_n)}{s_n^2} \leqslant \frac{1}{2} \pi\varkappa_2.$$

Consequently $t_n = o(s_n)$, and according to (7.36) $j(w_n) = -\kappa_2 s_n t_n + o(s_n t_n)$. The contradiction so obtained proves (7.46).

It follows from (7.40), (7.43) and (7.49) that

$$\left.\begin{aligned} g_w(z) &= -2\pi\theta p_\alpha^+(z) + t\hat{u}_\alpha(z) + o(\theta) + o(t) && \text{for } s > 0, \\ g_w(z) &= 2\pi\theta p_\alpha^-(z) + t\hat{u}_\alpha(z) + o(\theta) + o(t) && \text{for } s < 0. \end{aligned}\right\} \tag{7.50}$$

Let $w \to \alpha$ so that $-2\pi\theta t^{-1} \to \lambda$. Then

$$\lim \frac{g_w(z)}{|2\pi\theta|} = \begin{cases} p_\alpha^+(z) & \text{for } \lambda = +\infty, \\ p_\alpha^-(z) & \text{for } \lambda = -\infty \end{cases} \tag{7.51}$$

and

$$\lim \frac{g_w(z)}{t} = \begin{cases} \hat{u}_\alpha(z) + \lambda p_\alpha^+(z) & \text{for } 0 \leqslant \lambda < +\infty, \\ \hat{u}_\alpha(z) - \lambda p_\alpha^-(z) & \text{for } -\infty < \lambda \leqslant 0. \end{cases} \tag{7.52}$$

In particular, in the case of a strict non-tangential approach by w to α, $\lambda = 0$ and

$$\lim t^{-1} g_w(z) = \hat{u}_\alpha(z). \tag{7.53}$$

The restriction that $w \in [0, \alpha]$ can be removed exactly as in 7.8.

§8. The Martin boundary for the boundary value problem $\mathcal{A}$. A description of the non-negative solutions and the solutions bounded below

8.1. We shall now give a solution to the problems formulated in §3.

A description of the Martin boundary will be given in Theorems 8.1 and 8.2. In the case where the contour C is the unit circle, the statements of these theorems follow immediately from formulae (7.13), (7.45)–(7.46), (7.51)–(7.52). The general case reduces to the case of the unit circle by using a conformal mapping (see 4.1).

THEOREM 8.1. *Let D be a region bounded by a smooth closed contour C, let $v(z)$ be a Hölder continuous vector field on C with Γ as its set of exceptional points. The Martin boundary B for the boundary value problem $\mathcal{A}$ decomposes into connected components B^α $(\alpha \in \Gamma)$ that are in one-to-one correspondence with the points of Γ. For $\alpha \in \Gamma_-$, the component B^α consists of one point b^α. If $\alpha \in \Gamma_+$ and the field v is smooth in a neighbourhood of α, then the component B^α is a closed interval.*[1]

The points of this interval will be denoted by b_λ^α $(-\infty < \lambda < \infty)$.

Let s be the canonical parameter (arc length) for the contour C, starting from the point $\alpha \in \Gamma$ in the direction of the vector $v(\alpha)$. The point of the contour C corresponding to the value of the parameter s will be denoted by $c(s)$. Let $n(s)$ be the unit vector directed along the inward normal to C at $c(s)$, and let $w(s, t) = c(s) + t_n(s)$. If we restrict the values of s and t to a sufficiently small interval $(-\epsilon, +\epsilon)$, we obtain a local system of coordinates in a neighbourhood of the point α.

We denote by $\theta(s)$ the angle between the vector field and the positive tangent to C at the point $c(s)$ and write

---

[1] In our case $B = C \setminus \Gamma + \sum_{\alpha \in \Gamma} B_\alpha$. If $w \in C \setminus \Gamma$, then ·/ is defined as the set of all functions which are continuously differentiable near w and satisfy the condition $\partial f / \partial v = 0$ in a neighbourhood of w. For $w \in B_\alpha$, $\alpha \in \Gamma$, we denote by $/_w$ the set of all functions. Arguments of 4.7 imply that the sets $\cdot/_w$ satisfy 1.8.α. Obviously they satisfy also 1.8.β.

$$\left.\begin{aligned} \theta(s,\ t) &= \theta(s), \\ \varkappa &= \theta'(0), \end{aligned}\right\} \tag{8.1}$$

$$\zeta = \begin{cases} \dfrac{2\pi s}{t\,|\ln(s^2+t^2)|} & \text{when } \varkappa \neq 0, \\[2ex] \dfrac{-2\pi\theta(s,\ t)}{t} & \text{when } \varkappa = 0. \end{cases} \tag{8.2}$$

Observe that $\theta(0) = 0$, and that $\theta(s)$ changes sign from plus to minus if $\alpha \in \Gamma_+$, and from minus to plus if $\alpha \in \Gamma_-$.

THEOREM 8.2. *If $\alpha \in \Gamma_-$, then for w to converge to b^α (in the Martin topology) it is necessary and sufficient that $w \to \alpha$ in the ordinary topology of the plane. Hence the point b^α can be identified with α.*

If $\alpha \in \Gamma_+$ and the vector field is smooth in a neighbourhood of α, then for w to converge to b_λ^α (in the Martin topology) it is necessary and sufficient that $w \to \alpha$ and $\zeta \to \lambda$.

8.2. We shall now study the harmonic functions corresponding to each point of the Martin boundary. Denote them by $k_\alpha(z)$ $(\alpha \in \Gamma_-)$ and $(\alpha \in \Gamma_+\ -\infty \leqslant \lambda \leqslant +\infty)$. It will be assumed that the region D is represented by the unit disc. Consider on this disc the harmonic functions

$$\left.\begin{aligned} \varphi_\alpha(z) &= \operatorname{Im}\ln\left(1-\frac{z}{\alpha}\right) = \arg\left(1-\frac{z}{\alpha}\right) = \operatorname{arctg}\frac{1-x}{y}, \\ \psi_\alpha(z) &= \operatorname{Re}\ln\left(1-\frac{z}{\alpha}\right) = \ln\left|1-\frac{z}{\alpha}\right| = \frac{1}{2}\ln\left[(1-x)^2+y^2\right], \\ \omega_\alpha(z) &= \operatorname{Re}\left(1-\frac{z}{\alpha}\right)^{-1} = \frac{1-x}{(1-x)^2+y^2} \end{aligned}\right\} \tag{8.3}$$

$\left(x+iy = \frac{z}{\alpha}\right)$. These functions are positive on D and continuous at all points of $D \cup C$, except at the point α where they have a singularity.

We denote by $\mathfrak{H}$ the set of all functions of the form

$$\sum_{\gamma \in \Gamma_-} a_\gamma \varphi_\gamma(z) + h(z),$$

where α_γ are real constants and $h(z)$ is a harmonic function continuous on $D \cup C$. (By 4.8 and Theorem 5.2 all bounded solutions to Problem $\mathfrak{A}$ that are continuous on $D^* \cup \Gamma_+$ can be represented in the above form.) Let us write $f \equiv g$ whenever $f - g \in \mathfrak{H}$.

THEOREM 8.3. *Let the vector field $v(z)$ be smooth in a neighbourhood of the point $\alpha \in \Gamma$. Then*

a) *if $\alpha \in \Gamma_-$*

$$k_\alpha(z) \equiv c'\left[\omega_\alpha(z) - \varkappa\psi_\alpha(z)\right], \tag{8.4}$$

b) *if* $\alpha \in \Gamma_+$, $\kappa \neq 0$

$$k_\alpha^\infty(z) \equiv -c\varphi_\alpha(z), \quad k_\alpha^{-\infty}(z) \equiv c\varphi_\alpha(z); \tag{8.5}$$

$$k_\alpha^\lambda(z) = \frac{1+\lambda^+}{2+|\lambda|} k_\alpha^\infty + \frac{1+\lambda^-}{2+|\lambda|} k_\alpha^{-\infty}; \tag{8.6}$$

c) *if* $\alpha \in \Gamma_+$, $\kappa = 0$, (8.5) *holds for* $k_\alpha^\infty(z)$ *and* $k_\alpha^{-\infty}(z)$, *while for finite values of* λ *the following formulae hold*:

$$k_\alpha^0(z) \equiv c''\omega_\alpha(z), \tag{8.7}$$

$$k_\alpha^\lambda(z) = \begin{cases} \frac{1}{1+\lambda} k_\alpha^0(z) + \frac{\lambda}{1+\lambda} k_\alpha^\infty(z) & \text{for } 0 \leqslant \lambda < +\infty, \\ \frac{1}{1+|\lambda|} k_\alpha^0(z) + \frac{|\lambda|}{1+|\lambda|} k_\alpha^{-\infty}(z) & \text{for } -\infty < \lambda \leqslant 0. \end{cases} \tag{8.8}$$

Here c, c', c'' *are positive constants (dependent of course on the field* $v(z)$ *and the point* α).

PROOF. For each $\alpha \in \Gamma_+$, $p_\alpha \equiv 0$ (see 5.7) and $h_\alpha \equiv \varphi_\alpha$ (4.8). From (7.19), (7.42) and (7.44) we have

$$\tilde{h}_\alpha \equiv h_\alpha \equiv \varphi_\alpha, \quad p_\alpha^+ \equiv -\hat{h}_\alpha \equiv \frac{1}{\pi}\varphi_\alpha, \quad \bar{p}_\alpha \equiv \hat{h}_\alpha \equiv -\frac{1}{\pi}\varphi_\alpha,$$

and (8.5) follows from (7.45) for the case $\kappa \neq 0$. The validity of the same formula for $\kappa = 0$ follows from (7.51).

Furthermore, according to (7.14), (7.19) and (7.42),

$$\left.\begin{aligned} u_\alpha^0 &\equiv u_\alpha \quad (\alpha \in \Gamma_-), \\ \hat{u}_\alpha &\equiv \tilde{u}_\alpha + 2\varkappa_2\tilde{h}_\alpha \equiv u_\alpha + 2\varkappa_2 h_\alpha \equiv u_\alpha + 2\varkappa_2\varphi_\alpha \quad (\alpha \in \Gamma_+). \end{aligned}\right\} \tag{8.9}$$

Let us now investigate the function $u_\alpha(z)$. From (4.23) it is not difficult to conclude that the numbers $r_{\alpha\gamma} = i\alpha P'_\gamma(\alpha)$ $(\alpha, \gamma \in \Gamma_1)$ are real. Hence (7.11) can be put in the form

$$u_\alpha = 2u_\alpha^1(z) - 2\sum_{\gamma \in \Gamma_1} r_{\alpha\gamma} u_\gamma^2(z) + c_1, \tag{8.10}$$

where

$$u_\alpha^1(z) = \operatorname{Re}\int_{z_0}^{z} \frac{\alpha T(z)}{(z-\alpha)^2}\,dz, \quad u_\gamma^2(z) = \operatorname{Re}\int_{z_0}^{z} \frac{T(z)}{iz}\,\Phi_\gamma(z)\,dz$$

(z_0 is any point of the disc D other than the centre; c_1 is a constant).

Let us denote by $H_1, H_2, H_3, \ldots$ functions that are harmonic on the unit disc punctured at the centre, and continuous on the circle C enclosing the disc. It is not difficult to deduce that

$$u_\gamma^2 = \varphi_\gamma + H_1 \tag{8.11}$$

(see 4.8). Furthermore, if we put $A(z) = T(z) - T(\alpha) - T'(\alpha)(z-\alpha)$, we obtain

$$u_\alpha^1 = u_\alpha^{11} + u_\alpha^{12}, \tag{8.12}$$

where

$$u_\alpha^{11}(z) = \operatorname{Re}\int_{z_0}^{z}\left[\frac{\alpha T(\alpha)}{(z-\alpha)^2} + \frac{\alpha T'(\alpha)}{z-\alpha}\right]dz, \qquad u^{12}(z) = \operatorname{Re}\int_{z_0}^{z}\frac{\alpha A(z)}{(z-\alpha)^2}\,dz.$$

Bearing in mind that $T(\alpha) = 1$ and $\alpha T'(\alpha) = -\kappa_1' + \kappa_2 i$ (see (7.34)), we have

$$u_\alpha^{11}(z) = \tilde{\omega}_\alpha(z) - \varkappa_2\varphi_\alpha(z) + c_2, \quad \text{where} \quad \tilde{\omega}_\alpha = \omega_\alpha - \varkappa_1\psi_\alpha = \omega_\alpha - \varkappa\psi_\alpha \tag{8.13}$$

and c_2 is a constant. On the other hand, by the Mean Value Theorem $A(z) = [T'(\tilde{z}) - T'(\alpha)](z-\alpha)$, where $\tilde{z}$ is a point of the interval $[z, \alpha]$. Hence $|A(t)| \leqslant c_3 |\tilde{z}-\alpha|^\epsilon |z-\alpha| \leqslant |z-\alpha|^{1+\epsilon}$ $(c_3 > 0, \epsilon > 0)$, and

$$u^{12}(z) = H_2. \tag{8.14}$$

Combining (8.10)–(8.14) we have $u_\alpha = 2\tilde{\omega}_\alpha - 2\kappa_2\varphi_\alpha - 2\sum_{\gamma\in\Gamma_-} r_{\alpha\gamma}\varphi_\gamma + H_3$.

Clearly $H_3 \in \mathfrak{H}$, and therefore

$$\left.\begin{aligned} u_\alpha &\equiv 2\tilde{\omega}_\alpha && \text{for} \quad \alpha\in\Gamma_-, \\ u_\alpha &\equiv 2\tilde{\omega}_\alpha - 2\varkappa_2\varphi_\alpha && \text{for} \quad \alpha\in\Gamma_+. \end{aligned}\right\} \tag{8.15}$$

It follows from (8.9) and (8.15) that $u_\alpha^0 \equiv 2\tilde{w}_\alpha$ $(\alpha\in\Gamma_-)$ and $\hat{u}_\alpha \equiv 2\tilde{w}_\alpha$ $(\alpha\in\Gamma_+)$. Taking into account (7.13) and (7.52) we arrive at (8.4) and (8.7).

We now recall that by definition (see 1.7) the functions k_α, k_α^λ all take the value 1 at some point z_0. Therefore (8.6) and (8.8) follow from (7.46) and (7.52).

8.3. We can now give a description of the class of all non-negative solutions to Problem $\mathscr{A}$.

THEOREM 8.4. *Suppose that the vector field $v(z)$ is smooth in a neighbourhood of each point $\alpha\in\Gamma$ and that the set Γ_+ is non-empty. Put $\alpha\in\Gamma_+^0$ if $\alpha\in\Gamma_+$ and $\kappa = \theta'(0) = 0$. Then every non-negative solution h of Problem $\mathscr{A}$ can be uniquely written in the form*

$$h(z) = \sum_{\alpha \in \Gamma_-} a_\alpha k_\alpha(z) + \sum_{\alpha \in \Gamma_+} \{c'_\alpha k_\alpha^{-\infty}(z) + c''_\alpha k_\alpha^{+\infty}(z)\} + \sum_{\alpha \in \Gamma_+^0} a'_\alpha k_\alpha^0(z), \quad (8.16)$$

where a_α, c'_α c''_α, a'_α are non-negative constants.

PROOF. Denote by B_e the subset of the Martin boundary B consisting of the points b_α $(\alpha \in \Gamma_-)$, $b_\alpha^{-\infty}$, $b_\alpha^{+\infty}$ $(\alpha \in \Gamma_+)$ and b_α^0 $(\alpha \in \Gamma_+^0)$. It follows easily from (8.4), (8.5) and (8.7) that B_e fulfills condition 1.8.A. Also, by (8.6) and (8.8) B_e also satisfies condition 1.8.B. According to 1.8 every solution of Problem $\mathcal{A}$ is uniquely representable in the form (1.14). But equation (1.14) is equivalent to (8.16).

Received by the editors 28th February 1964.

References

[1] F. D. Gakhov, *Kraevye zadachi*, (Boundary value problems), Izd. Fiz.-Mat. Lit., Moscow 1963.

[2] G. M. Goluzin, *Geometricheskaya teoriya funktsii kompleksnogo peremennogo*, Izd. Fiz.-Mat. Lit., Moscow–Leningrad 1952.
Translation: Geometrische Funktiontheorie, Deutscher Verlag der Wissenschaften, Berlin 1957.

[3] E. Goursat, Cours d'analyse mathematique, (vol.III, Part I), 5th ed. Gauthier–Villars, Paris 1942.
Translation: *Kurs matematicheskogo analiza*, Izd. Fiz.-Mat. Lit., Moscow–Leningrad 1933.

[4] E. B. Dynkin, *Markovskie protsessy*, Izd. Fiz.-Mat. Lit., Moscow–Leningrad 1963.
Translation: Markov processes, Springer Verlag, Berlin–Göttingen–Heidelberg 1965.

[5] M. V. Keldysh, On the solvability and stability of the Dirichlet problem, Uspehi Mat. Nauk 8 (1941), 170–231.

[6] A. Hurwitz and R. Courant, Allgemeine Funktiontheorie und Elliptische Funktionen, Part II, Springer, Berlin 1929.
Translation: *Geometricheskaya teoriya funktsii kompleksnoi peremennoi*, G.T.T.I., Leningrad–Moscow 1934.

[7] A. G. Kurosh, *Lektsii po obshchei algebre*, Izd. Fiz.-Mat. Lit., Moscow 1962.
Translation: Lectures on general algebra, Chelsea, New York 1963.

[8] L. A. Lyusternik and V. I. Sobolev, *Elementy funktsional'nogo analiza*, Gostehizdat, Moscow–Leningrad 1951.
Translation: Elements of Functional Analysis, Ungar, New York 1961.

[9] M. B. Malyutov, Brownian motion with reflection, and a problem with a directional derivative, Doklady Akad. Nauk 156 (1964).

[10] I. G. Petrovskii, *Lektsii ob uravneniyakh s chastnymi proizvodnymi*, Izd. Fiz.-Mat. Lit., Moscow 1961.
Translation of 1st ed.: Lectures on partial differential equations, Interscience, New York 1954.

[11] I. I. Privalov, *Subgarmonicheskie funktsii* (Subharmonic functions), ONTI Moscow–Leningrad 1937.

[12] V. I. Smirnov, *Kurs vishchei matematiki*, t.3, G.T.T.I., Moscow–Leningrad 1933. Translation of 16th ed.: Course of higher mathematics, Addison Wesley, Reading, Mass. 1964.
[13] S. L. Sobolev, *Uravneniya matematicheskoi fiziki*, (Equations of mathematical physics), Gostehizdat 1950.
[14] P. Halmos, Measure Theory, Van Nostrand, New York 1950. Translation: *Teoriya mery*, Izdat. Inost. Lit., Moscow–Leningrad 1953.
[15] M. G. Shur, Martin boundaries for a second order linear elliptic operator, Izv. Akad. Nauk, SSSR Ser. Mat. **27** (1963), 45–60.
[16] M. Brelot, Sur le principe des singularités positives et la topologie de R. S. Martin, Ann. Univ. Grenoble, Sect. Math. Phys. **23** (1948), 113–138.
[17] M. Brelot, Le probleme de Dirichlet. Axiomatique et frontière de Martin Journ. de Math. **35** (1956), 297–335.
[18] M. Brelot, Éléments de la théorie classique du potentiel, Paris, 1961.
[19] G. Choquet, Unicité des représentations intégrales au moyen des points extrémaux dans les cônes convexes réticulés, C. R. Acad. Sci. (Paris) **245** (1956), 555–557.
[20] G. Choquet, Existence des représentations intégrales au moyen des points extrémaux dans les cônes convexes, C. R. Acad. Sci. (Paris) **245** (1956), 699–702.
[21] G. Choquet, Les cônes convexes faiblement complets dans l'Analyse, Proceedings of the International Congress of Mathematicians Stockholm 1962. Uppsala 1963, 317–330.
[22] J. L. Doob, Discrete potential theory and boundaries, Journ. Math. Mech. **8** (1959), 433–458.
[23] G. A. Hunt, Markoff chains and Martin boundaries, Illinois Journ. Math. **4** (1960), 313–340.
= Matematika **5**:5 (1961), 121–149.
[24] R. S. Martin Minimal positive harmonic functions, Trans. Amer. Math. Soc. **49** (1941), 137–172.
[25] T. Watanabe, On the theory of Martin boundaries induced by countable Markov processes, Mem., Coll. Sci. Univ. Kyoto **A33**:1 (1960), 39–108.

Translated by G. G. Gould

BOUNDARY THEORY OF MARKOV PROCESSES
(THE DISCRETE CASE)

The paper contains a detailed account of the theory of Martin boundaries for Markov processes with a countable number of states and discrete time. The probabilistic method of Hunt is used as a basis. This method is modified so as not to go outside the limits of the usual notion of a Markov process. The generalization of this notion due to Hunt is discussed in the concluding section.

Contents

Introduction

The boundary theory of Markov processes permits the investigation of the "final" behaviour of the paths of such processes, that is, the behaviour as the time t tends to infinity (or to the death time). Knowledge of the final behaviour is in its turn a prerequisite for the investigation of general boundary conditions (from the probabilistic point of view this is reduced to the study of possible continuations of the process after the death time). Another important application of boundary theory is the description of all positive harmonic and superharmonic (excessive) functions connected with the process. This problem motivated the creation of the theory of the Martin boundary in 1941 [4]. Martin investigated the set of positive solutions of Laplace's equation in an arbitrary domain of Euclidean space.

The probability interpretation of Martin's results was proposed by Doob [1]: these results are directly related to the Wiener process, but Doob proved that they can also be extended to discrete Markov chains.

A new approach to the theory of the Martin boundary was proposed by Hunt [6]. In the Martin–Doob theory first an integral representation of excessive functions is deduced by probability methods, and then from it a theorem on the final behaviour of the paths is obtained. Hunt proved a theorem on the final behaviour directly by means of probability arguments, and then, applying this theorem to h-processes, he obtained a simple derivation of the integral representation of excessive functions.

The reading of Hunt's important paper is made more difficult because it is written in terms of a generalization, due to the author, of the idea of a Markov process (approximate Markov chains).[1] This may give the impression that the success of the methods applied depends significantly on this generalization. Actually this is not so, and in this paper Hunt's method is modified so that we need not go outside the classes of usual Markov chains.

Problems of boundary theory admit a natural dual formulation.

Instead of harmonic (excessive) functions we can investigate harmonic (excessive) measures. In view of the self-adjointness of Laplace's operator in the case considered by Martin, this dual problem does not contain in itself anything really new. The situation changes in the general case, and now, instead of one Martin boundary, two are constructed in Doob's theory: the exit boundary and the entrance boundary. The role played by the exit boundary in the study of the final behaviour of the paths must now be played by its dual, the entrance boundary, in investigating the "initial" behaviour. However, to give this latter term a meaning, we have to widen the usual interpretation of a Markov chain. One of the possible extensions consists in considering stationary processes defined for values of the time from $-\infty$ to $+\infty$. For such processes the "initial" behaviour means the behaviour as $t \to -\infty$. Another possibility is to consider the generalization of Markov processes proposed by Hunt: Hunt processes begin at a random instant $\xi \geqslant -\infty$, and the "initial" behaviour for these is the behaviour as $t \to \xi$. It is not necessary to construct the dual boundary again,

[1] Chapter 10 of the recent book of Kemeny, Snell and Knapp [3] is written in these terms. This chapter contains a well-considered and polished account of Hunt's paper.

since it can be obtained from the previously constructed one by time reversal.

These questions are dealt with in the concluding sections of this paper.

Thus, the reader can gain a first acquaintance with the idea of time reversal and its application to boundary theory from the simpler and more usual material of stationary processes. At the same time it must be emphasized that stationary processes are not exhausted by Hunt's theory, since they do not satisfy the Hunt requirement of finiteness of the mean number of hits on each state. The construction of a theory including both stationary processes and Hunt processes remains an open problem.

The present paper treats (as also the paper of Doob and Hunt) only discrete Markov chains. It can serve as an introduction to boundary theory for general Markov processes, to which the author intends to devote a subsequent paper.

For an understanding of this paper only a knowledge of elementary probability and measure theory is needed.

§1. Harmonic and excessive functions and measures

We take as starting point a *transition function* in a countable space E. This is a non-negative function $p(x, y)$, $(x, y \in E)$, satisfying the condition

$$\sum_{y} p(x, y) \leqslant 1 \qquad (x \in E). \tag{1}$$

Let f and μ be any functions on E. We denote by $\mathsf{P}f$ and $\mu\mathsf{P}$ functions given by the formulae[1]

$$\left.\begin{aligned} \mathbf{P}f(x) &= \sum_{y} p(x, y) f(y) \qquad (x \in E), \\ (\mu\mathbf{P})(y) &= \sum_{x} \mu(x) p(x, y) \qquad (y \in E). \end{aligned}\right\} \tag{2}$$

Since the right-hand sides contain infinite series, these formulae do not have a meaning for all f and μ. However, they have a meaning if f and μ are non-negative. (By a non-negative function we always mean one with values in the extended number half-line $[0, +\infty]$.)

The transition function $p(x, y)$ can be interpreted as a matrix of countable order. Here $\mathsf{P}f$ is the product of this matrix by a countable vector column f, and $\mu\mathsf{P}$ is the product of a vector row by P. From another point of view, the first formula in (2) describes the effect of the kernel $p(x, y)$ on functions, and the second describes its effect on measures in E.[2] The integral of f with respect to the measure μ is denoted by the inner product

---

[1] If the domain of summation is not indicated, this means that it is E.

[2] We have to deal almost exclusively with non-negative f and μ. Note that in the general case the first formula in (2) has a meaning if f is bounded, and the second if $\sum_{x} |\mu(x)| < \infty$, that is, if the signed measure has bounded variation.

$$(f, \mu) = \sum_{y} f(y)\mu(y).$$

If $(f, \mu) < \infty$, then we say that f is *μ-integrable*, and also that μ is *f-finite*.

A non-negative function f is called *excessive* if $\mathsf{P}f \leqslant f$,[1] and *harmonic* if $f(x) < \infty$ for all x and $\mathsf{P}f = f$. Similarly, a measure is called *excessive* if $\mu\mathsf{P} \leqslant \mu$, and *harmonic* if $\mu(x) < \infty$ for all x and $\mu\mathsf{P} = \mu$.

One of the central problems before us is the description of all harmonic and excessive functions and measures connected with the transition function $p(x, y)$. It is expedient here to consider only γ-integrable functions h and l-finite measures ν, where γ and l are, respectively, a previously selected reference measure and function on E. Of fundamental interest is the case when $p(x, y)$ is transient (see the definition in §3). In this case we are able to attach to each point $y \in E$ a γ-integrable excessive function k_y and a l-finite excessive measure κ_y. By means of the kernels $k_y(x)$ and $\kappa_y(x)$ we construct two compactifications E^* and $\hat{E}$ of E, such that $k_y(x)$ is extended for each $x \in E$ continuously to E^*, and $\kappa_y(x)$ is extended continuously to $\hat{E}$. From the sets $E^* \setminus E$ and $\hat{E} \setminus E$ we single out Borel sets B and $\hat{B}$, where k_y is a harmonic function and $(k_y, \gamma) = 1$, for $y \in B$, and κ_y is a harmonic measure and $(l, \kappa_y) = 1$ for $y \in \hat{B}$.

It can be proved that every γ-integrable excessive function h is representable uniquely in the form

$$h(x) = \int_{E \cup B} k_y(x)\, \mu_h(dy),$$

and any l-finite excessive measure ν is expressible uniquely in the form

$$\nu(x) = \int_{E \cup \hat{B}} \kappa_y(x)\, \mu^{\nu}(dy).$$

Here μ_h and μ^{ν} are finite measures which are determined uniquely by h and ν, respectively. We call them *spectral measures.*

The set B is called the *exit space*, and the set $\hat{B}$ the *entrance space*. The origin of these terms becomes clear in the following section.

§2. Markov processes

Suppose that a particle moves in a space E, going through a sequence of states $a_0, a_1, a_2, \ldots$ The path $a_0 a_1 a_2 \ldots$ may be terminate or may continue unboundedly. The set of all (terminating or non-terminating) paths is denoted by Ω. The set of all non-terminating paths is denoted by Ω_∞.

[1] By $f \leqslant g$ we mean that $f(x) \leqslant g(x)$ for all $x \in E$.

Among the subsets of Ω the so-called simple sets play a special role. A *simple set* $[a_0 a_1 a_2 \ldots a_n]$ is composed of all paths beginning with the states $a_0, a_1, \ldots, a_n$ and continuing in any manner after the moment n. We denote by $\mathcal{F}$ the σ-algebra in Ω generated by all simple sets.

We depend on the following theorem on measures in Ω.

THEOREM A. *Suppose that for any n and any $a_0, a_1, \ldots, a_n \in E$ a non-negative number $p(a_0, a_1, \ldots, a_n)$ is given, where*

$$\sum_{a_n} p(a_0, a_1, \ldots, a_n) \leqslant p(a_0, a_1, \ldots, a_{n-1}). \tag{3}$$

Then there exists a measure P, *which is moreover unique, on the σ-algebra $\mathcal{F}$ such that*

$$\mathbf{P}[a_0 a_1 \ldots a_n] = p(a_0, a_1, \ldots, a_n).$$

This theorem will be proved in the Appendix.[1]

From Theorem A it follows that for any measure ν on E with $\nu(x) < \infty$ for all x, there exists a measure P_ν on Ω, such that

$$\mathbf{P}_\nu [a_0 a_1 \ldots a_n] = \nu(a_0)\, p(a_0, a_1) \ldots p(a_{n-1}, a_n). \tag{4}$$

An important role is played by the particular case when ν is the unit measure concentrated at the point x (when $\nu(y) = \delta(x, y)$, where $\delta(x, y) = 1$ if $x = y$, $\delta(x, y) = 0$ if $x \neq y$). The corresponding measure in Ω is denoted by P_x, so that

$$P_x [a_0 a_1 \ldots a_n] = \delta(x, a_0)\, p(a_0, a_1) \ldots p(a_{n-1}, a_n), \tag{4'}$$

P_x is the probability measure concentrated on the paths starting from x. We note that for any ν.

$$\mathbf{P}_\nu = \sum_x \nu(x)\, \mathbf{P}_x$$

and that $\mathsf{P}_\nu(\Omega) = \nu(E)$.

Each measure in the space of paths Ω determines a random process.[2] The process determined by the measure P_ν is called the *Markov process with initial distribution ν and transition function $p(x, y)$*. The process corresponding to the measure P_x is called the *Markov process with initial state x and transition function $p(x, y)$*.

One of the basic results of boundary theory states that almost every non-

---

[1] The necessity of the condition (3) is obvious, since $[a_0, a_1, \ldots, a_n] \subseteq [a_0, a_1, \ldots, a_{n-1}]$ for any a_n, and since distinct $[a_0, a_1, \ldots, a_n]$ do not intersect.

[2] If $p(\Omega) = 1$, then $\mathsf{P}(A)$ can be interpreted as the probability that the trajectories of motion belong to A. In the general case $\mathsf{P}(A)$ may prove to be greater than 1 and even equal to ∞.

terminating path tends to some point of the exit space B. The measure of the set of paths for which this limit belongs to the Borel set $\Gamma \subseteq B$ is

$$\int_{\Gamma} (k_y, \nu)\, \mu_1 (dy),$$

where k_y is the harmonic function corresponding to the point $y \in B$, and μ_1 is the spectral measure of the excessive function 1.

To explain the role of the exit space, we have to introduce into the discussion paths without beginning or end. These are functions a_t with range in E defined for all integers t from $-\infty$ to $+\infty$. The set of such paths is denoted by $\hat{\Omega}$. We use the term simple sets in $\hat{\Omega}$ for the sets $[a_m, a_{m+1}, \ldots, a_n]_m^n$ consisting of all paths passing at the moments $m, m+1, \ldots, n$ through the points $a_m, a_{m+1}, \ldots, a_n$. (Before the moment m and after the moment n they can behave arbitrarily.) We denote by $\hat{\mathscr{F}}$ the σ-algebra in $\hat{\Omega}$ generated by all simple sets.

For the construction of measures in the space $\hat{\Omega}$ we can use the following modification of Theorem A.

THEOREM B. *Suppose that for any integers $m \leqslant n$ and any $a_m, a_{m+1}, \ldots, a_n \in E$, non-negative numbers $p_m^n (a_m, a_{m+1}, \ldots, a_n)$ are given, where*

$$\sum_{a_n} p(a_m, a_{m+1}, \ldots, a_n) = p(a_m, a_{m+1}, \ldots, a_{n-1}), \tag{3'}$$

$$\sum_{a_m} p(a_m, a_{m+1}, \ldots, a_n) = p(a_{m+1}, \ldots, a_n). \tag{3''}$$

Then there exists a unique measure P *on the σ-algebra $\hat{\mathscr{F}}$ such that*

$$\mathsf{P}\, [a_m a_{m+1} \ldots a_n]_m^n = p_m^n (a_m, a_{m+1}, \ldots, a_n).$$

The necessity of the conditions (3′), (3″) is evident. For the proof of Theorem B see the Appendix.

We suppose that the transition function $p(x, y)$ satisfies (1) with the equality sign and that ν is a harmonic measure. Then the function

$$p_m^n (a_m, a_{m+1}, \ldots, a_n) = \nu (a_m)\, p (a_m, a_{m+1}) \ldots p (a_{n-1}, a_n)$$

satisfies the conditions (3′)–(3″), and by Theorem B there exists a measure P_ν on $\hat{\mathscr{F}}$ such that

$$\mathsf{P}_\nu\, [a_m a_{m+1} \ldots a_n]_m^n = \nu (a_m)\, p (a_m, a_{m+1}) \ldots p (a_{n-1}, a_n). \tag{4''}$$

A random process determined by P_ν in $\hat{\Omega}$ is called *a stationary Markov process with stationary distribution ν and transition function $p(x, y)$*.

In boundary theory it is proved that for such a process almost all paths converge as $t \to -\infty$ to some point of the entrance space $\hat{B}$. The measure of the set of paths for which this limit belongs to a Borel set $\Gamma \subseteq \hat{B}$ is

$$\int_{\Gamma} (1, \varkappa_y)\, \mu^{\nu}(dy),$$

where κ_y is the harmonic measure corresponding to the point $y \in \hat{B}$ and μ^{ν} is the spectral measure for ν.

This result can be further generalized in several directions.

Random variables connected with Markov processes are $\hat{\mathscr{F}}$-measurable functions defined on Ω or on a subset of this space (or $\hat{\mathscr{F}}$-measurable functions on $\hat{\Omega}$ or a subset of $\hat{\Omega}$). The integral of such a function ξ over its domain with respect to the measure $\mathbf{P}_{\nu}$ is denoted by $M_{\nu}\xi$, and with respect to $\mathbf{P}_x$ by $M_x \xi$.

Here are some examples.

ζ is the terminal moment of a path: if the last moment at which the path ω is defined is n, then $\zeta(\omega) = n$; if the path does not terminate, then $\zeta(\omega) = +\infty$.

x_n is the position of a particle at the moment n. This function is defined on the set $\{\omega \colon \zeta(\omega) \geqslant n\}$. In the case of a stationary process

$$\mathbf{P}_{\nu}\,[x_n = y] = \nu\,(y)$$

for any n. For a process with the initial distribution ν

$$\mathbf{P}_{\nu}\,[x_n = y] = \sum_{z} \mathbf{P}_{\nu}\,[x_{n-1} = z]\, p\,(z,\ y). \tag{5}$$

To prove this equation it suffices to note that $\{\omega \colon x_n = y\} = \bigcup_{z} \{\omega \colon x_{n-1} = z,\ x_n = y\}$, to decompose the sets occurring here into simple sets, and to use (4).

We put

$$p\,(n,\ x,\ y) = \mathbf{P}_x\,\{x_n = y\}.$$

From (5) it follows that

$$p\,(n,\ x,\ y) = \sum_{z} p\,(n-1,\ x,\ z)\, p\,(z,\ y),$$

and in view of the obvious relation

$$\sum_{y} \delta\,(x,\ y) = 1$$

we have

$$\begin{aligned} M_x f(x_n) = M_x \sum_{y} \delta\,(x_n,\ y)\, f\,(x_n) = \sum_{y} M_x \delta\,(x_n,\ y)\, f\,(y) = \\ = \sum_{y} p\,(n,\ x,\ y)\, f\,(y) = \mathbf{P}^n f\,(x), \end{aligned} \tag{6}$$

where $\mathbf{P}$ is an operator given by the first formula in (2). (6) is valid also for $n = 0$, if $\mathbf{P}^0$ is taken as the unit operator.

§3. The Green's function

We return to the problem raised in §1 of describing all excessive functions corresponding to a transition function $p(x, y)$. From (1) it follows that the non-negative constants always belong to the set of excessive functions. It may happen that no other excessive functions exist.

For example, let E be the set of all integers and $p(x, y) = \frac{1}{2}$ if $|x - y| = 1$, and $p(x, y) = 0$ for remaining pairs x, y. (The corresponding Markov processes are called *simple random walks*.) It is evident that

$$\mathbf{P}f(x) = \frac{1}{2} f(x+1) + \frac{1}{2} f(x-1),$$

and the condition that f is an excessive function can be written

$$\varphi(x+1) \leqslant \varphi(x),$$

where $\varphi(x) = f(x + 1) - f(x)$. For any natural number k

$$f(x+k) = f(x) + \varphi(x) + \varphi(x+1) + \ldots + \varphi(x+k-1) \leqslant \leqslant f(x) + k\varphi(x),$$

$$f(x) = f(x-k) + \varphi(x-k) + \ldots + \varphi(x-1) \geqslant f(x-k) + k\varphi(x).$$

Since f is non-negative, it follows from the first inequality that $f(x) \geqslant -k\varphi(x)$ and from the second that $f(x) \geqslant k\varphi(x)$. Since k is arbitrary, it follows that $\varphi(x) = 0$, and therefore f is a constant.

Replying on the notion of a Green's function we derive a class of processes for which sufficiently many excessive functions exist. The *Green's function* is defined by the series

$$g(x, y) = \sum_{n=0}^{\infty} p(n, x, y). \tag{7}$$

The process is called *transient* if $g(x, y) < \infty$ for arbitrary x and y.

We note that by (5) $p(n, x, y) = \mathbf{P}_x\{x_n = y\} = M_x \delta(x_n, y)$. Hence

$$g(x, y) = M_x \sum_{n=0}^{\zeta} \delta(x_n, y). \tag{8}$$

Under the sign of mathematical expectation there stands the number of times the path hits the point y. The condition of being transient implies that this number is almost certainly finite. Thus, for a transient process almost all paths go only a finite number of times through one and the same state. Hence, if the states are enumerated in any order, for almost all non-terminating paths the number of the state tends to infinity.

The simple random walk considered above has the property: almost all paths go infinitely often through any point.[1] Processes with this property are called *recurrent*. It can be proved that every connected Markov process is either transient or recurrent. (We say that a Markov process is connected if for any two states x and y there exists n such that $p(n, x, y) > 0$; in other words, if $g(x, y) > 0$ for any x and y.) For any recurrent process, as for the simple random walk, there do not exist non-constant excessive functions. *Henceforth, without saying so each time, we only discuss transient processes.*

The Green's function corresponds to the operators

$$Gf(x) = \sum_y g(x, y) f(y),$$

$$\mu G(y) = \sum_x \mu(x) g(x, y).$$

From (6) and (7) it is clear that

$$G = \sum_{n=0}^{\infty} \mathbf{P}^n. \tag{9}$$

Hence, for non-negative f and μ

$$f + \mathbf{P}Gf = Gf, \quad \mu + \mu G\mathbf{P} = \mu G. \tag{10}$$

Therefore it is evident that Gf is an excessive function and μG an excessive measure.

We put $\delta_y(x) = \delta(x, y)$. It is obvious that

$$g(x, y) = (G\delta_y)(x) = (\delta_x G)(y).$$

Hence $g(x, y)$ is an excessive function of x for fixed y and an excessive measure with respect to y for fixed x. Thus, the Green's function permits us to connect an excessive measure and an excessive function with each point of E. It is this initial store of excessive functions and measures from which subsequently all excessive functions and measures are obtained.

We derive one important property of the Green's function.

LEMMA 1. *For any states x and y*

$$g(x, y) = \pi(x, y) g(y, y), \tag{11}$$

where $\pi(x, y) = \mathbf{P}_x\{x_n = y$ for some $n\}$ is the probability of hitting y starting from x.

PROOF. We put

$$A_m = \{x_0 \neq y, x_1 \neq y, \ldots, x_{m-1} \neq y, x_m = y\}.$$

---

[1] See, for example, [5]. Ch. XIII, §3.

We note that

$$\mathbf{P}_x \{A_m, x_{m+k} = y\} = \mathbf{P}_x (A_m)\, p\,(k, y, y). \tag{12}$$

To see this we have to decompose the set $\{A_m, x_{m+k} = y\}$ into simple sets and use (4).

The sum on the right in (8) is evidently equal to[1]

$$\sum_{m=0}^{\infty} \chi_{A_m} \sum_{n=m}^{\zeta} \delta\,(x_n, y)$$

Hence

$$g\,(x, y) = \sum_{m=0}^{\infty} \sum_{n=m}^{\infty} M_x \chi_{A_m} \delta\,(x_n, y) = \sum_{m=0}^{\infty} \sum_{k=0}^{\infty} \mathbf{P}_x \{A_m, x_{n+k} = y\}.$$

Bearing (12) and (7) in mind we obtain (11).

REMARK. We put $B_k^m = \{x_n = z \text{ for some } n \in [m, m+k]\}$. Decomposing the set $A_m \cap B_k^m$ into simple sets, we can prove that $\mathbf{P}_x \{A_m, B_k^m\} = \mathbf{P}_x (A_m)\, \mathbf{P}_y (B_k^0)$. Letting $k \to \infty$, we obtain $\mathbf{P}_x (A_m, B_\infty^m) = \mathbf{P}_x (A_m)\, \mathbf{P}_y (B_\infty^0)$. Hence

$$\pi\,(x, z) \geqslant \mathbf{P}_x \{ \bigcup_{m=0}^{\infty} [A_m \cap B_\infty^m]\} = \sum_{m=0}^{\infty} \mathbf{P}_x (A_m)\, \mathbf{P}_y (B_\infty^0) = \pi\,(x, y)\, \pi\,(y, z). \tag{13}$$

This remark will be used in §9.

§4. Supermartingales

The investigation of excessive functions and paths of Markov processes is conducted most conveniently by means of the apparatus of supermartingales. In this section we introduce the notion and present some properties of supermartingales. The presentation will be in a most elementary form, fully sufficient, however, for our purpose.

DEFINITION. Let $\mathbf{P}$ be a measure on the σ-algebra F in the space Ω. Suppose that in Ω there are given F-measurable functions $y_0, y_1, \ldots, y_N$ with values belonging to a countable set E and real-valued functions $z_0, z_1, \ldots, z_N$. We say that $z_0, z_1, \ldots, z_N$ is *a supermartingale with respect to* $y_0, y_1, \ldots, y_N$ if for any $n = 0, 1, \ldots, N$,

1) z_n is a function of $y_0, y_1, \ldots, y_n$: $z_n = f_n(y_0, y_1, \ldots, y_n)$;

2)[1] for any $a_0, a_1, \ldots, a_{n-1}$ of E

---

[1] χ_A denotes the indicator of A, that is, the function equal to 1 on A and to 0 outside A.

[2] In terms of conditional expectation condition 2) can be restated in the form $M(z_n \mid y_0, y_1, \ldots, y_{n-1}) \leqslant z_{n-1}$ almost surely (a.s.).

$$\sum_{a_n} \mathbf{P}\{y_0 = a_0, \ldots, y_{n-1} = a_{n-1}, y_n = a_n\} f_n(a_0, a_1, \ldots, a_n) \leqslant \leqslant \mathbf{P}\{y_0 = a_0, \ldots, y_{n-1} = a_{n-1}\} f_{n-1}(a_0, a_1, \ldots, a_{n-1}). \tag{14}$$

From this definition the following two properties follow at once:

4.A. If d is a constant, then, together with $\{z_n\}$ the sequence[1] $\{z_n \wedge d\}$ is also a supermartingale with respect to $\{y_n\}$.

4.B. For any non-negative function φ

$$M\varphi(y_0, \ldots, y_{n-1}) z_n \leqslant M\varphi(y_0, \ldots, y_{n-1}) z_{n-1}.$$

To deduce 4.B from 2) it is sufficient to note that

$$\varphi(y_0, \ldots, y_{n-1}) = \sum_{a_0, \ldots, a_{n-1}} \varphi(a_0, \ldots, a_{n-1}) \delta(a_0, y_0) \ldots \delta(a_{n-1}, y_{n-1}).$$

The most important property of supermartingales is stated in terms of Markov moments. A random variable[2] τ, taking the values 0, 1, 2, . . ., is called a Markov moment (with respect to the sequence $y_0, y_1, y_2, \ldots$) if for any n

$$\delta(\tau, n) = \varphi_n(y_0, \ldots, y_n) \tag{15}$$

(φ_n is some function). Intuitively this definition means that observing the values $y_0, y_1, \ldots, y_n, \ldots$, we can answer until the moment n the question whether the equation $\tau = n$ is true. It is easy to verify that along with τ the function $\tau \vee m$ is also a Markov moment, where m is a non-negative integer.

LEMMA 2. *Let $z_0, z_1, \ldots, z_N$ be a supermartingale with respect to $y_0, y_1, \ldots, y_N$ and let two Markov moments $\sigma \leqslant \tau \leqslant N$ be given. Then*

$$Mz_\sigma \leqslant Mz_\tau. \tag{16}$$

PROOF. First we prove that if the Markov moment satisfies $n \leqslant \tau \leqslant N$, then for any non-negative function φ

$$M\varphi(y_0, \ldots, y_n) z_\tau \leqslant M\varphi(y_0, \ldots, y_n) z_n. \tag{17}$$

This is obvious for $n = N$. Hence it is sufficient to verify that if it holds for $n = m$, then it holds for $n = m - 1$. Thus, let $m - 1 \leqslant \tau \leqslant N$. We have

$$M\varphi(y_0, \ldots, y_{m-1}) z_\tau = M\varphi(y_0, \ldots, y_{m-1}) \delta(\tau, m-1) z_{m-1} + + M\varphi(y_0, \ldots, y_{m-1}) [1 - \delta(\tau, m-1)] z_{\tau \vee m}. \tag{18}$$

By (15) we have $\delta(\tau, m-1) = \tilde{\varphi}(y_0, \ldots, y_{m-1})$, and applying the inductive hypothesis to the Markov moment $\tau \vee m \geqslant m$, we find that the second term in (18) does not exceed

[1] We denote by $a \wedge b$ the smaller of the two numbers a and b, and by $a \vee b$ the larger.

[2] In certain cases it is useful also to allow the value $+\infty$ for τ. Here, as before, it is required that (15) be satisfied for all finite n.

$$M\varphi(y_0, \ldots, y_{m-1})[1-\delta(\tau, m-1)] z_m.$$

By 4.B the last expression is not diminished if z_m is replaced by z_{m-1}. Making this change and substituting the estimate so obtained in (18) we see that (17) holds for $n = m - 1$.

To complete the proof of the lemma we note that by (17) and (18)

$$M\delta(\sigma, n) z_\tau = M\delta(\sigma, n) z_{\tau \vee n} \leqslant M\delta(\sigma, n) z_n = M\delta(\sigma, n) z_\sigma.$$

Summing this inequality for $n = 0, 1, \ldots, N$, we obtain (16).

Relying on Lemma 2 we now prove a fundamental lemma about the number of crossings of a fixed interval $[c, d]$ for a positive supermartingale.

The *number of down-crossings of* $[c, d]$ by the sequence $z_0, z_1, \ldots, z_N$ is the largest number k for which numbers $0 \leqslant t_1 < t_2 < \ldots < t_{2k-1} < t_{2k} \leqslant N$ can be chosen so that $z_{t_1} \geqslant d$, $z_{t_2} \leqslant c$, $z_{t_3} \geqslant d$, $z_{t_4} \leqslant c, \ldots, z_{t_{2k-1}} \geqslant d$, $z_{t_{2k}} \leqslant c$.

LEMMA 3. *Suppose that the non-negative random variables $z_0, z_1, \ldots, z_N$ form a supermartingale with respect to $y_0, y_1, \ldots, y_N$. Then the number of down-crossings of $[c, d]$ by the sequence $z_0, z_1, \ldots, z_N$ satisfies the inequality*

$$M\nu \leqslant \frac{1}{d-c} M z_0. \tag{19}$$

PROOF. We put $\tau_0 = 0$ and define τ_n $(n = 1, 2, \ldots)$ inductively as follows: τ_n for odd n is the smallest value $k \geqslant \tau_{n-1}$ for which $z_k \geqslant d$, or, if there are no such values of k, then $\tau_n = N$; τ_n for even n is the smallest value $k \geqslant \tau_{n-1}$ for which $z_k \leqslant c$, or, if there are no such values of k, then $\tau_n = N$. It is easily verified that $\tau_0, \tau_1, \ldots, \tau_n, \ldots$ are Markov moments and $\tau_n = N$ for $n \geqslant 2\nu + 2$.

According to 4.A. $\tilde{z}_n = z_n \wedge d$ is a supermartingale. We choose m so that $2m \geqslant N$, and put

$$S = (\tilde{z}_{\tau_1} - \tilde{z}_{\tau_2}) + (\tilde{z}_{\tau_3} - \tilde{z}_{\tau_4}) + \ldots + (\tilde{z}_{\tau_{2\nu-1}} - \tilde{z}_{\tau_{2\nu}}) + \\ + (\tilde{z}_{\tau_{2\nu+1}} - \tilde{z}_{\tau_{2\nu+2}}) + \ldots + (\tilde{z}_{\tau_{2m-1}} - \tilde{z}_{\tau_{2m}}).$$

We note that

$$\tilde{z}_{\tau_1} = d, \quad \tilde{z}_{\tau_2} \leqslant c, \quad \tilde{z}_{\tau_3} = d, \quad \tilde{z}_{\tau_4} \leqslant c, \ \ldots, \ \tilde{z}_{\tau_{2\nu-1}} = d, \quad \tilde{z}_{\tau_{2\nu}} \leqslant c,$$
$$\tilde{z}_{\tau_{2\nu+1}} \geqslant \tilde{z}_{\tau_{2\nu+2}} = \tilde{z}_{\tau_{2\nu+3}} = \ldots = \tilde{z}_{\tau_{2m}}.$$

Therefore

$$S \geqslant \nu(d-c). \tag{20}$$

On the other hand,

$$S = \tilde{z}_{\tau_1} + (\tilde{z}_{\tau_3} - \tilde{z}_{\tau_2}) + \ldots + (\tilde{z}_{\tau_{2m-1}} - \tilde{z}_{\tau_{2m-2}}) - \tilde{z}_{\tau_{2m}}.$$

By Lemma 2,

$$M\tilde{z}_0 = M\tilde{z}_{\tau_0} \geqslant M\tilde{z}_{\tau_1} \geqslant M\tilde{z}_{\tau_2} \geqslant \ldots \geqslant M\tilde{z}_{\tau_{2m-1}}.$$

Noting that $\tilde{z}_{\tau_{2m}} \geqslant 0$, we find

$$MS \leqslant M\tilde{z}_{\tau_1} \leqslant M\tilde{z}_0 \leqslant Mz_0. \tag{21}$$

Now (19) follows from (20) and (21).

§5. Excessive functions and supermartingales

Let f be a non-negative function in the space E and let $x_0, x_1, x_2, \ldots$ be a path of a Markov process with initial state x and transition function $p(x, y)$. We add to E one further point, which we denote by $*$, and we take $x_n = *$ if $n > \zeta$. We put $f(*) = 0$. With these conventions, the functions $x_n(\omega)$ and $f(x_n(\omega))$ are defined for all n and ω. We show that if f is excessive, then the sequence $f(x_0), f(x_1), \ldots, f(x_N)$ is a supermartingale with respect to $x_0, x_1, \ldots, x_N$. For if at least one of the points $a_0, a_1, \ldots, a_n$ is $*$, then all subsequent ones also are $*$. In this case $f(a_n) = 0$. Hence the left-hand side of (14) is zero, while the right is non-negative. If all the points $a_0, a_1, \ldots, a_n \in E$, then by (4) we can write (14) in the form

$$\sum_{a_n} \delta(x, a_0) p(a_0, a_1) \ldots p(a_{n-1}, a_n) f(a_n) \leqslant$$
$$\leqslant \delta(x, a_0) p(a_0, a_1) \ldots p(a_{n-2}, a_{n-1}) f(a_{n-1}).$$

For $n = 1$ these inequalities coincide with the condition that f is excessive, and from their validity for $n = 1$ the validity for all n follows.

Let ν_N be the number of down-crossings of $[c, d]$ by the sequence $f(x_0), f(x_1), \ldots, f(x_N)$. By Lemma 3 of §4

$$M_x \nu_N \leqslant \frac{1}{d-c} M_x f(x_0) = \frac{f(x)}{d-c}. \tag{22}$$

Now let ν be the number of down-crossings of $[c, d]$ by the infinite sequence $f(x_0), f(x_1), \ldots$ Evidently, $\nu_N \uparrow \nu$, and hence it follows from (22) that

$$M_x \nu \leqslant \frac{f(x)}{d-c}.$$

We assume that $f(x) < \infty$. Then $M_x \nu < \infty$ and consequently $\nu < \infty$ ($\mathbf{P}_x$.a.e.).

However, it is easy to see that the following elementary proposition is true.

If a numerical sequence makes only a finite number of down-crossings of any interval $[c, d]$ with rational ends, then this sequence tends to a finite or infinite limit.

By what has been proved this theorem is applicable to $f(x_0), f(x_1), \ldots$, along almost all paths $x_0, x_1, \ldots$ Thus, almost surely there exists the limit

$$\xi = \lim_{n\to\infty} f(x_n).$$

By Fatou's lemma, from the inequality

$$M_x f(x_n) = \mathbf{P}^n f(x) \leqslant f(x)$$

it follows that

$$M_x \xi \leqslant f(x).$$

Hence ξ is almost surely finite.

Evidently $\xi = 0$ if $\zeta < \infty$. Hence there is interest only in the value of ξ on the set Ω_∞ of all non-terminating paths.

So we have proved the following theorem.

THEOREM 1. *If f is an excessive function and if $f(x) < \infty$, then the finite limit*

$$\lim_{n\to\infty} f(x_n)$$

exists $\mathbf{P}_x$*-a.e. on* Ω_∞.

We leave it to the reader to show that if $f(x) = \infty$, then $\mathbf{P}_x$-a.e. on Ω_∞ one of two possibilities holds: either $f(x_n) = +\infty$ for all n, or $f(x_n)$ tends to a finite limit.

In the following sections analogous properties will be established for the densities of excessive measures.

§6. Position of a particle at the last exit time from a set D

Let $D \subseteq E$. We define *the last exit time from the set D as*

$$\tau = \sup\{t : x_t \in D\}.$$

The random variable τ takes the values $0, 1, 2, \ldots$ and the value $+\infty$. It is taken to be undefined if $x_t \notin D$ for all t.

We put

$$L_D(x) = \mathbf{P}_x\{\tau = 0\} = \mathbf{P}_x\{x_0 \in D,\ x_t \notin D \text{ for } t > 0\}$$

We note that

$$\mathbf{P}_x\{x_\tau = y\} = \sum_{m=0}^{\infty} \mathbf{P}_x\{\tau = m,\ x_m = y\} = \\ = \sum_{m=0}^{\infty} p(m, x, y) L_D(y) = g(x, y) L_D(y). \quad (23)$$

It is clear that

$$\sum_y g(x, y) L_D(y) = \sum_y \mathbf{P}_x\{x_\tau = y\} \leq 1. \tag{24}$$

Let n be a non-negative integer. We investigate the distribution of the point $x_{\tau-n}$. This point is not defined if $\tau < n$ or $\tau = +\infty$ or if τ is not defined. In all three cases we put $x = *$. Let $a_0, a_1, \ldots, a_n \in E$. We have

$$\begin{aligned}
\mathbf{P}_x\{x_\tau = a_0,\ x_{\tau-1} = a_1,\ \ldots,\ x_{\tau-n} = a_n\} = \\
= \sum_{m=n}^{\infty} \mathbf{P}_x\{\tau = m,\ x_m = a_0,\ x_{m-1} = a_1,\ \ldots,\ x_{m-n} = a_n\} = \\
= \sum_{m=n}^{\infty} p(m-n, x, a_n)\, p(a_n, a_{n-1}) \ldots p(a_1, a_0) L_D(a_0) = \\
= g(x, a_n)\, p(a_n, a_{n-1}) \ldots p(a_1, a_0) L_D(a_0).
\end{aligned}$$

Multiplying this equation by $\gamma(x)$ and summing over x we obtain

$$\mathbf{P}_\gamma\{x_\tau = a_0,\ x_{\tau-1} = a_1,\ \ldots,\ x_{\tau-n} = a_n\} = \eta(a_n)\, p(a_n, a_{n-1}) \ldots p(a_1, a_0) L_D(a_0), \tag{25}$$

where

$$\eta = \gamma G. \tag{26}$$

In particular,

$$\mathbf{P}_\gamma\{x_\tau = y\} = \eta(y) L_D(y). \tag{27}$$

§7. Densities of excessive measures and supermartingales

Let τ be the last exit time from D and let f be a non-negative function in E. We put $f(*) = 0$. We ask the question: when is $f(x_\tau), f(x_{\tau-1}), \ldots, f(x_{\tau-N})$ a supermartingale with respect to $x_\tau, x_{\tau-1}, \ldots, x_{\tau-N}$ (for measure $\mathbf{P}\gamma$)?

By (25) we can write (14) as follows:

$$\sum_{a_n} \eta(a_n)\, p(a_n, a_{n-1}) \ldots p(a_1, a_0) L(a_0) f(a_n) \leq \eta(a_{n-1})\, p(a_{n-1}, a_{n-2}) \ldots p(a_1, a_0) L(a_0) f(a_{n-1}).$$

Obviously it is sufficient that

$$\sum_{a_n} \eta(a_n) f(a_n)\, p(a_n, a_{n-1}) \leq \eta(a_{n-1}) f(a_{n-1}),$$

that is, that the measure $f\eta$ is excessive.

Let ν_N be the number of down-crossings of $[c, d]$ by the sequence $f(x_\tau), f(x_{\tau-1}), \ldots, f(x_{\tau-N})$ or, what is equivalent, the number of up-crossings of $[c, d]$ by the sequence $f(x_{\tau-N}), \ldots, f(x_\tau)$. By Lemma 3 of §4

$$M_\gamma \nu_N \leqslant \frac{1}{d-c} M_\gamma f(x_\tau).$$

We denote by ν_D the number of up-crossings by the sequence $f(x_0), f(x_1), \ldots, f(x_\tau)$. If D is finite, then almost certainly $\tau < \infty$ and $\nu_N \uparrow \nu_D$ as $N \to \infty$. Hence $M_\gamma \nu_D \leqslant \dfrac{1}{d-c} M_\gamma f(x_{\tau_D})$. Now we consider an expanding sequence of finite sets D_n whose sum is the whole of E. Then $\nu_{D_n} \uparrow \nu$, where ν is the number of up-crossings of $[c, d]$ by the infinite sequence $f(x_0), f(x_1), \ldots$ It is evident that

$$M_\gamma \nu \leqslant \frac{1}{d-c} \sup_D M_\gamma f(x_{\tau_D}). \tag{28}$$

We say that f belongs to the class K_γ if

$$Q = \sup_D M_\gamma f(x_{\tau_D}) < \infty. \tag{29}$$

From (28) it is clear that if $f \in K_\gamma$, then $\nu < \infty$ (P_γ-a.e.). Hence, as in §5, there follows the existence P_γ-a.e. of the finite or infinite limit

$$\xi = \lim_{n \to \infty} f(x_n).$$

Let $\nu(c)$ be the number of up-crossings of $[c, 2c]$ by the infinite sequence $f(x_0), f(x_1), \ldots$ and let $\lim\limits_{c \to +\infty} \nu(c) = \bar{\nu}$. Evidently $\{\xi = \infty\} \subseteq \{\bar{\nu} \geqslant 1\}$. Hence

$$\mathsf{P}_\gamma\{\xi = \infty\} \leqslant \mathsf{P}_\gamma\{\bar{\nu} \geqslant 1\} \leqslant M_\gamma \bar{\nu}.$$

By Fatou's lemma and (28)

$$\mathsf{P}_\gamma\{\xi = \infty\} \leqslant \varliminf_{c \to \infty} M_\gamma \nu(c) \leqslant \varliminf_{c \to \infty} \frac{Q}{c} = 0.$$

So we have proved the following theorem:

THEOREM 2. *If the density f of the excessive measure μ with respect to the measure $\eta = \gamma G$ belongs to the class K_γ, then P_γ-a.e. on Ω_∞ there exists the finite limit*

$$\lim_{n \to \infty} f(x_n).$$

§8. Excessive measures with densities of class K_γ. The Martin kernel

Let γ be a finite measure, that is, $(1, \gamma) < \infty$. Then, by Lemma 1, for any $y \in E$

$$\eta(y) = \sum_x \gamma(x) g(x, y) \leqslant \sum_x \gamma(x) g(y, y) = (1, \gamma) g(y, y) < \infty.$$

We put $E_\gamma = \{y: \eta(y) > 0\}$. It is easily seen that E contains the set $\{y: \gamma(y) > 0\}$ and consists of all points that a particle starting from a point of this set hits with positive probability. The probability of going out of E_γ is zero. Hence the Markov process may be considered only on the set E_γ. The most interesting case is when $E_\gamma = E$. In this case we say that γ is a *reference measure.*

Henceforth we assume that γ is a reference measure.

According to §3 the measure μG is excessive if $\mu \geqslant 0$. The density of this measure with respect to $\eta = \gamma G$ is given by[1]

$$f(y) = \sum_x \mu(x) k_y(x) = (k_y, \mu), \tag{30}$$

where

$$k_y(x) = \frac{g(x, y)}{\eta(y)}. \tag{31}$$

The kernel $k(x, y) = k_y(x)$ is called the *Martin kernel.*

When does the density (30) belong to K_ν? By (27)

$$M_\gamma f(x_\tau) = \sum_y f(y) \mathbf{P}_\gamma \{x_\tau = y\} = \sum_y f(y) \eta(y) L(y) = (\mu G, L) = \sum_{x, y} \mu(x) g(x, y) L(y),$$

and by (24)

$$M_\gamma f(x_\tau) \leqslant \sum_x \mu(x).$$

Thus, if μ is a finite measure, then the measure μG has with respect to η a density of class K_γ. The following proposition follows from Theorem 3.

THEOREM 3. *For any finite measure μ there exists $\mathbf{P}_\gamma$-a.e. on Ω_∞ the finite limit*

$$\lim_{n \to \infty} (k_{x_n}, \mu).$$

In particular, for any y there exists $\mathbf{P}_\gamma$-a.e. on Ω_∞ the finite limit

$$\lim_{n \to \infty} k_{x_n}(y).$$

[1] If γ is not a reference measure, then μG has a density with respect to γG if and only if $E_\mu \subseteq E_\gamma$. Formula (30) remains valid if $k_y(x)$ is defined by (31) for $y \in E_\gamma$ and $k_y(x)$ is given arbitrarily for $y \notin E_\gamma$.

§9. Martin compactification

As already stated in §3, the Green's function $g(x, y)$ determines for each point $y \in E$ an excessive function $g(x, y)$. The function $k_y(x)$ differs from it only by a factor not depending on x and satisfies the normalizing relation

$$(k_y, \gamma) = 1. \tag{32}$$

By Lemma 1

$$k_y(x) = \frac{\pi(x, y)\, g(y, y)}{\sum_z \gamma(z)\, \pi(z, y)\, g(y, y)} = \frac{\pi(x, y)}{\sum_z \gamma(z)\, \pi(z, y)}$$

(because $g(y, y) \geqslant p(0, y, y) = 1$). According to (13) $\pi(z, y) \geqslant \pi(z, x)\, \pi(x, y)$. Hence for all y

$$k_y(x) \leqslant \frac{1}{a(x)}, \tag{33}$$

where

$$a(x) = \sum_z \gamma(z)\, \pi(z, x)$$

(since γ is a reference measure, $a(x) > 0$ for all x).

We enumerate the points of E in arbitrary order by the integers and let $N(x)$ be the number of the point x. We put

$$\rho(y, z) = |2^{-N(y)} - 2^{-N(z)}| + \sum_x |k_y(x) - k_z(x)|\, a(x)\, 2^{-N(x)}.$$

This defines a metric in E, and the distance between any two points does not exceed 3. Forming the completion of E with respect to this metric we obtain the compactum E^*. The set E is open in E^*, so that the boundary ∂E of E in E^* is $E^* \setminus E$. The compactification so constructed is called the *Martin compactification* and the boundary ∂E the *Martin boundary*.

It should be noted that the metric $\rho(y, z)$ can be chosen with a certain degree of arbitrariness. It is essential only that y_n is a Cauchy sequence if and only if:

a) the functions $k_{y_n}(x)$ converge at any point x and

b) $N(y_n) \to \infty$ or y_n remains constant from some number n_0 on.

For each $x \in E$, $k_y(x)$ as a function of y can be extended continuously to E^*. Suppose that the sequence $y_n \in E$ converges to $y \in \partial E$. Then $k_{y_n}(x) \to k_y(x)$ for any x. Hence it follows that for $y \in \partial E$, k_y is excessive and satisfies the condition

$$(k_y, \gamma) \leqslant 1 \tag{34}$$

(equality in (32) need not hold).

The topology in E^* induced by the metric $\rho(x, y)$ is called the M_+-topology.

The fundamental role of Martin boundaries in the theory of Markov processes is determined by the following theorem.

THEOREM 4. *With any initial state x, for almost all non-terminating paths there exists in the topology M_+ the limit*

$$\lim_{n\to\infty} x_n = x_\infty \in \partial E.$$

PROOF. For a process with initial distribution γ this statement follows at once from Theorem 3 and the Remark in §3, according to which for a transient chain $N(x_n) \to \infty$ almost surely.

We denote by A the set of all non-terminating paths for which the theorem does not hold and put $h(x) = \mathbf{P}_x(A)$. It will be proved later (see Corollary to Theorem 8), that h is harmonic. By what has been proved, $(h, \gamma) = \mathbf{P}_\gamma(A) = 0$. Hence Theorem 4 follows from the following lemma.

LEMMA 4. *If γ is a reference measure and h an excessive function, then from $(h, \gamma) = 0$ if follows that $h = 0$ everywhere.*

PROOF OF LEMMA 4. For any n we have $\mathbf{P}^n h \leqslant h$ and therefore

$$0 \leqslant (\gamma \mathbf{P}^n, h) = (\gamma, \mathbf{P}^n h) \leqslant (\gamma, h) = 0.$$

Thus, $(\gamma\mathbf{P}^n, h) = 0$. Summing over n we have $(\eta, h) = 0$, where $\eta = \gamma G$. By definition of a reference measure η is everywhere positive, hence h is zero everywhere.

§10. Distribution of x_ζ

If $\zeta < \infty$, then x_ζ is a point of E at which a path terminates. If $\zeta = \infty$, then $x_\zeta = x_\infty$ is defined in Theorem 4 and belongs to ∂E.

Let τ be the last exit time from D. Comparing (23), (27) and (31) we have

$$\mathbf{P}_x\{x_\tau = y\} = k_y(x)\, \mathbf{P}_\gamma\{x_\tau = y\}.$$

Hence for any function

$$M_x f(x_\tau) = M_x \sum_y \delta(x_\tau, y) f(y) = \sum_y f(y)\, \mathbf{P}_x\{x_\tau = y\} = \sum_y f(y)\, k_y(x)\, \mathbf{P}_\gamma\{x_\tau = y\} =$$
$$= M_\gamma \sum_y f(y)\, k_y(x)\, \delta(x_\tau, y) = M_\gamma f(x_\tau)\, k(x, x_\tau) \quad (35)$$

(we recall that $k(x, y) = k_y(x)$). We consider now a sequence of finite sets $D_n \uparrow E$ and denote by τ_n the last exit time from D_n. On the set Ω_∞ we have $\tau_n \to \infty$, and hence $x_{\tau_n} \to x_\infty$ and $k(x, x_{\tau_n}) \to k(x, x_\infty) = k(x, x_\zeta)$ almost surely. On the set $\{\zeta < \infty\}$, we have $\tau_n = \zeta$ beginning with some $n_0(\omega)$; hence $k(x, x_{\tau_n}) \to k(x, x_\zeta)$. Suppose that[1] $f \in C(E^*)$. Then $f(x_{\tau_n}) \to f(x_\zeta)$.

[1] $C(E^*)$ is the space of all continuous functions on the compactum E^*.

Bearing (33) in mind, we can take the limit in (35). Thus,

$$M_x f(x_\zeta) = M_\gamma f(x_\zeta)\, k(x, x_\zeta). \tag{36}$$

On the Borel sets of the compactum E^* we consider the measure μ_1 defined by

$$\mu_1(\Gamma) = \mathbf{P}_\gamma \{x_\zeta \in \Gamma\}. \tag{37}$$

By (36)

$$M_x f(x_\zeta) = \int_{E^*} k_y(x)\, f(y)\, \mu_1(dy). \tag{38}$$

The formula (38), which has been proved for continuous functions f, extends in an obvious way to all Borel non-negative functions. Putting $f = \chi_\Gamma$, we get

$$\mathbf{P}_x \{x_\zeta \in \Gamma\} = \int_\Gamma k_y(x)\, \mu_1(dy). \tag{39}$$

Thus, $k_y(x)$ ($x \in E$, $y \in E^*$) can be interpreted as the density of the distribution for the point x_ζ (with respect to μ_1) for the initial state x.

We note that by (4′)

$$\mathbf{P}_x \{x_n = y, \zeta = n\} = p(n, x, y)\, [1 - \mathbf{P}1](y).$$

Hence

$$\mathbf{P}x \{x_\zeta = y\} = \sum_{n=0}^{\infty} p(n, x, y)(1 - \mathbf{P}1)(y) = g(x, y)\, [1 - \mathbf{P}1](y). \tag{40}$$

From (37) and (40)

$$\mu_1(y) = \eta(y)\, [1 - \mathbf{P}1](y) \qquad (y \in E). \tag{41}$$

Next, if $f \in C(E^*)$, then

$$M_x f(x_\infty) = \lim_{n\to\infty} M_x f(x_n) = \lim_{n\to\infty} \mathbf{P}^n f(x)$$

and by (37) and (41)

$$\int_{E^*} f(y)\, \mu_1(dy) = M_\gamma f(x_\zeta) = (1 - \mathbf{P}1, f\eta) + \lim_{n\to\infty} \sum_{x, y} \gamma(x)\, p(n, x, y)\, f(y). \tag{42}$$

We note that by (38) and (41)

$$M_x f(x_\zeta)\, \chi_{\zeta<\infty} = M_x f(x_\zeta)\, \chi_E(x_\zeta) = \int_E k_y(x)\, f(y)\, \mu_1(dy) = G[f(1 - \mathbf{P}1)](x) \tag{43}$$

and by (38)

$$M_x f(x_\infty) = M_x f(x_\zeta) \chi_{\partial E}(x_\zeta) = \int_{\partial E} k_y(x) f(y) \mu_1(dy). \tag{44}$$

§11. h-processes. Martin representation of excessive functions

Let γ be a reference measure and h a γ-integrable excessive function. We prove that h is everywhere finite. For each $y \in E$ we can find n and x such that $\gamma(x)p(n, x, y) > 0$. Since $\mathbf{P}^n h \leqslant h$, we have

$$\gamma(x)\, p(n, x, y)\, h(y) \leqslant (\mathbf{P}^n h, \gamma) \leqslant (h, \gamma) < \infty.$$

Hence $h(y) < \infty$. We put $E^h = \{x\colon 0 < h(x)\}$ and define on E^h the transition function

$$p^h(x, y) = \frac{1}{h(x)} p(x, y) h(y). \tag{45}$$

A Markov process corresponding to the transition function $p^h(x, y)$ is called an *h-process*. All characteristics of this process are denoted by the same letters as for the initial process but with the upper suffix h: $\mathbf{P}_x^h$, $g^h(x, y)$, etc. For the Green's function the following relation holds:

$$g^h(x, y) = \frac{1}{h(x)} g(x, y) h(y).$$

γh is a reference measure for the h-process. We have $(\gamma h)G^h = \eta h$ and the Martin kernel corresponding to γh is given by

$$k_y^h(x) = \frac{g^h(x, y)}{(\eta h)(y)} = \frac{k_y(x)}{h(x)}. \tag{46}$$

From this it is evident that the Martin topology for an h-process coincides with the Martin topology of the initial process, and the Martin compactification E^{h*} for the h-process leads to the closure of E^h in the space E^*. The Martin boundary ∂E^h is simply the boundary of E^h in E^*.

We put

$$\mu_h(\Gamma) = \mathbf{P}_{\gamma h}^h \{x_\zeta \in \Gamma\}. \tag{47}$$

Evidently for any Borel function $f \geqslant 0$

$$\int_{E^*} f(y) \mu_h(dy) = M_{\gamma h}^h f(x_\zeta). \tag{48}$$

Hence

$$M_{\gamma h}^h f(x_\infty) = M_{\gamma h}^h (f \chi_{\partial E})(x_\zeta) = \int_{\partial E} f(y) \mu_h(dy). \tag{49}$$

Applying to the h-process the formulae (38), (44), (43), (41) and (42) we have

$$M_x^h f(x_\zeta) = \frac{1}{h(x)} \int_{E^*} k_y(x) f(y) \mu_h(dy), \tag{50}$$

$$M_x^h f(x_\infty) = \frac{1}{h(x)} \int_{\partial E} k_y(x) f(y) \mu_h(dy), \tag{51}$$

$$M_x^h f(x_{\zeta-}) \chi_{\zeta<\infty} = \frac{1}{h(x)} G[(h - \mathbf{P}h) f], \tag{52}$$

$$\mu_h(y) = \eta(y) [h(y) - \mathbf{P}h(y)], \tag{53}$$

$$\int_{E^*} f(y) \mu_h(dy) =$$

$$= (h - \mathbf{P}h, f\eta) + \lim_{n\to\infty} \sum_{x,y} \gamma(x) p(n, x, y) h(y) f(y) \qquad (f \in C(E^*)). \tag{54}$$

(In (50)–(51) we enlarge the domain of integration by taking $\mu_h(E^* \setminus E^{h*}) = 0$).

Putting $f = 1$ in (50) and noting (46), we observe that for any $x \in E^h$

$$h(x) = \int_{E^*} k_y(x) \mu_h(dy). \tag{55}$$

Outside the set E^h both sides of this equation are zero. (If $x \notin E^h$, $y \in E^h$, then $p(n, x, y) = 0$ for all n; hence, $g(x, y) = 0$ and $k_y(x) = 0$. Thus, $k_y(x) = 0$ also for y in the set E^{h*} on which the measure μ_h is concentrated). Thus, the representation (55) holds for all $x \in E$. It is called the *Martin representation of the excessive function h*. The measure μ_h is called *the spectral measure of the function h.*

By (53)

$$\int_E k_y(x) \mu_h(dy) = G(h - \mathbf{P}h).$$

Hence the Martin decomposition can be put in the form

$$h(x) = G(h - \mathbf{P}h)(x) + \int_{\partial E} k_y(x) \mu_h(dy). \tag{56}$$

§12. The spectral measure of k_z. The exit space

First let $z \in E$. Then

$$\mathbf{P}k_z(y) = \sum_u p(y, u) \frac{g(u, z)}{\eta(z)} = \frac{g(y, z) - \delta(y, z)}{\eta(z)} = k_z(y) - \frac{\delta(y, z)}{\eta(z)}, \tag{57}$$

and according to (54)

$$\mu_{k_z}(y) = \eta(y) [k_z(y) - \mathbf{P}k_z(y)] = \delta_z(y), \tag{58}$$

where $\delta_z(y) = \delta(y, z)$ is the unit measure concentrated at z. Thus, $\mu_{k_z} = \delta_z$ for all $z \in E$.

The set of all $z \in \partial E$ for which $\mu_{k_z} = \delta_z$ is called *the exit space* and is denoted by B.

THEOREM 5. *The space of exits B is a Borel subset of ∂E. For any γ-integrable excessive function h we have $\mu_h(\partial E \setminus B) = 0$. If $z \in B$, then k_z is a harmonic function and $(k_z, \gamma) = 1$.*

PROOF. If $z \in B$, then evidently

$$\mu_{k_z}\{z\} = 1. \tag{59}$$

On the other hand, for any $z \in \partial E$ by (47) and (34)

$$\mu_{k_z}(E^*) = (k_z, \gamma) \leqslant 1. \tag{60}$$

Hence, if (59) is satisfied, then $\mu_{k_z} = \delta_z$ and $z \in B$. Thus, B is given by (59). By (54), for $z \in \partial E$,

$$\mu_{k_z}\{z\} = \lim_{m\to\infty} \int_{E^*} e^{-m\rho(x,z)} \mu_{k_z}(dx) = \lim_{m\to\infty} \lim_{n\to\infty} \sum_{x,y} \gamma(x) p(n, x, y) k_z(y) e^{-m\rho(x,y)}$$

which implies that B is a Borel set.

We write $k = k_z$, omitting the subscript z when no confusion can arise.

Let $z \in \partial E$ and $\varphi, \psi \in C(E^*)$. For any $n \geqslant 0, m \geqslant 0$

$$M^h_{h\gamma}\varphi(x_n)\psi(x_{n+m}) = \sum_{x,y,z} \gamma(x) h(x) p^h(n, x, y) \varphi(y) p^h(m, y, z) \psi(z) =$$
$$= \sum_{x,y} \gamma(x) p(n, x, y) h(y) \varphi(y) M^h_y \psi(x_m).$$

Taking the limit as $m \to \infty$ and using (51) we have

$$M^h_{h\gamma}\varphi(x_n)\psi(x_\infty) = \sum_{x,y} \gamma(x) p(n, x, y) h(y) \varphi(y) \int_{\partial E} k^h_z(y) \psi(z) \mu_h(dz) =$$
$$= \int_{\partial E} \sum_{x,y} \gamma(x) h(x) p^k(n, x, y) k(y) \varphi(y) \psi(z) \mu_h(dz) = \int_{\partial E} M^k_{k\gamma}\varphi(x_n) \psi(z) \mu_h(dz).$$

Now letting $n \to \infty$ and noting (48) we have

$$\int_{\partial E} \varphi(z) \psi(z) \mu_h(dz) = \int_{\partial E} \left[\int_{\partial E} \varphi(u) \mu_k(du) \right] \psi(z) \mu_h(dz).$$

Since ψ is an arbitrary continuous function, it follows that μ_h – a.e.

$$\varphi(z) = \int_{\partial E} \varphi(u) \mu_{k_z}(du).$$

It is clear that for μ_h-almost all z this equation holds simultaneously for the sequence of functions

$$\varphi_m(y) = e^{-m\rho(y,z)} \quad (m = 1, 2, \ldots)$$

and hence in the limit as $n \to \infty$ the resulting equation (59) is satisfied. Hence

$$\mu_h(\partial E \setminus B) = 0.$$

For $z \in B$ it follows from (58) that k_z is harmonic, and from (60) that $(k_z, \gamma) = 1$.

REMARK. From (50) it is clear that if $z \in E \cup B$, then $\mathsf{P}_x^{k_z}\{x_\zeta = z\} = 1$, so that for any initial state x almost all paths of a k_z-process terminate at z.

§13. The Uniqueness Theorem

THEOREM 6. *Every γ-integrable excessive function h has a unique representation of the form*

$$h(x) = \int_{E \cup B} k_z(x)\, \mu(dz), \tag{61}$$

where μ is a measure on the Borel subsets of $E \cup B$. The measure μ is finite.

For any finite measure μ (61) defines a γ-integrable excessive function. This function is harmonic if and only if $\mu(E) = 0$.

From Theorem 6 it follows, in particular, that if h has a representation of the form (61), then μ coincides with the spectral measure μ_h, and (61) coincides with the Martin representation.

PROOF. From Theorem 5, the Martin representation (55) of the γ-integrable excessive function h can be rewritten in the form (61), where $\mu = \mu_h$. Since $\mathsf{P}k_z \leqslant k_z$ for $z \in E$ and $\mathsf{P}k_z = k_z$ for $z \in B$, every function h obtained by (61) is excessive, and if $\mu(E) = 0$, it is harmonic. Since $(k_z, \gamma) = 1$ for all $z \in E \cup B$, we see that $(h, \gamma) = \mu(E \cup B) < \infty$.

We show now that if h is given by (61), then μ coincides with the spectral measure μ_h. Let $f \in C(E)$. Applying (54) to h and to k_z we note that

$$\int_{E^*} f(y)\, \mu_h(dy) = \int_{E \cup B} \left[\int_{E^*} f(y)\, \mu_{k_z}(dy) \right] \mu(dz).$$

But $\mu_{k_z} = \delta_z$ for $z \in E \cup B$. Therefore

$$\int_{E^*} f(y)\, \mu_h(dy) = \int_{E \cup B} f(z)\, \mu(dz).$$

Since μ_h is concentrated on $E \cup B$, it follows that $\mu_h = \mu$.

We observe finally that if $\mathsf{P}h = h$, then by (53) $\mu_h(y) = 0$ for all $y \in E$.

§14. Minimal excessive functions

A non-zero excessive function h is called *minimal* if from $h = h_1 + h_2$, where h_1 and h_2 are excessive functions, it follows that $h_1 = c_1 h$, $h_2 = c_2 h$ (c_1 and c_2 are constants). It is easily proved that a harmonic function h is minimal if and only if every harmonic function h_1 satisfying $0 \leqslant h_1 \leqslant h$, is proportional to h.

THEOREM 7. *The general form of γ-integrable minimal excessive functions is ck_z, where $z \in E \cup B$ and c is a positive constant.*

PROOF. From (54) it is clear that $\mu_{h_1+h_2} = \mu_{h_1} + \mu_{h_2}$. Let $z \in E \cup B$ and $k_z = h_1 + h_2$, where h_1 and h_2 are excessive. Then $\mu_{h_1} + \mu_{h_2} = \mu_{k_z} = \delta_z$. Hence $0 = \delta(E^* \setminus z) = \mu_{h_1}(E^* \setminus z) + \mu_{h_2}(E^* \setminus z)$ and $\mu_{h_i}(E^* \setminus z) = 0$. From (55)

$$h_i(z) = \int_{E^*} k_y \mu_h(dy) = k_z \mu_{h_i}(z).$$

Thus, k_z is minimal.

Next, let h be any γ-integrable minimal excessive function and μ_h its spectral measure. Then $\mu_h(E \cup B) = (h, \gamma)$. By Lemma 5 this quantity is positive if $h \neq 0$. Hence there exists a point $z \in E \cup B$, any neighbourhood of which has positive measure μ_h. We put $U_n = \{y: \rho(y, z) < 1/n\}$,

$$h_n = \int_{U_n} k_y \mu_h(dy).$$

It is obvious that h_n and $h - h_n$ are excessive. Since h_n is minimal, we have $h_n = c_n h$. Since $(h_n, \gamma) = \mu_h(U_n)$ and $(h, \gamma) = \mu_h(E \cup B) = (h, \gamma)$, we have $c_h = \mu_h(U_n) / \mu_h(E \cup B)$ and consequently

$$h = \frac{(h, \gamma)}{\mu_h(U_n)} \int_{U_n} k_y \mu_h(dy).$$

Taking the limit as $n \to \infty$ we have $h = (h, \gamma) k_z$.

§15. The operator θ. Final random variables

We consider in the space of paths Ω the mapping θ that is defined on the set $\Omega_1 = \{\omega: \zeta(\omega) \geqslant 1\}$ and carries the path $a_0 a_1 a_2 \ldots$ into $a_1 a_2 a_3 \ldots$ For any function $\xi(\omega)$ we put

$$\theta\xi(\omega) = \begin{cases} \xi(\theta\omega) & \text{if} \quad \omega \in \Omega_1; \\ 0 & \text{if} \quad \omega \notin \Omega_1. \end{cases} \tag{62}$$

The random variable ξ is called *final* if $\theta\xi = \xi$. We note that final random variables are different from zero only on Ω_∞ and do not change their value if an arbitrary initial interval of the path is changed.

We denote by A the set of non-terminating paths for which the limit $x_\infty = \lim_{n \to \infty} x_n$ does not exist or does not belong to ∂E. An example of a final random variable is the function that is equal to some constant b on A, to $f(x_\infty)$ on $\Omega_\infty \setminus A$ (f is an arbitrary Borel function on ∂E), and to 0 outside Ω_∞. In particular, the random variable χ_A is final.

LEMMA 5. *For any non-negative random variable* ξ

$$M_x\{\delta(x_0, a_0)\,\delta(x_1, a_1)\ldots\delta(x_n, a_n)\,\theta^n\xi\} = \\ = \delta(x, a_0)\,p(a_0, a_1)\ldots p(a_{n-1}, a_n)\,M_{a_n}\xi. \quad (63)$$

PROOF. We denote by $\mu_1(A)$ and $\mu_2(A)$ the values of the right- and left-hand sides of (63) for $\xi = \chi_A$. Obviously it is sufficient to prove that $\mu_1(A) = \mu_2(A)$ for any $A \in \mathcal{F}$. The functions μ_1 and μ_2 are measures. By (63) for the simple set $A = [a_0a_1\ldots a_n]$

$$\mu_1(A) = \mu_2(A) = \delta(x, a_0)\,p(a_0, a_1)\ldots \\ \ldots p(a_{n-1}, a_n)\,\delta(a_n, b_0)\,p(b_0, b_1)\ldots p(b_{m-1}, b_m).$$

In accordance with Theorem A, from the fact that two measures coincide on simple sets it follows that they coincide on the σ-algebra $\mathcal{F}$.

THEOREM 8. *If* $\xi \geqslant 0$ *is a final random variable and* $h(x) = M_x\xi$ *is finite for each* x, *then* $h(x)$ *is a harmonic function, and measures corresponding to an* h-*process are given by the formula*

$$\mathbf{P}_x^h(A) = \frac{\int_A \xi\, d\mathbf{P}_x}{h(x)} \qquad (x \in E^h). \quad (64)$$

PROOF. By (62) and (63)

$$h(x) = M_x\xi = M_x\sum_y \delta(x, y)\,\theta\xi = \sum_y p(x, y)\,M_y\xi = \sum_y p(x, y)\,h(y).$$

Hence h is harmonic.

To prove the second statement it is sufficient to note that (64) defines a measure on the σ-algebra $\mathcal{F}$, such that the measure of a simple set $[a_0a_1a_2\ldots a_n]$ is

$$\frac{1}{h(z)}M_x\delta(x_0, a)\ldots\delta(x_n, a_n)\,\xi = \frac{1}{h(x)}M_x\delta(x_0, a)\ldots\delta(x_n, a)\,\theta^n\xi = \\ = \frac{1}{h(x)}\delta(x, a_0)\,p(a_0, a_1)\ldots p(a_{n-1}, a_n)\,h(a_n),$$

that is, coincides with $\mathbf{P}_x^h$.

COROLLARY. *If* $M_x\xi = M_x\eta$ *for any two final random variables* ξ *and* η, *then* $\xi = \eta$ ($\mathbf{P}_x$-*a.e.*).

PROOF. From (64), for any $A \in \mathcal{F}$

$$\int_A \xi\, d\mathbf{P}_x = \int_A \eta\, d\mathbf{P}_x.$$

THEOREM 9. *Let* h *be a bounded harmonic function. Then*

$$h(x) = \int_B k_y(x) \varphi(y) \mu_1(dy), \tag{65}$$

where φ is a bounded Borel function. Also P_x-a.e. on Ω_∞,

$$\lim_{n\to\infty} h(x_n) = \varphi(x_\infty) \tag{66}$$

and

$$M_x \varphi(x_\infty) = h(x). \tag{67}$$

PROOF. Let $0 \leqslant h \leqslant a$, where a is a constant. The functions h and $f = a - h$ are excessive. By Lemma 4 and Theorem 7

$$a = \int_B k_y(x) \mu_a(dy) = \int_B k_y(x) (\mu_h + \mu_f)(dy).$$

We assume that $a > 0$. Dividing the equation by a we find by Theorem 7 that $\mu_1 = (1/a)(\mu_h + \mu_f)$. Hence μ_h has a bounded density with respect to $\mu_1 : \mu_h(dy) = \varphi(y) \mu_1(dy)$. Therefore (61) can be written in the form (65). By (44) and (65) $M_x \varphi(x_\infty) = h(x)$. On the other hand, by Theorem 1 there exists P_x-*a.e.* the limit

$$\xi = \lim_{n\to\infty} h(x_n).$$

By (6)

$$M_x \xi = \lim_{n\to\infty} M_x h(x_n) = \lim_{n\to\infty} \mathbf{P}^n h(x) = h(x).$$

Thus, $M_x \varphi(x_\infty) = M_x \xi$. But $\varphi(x_\infty)$ and ξ are final random variables, and by the Corollary to Theorem 8 $\varphi(x_\infty) = \xi$ P_x-*a.e.* (we must take $\varphi(x_\infty) = 0$ outside Ω_∞).

COROLLARY. *Every final random variable ξ coincides almost surely with $\varphi(x_\infty)$, where φ is some Borel function on B.*

PROOF. Let ξ be bounded. Then $h(x) = M_x \xi$ is a bounded harmonic function, and by Theorem 9 P_x-*a.e.* on Ω_∞ we have $\lim h(x_n) = \varphi(x_\infty)$ and $M_x \varphi(x_\infty) = h(x) = M_x \xi$. By the Corollary to Theorem 8, P_x-*a.e.* $\xi = \varphi(x_\infty)$. If ξ is arbitrary, then $\xi \wedge a$ is bounded, a being a constant. By what has been proved, $\xi \wedge a = \varphi_a(x_\infty)$ P_x-*a.e.* Obviously it follows from this that $\xi = \varphi(x_\infty)$ P_x-*a.e.*, where $\varphi(x) = \lim_{a\to\infty} \varphi_a(x)$.

THEOREM 10. *Let φ be a non-negative Borel μ_1-integrable function on ∂E. Then*

$$h(x) = \int_B k_y(x) \varphi(y) \mu_1(dy). \tag{68}$$

defines a harmonic function h such that P_x-a.e. on Ω_∞,

$$\lim_{n\to\infty} h(x_n) = \varphi(x_\infty).$$

PROOF. The function $h(x)$ is finite everywhere by (33). We select a positive constant c and put

$$\varphi_1(x) = \begin{cases} \varphi(x) & \text{if } \varphi(x) \leqslant c, \\ 0 & \text{if } \varphi(x) > c, \end{cases} \qquad \varphi_2(x) = \begin{cases} 0 & \text{if } \varphi(x) \leqslant c, \\ \varphi(x) & \text{if } \varphi(x) > c, \end{cases}$$

$$h_i(x) = \int_B k(x, y)\varphi_i(y)\mu_1(dy).$$

From (44)

$$M_x\varphi(x_\infty) = h(x), \quad M_x\varphi_i(x_\infty) = h_i(x), \tag{69}$$

and by Theorem 8 h_1, h_2, h are harmonic functions. h_1 is bounded, and by Theorem 9 $\mathbf{P}_x$-*a.e.*

$$\lim_{n\to\infty} h_1(x_n) = \varphi_1(x_\infty).$$

By Theorem 1 $\mathbf{P}_x$-*a.e.* there exist also the limits

$$\xi = \lim_{n\to\infty} h_1(x_n), \quad \xi_2 = \lim_{n\to\infty} h_2(x). \tag{70}$$

By Fatou's lemma

$$M_x\xi_2 \leqslant \lim_{n\to\infty} M_x h_2(x_n) = \lim \mathbf{P}^n h_2(x) = h_2(x).$$

Hence by (69) and (70) we find

$$M_x \mid \xi - \varphi(x_\infty) \mid = M_x \mid \xi_2 - \varphi_2(x_\infty) \mid \leqslant M_x\xi_2 + M_x\varphi_2(x_\infty) \leqslant 2h_2(x).$$

The left-hand side does not depend on c and the right-hand side tends to zero as $c \to \infty$. Hence $\xi = \varphi(x_\infty)$, $\mathbf{P}_x$-*a.e.*

REMARK 1. According to Theorem 10, (68) gives a generalized solution of Dirichlet's problem with boundary function $\varphi(x)$; the boundary values are taken along almost all non-terminating paths.

REMARK 2. Theorems 9 and 10 can be "relativized" in an obvious way; selecting any harmonic function H we can replace the measure μ_1 by μ_H, the condition of boundedness of h by $h \leqslant H$ and the measure $\mathbf{P}_x$ by $\mathbf{P}_x^H$.

§16. The entrance space. Decomposition of excessive measures

We say that $l(x) \geq 0$ is a reference function $l(x)$ if $s = Gl$ is everywhere positive and finite. We put

$$\varkappa_y(x) = \frac{g(y, x)}{s(y)}. \tag{71}$$

We note that κ_y is an excessive measure and $(l, \kappa_y) = 1$. According to Lemma 1 $g(y, x) = \pi(y, x)\, g(x, x)$. By (13) we have $g(y, z) = \pi(y, z)\, g(z, z) \geq$ $\geq \pi(y, x)\, \pi(x, z)\, g(z, z) = \pi(y, x)\, g(x, z)$. Hence $s(y) \geq \pi(y, x)\, s(x)$ and $\kappa_y(x) \leq 1/c(x)$, where $c(x) = s(x)/g(x, x) > 0$.

We consider in E the metric

$$\hat{\rho}(y, z) = |2^{-N(y)} - 2^{-N(z)}| + \sum_x |\varkappa_y(x) - \varkappa_z(x)|\, c(x)\, 2^{-N(x)},$$

where $N(x)$ has the same meaning as in §9. Forming the completion of E with respect to this metric we obtain a compactum $\hat{E}$. The topology of $\hat{E}$ determined by the metric $\hat{\rho}$ is called M_-. The boundary of E in the topology M_- is $\hat{E} \setminus E$. Repeating the arguments of §9 we extend $\kappa_y(x)$ continuously to $y \in \hat{E} \setminus E$.

We now prove the following theorem.

THEOREM 11. *With each l-finite excessive measure α we can associate a finite measure μ^α on the Borel subsets of $\hat{E}$ such that for $f \in C^*(E)$*

$$\int_{\hat{E}} f(y)\, \mu^\alpha(dy) = (f_s, \alpha - \alpha \mathbf{P}) + \lim_{n \to \infty} \sum_{x, y} f(y)\, \alpha(y)\, p(n, x, y)\, l(x). \tag{72}$$

The entrance space $\hat{B}$ is the set of all $z \in \hat{E} \setminus E$ for which the spectral measure of the excessive measure κ_z is δ_z. For $z \in \hat{B}$ the measure κ_z is harmonic and $(l, \kappa_z) = 1$.

Every l-finite excessive measure α has a unique representation in the form

$$\alpha(x) = \int_{E \cap \hat{B}} \varkappa_y(x)\, \mu(dy). \tag{73}$$

Here μ coincides with the spectral measure μ^α. For α to be harmonic it is necessary and sufficient that $\mu^\alpha(E) = 0$.

Let ν be an l-finite excessive measure[1] *and $0 < \nu(x) < \infty$ for all x. The formula*

---

[1] If the reference measure γ and the reference function l are chosen so that

$$\sum_{x, y} g(x, y)\, \gamma(x)\, l(y) < \infty,$$

then the measure $\nu = \gamma G$ satisfies the required conditions.

$$p_\nu(x, y) = \frac{\nu(y)\, p(y, x)}{\nu(x)} \tag{74}$$

defines in E a transition function. The measure $\hat{\gamma} = l\nu$ is a reference measure, and the corresponding Martin kernel is $\kappa_y(x)/\nu(x)$. The Martin compactification corresponding to this kernel coincides with $\hat{E}$, and the exit space coincides with $\hat{B}$. h is excessive (harmonic) with respect to $p_\nu(x, y)$ if and only if the measure $\alpha = h\nu$ is excessive (harmonic) with respect to $p(x, y)$. Here $(l, \alpha) = (h, \hat{\gamma})$, so that α is l-finite if and only if h is $\hat{\gamma}$-integrable. The spectral measure of the excessive function $h = \alpha/\nu$ with respect to $p_\nu(x, y)$ coincides with μ^α.

PROOF. The statements about the Martin kernel for the transition function $p_\nu(x, y)$ and the corresponding compactification are verified directly. It is also evident that if $\alpha = h\nu$, then $(l, \alpha) = (h, \hat{\gamma})$ and $\mathsf{P}h \leqslant h(\mathsf{P}h = h)$ if and only if $\alpha\mathsf{P}_\nu \leqslant \alpha$ $(\alpha\mathsf{P}_\nu = \alpha)$. We define the spectral measure of the excessive measure α as that of the excessive function $h = \alpha / \nu$ (with respect to the transition function $p_\nu(x, y)$). (72) follows from (42). From the definition of $\hat{B}$ it is clear that $\hat{B}$ is the exit space for $p_\nu(x, y)$, and from (72) that μ^α and B do not depend on the choice of ν. The representation (73) follows from the Martin representation for the p_ν-excessive function $h = \alpha / \nu$. The remaining parts of the theorem are verified with difficulty.

REMARK. If $\nu(x)$ becomes zero or infinite at some points, then (74) defines a transition function on $E^\nu = \{x : 0 < \nu(x) < +\infty\}$. In this case the Martin compactification for $p_\nu(x, y)$ coincides with the closure of E^ν in $\hat{E}$.

§17. The behaviour of a stationary process as $t \to -\infty$

We consider a stationary Markov process with transition function $p(x, y)$ and stationary distribution ν. The formula $a_t = \hat{a}_{-t} (t = 0, 1, \ldots)$ defines a mapping r of $\hat{\Omega}$ into Ω_∞. The inverse image of A under r is denoted by A^r. We define in Ω_∞ the measure

$$\mathbf{P}(A) = \mathbf{P}_\nu(A^r) \qquad (A \in F).$$

If $A = [a_0 a_1 \ldots a_n]$, then $A^r = [a_n a_{n-1} \ldots a_0]_{-n}^0$ and by (4″)

$$\mathbf{P}(A) = \mathbf{P}_\nu(A^r) = \nu(a_n)\, p(a_n, a_{n-1}) \ldots p(a_1, a_0).$$

On the other hand, let $p_\nu(x, y)$ be a transition function defined by (74). We denote by ${}^\nu\mathsf{P}_\alpha$ the measure in Ω corresponding to the Markov process with transition function $p_\nu(x, y)$ and initial distribution α. (Since $\sum_y p_\nu(x, y) = 1$, the measure ${}^\nu\mathsf{P}_\alpha$ is concentrated on Ω_∞.) We note that

$${}^\nu\mathbf{P}_\nu[a_0 a_1 \ldots a_n] = \nu(a_n)\, p(a_n, a_{n-1}) \ldots p(a_1, a_0).$$

Therefore $\mathbf{P}(A) = {}^{\nu}\mathbf{P}_{\nu}(A)$ for all simple sets A. By Theorem A this equation is satisfied on all $A \in \mathcal{F}$. Thus,

$$\mathbf{P}_{\nu}(A^{r}) = {}^{\nu}\mathbf{P}_{\nu}(A) \qquad (A \in \mathcal{F}). \tag{75}$$

We denote by A the set of all paths $\{a_n\}$ in Ω_∞ for which $\lim\limits_{n\to\infty} a_n$ does not exist, or exists but is not in $\hat{B}$, and by $\hat{A}$ the set of all paths $\hat{a}_t$ in $\hat{\Omega}$ for which $\lim\limits_{t\to-\infty} \hat{a}_t$ does not exist, or exists but is not in $\hat{B}$. It is easy to see that $A^r = \hat{A}$, and by (75)

$$\mathbf{P}_{\nu}(\hat{A}) = {}^{\nu}\mathbf{P}_{\nu}(A).$$

By Theorems 4 and 5 and (39) we have ${}^{\nu}\mathbf{P}_x(A) = 0$ for any $x \in E$. Hence $\mathbf{P}_{\nu}(\hat{A}) = {}^{\nu}\mathbf{P}_{\nu}(A) = 0$.

Next, from (44) and Theorem 5

$${}^{\nu}M_x f(x_\infty) = \int_{\hat{B}} \frac{\varkappa_y(x)}{\nu(x)} f(y)\,\mu(dy), \tag{76}$$

where μ is the spectral measure of the function 1 with respect to $p_{\nu}(x, y)$. By Theorem 11 it coincides with the measure μ^{ν}. From (75) and (76)

$$M_{\nu} f(x_{-\infty}) = {}^{\nu}M_{\nu} f(x_\infty) = \int_{\hat{B}} (1, \varkappa_y)\, f(y)\, \mu^{\nu}(dy). \tag{77}$$

Putting $f = \chi_\Gamma$, where $\Gamma \subseteq \hat{B}$, we have

$$\mathbf{P}_{\nu}\{x_{-\infty} \in \Gamma\} = \int_{\Gamma} (1, \varkappa_y)\, \mu^{\nu}(dy). \tag{78}$$

So we have proved the following theorem.

THEOREM 12. *For a stationary process with l-finite stationary measure ν the limit*

$$x_{-\infty} = \lim_{t\to-\infty} x_t \tag{79}$$

in the topology M_- exists almost surely and belongs to the entrance space $\hat{B}$. The distribution $x_{-\infty}$ is given by (78). *In particular, if $\nu = \kappa_y$, where $y \in B$, then almost surely $x_{-\infty} = y$.*

Applying Theorem 12 to an h-process we can obtain a more general statement in which the function 1 ceases to play an exceptional role.

THEOREM 13. *Let ν be an l-finite harmonic measure, h a γ-integrable harmonic function. Then for the stationary process with transition function*

$p^h(x, y) = (h(x))^{-1} p(x, y) h(y)$ and stationary measure $h\nu$, the limit (79) in the topology M_- exists almost surely and belongs to the entrance space $\hat{B}$. The measure of the set of paths for which $x_{-\infty} \in \Gamma \subseteq \hat{B}$ is

$$\int_\Gamma (h, \varkappa_y)\, \mu^\nu(dy). \tag{80}$$

REMARK. From Theorems 4 and 5 and (51) it follows that almost all paths of the process considered in Theorem 13 have in the topology M_+ the limit

$$\lim_{t \to +\infty} x_t = x_{+\infty} \in B,$$

and the measure of the set of paths for which $x_{+\infty} \in \Gamma \subseteq B$ has the value

$$\int_\Gamma (k_y, \nu)\, \mu_h(dy) \tag{81}$$

(μ_h is the spectral measure of the excessive function h).

§ 18. Stationary processes with random birth and death times

We attempt to extend the results of the previous section to the case where the measure ν and the function h are excessive, but not necessarily harmonic. The transition from $h = 1$ to an arbitrary excessive function h does not cause serious difficulties. Therefore we first take $h = 1$.

We construct in some countable space $\bar{E} \supset E$ a transition function and harmonic measure coinciding on E with $p(x, y)$ and ν and defining in $\bar{E}$ a stationary process. Almost all paths of this process remain in E from the moment ξ of first hitting E to the moment ζ of the last exit from E. In E there arises a random process with random birth and death times. We call this a stationary process with transition function $p(x, y)$ and stationary measure ν.

Let Z be the set of all integers. We define in the space $\bar{E} = Z \times E$ the transition function $\bar{p}$ for which

$$\begin{aligned} \bar{p}(0 \times x, 0 \times y) &= p(x, y), \\ \bar{p}(0 \times x, 1 \times x) &= 1 - \sum_y p(x, y), \\ \bar{p}(m \times x, (m+1) \times x) &= 1 \quad \text{if} \quad m \neq 0, \end{aligned}$$

and the remaining values are zero. It is clear that (1) is satisfied for $\bar{p}$ with equality. Next, let ν be an excessive measure for $p(x, y)$. The formulae

$$\bar{\nu}(0\times x)=\nu(x),\ \bar{\nu}(m\times x)=\nu(x)-\sum_y \nu(y)\,p(y,x)\ \text{for}\ m<0,$$

$$\bar{\nu}(m\times x)=\nu(x)-\nu(x)\sum_y p(x,y)\ \text{for}\ m>0,$$

define in $\bar{E}$ a harmonic measure with respect to $\bar{p}$. Identifying $0\times x$ with x we may assume that $E\subset\bar{E}$. Note that $\bar{p}$ coincides with p and $\bar{\nu}$ with ν on E. We can imagine that the points $m\times x$ lie above x if $m>0$ and below if $m<0$. We consider in $\bar{E}$ the process $\bar{x}_t$ with transition function $\bar{p}$ and stationary measure $\bar{\nu}$. Subtracting from $\hat{\Omega}$ a set of measure zero we may assume that the paths behave as follows. After unit time a particle not lying in E moves one unit upwards; a particle lying in E moves in E in correspodnence with the transition probability $p(x, y)$; if the process in E guided by this law must terminate, then instead of this there is a move in $\bar{E}$ of one unit upwards. Let ξ be the first hitting time of E and ζ the last exit time from E. Removing from each path the part not belonging to E we obtain in E a random process x_t with birth time ξ and death time ζ. For this process the measure of the set $\{x_m=a_m,\ldots,x_n=a_n\}$ is

$$\nu(a_m)\,p(a_m,a_{m+1})\ldots p(a_{n-1},a_n). \tag{82}$$

We have the same expression as in (4″). It is natural therefore to call the process so constructed *stationary with transition function $p(x, y)$ and stationary measure ν.*

The paths of this process are functions with values in E defined on all possible intervals of the form $[m, n]$, $(-\infty, n]$, $[m, +\infty)$ and $(-\infty, +\infty)$. The set of all paths is denoted by Ω', and the σ-algebra generated by all simple sets $[a_m\ldots a_n]_m^n$ by $\mathcal{F}'$. We have constructed on $\mathcal{F}'$ a measure $\mathbf{P}_\nu$ that is equal to (82) on the simple set $[a_m\ldots a_n]_m^n$. The measure $\mathbf{P}_\nu$ is defined uniquely by this condition, as follows from the following lemma, which we prove in the Appendix.

LEMMA A. *If two measures on the σ-algebra $\mathcal{F}'$ in Ω' coincide and are finite on all simple sets, then they coincide everywhere.*

From

$$\begin{aligned}\nu(a_m)\,h(a_m)\,p^h(a_m,a_{m+1})\ldots p^h(a_{n-1},a_n) &=\\ =\nu(a_m)\,p(a_m,a_{m+1})\ldots p(a_{n-1},a_n)\,h(a_n) &=\\ =h(a_n)\,\nu(a_n)\,p_\nu(a_n,a_{n-1})\ldots p(a_{m+1},a_m)&\end{aligned}$$

it follows that if x_t is a stationary process with transition function p^h and stationary measure $h\nu$, then $\hat{x}_t=x_{-t}$ is a stationary process with transition function p_ν and stationary measure $h\nu$. Using these remarks it is easy to prove the following theorem, a generalization of Theorem 13.

THEOREM 14. *Let ν be an l-finite excessive function, and h a γ-integrable excessive function. Then for the stationary process with transition function*

$p^h(x, y)$ and stationary measure $h\nu$ almost all non-terminating paths have in the topology M_+ a limit $x_{+\infty} \in B$. The measure of the set of paths for which $x_\zeta \in \Gamma \subseteq E \cup B$ is

$$\int_\Gamma (k_y, \nu)\, \mu_h(dy). \tag{83}$$

Almost all paths without a beginning have in M_- a limit $x_{-\infty} \in \hat{B}$. The measure of the set $\{x\xi \in \Gamma\}$, where $\Gamma \subseteq E \cup \hat{B}$, is

$$\int_\Gamma (h, \varkappa_y)\, \mu^\nu(dy). \tag{84}$$

PROOF. We consider only the case $h = 1$. (The passage to the general case is effected just as in the derivation of Theorem 13 from Theorem 12.) Apart from the transition function p in E we consider the transition function $\bar{p}$ in $\bar{E}$, all entries related to $\bar{p}$ are distinguished by a bar on top. We introduce in $\bar{E}$ a reference measure $\bar{\gamma}$, putting $\bar{\gamma}(m \times x) = 0$ for $m > 0$ and selecting the value of $\bar{\gamma}(m, x)$ for $m \leqslant 0$ so that

$$\sum_m \bar{\gamma}(m \times x) = \gamma(x).$$

It is easy to calculate that

$$\bar{k}_{n\times y}(m \times x) = k_y(x) \quad \text{for} \quad n \geqslant 0. \tag{85}$$

We denote by E^+ the set of all points $n \times y$, where $n \geqslant 0$. By (85) the "divergent to infinity" sequence $n_r \times y_r \in E^+$ converges in the M^+-topology corresponding to $\bar{p}(x, y)$ if and only if y_r converges in the M^+-topology connected with $p(x, y)$, or $n_r \to \infty$ and $y_r = y$ beginning with some r. Hence the Martin boundary ∂E^+ is contained in $\partial\bar{E}$ and $\partial E^+ = \partial E \cup E'$, where E' is in natural correspondence with E.

Let $\bar{h}$ be a $\bar{p}$-harmonic function and $\mu_{\bar{h}}$ its spectral measure. By (54)

$$\int_{\bar{E}^*} f(y)\, \mu_{\bar{h}}(dy) = \lim_{n\to\infty} M_{\bar{\gamma}} f(\bar{x}_n)\, \bar{h}(\bar{x}_n) \qquad (f \in C(\bar{E}^*)).$$

We put $f(y) = \rho(y, E^+)$. Then the right-hand side is evidently zero. Hence $\mu_{\bar{h}}$ is concentrated on $E^+ \cup \partial E^+$. On the other hand, $\mu_{\bar{h}}$ is concentrated on $\partial\bar{E}$. This means that it is concentrated on ∂E^+. Hence it is easy to deduce that the exit space $\bar{B}$ is $B \cup E'$ and may be naturally identified with $B \cup E$. We apply to the process $\bar{x}_t$ the remark at the end of §17. It is obvious that

$\{\bar{x}_\infty = y\} = \{x_\zeta = y\}$. Hence the first part of Theorem 13 follows.

To prove the second part it is sufficient to apply the part just proved to the reversed process $\hat{x}_t = x_{-t}$.

§19. Hunt processes

Hunt noticed that boundary theory is applicable to a class of processes wider than Markov processes. Roughly speaking, these are processes which behave like Markov processes with transition function $p(x, y)$ after the moment of first reaching any finite set D.

We denote (as in §18) by Ω' the set of all functions a_t with values in E, defined on all intervals $(-\infty, n]$, $[m, n]$, $[m, +\infty)$ and $(-\infty, +\infty)$. The left end of an interval is called the birth time and the right the death time. In the space Ω' we consider the δ-algebra F' generated by all simple sets. Important examples of F'-measurable functions are: a) the birth time ξ; b) the death time ζ; c) the position x_t of the path at the moment t (x_t is defined on the set $\{\xi \leqslant t \leqslant \eta\}$; d) the moment σ_D of first hitting D and the moment τ_D of the last exit from D (defined on the set Ω'_D of paths hitting D); e) $\mathbf{N}(x)$, the number of hits in the point x (the domain is Ω').

Let $\mathbf{P}$ be a measure on the σ-algebra F'. The process determined by $\mathbf{P}$ is called *a Hunt process with transition function $p(x, y)$ and characteristic measure β if*:

19.A. For any finite D, any $n = 0, 1, \ldots$, and any $a_0, a_1, \ldots, a_n \in E$

$$\mathbf{P}\{x_{\sigma_D} = a_0, x_{\sigma_D+1} = a_1, \ldots, x_{\sigma_D+n} = a_n\} = \nu_D(a_0)\, p(a_0, a_1) \ldots p(a_{n-1}, a_n).$$

19.B. $\mathbf{MN}(x) = \beta(x) < \infty$ for any $x \in E$.

For $n = 0$ from 19.A we have $\nu_D(a_0) = \mathbf{P}\{x_{\sigma_D} = a_0\}$. Condition 19.A means that $y_t = x_{\sigma_D+t}$ is a Markov process with transition function $p(x, y)$ and initial distribution ν_D.

We note that by 19.B for almost all paths $\mathbf{N}(x)$ is finite, hence x_{σ_D} and x_{τ_D} are defined[1] on Ω'_D. We denote by $\mathbf{N}_D(x)$ the number of hits of the state x, starting with the moment σ_D. By 19.A, $\mathbf{MN}_D(x)$ is the mean number of hits of the point x for the Markov process with transition function $p(x, y)$ and initial distribution ν_D. Hence from (3) it follows that

$$\mathbf{MN}_D(x) = (\nu_D G)(x). \tag{86}$$

Evidently $\mathbf{N}_D(x) \uparrow \mathbf{N}(x)$ for $D \uparrow E$. Hence

$$(\nu_D G)(x) \uparrow \beta(x) \tag{87}$$

[1] Here, as in the whole of §19, D is taken as a finite set.

as $D \uparrow E$. From (87) it is clear that β is an excessive measure (with respect to $p(x, y)$). We note that

$$\beta(x) = \sum_t \mathbf{P}\{x_t = x\}.$$

Since $g(x, x) \geqslant 1$,

$$\nu_D(x) \leqslant (\nu_D G)(x) \leqslant \beta(x) < \infty.$$

If a state is attained only by a set of paths of measure zero, such a state may be removed from E. Hence, without restricting significantly the generality, it may be assumed that the following additional condition is satisfied:

19.C. $\beta(x) > 0$ for any $x \in E$.

EXAMPLE 1. The Markov process with transition function $p(x, y)$ and initial distribution ν can be regarded as a process with birth time $\xi = 0$ and death time ζ. Condition 19.A is always satisfied.[1] Condition 19.B is satisfied if the process is transient.

EXAMPLE 2. The stationary process with transition function $p(x, y)$ and stationary distribution ν satisfies condition 19.A, but not 19.B (if $\nu \neq 0$).

EXAMPLE 3. Let x_t be a Hunt process and $\nu(\omega)$ any integer-valued random variable. Then $\widetilde{x}_t = x_{\nu+t}$ is a Hunt process with the same transition function and characteristic measure. We call this transformation a *random translation in time.*

For the process x_t the reversed process is defined by the formula $\hat{x}_t = x_{-t}$. Its birth time is $-\zeta$, and death time $-\xi$.

THEOREM 15. *By reversing a Hunt process with transition function $p(x, y)$ and characteristic measure β a Hunt process is obtained with the same characteristic measure and the transition function*

$$p_\beta(x, y) = \frac{\beta(y)\, p(y, x)}{\beta(x)}.$$

PROOF. Let σ' be the moment of first hitting the finite set D' for the process $\hat{x}_t$. Then $\tau = -\sigma'$ is the last exit time from D' for the process x_t. We have

$$\mathbf{P}\{\hat{x}_{\sigma'} = a_0, \hat{x}_{\sigma'+1} = a_1, \ldots, \hat{x}_{\sigma'+n} = a_n\} = \mathbf{P}\{x_\tau = a_0, x_{\tau-1} = a_1, \ldots, x_{\tau-n} = a_n\}. \tag{88}$$

Let σ_D be the moment of first hitting D for the process x_t. By 19.A $y_t = x_{\sigma_D+t}$ is a Markov process with transition function $p(x, y)$ and initial distribution ν_D. Let $\bar{\tau}$ be the last exit time from D' for the process y_t. According to (25)

[1] It is left to the reader to verify this.

$$\mathbf{P}\{y_{\bar{\tau}} = a_0,\ y_{\bar{\tau}-1} = a_1,\ \ldots,\ y_{\bar{\tau}-n} = a_n\} =$$
$$= (\nu_D G)(a_n)\, p(a_n, a_{n-1}) \ldots p(a_1, a_0)\, L_D(a_0). \quad (89)$$

Obviously outside the set $A_D = \{\sigma_D > \tau\}$ we have $y_{\bar{\tau}} = x_\tau, \ldots, y_{\bar{\tau}-n} = x_{\tau-n}$. Hence the left-hand side of (89) differs by not more than $\mathbf{P}(A_D)$ from the probability (88). But as $D \uparrow E$, $\mathbf{P}(A_D) \downarrow 0$ and $\nu_D G \uparrow \beta$ by (87). Hence, taking the limit in (89), we have

$$\mathbf{P}\{x_\tau = a_0,\ x_{\tau-1} = a_1,\ \ldots,\ x_{\tau-n} = a_n\} =$$
$$= \beta(a_n)\, p(a_n, a_{n-1}) \ldots p(a_1, a_0)\, L_D(a_0). \quad (90)$$

The right-hand side is equal to $L_D(a_0)\,\beta(a_0)\,p_\beta(a_0, a_1) \ldots p_\beta(a_n, a_{n-1})$ and by (88) the process $\hat{x}_t$ satisfies 19.A. That 19.B is satisfied for this process is obvious.

COROLLARY. *Let* ${}^\beta\mathbf{P}_x$ *be the measure corresponding to the Markov process with transition function* $p_\beta(x, y)$ *and initial state x. Let*

$$L_D^\beta(x) = {}^\beta\mathbf{P}_x\{x_0 = x,\ x_t \,\bar{\in}\, D \text{ for } t > 0\}.$$

Then for the Hunt process with transition function $p(x, y)$ *and characteristic function* β

$$\nu_D(x) = \beta(x)\, L_D^\beta(x). \quad (91)$$

PROOF. Putting $n = 0$ in (91), (88) and (90), we have

$$\mathbf{P}\{\hat{x}_{\sigma'} = a_0\} = \mathbf{P}\{x_\tau = a_0\} = \beta(a_0)\, L_D(a_0). \quad (92)$$

Considering now $\hat{x}_t$ as an initial process and x_t as a reversed one, on applying (92) we get (91).

(91) shows that the distribution ν_D can be uniquely reconstructed from the transition function $p(x, y)$ and the characteristic measure β.

LEMMA 6. *For* $D \uparrow E$ *there exist the limits*

$$(k_y, \nu_D) \uparrow \mathscr{H}_\beta(y), \qquad (y \in E^*),$$
$$(L_D, \varkappa_y) \uparrow S(y), \qquad (y \in \hat{E}).$$

Here for $y \in E$

$$\mathscr{H}_\beta(y) = \frac{\beta(y)}{\eta(y)},$$
$$S(y) = \frac{1}{s(y)}.$$

(Outside E $\mathscr{H}_\beta$ *and S may equal to* $+\infty$*. S depends only on the transition function* $p(x, y)$*, and* $\mathscr{H}_\beta$ *depends only on* $p(x, y)$ *and* β*.)*

PROOF. By (31) and (71) for $y \in E$

$$(k_y, \nu_D) = \frac{(\nu_D G)(y)}{\eta(y)}, \qquad (L_D, \varkappa_y) = \frac{GL_D(y)}{s(y)}.$$

By (87) it follows from this that

$$(k_y, \nu_D) \uparrow \frac{\beta(y)}{\eta(y)} \quad \textit{for} \quad y \in E.$$

By (24), $GL_D(y)$ is the probability that the path of the Markov process with initial state y and transition function $p(x, y)$ hits[1] D. Hence it is clear that $GL_D(y) \uparrow 1$ and $(L_D, \kappa_y) \uparrow 1/s(y)$. It remains to note that

$$(k_y, \nu_D) = \sum_{x \in D} \nu_D(x)\, k_y(x)$$

is a continuous function of y on E^* and hence, for all $y \in E^*$, this function increases as $D \uparrow E$. Similar arguments are applicable to (L_D, κ_y).

THEOREM 16. *For the Hunt process with transition function $p(x, y)$ and characteristic measure β almost all non-terminating paths have in the topology M_+ a limit $x_{-\infty} \in B$. Almost all paths not having a beginning have in the topology M_- a limit $x_{-\infty} \in \hat{B}$. Here*

$$\mathbf{P}\{x_{\zeta} \in \Gamma\} = \int_{\Gamma} \mathscr{H}_B(y)\, \mu_1(dy), \qquad (\Gamma \subseteq E \cup B), \tag{93}$$

$$\mathbf{P}\{x_{\zeta} \in \Gamma\} = \int_{\Gamma} S(y)\, \mu^{\beta}(dy), \qquad (\Gamma \subseteq E \cup \hat{B}) \tag{94}$$

(μ_1 *is the spectral measure of the excessive function* 1, *and* μ^{β} *is the spectral measure of the excessive measure* β).

PROOF. By 19.A $y_t = x_{\sigma_D + t}$ is a Markov process with transition function $p(x, y)$ and initial distribution ν_D. By Theorems 4 and 5 for almost all non-terminating paths of this process the limit $y_{+\infty}$ exists and is in B. By (83) the probability that $y_{\bar{\zeta}} \in \Gamma$ is

$$\int_{\Gamma} (k_y, \nu_D)\, \mu_h(dy) \tag{95}$$

($\bar{\zeta}$ is the death time for y_t). We denote by C_D the set of paths of the process x_t not meeting D. Evidently outside D $y_{\bar{\zeta}} = x_{\zeta}$. Hence (95) differs from $\mathbf{P}\{x_{\zeta} \in \Gamma\}$ by not more than $\mathbf{P}(C_D)$. But for $D \uparrow E$ we have $\mathbf{P}(C_D) \downarrow 0$, and hence the limit (95) is equal to $\mathbf{P}\{x_{\zeta} \in \Gamma\}$. This proves (93).

The remaining statements of the theorem are obtained by applying the part already proved to the reversed process. Here we have to use Theorem 11 and the Corollary to Theorem 15.

---

[1] We recall that being transient almost all paths hitting D leave D.

So far the question of the existence and uniqueness of a Hunt process with given transition function $p(x, y)$ and characteristic measure β has remained open. First we prove the uniqueness theorem.

THEOREM 17. *The transition function and characteristic measure determine a Hunt process uniquely to within a random translation of time.*

PROOF. As in §9, let $N(x)$ denote the number of the state x. For each path $\{a_t\}$ we consider the smallest of the numbers $N(a_t)$ and call the first moment for which this smallest number is attained canonical. The canonical moment υ is a function of the path that can be defined by the conditions

$$N(x_t) \geqslant N(x_\upsilon) \quad \text{for all} \quad t,$$
$$N(x_t) > N(x_\upsilon) \quad \text{for} \quad t < \upsilon.$$

By means of a random translation of time we can make sure that $\upsilon = 0$. We assume that this condition is satisfied and put $D_k = \{x: N(x) \leqslant k\}$, $\sigma_k = \sigma_{D_k}$, $\nu_k = \nu_{D_k}$. We note that $\sigma_k \leqslant \upsilon = 0$ for any k. Next, $N(x_t) > N(x_{\sigma_k})$ for $t < \sigma_k$. Hence

$$\{\sigma_k = -m\} = \{\sigma_k + m = 0\} = \{N(x_{\sigma_k+t}) > N(x_{\sigma_k+m}) \text{ for } t = 0, 1, \ldots,$$
$$\ldots, m-1, \; N(x_{\sigma_k+t}) \geqslant N(x_{\sigma_k+m}) \text{ for } t > m\} = \{\upsilon' = m\},$$

where υ' is the canonical moment for the process $y_t = x_{\sigma_k+t}$ $(t \geqslant 0)$. By 19.A

$$\mathbf{P}\{\sigma_k = -m, \; x_{\sigma_k} = x, \; x_n = a_n, \; x_{n+1} = a_{n+1}, \ldots, x_r = a_r\} =$$
$$= \mathbf{P}\{y_0 = x, \; y_{n+m} = a_n, \ldots, y_{r+m} = a_r, \; \upsilon' = m\}. \quad (96)$$

Let $a_n \in D_k$. Then the left-hand side of (96) is zero for $n < -m$. Hence, summing over all $x \in D_k$ and over all $m \geqslant 0 \vee (-n)$, we get

$$\mathbf{P}\{x_n = a_n, \; x_{n+1} = a_{n+1}, \ldots, x_r = a_r\} =$$
$$= \mathbf{P}_{\nu_k}\{\upsilon' + n \geqslant 0, \; y_{n+\upsilon'} = a_n, \ldots, y_{r+\upsilon'} = a_r\}. \quad (97)$$

Thus, the measures of simple sets in Ω' can be reconstructed from the transition function $p(x, y)$ and the measures ν_k. By the Corollary to Theorem 15 the latter is defined uniquely by $p(x, y)$ and the measure β. It remains to use Lemma A of §18.

The following existence theorem holds.

THEOREM 18. *Let β be an excessive measure for the transition function $p(x, y)$, where $\beta(x) < \infty$ for all $x \in E$. Then there exists a Hunt process corresponding to $p(x, y)$ and β.*

To prove this theorem measures ν_D can be defined by (92) and then the measures of simple sets can be given by (97). The detailed execution of this plan is somewhat unwieldy (it is carried through in [3], Chapter 10, §12).

If $\beta = \alpha G$, the required Hunt process is obtained by considering a Markov process with initial distribution α.

The general case can be treated as follows: It is easy to verify that if $D_k \uparrow E$ and $\nu_k = \nu_{D_k}$ are defined by (92), then $\nu_k G \uparrow \beta$. We construct the Hunt process with characteristic measures $\beta_k = \nu_k G$, produce in each of them a random translation of time, making the canonical moment zero, and then take the limit as $k \to \infty$.

Appendix
Measures in spaces of paths

We prove Theorems A and B stated in §2 and Lemma A of §18. First we prove the propositions on the *uniqueness* of a measure. Here we depend on a simple lemma from set theory.

A system $\mathscr{C}$ of subsets of a set Ω is called a π-system if the intersection of two sets of $\mathscr{C}$ also belongs to $\mathscr{C}$. The system $\mathscr{H}$ is called a λ-system if: λ_1) the sum of two disjoint sets of $\mathscr{H}$ also belongs to $\mathscr{H}$; λ_2) if $A, B \in \mathscr{H}$ and $A \supseteq B$, then $A \setminus B \in \mathscr{H}$; λ_3) if $A_1, \ldots, A_n, \ldots \in \mathscr{H}$ and $A_n \uparrow A$, then $A \in \mathscr{H}$; λ_4) $\Omega \in \mathscr{H}$.

LEMMA B. *If the λ-system $\mathscr{H}$ contains the π-system $\mathscr{C}$, then $\mathscr{H}$ contains the σ-algebra $\sigma(\mathscr{C})$, generated by $\mathscr{C}$.*

This lemma is proved in the first few pages of [2] (see Lemma 1.1).

COROLLARY. *Suppose that two measures given on a σ-algebra $\mathscr{F}$ in a space Ω coincide and are finite on a π-system $\mathscr{C}$ generating $\mathscr{F}$. If Ω can be partitioned into the sum of a countable number of pairwise disjoint sets $\Omega_n \in \mathscr{C}$, then the two measures coincide everywhere on $\mathscr{F}$.*

PROOF. We denote by $\mathscr{H}$ the family of all sets $A \in \mathscr{F}$ on which the measures coincide and are finite. We put $A \in \mathscr{H}_n$ if $A \in \mathscr{H}$ and $A \subseteq \Omega_n$; $A \in \mathscr{C}_n$ if $A \in \mathscr{C}$ and $A \subseteq \Omega_n$. Let $\mathscr{S}_n$ be the σ-algebra in the space Ω_n generated by $\mathscr{C}_n$. Evidently $\mathscr{C}_n$ is a π-system, $\mathscr{H}_n$ is a λ-system in Ω_n and $\mathscr{C}_n \subseteq \mathscr{H}_n$. By Lemma B $\mathscr{S}_n \subseteq \mathscr{H}_n$.

We put $A \in \bar{\bar{\mathscr{F}}}$ if $A_n \cap \Omega_n \in \mathscr{S}_n$ for all n. Evidently $\bar{\bar{\mathscr{F}}}$ contains $\mathscr{C}$ and is a σ-algebra. Hence $\bar{\bar{\mathscr{F}}} \supseteq \mathscr{F}$. Thus, if $A \in \mathscr{F}$, then for all n, $A \cap \Omega_n \in$ $\in \mathscr{S}_n \subseteq \mathscr{H}_n \subseteq \mathscr{H}$. But if two measures coincide on $A \cap \Omega_n$ for all n, they coincide on A.

It is now quite simple to prove the uniqueness of the measure in Theorems A and B. It is sufficient to apply the Corollary just proved to the π-system of all simple sets and to note that the simple sets $[a_0]$ ($a_0 \in E$) are pairwise disjoint and that their sum is the entire space of paths.

To prove Lemma A of §18 we denote by $\mathscr{C}$ the family of sets of the form

$$\{\xi = s,\ x_{m_1} = a_m,\ x_{m_2} = a_{m_2},\ \ldots,\ x_{m_k} = a_{m_k},\ \zeta = t\}, \tag{0.1}$$

where $-\infty \leqslant s \leqslant m_1 < m_2 < \ldots < m_n \leqslant t \leqslant +\infty$; $k = 1, 2, \ldots$ We note that the sets

$$\{\xi = s,\ x_s = x,\ \zeta = t\} \qquad (-\infty < s \leqslant t \leqslant +\infty,\ x \in E),$$
$$\{\xi = -\infty,\ x_t = x,\ \zeta = t\} \qquad (-\infty < t < +\infty,\ x \in E),$$
$$\{\xi = -\infty,\ x_0 = x,\ \zeta = +\infty\} \qquad (x \in E)$$

belong to $\mathscr{C}$, are pairwise disjoint, and that their sum is the whole space of paths. It remains to verify that the given measures are finite and coincide on $\mathscr{C}$. For the sets

$$[a_m \ldots a_n]_m^n = \{\xi \leqslant m,\ x_m = a_m, \ldots, x_n = a_n, \zeta \geqslant n\} \tag{0.2}$$

this is true by the conditions of the Lemma. Hence this is true also for the sets

$$\{\xi \leqslant s,\ x_{m_1} = a_{m_1},\ x_{m_2} = a_{m_2}, \ldots, x_{m_k} = a_{m_k},\ \zeta \geqslant t\} \tag{0.3}$$
$$(-\infty < s \leqslant m_1 < m_2 < \ldots < m_k \leqslant t < +\infty),$$

which can be expressed as a countable sum of pairwise disjoint sets (0.2). Letting $s \downarrow -\infty$ or $t \uparrow +\infty$ we conclude that the measures coincide on the sets (0.3) also for $s = -\infty$ and for $t = +\infty$. Hence it is clear that the measures coincide on all sets (0.1).

The proof of the existence of the measures described in Theorems A and B is based on the following general theorem from measure theory.

THEOREM B. *Let $\mathscr{A}$ be an algebra of sets in the space Ω (that is, a family of sets containing together with any two sets their sum and together with any set its complement). Let* $\mathbf{P}(A)$ *be a non-negative function on $\mathscr{A}$, satisfying the conditions*:

α. *If A_1, A_2 belong to $\mathscr{A}$ and are disjoint, then*

$$\mathbf{P}(A_1 \cup A_2) = \mathbf{P}(A_1) + \mathbf{P}(A_2).$$

β. *The space Ω can be partitioned into a countable union of disjoint subsets $\Omega_n \in A$ such that $P(\Omega_n) < \infty$.*

γ. *If $A_1 \supset A_2 \supset \ldots \supset A_n \supset \ldots$ are sets in $\mathscr{A}$ and* $\lim \mathbf{P}(A_n) > 0$, *then $\bigcap_n A_n$ is non-void.*

Then, on the σ-algebra $\mathscr{F}$ generated by $\mathscr{A}$ there exists a measure coinciding with $\mathbf{P}$ *on $\mathscr{A}$.*

The sequence of sets $A_n \in \mathscr{A}$ is called a *nest*, if $A_0 \supseteq A_1 \supseteq \ldots \supseteq A_n \supseteq \ldots$, $\lim \mathbf{P}(A_n) > 0$ and $\mathbf{P}(A_1) < \infty$. By virtue of β it is easy to prove that the condition γ is equivalent to the following:

γ′. *Every nest has a non-void intersection.*

Theorem A will be deduced from the following theorem:

THEOREM D. *Let $\mathcal{F}$ be a σ-algebra in the space Ω_∞ of non-terminating paths generated by the simple sets $[a_0 \ldots a_n]$. For any n and any $a_0 \ldots a_n \in E$ suppose that a non-negative number $p(a_0, a_1, \ldots, a_n)$ is given, where*

$$\sum_{a_n} p(a_0, a_1, \ldots, a_n) = p(a_0, a_1, \ldots, a_{n-1}). \tag{0.4}$$

Then there exists a measure **P** *on the σ-algebra $\mathcal{F}$ for which*

$$\mathbf{P}[a_0, a_1 \ldots a_n] = p(a_0, a_1, \ldots, a_n). \tag{0.5}$$

PROOF. We call a simple set in Ω_∞ of the form $[a_0, a_1, \ldots, a_n]$ a *simple n-set*. A set that can be represented as the sum of simple n-sets is called a *cylinder n-set*.[1]

We note that for $m \geqslant n$:

a) If a simple m-set A intersects a simple n-set B, then $A \subseteq B$. (Hence it follows that for $m = n$ $A = B$.)

b) If a simple m-set intersects a cylindrical n-set, then it is contained in it.

c) A simple n-set is a cylindrical m-set.

d) A cylindrical n-set is a cylindrical m-set.

e) The sum and the complement of a cylindrical n-set is also á cylindrical n-set.

Properties a), c), e), are evident, b) follows from a), and d) from c). By d) and e) the family A of all cylindrical sets is an algebra. We define on this algebra a measure, relating to each cylindrical n-set A the sum of the values $p(a_0, \ldots, a_n)$ over all simple sets $[a_0, \ldots, a_n]$ contained in A. Although every cylindrical n-set A is at the same time an $(n + 1)$-set, by (0.4) the number associated with A does not depend on whether it is regarded as an n-set or as an $(n + 1)$-set. It is easily seen that the function on A introduced in this way satisfies α and β. If we prove that it satisfies also condition γ', then Theorem D follows from Theorem C.

We require the following:

LEMMA C. *If the sets C_n form a nest, then for each m there exists a simple m-set B such that $B \cap C_n$ also forms a nest.*

We note that

$$0 < \lim \mathbf{P}(C_n) = \lim \sum \mathbf{P}(B \cap C_n) = \sum \lim \mathbf{P}(B \cap C_n) \tag{0.6}$$

(the sum is taken over all simple m-sets B; the signs of summation and limit can be interchanged, since $\mathbf{P}(B \cap C_n) \leqslant \mathbf{P}(B \cap C_0)$ for all n and $\Sigma\mathbf{P}(B \cap C_0) = \mathbf{P}(C_0) < \infty$). From (0.6) it follows that for some B $\lim \mathbf{P}(B \cap C_n) > 0$ and hence $B \cap C_n$ is a nest.

---

[1] Theorem D, as the Theorem E deduced below, is a particular case of a well known theorem of Kolmogorov on measures in products. In this particular case the general proof is simplified significantly. In our presentation we follow the book of Kemeny, Snell and Knapp [3].

We come to the proof of γ'. Let A_n be a nest. By d) we may regard A_n as a cylindrical n-set. By Lemma C a simple 0-set B_0 can be chosen so that $A_n^1 = A_n \cap B_0$ form a nest. Next, a simple 1-set B_1 can be chosen so that $A_n^2 = A_n^1 \cap B_1 = A_n \cap B_0 \cap B_1$ form a nest. Continuing this construction, for each m we construct the nest $A_n^m = A_n^{m-1} \cap B_m = A_n \cap B_0 \cap B_1 \cap \ldots \cap B_m$. Obviously $\mathbf{P}(A_n \cap B_0 \cap B_1 \cap \ldots \cap B_n) \neq 0$. Hence $A_n \cap B_0 \cap B_1 \cap \ldots \cap B_n$ is non-void. In view of a) $B_0 \supseteq B_1 \supseteq \ldots \supseteq B_n$, and in view of b) $B_n \subseteq A_n$. Obviously there exists a path $\omega = b_0 b_1 \ldots b_n \ldots$ such that $B_n = [b_0 b_1 \ldots b_n]$. It is clear that $\omega \in B_n \subseteq A_n$, and hence the intersection of A_n is non-void.

PROOF OF THEOREM A. We extend the space E to $\bar{E}$ adding to E one more point $*$. We put

$$\bar{p}(a_0, a_1, \ldots, a_n) = \begin{cases} p(a_0, a_1, \ldots, a_n), & \text{when } a_0, a_1, \ldots, a_n \in E, \\ p(a_0, a_1, \ldots, a_m), & \text{when } a_0, a_1, \ldots, a_m \in E,\ a_{m+1} = \ldots = a_n = *, \\ 0 & \text{, otherwise.} \end{cases}$$

It is easily seen that $\bar{p}$ satisfies (0.4). Let $\bar{\Omega}_\infty$ be a space of non-terminating paths and $\mathscr{F}$ a σ-algebra in this space generated by cylindrical sets. We remove from each path in $\bar{\Omega}_\infty$ containing the element $*$ the part of it beginning with the first asterisk (paths not containing $*$ are left unchanged). So we obtain a mapping α of $\bar{\Omega}_\infty$ into Ω.

By Theorem D there exists a measure $\bar{\mathbf{P}}$ on the σ-algebra $\mathscr{F}$ such that $\bar{\mathbf{P}}[a_0 a_1 \ldots a_n] = p(a_0, a_1, \ldots, a_n)$. The formula

$$\mathbf{P}(A) = \bar{\mathbf{P}}[\alpha^{-1}(A)], \qquad (A \in \mathscr{F})$$

defines a measure such that $\mathbf{P}[a_0 \ldots a_n] = \bar{p}(a_0, a_1, \ldots, a_n) = p(a_0, a_1, \ldots, a_n)$ for $a_0, a_1, \ldots, a_n \in E$.

PROOF OF THEOREM B. The formula

$$\hat{a}_n = \begin{cases} a_n & \text{if } n = 0, \\ a_{2n-1} & \text{if } n > 0, \\ a_{-2n} & \text{if } n < 0 \end{cases}$$

defines a mapping of Ω_∞ into $\hat{\Omega}$. (Under this mapping the path $a_0 a_1 a_2 a_3 a_4 \ldots$ *goes into* $\ldots a_4 a_2 a_0 a_1 a_3 \ldots$). We put

$$p'(a_0, a_1, a_2, \ldots, a_{2k}) = p_{-k}^{k}[a_{2k}, a_{2k-2}, \ldots, a_2, a_0, a_1, \ldots, a_{2k-1}],$$
$$p'(a_0, a_1, a_2, \ldots, a_{2k-1}) = p_{-k+1}^{k}[a_{2k-2}, \ldots, a_2, a_0, a_1, \ldots, a_{2k-1}].$$

By (3′)–(3″) the function p' satisfies (0.4). By Theorem D a measure $\mathbf{P}'$ can be constructed in Ω_∞ such that $\mathbf{P}'[a_0 a_1 \ldots a_n] = p'[a_0, a_1, \ldots, a_n)$. We define in $\hat{\Omega}$ the measure

$$\mathbf{P}(A) = \mathbf{P}'[\alpha^{-1}(A)].$$

It is easy to see that for it

$$\mathbf{P}[a_{-k}, \ldots, a_0, \ldots, a_k]_{-k}^{k} = p_{-k}^{k}(a_{-k}, \ldots, a_0, \ldots, a_k).$$

Using the additivity of p and the properties (3')–(3'') of p_m^n, it is easy to prove that for any $m \leqslant n$

$$\mathbf{P}[a_m \ldots a_n]_m^n = p_m^n(a_m, \ldots, a_n).$$

References

[1] J. L. Doob, Discrete potential theory and boundaries, J. Math. Mech. 8 (1959), 433–458.

[2] E. B. Dynkin, *Osnovaniya teorii markovskikh protsessov*, Izd. Fiz.-Mat., Moscow 1959. Translation: Theory of Markov Processes, Pergamon Press, Oxford 1960.

[3] J. G. Kemeny, J. L. Snell, and A. W. Knapp, Denumerable Markov chains, van Nostrand, New York 1966.

[4] R. S. Martin, Minimal positive harmonic functions, Trans. Amer. Math. Soc., 49 (1941), 137–172.

[5] W. Feller, An Introduction to probability theory and its applications, Wiley, New York 1950. Translation: *Vvedenie v teoriyu veroyatnostei i ee prilozheniya*, Izd. "Mir", Moscow 1964.

[6] G. A. Hunt, Markov chains and Martin boundaries, Illinois J. Math. 4 (1960), 313–340. = Sb. per. Matem. 5:5 (1961), 121–149.

Received by the Editors December 3, 1968.

Translated by P. G. Gormley

THE INITIAL AND FINAL BEHAVIOUR OF TRAJECTORIES OF MARKOV PROCESSES

The initial and final behaviour of the trajectories of Markov processes is studied within the theory of Martin boundaries. We propose a simpler approach, based on a direct investigation of the class $\mathscr{K}$ of Markov processes with a given transition function and the class $\mathscr{K}^*$ of Markov processes with a given cotransition function. In the class $\mathscr{K}(\mathscr{K}^*)$ processes with the following property are distinguished: the probability of any event, determined by an arbitrarily small initial (final) section of a trajectory, is equal to 0 or 1. Every process of $\mathscr{K}(\mathscr{K}^*)$ decomposes uniquely into such "ergodic" processes, and the corresponding measure completely describes the initial (final) behaviour of trajectories. The theory is invariant with respect to reversal of time.

Based on the results of the present paper we shall study in a subsequent publication the excessive measures and excessive functions associated with a Markov process.

A brief account of the main ideas of this work (for processes with non-random births and deaths) was given in the author's invited address at the International Congress of Mathematicians in Nice (1970).

Contents

§1. Introduction

1.1. A Markov process is a random process satisfying the Markov condition of independence of the past and future, given the present.

We consider processes with a fixed state space $(E, \mathscr{B})$ and time set T. Here $(E, \mathscr{B})$ is an arbitrary measurable space, and T is any subset of the real line. E may be interpreted as a space in which motion is taking place, and T as the set of times when the motion is observed. The times $\alpha = \inf T$ and $\beta = \sup T$ are called the *initial* and *final times*, respectively (or the *birth and death times*).

By attaching to each time t the position $\omega(t)$ of a moving particle we define the trajectory ω. Instead of $\omega(t)$ we often write $x_t(\omega)$. We denote by Ω the set of all trajectories (that is, all maps of T to E). A random process is a probability measure $\mathbf{P}$ on Ω given on the Kolmogorov σ-algebra $\mathscr{N}$. The latter is defined as the minimal σ-algebra containing all the sets $\{\omega : x_t(\omega) \in \Gamma\}$ $(t \in T, \Gamma \in \mathscr{B})$.

We denote by $\mathscr{N}_\Lambda$ the minimal σ-algebra in Ω containing all the sets $\{\omega: x_t(\omega) \in \Gamma\} (t \in T \cap \Lambda, \Gamma \in \mathscr{B})$ and put $\mathscr{N}_t = \mathscr{N}_{(-\infty, t]}$, $\mathscr{N}^t = \mathscr{N}_{[t, +\infty)}$. Then $\mathscr{N}_t$ can be interpreted as the collection of events that are defined "by the past" and $\mathscr{N}^t$ as the collection of events that are defined "by the future". A random process $\mathbf{P}$ is called *Markov*[1] if for any $t \in T$ and any $A \in \mathscr{N}_t$, $B \in \mathscr{N}^t$

$$\mathbf{P}(AB \mid x_t) = \mathbf{P}(A \mid x_t)\mathbf{P}(B \mid x_t) \qquad \text{(a.s. } \mathbf{P}) \tag{1.1}$$

1.2. An important feature of the definition of a Markov process is its invariance relative to reversal of time. This invariance, however, is lost with the introduction of the notion of a transition function, without which the advanced theory of Markov processes is impossible.

A *transition function* is a family of stochastic kernels $p_t^s (s < t \in T)$ in the space E, connected by the relation $p_t^s * p_u^t = p_u^s$ for all $s < t < u \in T$ (the Kolmogorov – Chapman equations).[2] We will frequently write $p(s, x; t, \Gamma)$ instead of $p_t^s(x, \Gamma)$.

A Markov process $\mathbf{P}$ *has the transition function* $p(s, x; t, \Gamma)$ if for any $s < t \in T$, $\Gamma \in \mathscr{B}$

$$\mathbf{P}\{x_t \in \Gamma \mid x_s\} = p(s, x_s; t, \Gamma) \qquad \text{(a.s. } \mathbf{P}) \tag{1.2}$$

To restore the symmetry of past and future we consider in parallel with transition functions (which we could perhaps call "forward transitions") the dual notion of cotransition functions (or "backward transition" functions). A *cotransition function* is a family of stochastic kernels $p_t^s (s > t \in T)$ in E connected by the relation $p_t^s * p_u^t = p_u^s$ for any $s > t > u \in T$. We usually denote a cotransition function by $p^*(s, x; t, \Gamma)$. A *Markov process* $\mathbf{P}$ *has the cotransition function* $p^*(s, x; t, \Gamma)$ if for any $s > t$, $t \in T$, $\Gamma \in B$

$$\mathbf{P}\{x_t \in \Gamma \mid x_s\} = p^*(s, x_s; t, \Gamma). \tag{1.2*}$$

1.3. The notion of a measurable space is too broad, and from now on we assume that the state space $(E, \mathscr{B})$ is *standard*. This means that $(E, \mathscr{B})$ is isomorphic[3] to a space $(E_1, \mathscr{B}_1)$ where E_1 is a Borel subset in some

[1] The abbreviations (a.s. P), (a.s. P, A), (a.s. $\mathscr{K}$) mean, respectively, "almost surely with respect to P", "almost surely on the set A with respect to P", "almost surely with respect to all measures P of the class $\mathscr{K}$".

[2] A stochastic kernel in E is a function $p(x, \Gamma)$ $(x \in E, \Gamma \in B)$ which is measurable in x and a probability measure with respect to Γ. Writing $p = p' * p''$ means that p is the product of the kernels p' and p'' that is

$$p(x, \Gamma) = \int_E p'(x, dy)\, p''(y, \Gamma).$$

[3] A one-to-one mapping of E onto E_1 is called an isomorphism of $(E, \mathscr{B})$ and $(E_1, \mathscr{B}_1)$, if it and its inverse are measurable.

complete separable metric space X and $\mathscr{B}_1$ is the collection of all Borel sets of X contained in E_1.

We fix any transition function $p(s, x; t, \Gamma)$ and put[1]

$$(1.3) \quad p(s, x; t_1, \Gamma_1, \ldots, t_n, \Gamma_n) = \\ = \int_{\Gamma_1}\int_{\Gamma_2} \cdots \int_{\Gamma_n} p(s, x; t_1, dy_1)\, p(t_1, y_1; t_2, dy_2) \ldots p(t_{n-1}, y_{n-1}; t_n, dy_n) \\ (s \leqslant t_1 < \ldots < t_n \in T,\ \Gamma_1, \ldots, \Gamma_n \in \mathscr{B}).$$

These functions are consistent in the following sense: if we set $\Gamma_i = E$ or if we cross out the pair of arguments t_i, Γ_i, then we obtain the same value. By a wellknown theorem of Kolmogorov (see, for example, [4] Chapter III, 3) it then follows that there is a measure $\mathbf{P}_{s,x}$ on the σ-algebra $\mathscr{N}^s$ for which

$$(1.4) \quad \mathbf{P}_{s,x}\{x_{t_1} \in \Gamma_1, \ldots, x_{t_n} \in \Gamma_n\} = p(s, x; t_1, \Gamma_1, \ldots, t_n, \Gamma_n).$$

We denote by $\mathscr{K}$ the class of all Markov processes having the transition function $p(s, x; t, \Gamma)$. It is not difficult to show that for processes of this class and for $A \in \mathscr{N}^s$

$$(1.5) \qquad \mathbf{P}(A \mid \mathscr{N}_s) = \mathbf{P}_{s, x_s}(A) \qquad (\text{a.s. } \mathbf{P}).$$

(1.5) is equivalent to the system of conditions (1.1) and (1.2), so that $\mathscr{K}$ may be written as the collection of probability measures $\mathbf{P}$ on the σ-algebra $\mathscr{N}$ for which (1.5) is satisfied.

From an intuitive point of view (1.5) means that the probability of predicting the future for a fixed past depends only on the present and is identical for all processes of $\mathscr{K}$. It is natural to take these processes as differing only in the initial behaviour of trajectories.

A system of measures $\mathbf{P}^{s,x}$ $(s \in T, x \in E)$ on the σ-algebra $\mathscr{N}^s$ for the cotransition function $p^*(s, x; t, \Gamma)$ is constructed in an analogous way. The class $\mathscr{K}^*$ of all Markov processes corresponding to p^* may also be written as the collection of probability measures $\mathbf{P}$ on $\mathscr{N}$ such that for any $s \in T$ and $A \in \mathscr{N}^s$

$$(1.6) \qquad \mathbf{P}(A \mid \mathscr{N}^s) = \mathbf{P}^{s, x_s}(A) \qquad (\text{a.s. } \mathbf{P}).$$

It is natural to regard the processes of $\mathscr{K}^*$ as differing only in the final behaviour of their trajectories.

1.4. It is our aim to introduce convenient characteristics of the initial (final) behaviour of trajectories, in other words, characteristics which together with a transition (cotransition) function uniquely define a Markov process.

If the birth time α is contained in T, then the problem of initial behaviour is solved trivially: the required characteristic is the probability

[1] We assume that $p(s, x; s, \Gamma) = 1$ for $x \in \Gamma$ and 0 for $x \notin \Gamma$.

distribution of the points x_α (the *initial distribution*); there exists a one-to-one correspondence between processes of $\mathscr{K}$ and probability measures on E (initial distributions). We shall see that the answer is analogous even when $\alpha \notin T$, except that the initial distribution must be given not in E but in some new space $\mathscr{V}$, the space of entries.

More precisely, we construct a measurable map φ of the space $(\Omega, \mathscr{N})$ onto some standard space $(\mathscr{V}, \mathscr{B}_{\mathscr{V}})$ so that the formula

$$\mu(\Gamma) = \mathbf{P}\{\omega : \varphi(\omega) \in \Gamma\} \qquad (\Gamma \in \mathscr{B}_{\mathscr{V}})$$

defines a one-to-one correspondence between $\mathscr{K}$ and the class $\mathscr{M}$ of all probability measures on $(\mathscr{V}, \mathscr{B}_{\mathscr{V}})$. Here $\varphi(\omega)$ (denoted by $\chi_\alpha(\omega)$) is interpreted as the beginning of the trajectory ω, or the entrance through which ω gets into E, and μ is interpreted as the initial distribution of the process $\mathbf{P}$.

The role of the entrance space with respect to $\mathscr{K}$ is played by the exit space$(\mathscr{V}^*, \mathscr{B}_{\mathscr{V}^*})$with respect to $\mathscr{K}^*$. Under time reversal these spaces are transformed into each other.

1.5. We call the intersection of the σ-algebras $\mathscr{N}_t$ for all $t \in T$ the *initial σ-algebra* and denote it by $\mathscr{N}_{\alpha+}$. If $\alpha \in T$, then this σ-algebra is the same as $\mathscr{N}_\alpha$. We interpret $\mathscr{N}_{\alpha+}$ as the system of events that are determined by the initial behaviour of a trajectory. The dual formation is the final σ-algebra $\mathscr{N}^{\beta-}$, which is the intersection of the $\mathscr{N}^t$ for all $t \in T$.

We call $\mathbf{P}$ *$\mathscr{N}_{\alpha+}$-ergodic* if $\mathbf{P}(A)$ is equal to 0 or 1 for all $A \in \mathscr{N}_{\alpha+}$. It is natural to assume that the initial behaviour of such a process is non-random and that almost all its trajectories "get into the state space E through the same entrance". This leads to the idea of identifying entrances with $\mathscr{N}_{\alpha+}$-ergodic processes of $\mathscr{K}$.

The programme outlined will be implemented in §2–4. We study in §2 the general problem of decomposition into ergodic measures. The applicability of the results of §2 to $\mathscr{K}$ and $\mathscr{K}^*$ is proved in §3, the entry space and the exit space are constructed in §4.

1.6. Let us agree on brief notations. Let $\mathscr{F}$ be any σ-algebra. For a function f the symbol $f \in \mathscr{F}$ means that f is non-negative and $\mathscr{F}$-measurable. Thus, $A \in \mathscr{F}$ if and only if $\chi_A \in \mathscr{F}$ (where χ_A is the indicatrix of A, that is, the function with the value 1 on A and 0 outside A). If $f \in \mathscr{F}$ and μ is a measure on $\mathscr{F}$, then by $\mu(f)$ (or μf) we mean the integral of f with respect to μ. In particular, if $\xi \in \mathscr{N}$ and $\mathbf{P}$ is a probability measure on $\mathscr{N}$, then $\mathbf{P}\,\xi$ denotes the mathematical expectation of ξ. (Usually the letter $\mathbf{E}$ or $\mathbf{M}$ is used for mathematical expectation, but this is inconvenient when we have to deal simultaneously with several measures.)

It is easy to see that condition (1.5), which defines $\mathscr{K}$, is equivalent to either of the conditions:

1.6.A. For any $s \in T$ and all functions $\xi \in \mathcal{N}^{s}$

$$\mathbf{P}(\xi \mid \mathcal{N}_s) = \mathbf{P}_{s, x_s}\xi \quad \text{(a.s. } \mathbf{P})$$

1.6.A′. For any $s \in T$ and all functions, $\xi \in \mathcal{N}^{s}$, $\eta \in \mathcal{N}_s$

$$\mathbf{P}\eta\xi = \mathbf{P}(\eta \mathbf{P}_{s, x_s}\xi).$$

§ 2. Decomposition into ergodic measures

2.1. Let $(\Omega, \mathcal{F})$ be any measurable space. We consider various classes of probability measures in this space. We treat each such class as a measurable space $(\mathcal{M}, \mathcal{C}_{\mathcal{M}})$: $\mathcal{C}_{\mathcal{M}}$ is defined as the minimal σ-algebra with respect to which all the functions $\mathbf{P}\xi (\xi \in \mathcal{F})$ are measurable. (This is the σ-algebra generated by the system of functions $\mathbf{P}(A)$ $(A \in \mathcal{F})$.)

We call the class $\mathcal{M}$ *separable* if there exists a countable family W of functions $f \in \mathcal{F}$, separating the measures of $\mathcal{M}$ (this means that for any $\mu_1 \neq \mu_2$ of $\mathcal{M}$ there is an $f \in W$ such that $\mu_1(f) \neq \mu_2(f)$).

Let $\mathcal{A}$ be a σ-algebra contained in $\mathcal{F}$. A measure $\mathbf{P}$ is called $\mathcal{A}$-ergodic if $\mathbf{P}(A)$ is equal to 0 or 1 for all $A \in \mathcal{A}$. We are interested in conditions under which all measures from $\mathcal{M}$ uniquely decompose into $\mathcal{A}$-ergodic measures from $\mathcal{M}$.

A function $\mathbf{P}^{\omega}(A)$ $(\omega \in \Omega, A \in \mathcal{F})$ is called an $(\mathcal{M}, \mathcal{A})$*-kernel* if:

2.1.A. $\mathbf{P}^{\omega} \in \mathcal{M}$ for each $\omega \in \Omega$; $\mathbf{P}^{\omega}(A)$ is $\mathcal{A}$-measurable for any $A \in \mathcal{F}$.

2.1.B. For any $\mathbf{P} \in \mathcal{M}$, $A \in \mathcal{F}$

$$\mathbf{P}(A \mid \mathcal{A}) = \mathbf{P}^{\omega}(A) \quad \text{(a.s. } \mathbf{P})$$

(An $(\mathcal{M}, \mathcal{A})$-kernel gives a general conditional probability distribution relative to $\mathcal{A}$ for all measures $\mathbf{P} \in \mathcal{M}$).

Condition 2.1.A. can also be written in the form:

2.1.A′. $\omega \to \mathbf{P}^{\omega}$ is a measurable map of $(\Omega, \mathcal{A})$ to $(\mathcal{M}, \mathcal{C}_{\mathcal{M}})$.

The aim of this section is to prove the following theorem.

THEOREM 2.1. *Let $\mathcal{M}_e$ be the collection of all $\mathcal{A}$-ergodic measures of the separable class $\mathcal{M}$. If there exists an $(\mathcal{M}, \mathcal{A})$-kernel $\mathbf{P}^{\omega}$, then $\mathcal{M}_e \in \mathcal{C}_{\mathcal{M}}$ and each measure $\mathbf{P}$ from $\mathcal{M}$ can be uniquely represented in the form*

$$\mathbf{P}(A) = \int_{\mathcal{M}_e} \tilde{\mathbf{P}}(A)\, \mu(d\tilde{\mathbf{P}}) \qquad (A \in \mathcal{F}), \tag{2.1}$$

where μ is a probability measure on $(\mathcal{M}_e, \mathcal{C}_{\mathcal{M}_e})$. The measure μ is expressed in terms of $\mathbf{P}$ by the formula

$$\mu(\Gamma) = \mathbf{P}\{\omega : \mathbf{P}^{\omega} \in \Gamma\}. \tag{2.2}$$

2.2. The proof of theorem 2.1 is based on two lemmas.

LEMMA 2.1. *Let W be a countable family of functions separating the measures in* $\mathcal{M}$. *A measure* **P** *from* $\mathcal{M}$ *is* $\mathcal{A}$*-ergodic if and only if it satisfies any of the following equivalent conditions*:

2.2.A. $\mathbf{P}^{\omega}(\xi) = \mathbf{P}(\xi)$ *for all* $\xi \in W$ (a.s. **P**).

2.2.B. $\Phi_{\xi}(\mathbf{P}) = 0$ *for all* $\xi \in W$, *where*

$$\Phi_{\xi}(\mathbf{P}) = \int_{\Omega} \mathbf{P}^{\omega}(\xi)^2 \mathbf{P}(d\omega) - \mathbf{P}(\xi)^2 = \int_{\Omega} [\mathbf{P}^{\omega}(\xi) - \mathbf{P}(\xi)]^2 \mathbf{P}(d\omega).$$

2.2.C. $\mathbf{P}\{\omega: \mathbf{P}^{\omega} = \mathbf{P}\} = 1$.

PROOF. The equivalence of 2.2.A. and 2.2.B. is obvious. The equivalence of 2.2.A. and 2.2.C. follows from the equality

$$\{\omega: \mathbf{P}^{\omega} = \mathbf{P}\} = \{\omega: \mathbf{P}^{\omega}(\xi) = \mathbf{P}(\xi) \text{ for all } \xi \in W\}.$$

If $\mathbf{P} \in \mathcal{M}_e$, then $\eta = \mathbf{P}(\eta)$ (a.s. **P**) for all $\eta \in \mathcal{A}$. Putting $\eta = \mathbf{P}^{\omega}(\xi)$ $(\xi \in \mathcal{F})$ and considering 2.1.B., we have

$$\mathbf{P}^{\omega}(\xi) = \int_{\Omega} \mathbf{P}^{\omega}(\xi)\mathbf{P}(d\omega) = \mathbf{P}(\mathbf{P}(\xi \mid \mathcal{A})) = \mathbf{P}(\xi) \quad \text{(a.s. } \mathbf{P})$$

Hence 2.2.A. follows.

On the other hand, if 2.2.C. is satisfied, then for $A \in \mathcal{A}$

$$\mathbf{P}(A) = \mathbf{P}^{\omega}(A) = \mathbf{P}(A \mid \mathcal{A}) = \chi_A \text{ (a.s. } \mathbf{P}).$$

This means $\mathbf{P}(A)$ is equal to 0 or 1.

COROLLARY. $\mathcal{M}_e$ *is contained in the set of measures* $\mathbf{P}^{\omega}$ $(\omega \in \Omega)$. (This follows from 2.2.C.)

LEMMA 2.2. $\mathcal{M}_e \in \mathscr{C}_M$ *and* $\mathbf{P}\{\omega: \mathbf{P}^{\omega} \in \mathcal{M}_e\} = 1$ *for all* $\mathbf{P} \in \mathcal{M}$.

PROOF. The first assertion follows from 2.2.B. and the $\mathscr{C}_{\mathcal{M}}$-measurability of the function Φ_{ξ}.

Setting for brevity $\eta(\omega) = \mathbf{P}^{\omega}(\xi)^2$ we have

$$\Phi_{\xi}(\mathbf{P}^{\omega}) = \mathbf{P}^{\omega}(\eta) - \eta = \mathbf{P}(\eta \mid \mathcal{A}) - \eta \quad \text{(a.s. } \mathbf{P}).$$

Therefore the integral of $\Phi_{\xi}(\mathbf{P}^{\omega})$ with respect to **P** is 0 and since Φ_{ξ} is non-negative $\Phi_{\xi}(\mathbf{P}^{\omega}) = 0$ for **P**-almost all ω. Using 2.1.C. we conclude that $\mathbf{P}\{\omega: P^{\omega} \in \mathcal{M}_e\} = 1$.

2.3. PROOF OF THEOREM 2.1. *Let* $\mathbf{P} \in \mathcal{M}$. *By* 2.1.B.

$$\mathbf{P}(A) = \mathbf{P}[\mathbf{P}(A \mid \mathcal{A})] = \int_{\Omega} \mathbf{P}^{\omega}(A)\mathbf{P}(d\omega) \qquad (A \in \mathcal{F}).$$

If we replace the integration variable by $\mathbf{P}^{\omega} = \widetilde{\mathbf{P}}$, we have

$$\mathbf{P}(A) = \int_{\mathcal{M}} \widetilde{\mathbf{P}}(A)\,\mu(d\widetilde{\mathbf{P}}), \tag{2.3}$$

where μ is defined by (2.2). By Lemma 2.2 $\mu(\mathcal{M}_e) = 1$. Therefore (2.1) follows from (2.3).

To prove the uniqueness of (2.1) we note that by 2.2.C $\widetilde{\mathbf{P}}\{\omega: \mathbf{P}^{\omega} \in \Gamma\} = \chi_{\Gamma}(\widetilde{\mathbf{P}})$ for $\widetilde{\mathbf{P}} \in \mathcal{M}_e$. Therefore it follows from (2.1) that for any $\Gamma \in \mathscr{C}_{\mathcal{M}_e}$

$$\text{(2.4)} \quad \mathbf{P}\{\omega: \mathbf{P}^{\omega} \in \Gamma\} = \int_{\mathcal{M}_e} \tilde{\mathbf{P}}\{\omega: \mathbf{P}^{\omega} \in \Gamma\} \mu(d\tilde{\mathbf{P}}) = \int_{\mathcal{M}_e} \chi_{\Gamma}(\tilde{\mathbf{P}}) \mu(d\tilde{\mathbf{P}}) = \mu(\Gamma).$$

2.4. Any two $(\mathcal{M}, \mathcal{A})$-kernels $\mathbf{P}^{\omega}$ and $\mathbf{P}_1^{\omega}$ are equivalent in the following sense: a set $\Omega' \in A$ can be constructed so that $\mathbf{P}(\Omega') = 1$ for all $\mathbf{P} \in \mathcal{M}$ and $\mathbf{P}^{\omega} = \mathbf{P}_1^{\omega}$ for all $\omega \in \Omega'$. (Ω' is defined by the condition: $\mathbf{P}^{\omega}(\xi) = \mathbf{P}_1^{\omega}(\xi)$ for all ξ of any countable set of functions which separate measures of $\mathcal{M}$.)

On the other hand, if a function $\mathbf{P}^{\omega}$ satisfies 2.1.A′, and is equivalent to the $(\mathcal{M}, \mathcal{A})$-kernel $\mathbf{P}_1^{\omega}$, then it is also an $(\mathcal{M}, \mathcal{A})$-kernel.

An $(\mathcal{M}, \mathcal{A})$-kernel $\mathbf{P}^{\omega}$ is called *normal* if $\mathbf{P}^{\omega} \in \mathcal{M}_e$ for all $\omega \in \Omega$. Starting with any $(\mathcal{M}, \mathcal{A})$-kernel $\mathbf{P}_1^{\omega}$ we can construct a normal $(\mathcal{M}, \mathcal{A})$-kernel $\mathbf{P}^{\omega}$ by putting

$$\mathbf{P}^{\omega} = \begin{cases} \mathbf{P}_1^{\omega} & \text{for } \mathbf{P}_1^{\omega} \in \mathcal{M}_e, \\ \mathbf{P}_0 & \text{for } \mathbf{P}_1^{\omega} \notin \mathcal{M}_e, \end{cases}$$

where $\mathbf{P}_0$ is any element of $\mathcal{M}_e$.

From the corollary to Lemma 2.1 it follows that if the $(\mathcal{M}, \mathcal{A})$-kernel is normal then $\mathcal{M}_e$ coincides with the set of measures $\mathbf{P}^{\omega}$ $(\omega \in \Omega)$.

§ 3. Construction of a $(\mathcal{K}, \mathcal{N}_{\alpha+})$-kernel

3.1. If T contains α, then a $(\mathcal{K}, \mathcal{N}_{\alpha+})$-kernel can be obtained by the formula $\mathbf{P}^{\omega} = \mathbf{P}_{\alpha, x_{\alpha}(\omega)}$. Therefore we will assume, without saying so each time, that $\alpha \notin T$.

The construction of a $(\mathcal{K}, \mathcal{N}_{\alpha+})$-kernel is based on the formula

$$\text{(3.1)} \qquad \mathbf{P}(\xi \mid \mathcal{N}_{\alpha+}) = \lim_{t \downarrow \alpha,\, t \in \Lambda} \mathbf{P}(\xi \mid \mathcal{N}_t) \quad \text{(a.s. } \mathbf{P}\text{)}.$$

which holds for all $\xi \in \mathcal{N}$ and $\mathbf{P}$, if Λ is a countable subset of T having α as a limit point (see [1], Chapter 7, Theorem 4.3).

We say that ξ is *independent of the beginning of a trajectory* if $\xi \in \mathcal{N}^s$ for some $s > \alpha$. For such functions it follows from 1.6.A. and (3.1) that for $\mathbf{P} \in \mathcal{K}$

$$\text{(3.2)} \qquad \mathbf{P}(\xi \mid \mathcal{N}_{\alpha+}) = \lim_{\substack{t \downarrow \alpha \\ t \in \Lambda}} \mathbf{P}_{t, x_t} \xi \quad \text{(a.s. } \mathbf{P}\text{)}.$$

The right-hand side does not depend on $\mathbf{P}$. To construct a $(\mathcal{K}, \mathcal{N}_{\alpha+})$-kernel it is sufficient to find a measure $\mathbf{P}^{\omega} \in \mathcal{K}$, satisfying the condition

$$\text{(3.3)} \qquad \mathbf{P}^{\omega}\xi = \lim_{t \downarrow \alpha,\, t \in \Lambda} \mathbf{P}_{t, x_t}\xi \quad \text{(a.s. } \mathcal{K}\text{)}$$

for some countable family W of functions ξ independent of the beginning of a trajectory and separating measures in $\mathcal{K}$.

The required family W is given by the formula

$$\text{(3.4)} \qquad \xi = f(x_s) \qquad (s \in \Lambda,\ f \in W_q),$$

where W_q is a special family of functions on the state space $(E, \mathscr{B})$. The family W_q is introduced and studied in 3.2. In 3.3 the relation between the measures $\mathbf{P} \in \mathscr{K}$ and the systems of measures $\nu_t (t \in \Lambda)$ in $(E, \mathscr{B})$ is settled. After these preparations the theorem on the existence of a $(\mathscr{K}, \mathscr{N}_\alpha)$-kernel is proved in 3.4. At the end of the section it will be proved that the spaces $(\mathscr{K}, \mathscr{C}_{\mathscr{K}})$ and $(\mathscr{K}_e, \mathscr{C}_{\mathscr{K}_e})$ are standard.

3.2. It is well known (see, for example, [3], Chapter 3, § 37, II) that each standard measurable space is isomorphic to one of the spaces $(J, \mathscr{B}_J)$ or $(I, \mathscr{B}_I)$, where J is a finite or countable set, $\mathscr{B}_J$ is the system of all its subsets; $I = \{x\colon 0 \leqslant x \leqslant 1\}$, $\mathscr{B}_I$ is the collection of all Borel sets of I.

We realize J as a closed subset of the interval I and call a real function q on E a *coordinate* on $(E, \mathscr{B})$ if it gives an isomorphic map of $(E, \mathscr{B})$ onto $(I, \mathscr{B}_I)$ or $(J, \mathscr{B}_J)$. (It is not difficult to show that if q is any coordinate, then the general coordinate is $\varphi(q)$, where φ is any automorphism of $(I, \mathscr{B}_I)$.)

We write $x_n \overset{q}{\rightarrow} x$ if $q(x_n) \rightarrow q(x)$ and $\mu_n \overset{q}{\rightarrow} \mu$ if $\mu_n(q^k) \rightarrow \mu(q^k)$ for $k = 0, 1, 2, \ldots$ Relative to the q-topology the space E and the set $\mathscr{M}(E)$ of all probability measures on $(E, \mathscr{B})$ are compact. For if x is identified with $q(x)$, then E goes over in to J or I with their usual topology and $\mathscr{M}(E)$ into the space of all probability measures on J or I with the weak topology.[1]

We denote by W_q the family of functions $1, q, q^2, \ldots$ We need the following properties of this family.

3.2.A. Suppose that a system $\mathscr{H}$ of functions contains W_q and satisfies the conditions:

3.2.A$_1$. If $f_1, f_2 \in \mathscr{H}, c_1, c_2$ real, then $c_1 f_1 + c_2 f_2 \in \mathscr{H}$.

3.2.A$_2$. If the sequence $f_n \in \mathscr{H}$ is uniformly bounded and converges to f at each point, then $f \in \mathscr{H}$.

Then $\mathscr{H}$ contains all bounded $\mathscr{B}$-measurable functions.

3.2.B. W_q separates probability measures.

3.2.C. If for the probability measures μ_n the sequence $\mu_n(f)$ converges for any $f \in W_q$, then there is a probability measure μ such that $\mu_n(f) \rightarrow \mu(f)$ for $f \in W_q$.

Let us prove 3.2.A. Let $(E, \mathscr{B}) = (I, \mathscr{B}_I)$ and $q(x) = x$. If $\mathscr{H} \supseteq W_q$, then $\mathscr{H}$ contains all polynomials, consequently all continuous functions. Hence $\mathscr{H}$ contains all bounded $\mathscr{B}_I$-measurable functions (see, for example, [2], Lemma 1.8). The case $(E, \mathscr{B}) = (J, \mathscr{B}_J)$ is treated similarly.

If $\mu_1, \mu_2 \in \mathscr{M}(E)$, then the set $\mathscr{H}$ of bounded $\mathscr{B}$-measurable functions f for which $\mu_1(f) = \mu_2(f)$ satisfies conditions 3.2.A$_1$–3.2.A$_2$. Therefore 3.2.B follows from 3.2.A.

---

[1] If $\mu_n \overset{q}{\rightarrow} \mu$, then $\mu_n[f(q)] \rightarrow \mu[f(q)]$ for all polynomials and hence all continuous functions f (since by Weierstrass' theorem every continuous function is uniformly approximable by polynomials on I).

To prove 3.2.C we note that any measure μ which is a limit point for the sequence μ_n in the q-topology satisfies the condition $\mu(f) = \lim \mu_n(f)$ for all $f \in W_q$. By 3.2.B it follows that the sequence μ_n has a unique limit point and hence converges.

3.3 Put

$$(3.5)\qquad \begin{cases} P_t^s f(x) = \int_E p(s, x; t, dy) f(y) = \mathbf{P}_{s,x} f(x_t), \\ \nu P_t^s(\Gamma) = \int_E \nu(dx) p(s, x; t, \Gamma) \end{cases}$$

($p(s, x; t, \Gamma)$ is the transition function that defines $\mathcal{K}$).

LEMMA 3.1. *Let $\Lambda \subseteq T$ be a countable set and α a limit point of it. Denote by $\mathcal{L}$ the collection of functions $\nu_t (t \in \Lambda)$ with values in $\mathcal{M}(E)$ satisfying the condition*

$$(3.6)\qquad \nu_s P_t^s = \nu_t \qquad (s < t \in \Lambda).$$

The formula

$$(3.7)\qquad \nu_t(\Gamma) = \mathbf{P}\{x_t \in \Gamma\} \qquad (\Gamma \in \mathcal{B},\ t \in \Lambda)$$

defines a one-to-one map of $\mathcal{K}$ onto $\mathcal{L}$.

PROOF. It is clear that for any $f \in \mathcal{B}$

$$(3.8)\qquad \mathbf{P} f(x_t) = \nu_t(f).$$

By 1.6.A′ and (3.8) for $\mathbf{P} \in \mathcal{K}$ and any $\xi \in \mathcal{N}^s$

$$(3.9)\qquad \mathbf{P}\xi = \mathbf{P}(\mathbf{P}_{s,x_s}\xi) = \int_E \nu_s(dx) \mathbf{P}_{s,x}\xi.$$

Putting $\xi = \chi_{\Gamma_1}(x_{t_1}) \ldots \chi_{\Gamma_n}(x_{t_n})$ $(s < t_1 < \ldots < t_n \in T; \Gamma_1, \ldots \Gamma_n \in \mathcal{B})$ and bearing (1.4) in mind we have

$$(3.10)\qquad \mathbf{P}\{x_{t_1} \in \Gamma_1, \ldots, x_{t_n} \in \Gamma_n\} = \int_E \nu_s(dx) p(s, x; t_1, \Gamma_1, \ldots, t_n, \Gamma_n).$$

For $n = 1$ and $t_n = t \in \Lambda$ this formula is equivalent to (3.6). Consequently, the set $\nu = \{\nu_s\}$, corresponding to $\mathbf{P} \in \mathcal{K}$, belongs to L. From (3.10) it is obvious that the measure $\mathbf{P}$ is uniquely recoverable from $\{\nu_s\}$.

On the other hand, if $\{\nu_s\} \in \mathcal{L}$, then by (3.6) the right-hand side of (3.10) does not depend on the choice of s from $\Lambda \cap (\alpha, t_1)$. We denote it by $\mu_{t_1, \ldots, t_n}(\Gamma_1, \ldots, \Gamma_n)$. The functions $\mu_{t_1, \ldots, t_n}$ are obviously consistent, and by Kolmogorov's theorem there exists a measure $\mathbf{P}$ on the σ-algebra $\mathcal{N}$, satisfying (3.10). Using (1.3), (1.4), and (3.10) it is easy to verify that $\mathbf{P}$ satisfies 1.6.A′ and consequently belongs to $\mathcal{K}$.

3.4. THEOREM 3.1. *There exists a $(\mathcal{K}, \mathcal{N}_{\alpha+})$-kernel* $\mathbf{P}^\omega$

PROOF. The system of functions W defined by (3.4) separates measures from $\mathcal{K}$. For let $\mathbf{P}$, $\overline{\mathbf{P}} \in \mathcal{K}$, and let $\{\nu_s\}$, $\{\overline{\nu}_s\}$ be the elements of $\mathcal{L}$, corresponding to them by (3.7). If $\mathbf{P}_\xi = \overline{\mathbf{P}}_\xi$ for all $\xi \in W$, then $\nu_s(f) = \overline{\nu}_s(f)$ for all $f \in W_q$, $s \in \Lambda$. By 3.2.B $\{\nu_s\} = \{\overline{\nu}_s\}$ and by Lemma 3.1 $\mathbf{P} = \overline{\mathbf{P}}$.

We put $\omega \in \Omega_0$ if for all $\xi \in W$ the limit on the right-hand side of (3.2) exists. It is clear that $\Omega_0 \in \mathcal{N}_{\alpha+}$ and $\mathbf{P}(\Omega_0) = 1$ for all $\mathbf{P} \in \mathcal{K}$. Let

$$\nu_s^{\omega, t}(\Gamma) = \mathbf{P}_{t, x_t(\omega)}\{x_s \in \Gamma\} \qquad (\Gamma \in \mathcal{B}). \tag{3.11}$$

If $\omega \in \Omega_0$ then for any $s \in \Lambda$, $f \in W_q$ the limit

$$\lim_{t \downarrow \alpha,\, t \in \Lambda} \nu_s^{\omega, t}(f). \tag{3.12}$$

exists. By 3.2.B there is a probability measure ν_s^ω such that for any $f \in W_q$ the limit (3.12) is equal to $\nu_s^\omega(f)$. By (3.11)

$$\nu_s^\omega(f) = \lim_{t \downarrow \alpha,\, t \in \Lambda} \mathbf{P}_{t, x_t} f(x_s). \tag{3.13}$$

From (3.13) and (3.2) we conclude that for any $f \in W_q$, $s \in \Lambda$, the function $\nu_s^\omega(f)$ is $\mathcal{N}_{\alpha+}$-measurable and

$$\nu_s^\omega(f) = \mathbf{P}(f(x_s) \mid \mathcal{N}_{\alpha+}) \qquad (\text{a.s. } \mathcal{K}) \tag{3.14}$$

By 3.2.A, the validity of this assertion for all bounded $\mathcal{B}$-measurable functions f follows from its validity for $f \in W_q$.

Since $\mathcal{N}_{\alpha+} \subseteq \mathcal{N}_s$, for $\xi \in \mathcal{N}^s$

$$\mathbf{P}\{\xi \mid \mathcal{N}_{\alpha+}\} = \mathbf{P}\{\mathbf{P}(\xi \mid N_s) \mid \mathcal{N}_{\alpha+}\} = \mathbf{P}\{\mathbf{P}_{s, x_s}\xi \mid \mathcal{N}_{\alpha+}\} \quad (\text{a.s. } \mathbf{P}). \tag{3.15}$$

In particular, for $\alpha < s < t$ and $f \in \mathcal{B}$

$$\mathbf{P}\{f(x_t) \mid \mathcal{N}_{\alpha+}\} = \mathbf{P}\{P_t^s f(x_s) \mid \mathcal{N}_{\alpha+}\} \quad (\text{a.s. } \mathbf{P}). \tag{3.16}$$

From (3.14) and (3.16) it follows for $f \in \mathcal{B}$ that $\mathcal{K}$-almost surely

$$\nu_t^\omega(f) = \nu_s^\omega(P_t^s f), \tag{3.17}$$

where both sides are $\mathcal{N}_{\alpha+}$-measurable functions. We put $\omega \in \Omega_1$ if (3.17) is satisfied for all $f \in W_q$, $s < t \in \Lambda$. It is clear that $\Omega_1 \in \mathcal{N}_{\alpha+}$ and $\mathbf{P}(\Omega_1) = 1$ for all $\mathbf{P} \in \mathcal{K}$. If $\omega \in \Omega_1$, then $\{\nu_s^\omega\}$ satisfies (3.6) and consequently belongs to $\mathcal{L}$. We denote by $\mathbf{P}^\omega$ the corresponding element of $\mathcal{K}$. Outside Ω_1 we put $\mathbf{P}^\omega = \mathbf{P}^{\omega_1}$, where ω_1 is any fixed element of Ω_1. By (3.8) and (3.10) $\mathbf{P}^\omega$ satisfies condition 2.1.A. From (3.15), (3.14), (3.8), and 1.6.A′ we have for $\mathbf{P} \in \mathcal{K}$, $\xi \in \mathcal{N}^s$

$$\mathbf{P}(\xi \mid \mathcal{N}_{\alpha+}) = \mathbf{P}(\mathbf{P}_{s, x_s}(\xi) \mid \mathcal{N}_{\alpha+}) = \mathbf{P}^\omega(\mathbf{P}_{s, x_s}\xi) = \mathbf{P}^\omega(\xi) \quad (\text{a.s. } \mathbf{P})$$

Thus, $\mathbf{P}^\omega$ also satisfies 2.1.B.

3.5. THEOREM 3.2. *The spaces $(\mathcal{K}, \mathcal{C}_{\mathcal{K}})$ and $(\mathcal{K}_e, \mathcal{C}_{\mathcal{K}_e})$ are standard.*

PROOF. By Theorem 2.1 $\mathcal{K}_e \in \mathcal{C}_{\mathcal{K}}$. Therefore it is sufficient to prove that $(\mathcal{K}, \mathcal{C}_{\mathcal{K}})$ is standard. We construct a σ-algebra $\mathcal{F}$ in the space $\mathcal{L}$ of Lemma 3.1 such that $(\mathcal{L}, \mathcal{F})$ is standard, and we show that the map (3.7) is an isomorphism of $(\mathcal{K}, \mathcal{C}_{\mathcal{K}})$ onto $(\mathcal{L}, \mathcal{F})$.

The set $\mathscr{L}$ is embedded in the space $\overline{\mathscr{L}}$ of all functions $\nu_t (t \in \Lambda)$ with values in $\mathscr{M}(E)$. Choose in $(E, \mathscr{B})$ any coordinate q and assume that the sequence $\nu^n \in \overline{\mathscr{L}}$ converges to $\nu \in \mathscr{L}$ if $\nu_t^n \overset{q}{\to} \nu_t$ for all $t \in \Lambda$. It is clear that $\overline{\mathscr{L}}$ is compact. Denote by $\overline{\mathscr{F}}$ the family of its Borel sets. If f is a bounded $\mathscr{B}$-measurable function, then for each $t \in \Lambda$ the function $F(\nu) = \nu_t(f)$ is $\overline{\mathscr{F}}$-measurable: for $f \in W_q$ this follows from the continuity of $F(\nu)$, and the extension from W_q to $\mathscr{B}$ is achieved using 3.2.A.

Condition (3.6) is equivalent to the countable system of equalities

$$\nu_s(P_t^s f) = \nu_t(f) \qquad (f \in W_q,\ s < t \in \Lambda).$$

Since both sides of each equality are $\overline{\mathscr{F}}$-measurable, we have $\mathscr{L} \in \overline{\mathscr{F}}$.

Let $\mathscr{F}$ be the collection of sets of $\overline{\mathscr{F}}$, that are contained in $\mathscr{L}$. The sets

$$C = \{\nu\colon c_1 < \nu_t(f) < c_2\} \qquad (t \in \Lambda,\ f \in W_q,\ 0 \leqslant c_1 < c_2 \leqslant 1)$$

form a base of the topology in $\overline{\mathscr{L}}$. The inverse image of C in $\mathscr{K}$ is the set

$$\{\mathbf{P}\colon c_1 < \mathbf{P}f(x_t) < c_2\} \in \mathscr{C}_{\mathscr{K}}.$$

Consequently, the map (3.7) of $(\mathscr{K}, \mathscr{C}_{\mathscr{K}})$ to $(\mathscr{L}, \mathscr{F})$ is measurable. To prove that the inverse map is measurable it is sufficient to verify that the function defined by (3.10) is $\overline{\mathscr{F}}$-measurable. But this function can be written in the form $\nu_s(f)$, where $f(x) = p(s, x; t_1, \Gamma_1, \ldots, t_n, \Gamma_n)$. Therefore[1] $\overline{\mathscr{F}}$-measurability follows from the $\mathscr{B}$-measurability of f.

§4. The entrance space and the exit space

4.1. Let $(\mathscr{V}, \mathscr{B}_{\mathscr{V}})$ be any measurable space and let

$$(4.1) \qquad (\Omega, \mathscr{N}_{\alpha+}) \overset{\varphi}{\longrightarrow} (\mathscr{V}, \mathscr{B}_{\mathscr{V}}) \overset{\psi}{\longrightarrow} (\mathscr{K}, \mathscr{C}_{\mathscr{K}}),$$

be given measurable maps where:

4.1.A. The map $\psi\varphi$ defines a normal $(\mathscr{K}, \mathscr{N}_{\alpha+})$-kernel.

4.1.B. $\varphi(\Omega) = \mathscr{V}$.

4.1.C. $\psi(v_1) \neq \psi(v_2)$ for $v_1 \neq v_2$; $\mathscr{B}_{\mathscr{V}} = \psi^{-1}(\mathscr{C}_{\mathscr{K}})$[2].

Then we say that $(\mathscr{V}, \mathscr{B}_{\mathscr{V}})$ is an *entrance space*. We call the point $\varphi(\omega)$ the *beginning of the trajectory* and denote it by $x_{\alpha+}(\omega)$; the measure $\psi(v)$ we denote by $\mathbf{P}_{\alpha,v}$ and call it the *process starting at v at time α*.

---

[1] If Λ consists of one point, then $\overline{\mathscr{L}}$ coincides with the space $\mathscr{M} = \mathscr{M}(E)$ of all probability measures on $(E, \mathscr{B})$, and the arguments in the proof of Theorem 3.2. prove that $(\mathscr{M}, \mathscr{C}_{\mathscr{M}})$ is standard.

[2] Without violating 4.1.A it is always possible to satisfy 4.1.B–4.1.C. For this it is sufficient to remove from $\mathscr{V}$ points not belonging to $\varphi(\Omega)$, then identify points v with the same image $\psi(v)$ and contract $\mathscr{B}_{\mathscr{V}}$ to $\psi^{-1}(\mathscr{C}_{\mathscr{K}})$.

In the new notation $\psi\varphi(\omega) = \mathbf{P}_{\alpha,x_{\alpha+}}$. This map automatically satisfies 2.1.A. Therefore 4.1.A can be replaced by the requirement:

4.1.A′. For any $\mathbf{P} \in \mathscr{K}$, $\xi \in \mathscr{N}$

$$\mathbf{P}(\xi \mid \mathscr{N}_{\alpha+}) = \mathbf{P}_{\alpha,x_{\alpha+}}\xi \quad \text{(a.s. } \mathbf{P}).$$

Recalling the definition of the σ-algebra $\mathscr{C}_{\mathscr{K}}$ (see 2.1) we can write 4.1.C in the form.

4.1.C′. The system of functions $\mathbf{P}_{\alpha,v}(A)$ $(A \in \mathscr{N})$ separates points of $\mathscr{V}$ and generates the σ-algebra $\mathscr{B}_{\mathscr{V}}$.

When T contains α, then $x_t(\omega) \in E$ and $\mathbf{P}_{t,x}$ $(x \in E)$ is defined for $t = \alpha$. If we put $(\mathscr{V}, \mathscr{B}_{\mathscr{V}}) = (E, \mathscr{B})$, then the validity of 4.1.A′ follows from 1.6.A and that of 4.1.B and 4.1.C′ from the formula

$$\mathbf{P}_{s,x}\{x_s \in \Gamma\} = \chi_\Gamma(x) \qquad (s \in T,\ \Gamma \in \mathscr{B}),$$

which follows from (1.4). Consequently the state space $(E, \mathscr{B})$ is an entry space.

4.2. In the general case an entrance space can be constructed in the following way. By Theorem 3.1 there exists a $(\mathscr{K}, \mathscr{N}_{\alpha+})$-kernel. Consequently (see 2.4) there exists a normal $(\mathscr{K}, \mathscr{N}_{\alpha+})$-kernel $\mathbf{P}^\omega$. Consider the diagram

$$(4.2) \qquad (\Omega, \mathscr{N}_{\alpha+}) \xrightarrow{\bar\varphi} (\mathscr{K}_e, \mathscr{C}_{\mathscr{K}_e}) \xrightarrow{\bar\psi} (\mathscr{K}, \mathscr{C}_{\mathscr{K}}),$$

where $\bar\varphi(\omega) = \mathbf{P}^\omega$ and $\bar\psi$ is the embedding of $\mathscr{K}_e$ in $\mathscr{K}$. The measurability of $\bar\varphi$ follows from 2.1.A′, and that of $\bar\psi$ from the definitions of $\mathscr{C}_{\mathscr{K}}$ and $\mathscr{C}_{\mathscr{K}_e}$. The validity of 4.1.A and 4.1.B is obvious, that of 4.1.C follows from 2.4.

If another normal $(\mathscr{K}, \mathscr{N}_{\alpha+})$-kernel is chosen, then in (4.2) the map $\bar\psi$ does not change, but $\bar\varphi$ is replaced by a $\bar\varphi'$, which is equivalent to $\bar\varphi$ in the following sense: there exists a set $\Omega' \in \mathscr{N}_{\alpha+}$, such that $\mathbf{P}(\Omega') = 1$ for all $\mathbf{P} \in \mathscr{K}$ and $\bar\varphi(\omega) = \bar\varphi'(\omega)$ for all $\omega \in \Omega'$ (see 2.4).

Now we prove that for a fixed kernel $\mathbf{P}^\omega$ the entrance space $(\mathscr{V}, \mathscr{B}_{\mathscr{V}})$ is uniquely determined to within isomorphism. More precisely, if $\psi\varphi = \bar\psi\bar\varphi$, then we have a commutative diagram

$$(4.3) \qquad (\Omega, \mathscr{N}_{\alpha+}) \begin{array}{c} \overset{\varphi}{\nearrow} (\mathscr{V}, \mathscr{B}_{\mathscr{V}}) \overset{\psi}{\searrow} \\ \updownarrow \\ \underset{\bar\varphi}{\searrow} (\mathscr{K}_e, \mathscr{C}_{\mathscr{K}_e}) \underset{\bar\psi}{\nearrow} \end{array} (\mathscr{K}, \mathscr{C}_{\mathscr{K}}).$$

(The vertical arrow represents an isomorphism of $(\mathscr{V}, \mathscr{B}_{\mathscr{V}})$ and $(\mathscr{K}_e, \mathscr{C}_{\mathscr{K}_e})$.)

For by 4.1.A and 2.4 $\psi\varphi(\Omega) = \mathscr{K}_e$. Taking account of 4.1.B we have $\varphi(V) = \mathscr{K}_e$. The map

$$(4.4) \qquad (\mathscr{V}, \mathscr{B}_{\mathscr{V}}) \xrightarrow{\psi} (\mathscr{K}_e, \mathscr{C}_{\mathscr{K}_e}),$$

is obviously measurable. By 4.1.C it is one-to-one. Since $\mathscr{K}_e \in \mathscr{C}_{\mathscr{K}}$ (Theorem 2.1), it follows from 4.1.C that the inverse map is also measurable. Consequently (4.4) is an isomorphism. It is easily seen that it satisfies (4.3).

By Theorem 3.2 the space $(\mathscr{V}, \mathscr{B}_{\mathscr{V}})$ is standard.

The results obtained can be summarized as follows:

THEOREM 4.1. *For every transition function with standard space $(E, \mathscr{B})$ it is possible to construct an entrance space. In the diagram* (4.1) *the space $(E, \mathscr{B})$ and the map $\psi(v) = \mathbf{P}_{\alpha,v}$ are uniquely determined to within isomorphism. The map $\varphi(\omega) = x_{\alpha+}(\omega)$ is uniquely determined to within equivalence. The space $(\mathscr{V}, \mathscr{B}_{\mathscr{V}})$ is standard. It may be identified with $(\mathscr{K}_e, \mathscr{C}_{\mathscr{K}_e})$ (and when $\alpha \in T$ also with $(E, \mathscr{B})$).*

4.3. We now prove that the entrance space has the property announced in 1.4.

THEOREM 4.2. *The formulae*

$$\mu(\Gamma) = \mathbf{P}\{x_{\alpha+} \in \Gamma\} \qquad (\Gamma \in \mathscr{B}_{\mathscr{V}}), \tag{4.5}$$

$$\mathbf{P}(A) = \int_{\mathscr{V}} \mathbf{P}_{\alpha,v}(A)\,\mu(dv) \qquad (A \in \mathscr{N}) \tag{4.6}$$

establish a one-to-one correspondence between the measures $\mathbf{P}$ *from* $\mathscr{K}$ *and all probability measures* μ *on* $(\mathscr{V}, \mathscr{B}_{\mathscr{V}})$ *(the initial distributions).*

PROOF. We consider the $(\mathscr{K}, \mathscr{N}_{\alpha+})$-kernel $\mathbf{P}^{\omega} = \mathbf{P}_{\alpha, x_{\alpha+}}$. By Theorem 2.1 every $\mathbf{P} \in \mathscr{K}$ can be uniquely represented in the form

$$\mathbf{P}(A) = \int_{K_e} \mathbf{P}(A)\,\mu(d\mathbf{P}), \tag{4.7}$$

where

$$\mu(\Gamma) = \mathbf{P}\{\omega\colon \mathbf{P}_{\alpha, x_{\alpha+}} \in \Gamma\}. \tag{4.8}$$

The formulae (4.7)–(4.8) go over into (4.5)–(4.6) if $(\mathscr{V}, \mathscr{B}_{\mathscr{V}})$ is identified with $(\mathscr{K}_e, \mathscr{C}_{\mathscr{K}_e})$. On the other hand, the validity of 1.6.A′ for the measures $\mathbf{P}_{\alpha,v}(v \in \mathscr{V})$ implies its validity for any measure $\mathbf{P}$ defined by (4.6). Thus, to an arbitrary probability measure μ there corresponds a $\mathbf{P} \in \mathscr{K}$.

4.4. Using 4.1.A′ we can rewrite (3.2) in the form

$$\mathbf{P}_{\alpha, x_{\alpha+}}\xi = \lim_{t \downarrow \alpha,\, t \in \Lambda} \mathbf{P}_{t, x_t}\xi \quad (\text{a.s. } \mathscr{K}) \tag{4.9}$$

(Λ is a countable subset of T, ξ does not depend on the initial behaviour of the trajectory). Let W be any countable system of functions independent of the beginning of a trajectory and separating measures of $\mathscr{K}$. We introduce in the space of all probability measures on the σ-algebra $\mathscr{N}$ the W-topology, by taking $\mathbf{P}_n$ as converging to $\mathbf{P}$ if $\mathbf{P}_n(\xi) \to \mathbf{P}(\xi)$ for all $\xi \in W$. It is clear that in this topology[1]

[1] We assume that the measures $\mathbf{P}_{t,x}$ are extended in some way from the σ-algebra $\mathscr{N}^{t}$ to $\mathscr{N}$.

$$(4.10) \qquad \mathbf{P}_{\alpha, x_{\alpha+}} = \lim_{t \downarrow \alpha,\, t \in \Lambda} \mathbf{P}_{t, x_t} \qquad (\text{a.s. } \mathscr{K}).$$

Identifying (t, x) with $\mathbf{P}_{t,x}$ and (α, v) with $\mathbf{P}_{\alpha,v}$ we transfer the W-topology to the space $(T \times E) \cup \mathscr{V}$. Then (4.10) can be rewritten in the form

$$(\alpha, x_{\alpha+}) = \lim_{\substack{t \downarrow \alpha \\ t \in \Lambda}} (t, x_t) \quad (\text{a.s. } \mathscr{K}).$$

The exceptional set of trajectories for which convergence does not hold depends on the choice of the system W.

4.5. We note some properties of the initial point $x_{\alpha+}$:

4.5.A. $\mathbf{P}_{\alpha, v} \{x_{\alpha+} = v\} = 1$ for all $v \in \mathscr{V}$.

4.5.B. If $\Gamma \in \mathscr{B}_{\mathscr{V}}$, then $\{x_{\alpha+} \in \Gamma\} \in \mathscr{N}_{\alpha+}$. Any set A from $\mathscr{N}_{\alpha+}$ differs from some set $\{x_{\alpha+} \in \Gamma\}$ $(\Gamma \in \mathscr{B}_{\mathscr{V}})$ by a set of measure zero relative to the class $\mathscr{K}$.

Since $x_{\alpha+}$ is defined to within equivalence (see **4.2**), it is sufficient to verify 4.5.A in the case described by diagram (4.2). But then 4.5.A reduces to 2.2.C.

Of the two assertions contained in 4.5.B the first is obvious. To prove the second we observe that by 4.1.C′ for $A \in \mathscr{N}_{\alpha+}$

$$\chi_A = \mathbf{P}(A \mid \mathscr{N}_{\alpha+}) = \mathbf{P}_{\alpha, x_{\alpha+}}(A) \quad (\text{a.s. } \mathscr{K}),$$

so that it is sufficient to put $\Gamma = \{v : \mathbf{P}_{\alpha+, v}(A) = 1\}$.

4.6. Let $\mathscr{M}$ be any set of measures. An element $\mathbf{P}$ of $\mathscr{M}$ is called extremal if from the relation $\mathbf{P} = \lambda_1\mathbf{P}_1 + \lambda_2 \mathbf{P}_2$ $(\mathbf{P}_1, \mathbf{P}_2 \in \mathscr{M}; \lambda_1, \lambda_2 \geqslant 0; \lambda_1 + \lambda_2 = 1)$ it follows that $\mathbf{P}_1 = \mathbf{P}_2 = \mathbf{P}$.

THEOREM 4.3. *The set of extremal elements of $\mathscr{K}$ coincides with $\mathscr{K}_e$ (or, what is the same thing, with the set of measures $\mathbf{P}_{\alpha, v}(v \in \mathscr{V})$).*

PROOF We denote by $\mathscr{M}$ the set of all probability measures on $(\mathscr{V}, \mathscr{B}_{\mathscr{V}})$. By Theorem 4.2 the proposition stated is equivalent to the following: the collection of all extremal elements of $\mathscr{M}$ coincides with the set of all measures concentrated on one point. It is clear that a measure concentrated on a point is extremal. Let μ be any extremal element of $\mathscr{M}$. Assume that $0 < \mu(B) < 1$ for some $B \in \mathscr{B}_{\mathscr{V}}$ and put $\overline{B} = \mathscr{V} \setminus B$.

Obviously $\mu = \mu(B)\mu_1 + \mu(\overline{B})\mu_2$, where $\mu_1(A) = \dfrac{\mu(AB)}{\mu(B)}$; $\mu_2(A) = \dfrac{\mu(A\overline{B})}{\mu(\overline{B})}$.

Hence $\mu_1 = \mu_2 = \mu$. But $\mu_1(B) = 1$, $\mu_2(B) = 0$. This contradiction shows that $\mu(B)$ is equal to 0 or 1 for all $B \in \mathscr{B}_{\mathscr{V}}$. Let $r(v)$ be a coordinate in the standard space $(\mathscr{V}, \mathscr{B}_{\mathscr{V}})$ and let c_0 be the least upper bound of all c for which $\mu\{v : r(v) < c\} = 1$. It is clear that the measure μ is concentrated on the one point set $\{v : f(v) = c_0\}$.

4.7. If in the constructions of 4.1.–4.6 we replace each notion by its dual ($\mathscr{K}$ by $\mathscr{K}^*$, $\mathscr{N}_\alpha$ by $\mathscr{N}^\beta$, $\mathbf{P}_{t,x}$ by $\mathbf{P}^{t,x}$ etc.), then we obtain the definition and properties of the *exit space.* Diagram (4.1) is replaced by

$$(\Omega, \mathcal{N}^{\cdot\beta}) \xrightarrow{\varphi^*} (\mathcal{V}^{\cdot *}, \mathcal{B}_{\mathcal{V}^*}) \xrightarrow{\psi^*} (\mathcal{K}^*, \mathcal{C}_{K^*}),$$

the initial point $x_{\alpha+}$ by the final point $x_{\beta-}$, the measures $\mathbf{P}_{\alpha,x}$ by $\mathbf{P}^{\beta,x}$. To each proposition on the entrance space there corresponds to a dual proposition on the exit space. (We use the same numbers but with asterisks). For example, 4.1.A$'$ corresponds to the property:

4.1.A$'$*. For any $\mathbf{P} \in \mathcal{K}^*$, $\xi \in \mathcal{N}^{\cdot}$

$$\mathbf{P}(\xi \mid \mathcal{N}^{\cdot\beta-}) = \mathbf{P}^{\beta,\, x_\beta}\xi \quad \text{(a.s. } \mathbf{P}).$$

To (4.6) corresponds the formula

$$\mathbf{P}(A) = \int_{\mathcal{V}^*} \mathbf{P}^{\beta,\, v}(A)\, \mu(dv) \qquad (A \in \mathcal{N}^{\cdot}). \tag{4.6*}$$

4.8. Up to now we have assumed that the space of trajectories is the same as the space Ω of all maps from T to E. However, in the theory of Markov processes the set of trajectories is usually some subset Ω' of Ω. The most important cases are when E is a metric space and Ω' is the class Ω^c of all continuous maps from T to E or with the class Ω^d of all maps that are right-continuous and do not have discontinuities of the second kind.

The Kolmogorov σ-algebra $\mathcal{N}^{\cdot\prime}$ in Ω' coincides with the system of sets of the form $A \cap \Omega'$, where $A \in \mathcal{N}^{\cdot}$. We denote by $\mathcal{R}$ the collection of all probability measures $\mathbf{P}$ on $(\Omega', \mathcal{N}^{\cdot})$ with respect to which the outer measure of Ω' is equal to 1. It is not difficult to verify that the formula

$$\mathbf{P}'(A \cap \Omega') = \mathbf{P}(A) \tag{4.11}$$

establishes a one-to-one correspondence between $\mathbf{P} \in \mathcal{R}$ and all probability measures $\mathbf{P}'$ on $(\Omega', \mathcal{N}^{\cdot\prime})$ Here to the set $\mathcal{K} \cap \mathcal{R}$ there corresponds the class $\mathcal{K}'$ of processes $\mathbf{P}'$ with trajectories Ω' and transition function $p(s, x; t, \Gamma)$. If $\mathcal{K} \subseteq \mathcal{R}$, then (4.11) gives a one-to-one correspondence between $\mathcal{K}$ and $\mathcal{K}'$, moreover $\mathbf{P}(\xi) = \mathbf{P}'(\xi')$, where ξ' is the restriction of $\xi \in \mathcal{N}^{\cdot}$ to Ω'. It is clear that all the results of **4.1–4.6** are preserved under restriction of Ω to Ω'. When can it be asserted that $\mathcal{K} \subseteq \mathcal{R}$?

For each $\Lambda \subseteq T$ we put $\omega \in \Omega'(\Lambda)$, if there exists a trajectory $\omega' \in \Omega'$ that coincides with ω on Λ. It is clear that $\Omega'(\Lambda) \supseteq \Omega'(\bar{\Lambda})$ for $\bar{\Lambda} \supseteq \Lambda$. In particular, $\Omega'(\Lambda) \supseteq \Omega'(T) = \Omega'$.

We assume that the following conditions are satisfied.

4.8.A. For any countable $\Lambda \subseteq T$ the set $\Omega'(\Lambda)$ belongs to the σ-algebra $\mathcal{N}^{\cdot}_{\Lambda}$, generated by the sets $\{\omega : x_t \in \Gamma\}$ $(t \in \Lambda,\ \Gamma \in \mathcal{B})$.

4.8.B. $\Omega'(\Lambda_n) \downarrow \Omega'(\Lambda)$ if $\Lambda_n = \Lambda \cap [s_n, \infty)$ and $s_n \downarrow \alpha$.

4.8.C. For any $s \in T$, $x \in E$ and any countable $\Lambda \subseteq T \cap [s, \infty)$

$$\mathbf{P}_{s,x}(\Omega'(\Lambda)) = 1.$$

Let us prove that $\mathcal{K} \subseteq \mathcal{R}$. We note that for each $A \in \mathcal{N}^{\cdot}$ there is a countable set $\Lambda(A) \subseteq T$ such that $A \in \mathcal{N}^{\cdot}_{\Lambda(A)}$ (because the events A with the indicated property form a σ-algebra containing all the sets $\{\omega : x_t \in \Gamma\}$ ($t \in T$, $\Gamma \in \mathscr{B}$)). We have to prove that if $\mathbf{P} \in \mathscr{K}$ and $A \supseteq \Omega'$, then $\mathbf{P}(A) = 1$. Let $\Lambda = \Lambda(A)$. Then, as is easily seen, $A \supseteq \Omega'(\Lambda)$. Therefore it is sufficient to verify that $\mathbf{P}[\Omega'(\Lambda)] = 1$. By 4.8.A we have $\Omega'(\Lambda_n) \in \mathscr{N}_{\Lambda_n} \subseteq \mathscr{N}^{s_n}$, and from 1.6.A'

$$\mathbf{P}(\Omega'(\Lambda_n)) = \mathbf{P}\mathbf{P}_{s_n, x_{s_n}}(\Omega'(\Lambda_n)).$$

According to 4.8.C the right-hand side is equal to 1 and by 4.8.B

$$\mathbf{P}(\Omega'(\Lambda)) = \lim \mathbf{P}(\Omega'(\Lambda_n)) = 1.$$

Conditions 4.8.A–4.8.C are satisfied if $\Omega' = \Omega$, because then $\Omega'(\Lambda) = \Omega$ for all Λ. If Ω' is equal to Ω^c or Ω^d, then the requirements 4.8.A–4.8.B are satisfied and there are fairly broad conditions on a transition function under which 4.8.C is valid (see [2], Chapter 6, § 3).

One more example. Let $T = (r, \infty)$ and let Ω' be the set of trajectories that are absorbed at a fixed point b (this means that for some $t_0 \in (r, \infty]$, depending on ω, $\omega(t) = b$ for $t \geqslant t_0$ and $\omega(t) \neq b$ for $t < t_0$). For any countable everywhere dense set $\Lambda \subseteq T$ we then have $\Omega'(\Lambda) = \Omega'_1 \cap \Omega'_2 \cap \Omega'_3$, where

$$\Omega'_1 = \bigcup_{s \in \Lambda} \{\omega(s) \neq b\},$$

$$\Omega'_2 = \bigcap_{s \in \Lambda} [\omega(s) = b \text{ or } \omega(t) \neq b \quad \text{for some} \quad t \in \Lambda \cap (s, \infty)],$$

$$\Omega'_3 = \bigcap_{s \in \Lambda} [\omega(s) \neq b \text{ or } \omega(t) = b \quad \text{for all} \quad t \in \Lambda \cap (s, \infty)].$$

It is clear that 4.8.A and 4.8.B are satisfied. Condition 4.8.C is also satisfied if $p(s, b; t, b) = 1$ ($s < t \in T$), $\lim_{t \downarrow s} p(s, x; t, E) = 1$ for $x \neq b$.

§ 5. Markov processes with random birth and death times

5.1. To avoid a cumbersome exposition we assume that T coincides with the real line $(-\infty, +\infty)$. (The necessary changes for an arbitrary T will be mentioned at the end of the section.) We denote by $(E_a^b, \mathscr{B}_a^b)$ the measurable space obtained by adjoining[1] a pair of points a and b to the state space $(E, \mathscr{B})$. Let Ω be some set of maps from T to E_a^b, $\mathbf{P}$ a Markov process on $(E_a^b, \mathscr{B}_a^b)$ with Ω as trajectories, $\alpha < \beta$ functions in Ω with values in $[-\infty, +\infty]$. We say that $\mathbf{P}$ is a *process with birth time* α *and death time* β if for all $\omega \in \Omega$

$$\begin{cases} \omega(t) = a & \text{for} \quad t \leqslant \alpha(\omega), \\ \omega(t) \in E & \text{for} \quad \alpha(\omega) < t < \beta(\omega), \\ \omega(t) = b & \text{for} \quad t \geqslant \beta(\omega). \end{cases} \tag{5.1}$$

[1] $\mathscr{B}_a^b$ is the minimal σ-algebra on E_a^b containing $\mathscr{B}$ and the sets $\{a\}$, $\{b\}$. Obviously $(E_a^b, \mathscr{B}_a^b)$ is, together with $(E, \mathscr{B})$ a standard space.

We assume that Ω contains all maps ω for which (5.1) is satisfied.

5.2. Apart from $(E_a^b, \mathscr{B}_a^b)$ we also need the measurable spaces $(E_a, \mathscr{B}_a)$ and $(E^b, \mathscr{B}^b)$, which are obtained from $(E, \mathscr{B})$ by adjoining a and b, respectively. Transition functions are considered in $(E^b, \mathscr{B}^b)$, and cotransition functions in $(E_a, \mathscr{B}_a)$. In accordance with (5.1) we assume that a transition function satisfies the conditions

$$(5.2) \qquad \begin{cases} p(s, b; t, b) = 1 & (s < t), \\ \lim_{t \downarrow s} p(s, x; t, E) = 1 & \text{for } x \in E, \end{cases}$$

and a cotransition function the conditions

$$(5.3) \qquad \begin{cases} p^*(s, a; t, a) = 1 & (s > t), \\ \lim_{t \uparrow s} p^*(s, x; t, E) = 1 & \text{for } x \in E. \end{cases}$$

Under these conditions transition and cotransition functions[1] are uniquely recoverable from their values in $(E, \mathscr{B})$.

A Markov process $\mathbf{P}$ corresponds[2] to a transition function $p(s, x; t, \Gamma)$ if for all $s < t$, $\Gamma \in \mathscr{B}^b$

$$(5.4) \qquad \mathbf{P}\{x_t \in \Gamma \mid x_s\} = p(s, x_s; t, \Gamma) \quad (\text{a.s. } \mathbf{P}, \alpha < s).$$

Our task is to describe the class $\mathscr{K}$ of all Markov processes corresponding to the transition function $p(s, x; t, \Gamma)$. Here randomness of the birth time brings in an essentially new element. On the other hand, the randomness of the death time has no significance: nothing changes if b is included in the state space and the time of death is assumed to be ∞. (Of course, the roles of births and deaths are interchanged in the study of the class $\mathscr{K}^*$ of Markov processes with cotransition function $p^*(s, x; t, \Gamma)$.)

A process $\mathbf{P}$ belongs to $\mathscr{K}$ if and only if it satisfies any one of the two equivalent conditions:

5.2.A. For any s and for all $\xi \in \mathscr{N}^s$

$$\mathbf{P}(\xi \mid \mathscr{N}_s) = \mathbf{P}_{s, x_s}\eta \quad (\text{a.s. } \mathbf{P}, \alpha < s).$$

5.2.A′. For any s and for all $\xi \in \mathscr{N}^s$, $\eta \in \mathscr{N}_s$

$$\mathbf{P}(\eta\chi_{\alpha<s}\xi) = \mathbf{P}(\eta\chi_{\alpha<s}\mathbf{P}_{s, x_s}\xi).$$

(The measures $\mathbf{P}_{s,x}$ $(s \in T, x \in E^b)$ on the σ-algebra $\mathscr{N}^s$ are constructed starting from the transition function $p(s, x; t, \Gamma)$.)

5.3. One of the possible approaches to the investigation of $\mathscr{K}$ is to begin with processes whose birth times are not random.

We consider the class $\mathscr{K}(r, \infty)$ of all Markov processes with transition

---

[1] Transition and cotransition functions, if considered only on $(E, \mathscr{B})$, possess all the properties listed in **1.2** except one: instead of $p(s, x; t, E) = 1$ we have $p(s, x; t, E) \uparrow 1$ for $t \downarrow s$.

[2] By (5.2) we know that (5.4) is automatically satisfied (a.s. $\mathbf{P}$, $x_s = b$). Therefore it is sufficient to demand that it is satisfied (a.s. $\mathbf{P}$, $x_s \in E$).

function $p(s, x; t, \Gamma)$, defined on the time interval (r, ∞) and in the state space $(E^b, \mathscr{B}^b)$. Since p satisfies (5.2), by 4.8 $\mathscr{K}(r, \infty)$ does not change if we contract the space of trajectories to the set of trajectories that are absorbed at b. If each of these trajectories is continued by the formula $\omega(t) = a$ for $t \leqslant r$ we obtain a process of $\mathscr{K}$. We put $\mathbf{P} \in \mathscr{K}^r$, if $\mathbf{P} \in \mathscr{K}$ and $\mathbf{P}\{\alpha = r\} = 1$. It is easy to see that the construction described defines a one to-one correspondence between $\mathscr{K}(r, \infty)$ and $\mathscr{K}^r$.

It is natural to assume that any process of $\mathscr{K}$ is representable in the form

$$\mathbf{P} = \int_{-\infty}^{+\infty} \mathbf{P}^r \, dF(r), \tag{5.5}$$

where $F(r) = \mathbf{P}\{\alpha < r\}$ and $\mathbf{P}^r \in \mathscr{K}^r$. This formula could have been proved directly, and then we could make use of the results of the preceding sections concerning the construction of $\mathscr{K}(r, \infty)$. However, we prefer to construct the entrance space for $\mathscr{K}$ directly by the same scheme as for processes with non-random birth times.

5.4. The first step is the introduction of the *initial* σ*-algebra* $\mathscr{N}_{\alpha+}$, which is now defined by the condition: $A \in \mathscr{N}_{\alpha+}$ if $\{A, \alpha < t\} \in \mathscr{N}_t$ for all t. (We note that $\{\alpha < t\} = \{x_t \in E^b\} \in \mathscr{N}_t$.)

The second step is the derivation of a formula replacing (3.2).

LEMMA 5.1 *Let Λ be a countable everywhere dense subset of T and let $\Lambda_n = \{t_1^n < t_2^n < \ldots < t_n^n\} \uparrow \Lambda$. We put*

$$\alpha_n = \begin{cases} t_1^n & \text{for } \alpha < t_1^n, \\ t_i^n & \text{for } t_{i-1}^n \leqslant \alpha < t_i^n, \\ +\infty & \text{for } t_n^n \leqslant \alpha. \end{cases} \tag{5.6}$$

If $\mathbf{P} \in \mathscr{K}$, then for any u, $f \in \mathscr{B}^b$

$$\mathbf{P}\{f(x_u) \mid \mathscr{N}_{\alpha+}\} = \lim_{n\to\infty} \mathbf{P}_{\alpha_n, x_{\alpha_n}} f(x_u) \quad (\text{a.s. } \mathbf{P},\ \alpha < u). \tag{5.7}$$

PROOF. Put $A \in \mathscr{N}_{\alpha_n}$ if $\{A, \alpha_n \leqslant t\} \in \mathscr{N}_t$ for all t. It is easy to see that $\mathscr{N}_{\alpha_n} \downarrow \mathscr{N}_{\alpha+}$. Therefore

$$\mathbf{P}\{f(x_u) \mid \mathscr{N}_{\alpha+}\} = \lim_{n\to\infty} \mathbf{P}\{f(x_u) \mid \mathscr{N}_{\alpha_n}\} \qquad (\text{a.s. } \mathbf{P}). \tag{5.8}$$

If $\xi \in \mathscr{N}_{\alpha_n}$, then by 5.2.A′ for any $t < u$

$$\mathbf{P}(\xi \chi_{\alpha_n = t} f(x_u)) = \mathbf{P}(\xi \chi_{\alpha_n = t} \mathbf{P}_{t, x_t} f(x_u)) = \mathbf{P}(\xi \chi_{\alpha_n = t} \mathbf{P}_{\alpha_n, x_{\alpha_n}} f(x_u)).$$

Consequently, $\mathbf{P}(\xi \chi_{\alpha_n \leqslant u} f(x_u)) = \mathbf{P}(\xi \chi_{\alpha_n \leqslant u} \mathbf{P}_{\alpha_n, x_{\alpha_n}} f(x_u))$ and

$$\mathbf{P}(f(x_u) \mid \mathscr{N}_{\alpha_n}) = \mathbf{P}_{\alpha_n, x_{\alpha_n}} f(x_u) \qquad (\text{a.s. } \mathbf{P},\ \alpha_n \leqslant u). \tag{5.9}$$

(5.7) follows from (5.8) and (5.9).

5.5. THEOREM 5.1. *There exists a $(\mathscr{K}, \mathscr{N}_{\alpha+})$-kernel $\mathbf{P}^\omega$.*

PROOF. Let q be any coordinate in the standard space $(E^b, \mathscr{B}^b)$, and Λ the set of Lemma 5.1. We denote by W the system of functions

$f(x_u)$ $(u \in \Lambda, f \in W_q)$ [see (3.4)]. Put

$$\nu_t(\Gamma) = \mathbf{P}\{x_t \in \Gamma\} \qquad (\Gamma \in \mathscr{B}^b,\ t \in \Lambda).$$

By 5.2.A′, for any $t_1 < \ldots < t_n \in \Lambda$, $\Gamma_1, \ldots, \Gamma_n \in \mathscr{B}^b$

$$(5.10) \qquad \mathbf{P}\{x_{t_1} \in \Gamma_1, \ldots, x_{t_n} \in \Gamma_n\} = \lim_{s \uparrow t_1,\, s \in \Lambda} \mathbf{P}\{\alpha < s, x_{t_1} \in \Gamma_1, \ldots, x_{t_n} \in \Gamma_n\} = \lim_{s \uparrow t,\, s \in \Lambda} \int_{E^b} \nu_s(dx)\, p(s, x; t_1, \Gamma_1, \ldots, t_n, \Gamma_n).$$

Therefore the measure $\mathbf{P} \in \mathscr{K}$ is uniquely recoverable from $\mathbf{P}\xi$ $(\xi \in W)$. In other words, W separates measures in $\mathscr{K}$.

According to (5.7), for any f and u the set $\{\omega\colon \alpha(\omega) < u$ and $\lim \mathbf{P}_{\alpha_n, x_{\alpha_n}} f(x_u)$ does not exist$\}$ has measure zero. Form the sum of such sets and denote its complement by Ω_0. It is clear that $\Omega_0 \in \mathscr{N}_{\alpha+}$, $\mathbf{P}(\Omega_0) = 1$ $(\mathbf{P} \in \mathscr{K})$ and for $\omega \in \Omega_0$ the limit on the right-hand side of (5.7) exists for all $u > \alpha(\omega)$. Let $\nu_u^{\omega,n}(\Gamma) = \mathbf{P}_{\alpha_n, x_{\alpha_n}}\{x_u \in \Gamma\}$. If $\omega \in \Omega_0$ and $u > \alpha(\omega)$, then for any $f \in W_q$ the limit of $\nu_u^{\omega,n}(f)$ exists for $n \to \infty$. By 3.1.B there is a probability measure ν_u^ω on $(E^b, \mathscr{B}^b)$, such that $\nu_u^{\omega,n}(f) \to \nu_u^\omega(f)$ $(f \in W_q)$. As in the proof of Theorem 3.1 we note that for any $u \in \Lambda$ and any bounded $\mathscr{B}^b$-measurable function f the function $\nu_u^\omega(f)$ is $\mathscr{N}_{\alpha+}$-measurable and

$$(5.11) \qquad \nu_u^\omega(f) = \mathbf{P}\{f(x_u) \mid \mathscr{N}_{\alpha+}\} \qquad (\text{a.s. } \mathbf{P},\ \alpha < u)$$

If $\eta \in \mathscr{N}_\alpha$, $\xi \in \mathscr{N}^s$, then $\eta\chi_{\alpha<s} \in \mathscr{N}_s$ and by 5.2.A′ $\mathbf{P}(\eta\chi_{\alpha<s}\xi) = \mathbf{P}(\eta\chi_{\alpha<s}\mathbf{P}_{s, x_s}\xi)$. Hence

$$(5.12) \qquad \mathbf{P}(\xi \mid \mathscr{N}_{\alpha+}) = \mathbf{P}(\mathbf{P}_{s, x_s}\xi \mid \mathscr{N}_{\alpha+}) \qquad (\text{a.s. } \mathbf{P},\ \alpha < s)$$

In particular, for $\xi = f(x_t)$ $(t > s)$

$$(5.13) \qquad \mathbf{P}\{f(x_t) \mid \mathscr{N}_{\alpha+}\} = \mathbf{P}\{\mathbf{P}_{tf}^s(x_s) \mid \mathscr{N}_{\alpha+}\} \qquad (\text{a.s. } \mathbf{P},\ \alpha < s)$$

From (5.11) and (5.13) it follows that for $f \in W_q$, $s < t \in \Lambda$ $\mathscr{K}$-almost surely on the set $\{\alpha < s\}$

$$(5.14) \qquad \nu_t^\omega(f) = \nu_s^\omega(\mathbf{P}_{tf}^s).$$

Put $\omega \in \Omega_1$ if (5.14) is satisfied for all $t > s > \alpha(\omega)$ $(t, s \in \Lambda)$ and all $f \in W_q$. It is clear that $\Omega_1 \in \mathscr{N}_{\alpha+}$ and $\mathbf{P}(\Omega_1) = 1$ for all $\mathbf{P} \in \mathscr{K}$. Fix $\omega \in \Omega_1$ and assume that $\alpha(\omega) = r$, $\Lambda_r = \Lambda \cap (r, \infty)$. The set of measures ν_t^ω $(t \in \Lambda_r)$ in $(E^b, \mathscr{B}^b)$ satisfies (3.6), and by Lemma 3.1 there is a process $\mathbf{P}$ of the class $\mathscr{K}(r, \infty)$ for which

$$(5.15) \qquad \mathbf{P}\{x_t \in \Gamma\} = \nu_t^\omega(\Gamma) \qquad (\Gamma \in \mathscr{B}^b,\ t \in \Lambda_r).$$

Denote by $\mathbf{P}^\omega$ the corresponding process of $\mathscr{K}^r$. For $\omega \in \Omega_1$ we put $\mathbf{P}^\omega = \mathbf{P}^{\omega_1}$, where ω_1 is any element of Ω. By (5.15) and (5.11)

$$(5.16) \qquad \mathbf{P}^\omega\{x_t \in \Gamma\} = \nu_t^\omega(\Gamma) = \mathbf{P}\{x_t \in \Gamma \mid \mathscr{N}_{\alpha+}\} \qquad (\text{a.s. } \mathbf{P},\ \alpha < t)$$

From (5.12), (5.16) and (3.10) we have for any $t_1 < \ldots < t_n \in T, \Gamma_1, \ldots, \Gamma_n \in \mathscr{B}^n$

$$\mathbf{P}\{x_{t_1} \in \Gamma_1, \ldots, x_{t_n} \in \Gamma_n \mid \mathscr{N}_{\alpha+}\} = \lim_{s \uparrow t_1} \mathbf{P}\{\alpha < s, x_{t_1} \in \Gamma_1, \ldots, x_{t_n} \in \Gamma_n \mid \mathscr{N}_{\alpha+}\} =$$
$$= \lim_{s \uparrow t_1} \mathbf{P}(p(s, x_s; t_1, \Gamma_1, \ldots, t_n, \Gamma_n) \mid \mathscr{N}_{\alpha+}) =$$
$$= \lim_{s \uparrow t_1} \mathbf{P}^{\omega}(p(s, x_s; t_1, \Gamma_1, \ldots, t_n, \Gamma_n)) = \mathbf{P}^{\omega}\{x_{t_1} \in \Gamma_1, \ldots, x_{t_n} \in \Gamma_n\}.$$

Hence it follows that $\mathbf{P}^{\omega}$ has the properties 2.1.A–2.1.B.

5.6. The entrance space $(\mathscr{V}, \mathscr{B}_{\mathscr{V}^{\circ}})$ is defined as in §4. We also preserve our notation $x_{\alpha+}(\omega)$ for the beginning of ω. The space $(\mathscr{V}, \mathscr{B}_{\mathscr{V}^{\circ}})$ is isomorphic to the space $(\mathscr{K}_e, \mathscr{C}_{\mathscr{K}_e})$ of $\mathscr{N}_{\alpha+}$-ergodic measures of $\mathscr{K}$. Put $\mathscr{K}_e^r = \mathscr{K}_e \cap \mathscr{K}^r$ (it is easy to see that $\mathscr{K}_e^r$ coincides with the set of all $\mathscr{N}_{r+}$-ergodic measures of $\mathscr{K}^r$). Since $\alpha \in \mathscr{N}_{\alpha+}$, we see that α = const (a.s. $\mathbf{P}$) for each $\mathbf{P} \in \mathscr{K}_e$. Consequently, $\mathscr{K}_e = \bigcup_r \mathscr{K}_e^r$. We consider the corresponding decomposition $\mathscr{V} = \bigcup_r \mathscr{V}^r$ of $\mathscr{V}$. To a point $v \in \mathscr{V}^r$ there corresponds a measure concentrated on the trajectories with birth time r. We denote it by $\mathbf{P}_{r,v}$. (Thus, the measures $\mathbf{P}_{r,v}$ are defined for $r \in [-\infty, +\infty), v \in \mathscr{V}^r$). Put $\alpha(v) = r$ for $v \in \mathscr{V}^r$, so that $\alpha[x_{\alpha+}(\omega)] = \alpha(\omega)$. To each $v \in \mathscr{V}$ there corresponds a measure $\mathbf{P}_{\alpha(v), v}$ or, briefly, $\mathbf{P}_{\alpha,v}$. In this notation all the formulae of §4 are valid for random α. It only remains to prove that $(\mathscr{V}, \mathscr{B}_{\mathscr{V}^{\circ}})$ is standard. We shall prove this elsewhere.

We note that $\alpha(v)$ is $\mathscr{B}_{\mathscr{V}^{\circ}}$-measurable. In fact, for any r

$$\{v: \alpha(v) < r\} = \{v: \mathbf{P}_{\alpha(v), v}(\alpha < r) = 1\} \in \mathscr{B}_{\mathscr{V}^{\circ}}.$$

Since $\alpha(v)$ is measurable and $(\mathscr{V}^{\circ}, \mathscr{B}_{\mathscr{V}^{\circ}})$ is standard, for each probability measure μ on $(\mathscr{V}^{\circ}, \mathscr{B}_{\mathscr{V}^{\circ}})$ there is a conditional distribution $\mu_r(\Gamma)$ relative to $\alpha(v)$ (see, for example, [1], Chapter 1, §9). This is a Borel function of r and is a measure, concentrated on $\mathscr{V}^{\circ r}$, relative to Γ; moreover, for any $\Gamma \in \mathscr{B}_{\mathscr{V}^{\circ}}$

$$\mu(\Gamma) = \int_{-\infty}^{+\infty} \mu_r(\Gamma)\, dF(r),$$

where $F(r) = \mu\{v: \alpha(v) < r\}$. Putting $\mathbf{P}^r = \int_{\mathscr{V}^{\circ}} \mathbf{P}_{r,v} \mu_r(dv)$, we arrive at (5.5).

5.7. We turn now to the case when T is arbitrary. Changes are required if some points of T are isolated to the left or to the right (for example, when T is the set of integers).

We denote by $T_+(T_-)$ the set of t in T that are isolated on the right (left). These sets are at most countable. We keep the conditions (5.1) for t other than α and β, and for α and β we replace them by the following:

$$\omega(\alpha) \in E \quad \text{for } \alpha \in T_+, \qquad \omega(\alpha) = a \text{ for } \alpha \in T \setminus T_+;$$
$$\omega(\beta) \in E \text{ for } \beta \in T_-, \qquad \omega(\beta) = b \text{ for } \beta \in T \setminus T_-.$$

Note that always $\alpha = \inf \{t : \omega(t) \in E\}$, $\beta = \sup \{t : \omega(t) \in E\}$.

In (5.4), 5.2.A and 5.2.A′ we have to replace the set $\{\alpha < s\}$ by $\{x_s \in E^b\}$; in (5.7) to replace $\{\alpha < u\}$ by $\{\alpha < u, \alpha \in T - T_+\}$. Formula (5.7) is supplemented by

$$\mathbf{P}\{f(x_u) \mid \mathcal{N}_{\alpha+}\} = \mathbf{P}_{\alpha, x_\alpha} f(x_u) \quad (\text{a.s. } \mathbf{P}, \{\alpha < u, \alpha \in T_+\}).$$

We assume that the set Λ of Lemma 5.1 contains T_-. Formula (5.10) holds for $s \notin T_-$; if $s \in T_-$, then its right-hand side is replaced by

$$\int_{\Gamma_1} \nu_{t_1}(dx)\, p(t_1, x; t_2, \Gamma_2, \ldots, t_n, \Gamma_n).$$

For the proof of Theorem 5.1 we have to replace in the definition of Ω_0 the set $\{\alpha < u\}$ by the set $\{\alpha < u, \alpha \in T - T_+\}$ and to put $\nu_u^\omega(\Gamma) = \mathbf{P}_{\alpha, x_\alpha}\{x_u \in \Gamma\}$ for $\alpha \in T_+$. Slight changes that must be made in the subsequent part of the proof are obvious.

To what we have said in **5.6** we only have to add that for $\alpha \in T_+$ the initial point $x_{\alpha+}$ coincides with x_α.

References

[1] J. L. Doob, Stochastic processes, Wiley, New York, 1953. Translation: *Veroyatnostnye protsessy*, Inost. lit, Moscow, 1956.

[2] E. B. Dynkin, *Osnovaniya teorii markovskikh protsessov,* Fizmatgiz, Moscow, 1959. Translation: Die Grundlagen der Theorie der Markoffschen Prozesse, Springer Verlag, Berlin-Heidelberg-Göthingen, 1962.

[3] K. Kuratowski, *Topologie*, 4th ed. 2 vols., Polish Acad. Sci., Warsaw, 1958. English Translation: Topology, 2 vols. Academic Press, New York, 1967–9. Russian Translation: Topologiya vol. 1, Izdat. Mir, Moscow 1966.

[4] J. Neveu, Bases mathématiques du calcul des probabilités, Masson et Cie, Paris, 1964. English Translation: Mathematical Foundations of the Calculus of Probability, Holden-Day, San Francisco, 1965. Translation: *Matematicheskie osnovy teoriya veroyatnostei,* Izdat. "Mir", Moscow, 1969.

Received by the Editors, 26 February, 1971

Translated by D. Newton

INTEGRAL REPRESENTATION OF EXCESSIVE MEASURES AND EXCESSIVE FUNCTIONS

One of the central results of classical potential theory is the theorem on the representation of an arbitrary non-negative superharmonic function in the form of a sum of a Green's potential and a Poisson integral. We obtain similar integral representations for the excessive measures and functions connected with an arbitrary Markov transition function. Many authors have studied the homogeneous excessive measures connected with a homogeneous transition function. We begin with the inhomogeneous case and then reduce the homogeneous case to it. The method proposed gives a considerable gain in generality.

The investigation is carried out in the language of convex measurable spaces and in contrast to previous papers no topological arguments are used. Our basis are the results obtained in [3] (also without topology) on the integral representation of Markov processes with a given transition function. For the reduction of the homogeneous case to the inhomogeneous we use a theorem from the theory of dynamical systems due to Yu. I. Kifer and S. A. Pirogov (see the Appendix at the end of this paper).

Contents

§ 1. Plan of the paper. Discussion of results

1.1. We start from the idea of a *transition function*. From an intuitive point of view $p(s, x; t, \Gamma)$ gives the probability of being at a time t in a set Γ if you are at the time s at the point x. Here x is an element of some set E, Γ belongs to a fixed σ-algebra $\mathscr{B}$ of subsets of E, and s and t are chosen from some set of numbers T. The value of $p(s, x; t, \Gamma)$ is not defined for all pairs s, t. We usually consider "forward" transition functions defined for pairs $s < t$. We shall also consider "backward" transition functions defined for $s > t$. Side by side with these rather clumsy expressions we use the names "direct and reverse transition functions".[1] The meaning of expressions such as "the direction of a transition function", "similarly (or oppositely) directed transition functions" needs no explanation. The statements of any proposition about transition functions of the opposite direction are obtained from each other by replacing all inequalities between elements of T by the opposite inequalities. There is no point in duplicating these statements, and we shall only speak of direct transition functions provided that the discussion does not simultaneously contain transition functions in both directions.

1.2. A set E is said to have a *measurable structure* if there is given some σ-algebra of subsets in E. The elements of this σ-algebra are called measurable and E with a fixed measurable structure is called a *measurable space*. Each system of sets (or functions) in E generates a measurable structure: it is characterized as the minimal σ-algebra containing all the given sets (or with respect to which all the given functions are measurable). Two measurable spaces are called *isomorphic* if there is a one-to-one mapping between them that preserves measurability of sets. A measurable space is called *standard* if it is isomorphic to a Borel subset X of a complete separable metric space. (Measurability in X is relative to the Borel subsets).

We only consider transition functions on a standard space E. Furthermore, to avoid a cumbersome exposition we shall assume that T coincides with the real line.

A formal definition of a (direct) transition function consists of the following two conditions:

1.2.A. $p(s, x; t, \Gamma)$ is measurable in x and a measure with respect to Γ, and $p(s, x; t, E) \leqslant 1$.

1.2.B. For any $x \in E, \Gamma \in \mathscr{B}, s < t < u \in T$

$$\int_E p(s, x; t, dy)\, p(t, y; u, \Gamma) = p(s, x; u, \Gamma).$$

[1] In [3] the term "transition function" refers only to forward transition functions, and backward transition functions were called cotransition functions.

We complete the definition of $p(s, x; t, \Gamma)$ by taking it to be zero for $s \geqslant t$.

Connected with each transition function p there are families of operators P_t^s acting according to the formulae

$$\nu P_t^s(\Gamma) = \int_E \nu(dx)\, p(s, x; t, \Gamma), \tag{1.1}$$

$$P_t^s f(x) = \int_E p(s, x; t, dy)\, f(y). \tag{1.2}$$

The first family maps the class of all finite measures on E to itself (and also the class of measures that are representable as a sum of countably many finite measures); the second acts on the set of all non-negative measurable functions on E. Condition **1.2.B.** is equivalent to: $P_t^s P_u^t = P_u^s$ for $s < t < u$ (for $s \geqslant t$, $P_t^s = 0$).

A transition function is called *homogeneous* if for any δ $p(s, x; t, \Gamma) = p(s + \delta, x; t + \delta, \Gamma)$. For a homogeneous transition function put $p(t, x, \Gamma) = p(0, x; t, \Gamma)$, $P_t = P_t^0$. Clearly $p(s, x; t, \Gamma) = p(t - s, x, \Gamma)$, $P_t^s = P_{t-s}$. The operators P_t form a one-parameter semigroup.

1.3. A non-negative function $h^t(x)$ $(t \in T, x \in E)$ that is $\mathscr{B}$-measurable with respect to x for each $t \in T$ is called *p-excessive* if:

1.3.A. $P_t^s h^t(x) \leqslant h^s(x)$ $(x \in E)$,

1.3.B. $P_t^s h^t(x) \to h^s(x)$ for $t \downarrow s$ $(x \in E)$.

A σ-finite measure ν_t on E, depending on a parameter $t \in T$, is called *p-excessive* if:

1.3.A′ $\nu_s P_t^s(\Gamma) \leqslant \nu_t(\Gamma)$ $(\Gamma \in \mathscr{B})$,

1.3.B′ $\nu_s P_t^s(\Gamma) \to \nu_t(\Gamma)$ for $s \uparrow t$ $(\Gamma \in \mathscr{B})$.

p-excessive functions and measures are called *invariant* if we have equality in condition **1.3.A.** (or **1.3.A′**) (condition **1.3.B.** (or **1.3.B′**) is then automatically satisfied). We call *p-null-execessive* those p-exessive functions (measures) for which $\lim_{t \to \infty} P_t^s h^t(x) = 0$ for $h^s(x) < \infty$ (respectively, $\lim_{s \to -\infty} \nu_s P_t^s(\Gamma) = 0$ for $\nu_t(\Gamma) < \infty$).

p-excessive function and measures that do not depend on t are called *homogeneous*. These are interesting only when the transition function p is homogeneous. In this case **1.3.A.** – **1.3.B.** and **1.3.A′** – **1.3.B′** take the following form:

1.3.α. $P_t h(x) \leqslant h(x)$ $(x \in E)$.

1.3.β. $P_t h(x) \to h(x)$ for $t \downarrow 0$ $(x \in E)$.

1.3.α′. $\nu P_t(\Gamma) \leqslant \nu(\Gamma)$ $(\Gamma \in \mathscr{B})$.

1.3.β′. $\nu P_t(\Gamma) \to \nu(\Gamma)$ for $t \downarrow 0$ $(\Gamma \in \mathscr{B})$.

Fix a transition function p. Let ν be a p-excessive measure, h a p-excessive function, and let h^t be summable with respect to ν_t for any t. For any finite set $\Lambda = \{t_0 < t_1 < \ldots < t_n\}$ put[1]

$$(1.3) \quad \{\nu, h\}_\Lambda = \nu_{t_0} h^{t_0} + \sum_{k=1}^{n} [\nu_{t_k} h^{t_k} - \nu_{t_{k-1}} P_{t_k}^{t_{k-1}} h^{t_k}] = \\ = \sum_{k=1} [\nu_{t_{k-1}} h^{t_{k-1}} - \nu_{t_{k-1}} P_{t_k}^{t_{k-1}} h^{t_k}] + \nu_{t_n} h^{t_n}.$$

We denote the least upper bound of this expression over all Λ by $\{\nu, h\}$.
If ν or h is invariant, then

$$(1.3') \quad \{\nu, h\} = \sup_{t \in T} \nu_t h^t.$$

A wide class of p-null-excessive measures and functions can be constructed by the formulae

$$(1.4) \quad \nu_t(\Gamma) = \int_{(-\infty, t) \times E} \gamma(ds, dx)\, p(s, x; t, \Gamma),$$

$$(1.5) \quad h^s(x) = \int_{(s, \infty) \times E} p(s, x; t, dy)\, l^t(y)\, d\lambda(t) = \int_{(s, \infty)} P_t^s l^t(x)\, d\lambda(t).$$

Here γ is a measure on $\mathscr{B}_T \times \mathscr{B}$ [$\mathscr{B}_T$ denotes the σ-algebra of Borel sets of T], $\lambda(t)$ is a non-decreasing function on T, $l^t(x)$ is a non-negative measurable function on $T \times E$. (1.4) makes sense under the assumption that for any $t \in T$, $\Gamma \in \mathscr{B}$ $p(-, -; t, \Gamma)$ is measurable relative to the completion of $\mathscr{B}_T \times \mathscr{B}$ with respect to γ. (1.5) makes sense if for all $s \in T$, $x \in E$, $\Gamma \in \mathscr{B}$ the function $p(s, x; -, \Gamma)$ is measurable relative to the completion of $\mathscr{B}_T$ with respect to $d\lambda(t)$. By 1.2.B, for all $s < t$

$$(1.6) \quad \nu_t(\Gamma) = \nu_s P_t^s(\Gamma) + \int_{[s, t) \times E} \gamma(du, dx)\, p(u, x; t, \Gamma),$$

$$h^s(x) = P_t^s h^t(x) + \int_{(s, t]} P_u^s l^u(x)\, d\lambda(u).$$

If the integrals on the right-hand sides of these equalities converge for any finite $s < t$, then ν is a p-null-excessive measure and h is a p-null-excessive function.

It is not difficult to show that if ν is given by (1.4), then

$$(1.8) \quad \{\nu, h\} = \int_{T \times E} \gamma(ds, dx)\, h^s(x),$$

---

[1] If h is a non-negative function and ν a measure, then νh (or $\nu(h)$) denotes the integral of h over ν. The expression $\nu_s P_t^s h^t$ may be understood as the integral of $P_t^s h^t$ over ν_s or the integral of h^t over $\nu_s P_t^s$ (the two integrals are clearly equal).

and if h is given by (1.5), then

$$\{\nu, h\} = \int_T \nu_u l^u \, d\lambda(u). \tag{1.9}$$

Indeed, comparing (1.3) and (1.6) we observe that

$$\{\nu, h\}_\Lambda = \int_{(-\infty, t_n)\times E} \gamma(ds, dx) \, P^s_{\varphi(s)} h^{\varphi(s)}(x), \tag{1.10}$$

where $\varphi(s) = t_0$ for $s < t_0$, $\varphi(s) = t_k$ for $t_{k-1} \leqslant s < t_k$ $(k = 1, 2, \ldots, n)$. By **1.3.A.** the right-hand side of (1.10) does not exceed the right-hand side of (1.8), and by **1.3.B.** it converges to the right-hand side of (1.8) when Λ ranges over an expanding sequence of finite sets whose union is dense in T. Hence (1.8) is proved. (In passing we have also proved the measurability of $h^s(x)$ in s, x). (1.9) is proved similarly.

If the transition function p is homogeneous, then (1.4) and (1.5) for $\gamma(ds, dx) = ds\gamma(dx)$, $l^t = l$, $\lambda(t) = t$ define homogeneous p-excessive measures and functions

$$\nu(\Gamma) = \int_0^\infty du \int_E \gamma(dx) \, p(u, x, \Gamma) = \int_E \gamma(dx) \, g(x, \Gamma) = \gamma G(\Gamma) \tag{1.11}$$

$$h(x) = \int_0^\infty du \int_E p(u, x, dy) \, l(y) = \int_E g(x, dy) \, l(y) = Gl(x), \tag{1.12}$$

where

$$g(x, \Gamma) = \int_0^\infty p(u, x, \Gamma) \, du. \tag{1.13}$$

1.4. One of the central results of this paper is a theorem that associates a Markov process with a transition function p and each pair: a p-excessive measure ν and a p-excessive function h connected by the relation $\{\nu, h\} = 1$.

The Markov processes in question are those with random birth and death times. We recall the corresponding definitions (see [3], 5.1. – 5.2.). Consider the measure space $(E^b_a, \mathscr{B}^b_a)$ obtained from $(E, \mathscr{B})$ by adding two points a and b. A mapping ω of T to E^b_a is called a *trajectory* if for some $-\infty \leqslant \alpha < \beta \leqslant +\infty$ (depending on ω) $\omega(t) = a$ for $t \leqslant \alpha$, $\omega(t) \in E$ for $\alpha < t < \beta$, $\omega(t) = b$ for $t \geqslant \beta$. Here α is called the *birth time* and β the *death time.* The set of all trajectories is denoted by Ω. The image $\omega(t)$ of t under the mapping ω is also denoted by $x_t(\omega)$.

We introduce in Ω the measurable structure generated by the sets $A_{t,\Gamma} = \{\omega : x_t(\omega) \in \Gamma\}$ $(t \in T, \Gamma \in \mathscr{B})$. The σ-algebras $\mathscr{N}_t$ generated by the sets $A_{s,\Gamma}$ $(s \leqslant t, \Gamma \in \mathscr{B})$ and $\mathscr{N}^t$ generated by the sets $A_{u,\Gamma}$ $(u \geqslant t, \Gamma \in \mathscr{B})$ also play an important role. A Markov process is a probability measure $\mathbf{P}$ satisfying the condition: for any $t \in T$, $A \in \mathscr{N}_t$, $B \in \mathscr{N}^t$

(1.14) $$\mathsf{P}(AB \mid x_t) = \mathsf{P}(A \mid x_t)\, \mathsf{P}(B \mid x_t) \text{ (a.s. } \mathbf{P}\{x_t \in E\}).$$

In § 2 it will be proved that if h is a p-excessive function, ν a p-excessive measure, and $\{\nu, h\} < \infty$, then there exists a unique measure $\mathbf{P}_\nu^h$ on Ω such that for any $t_1 < t_2 < \ldots < t_n$

(1.15) $$\mathsf{P}_\nu^h \{x_{t_1} \in dy_1, \ldots, x_{t_n} \in dy_n\} = \\ = \nu_{t_1}(dy_1)\, p(t_1, y_1; t_2, dy_2) \ldots p(t_{n-1}; y_{n-1}; t_n, dy_n)\, h^{t_n}(y_n).$$

The measure $\mathbf{P}_\nu^h$ satisfies condition (1.14). Also $\mathbf{P}_\nu^h(\Omega) = \{\nu, h\}$. In particular, for $\{\nu, h\} = 1$ $\mathbf{P}_\nu^h$ is a probability and defines a Markov process.

1.5. With the help of the theorem stated in **1.4.** the study of p-excessive functions and p-excessive measures can be reduced to the study of Markov processes. Furthermore, we make use of the results of [3] concerning the construction of the class of Markov processes corresponding to a given transition function.

We say that a Markov process $\mathbf{P}$ corresponds to a transition function p and write $\mathbf{P} \in \mathscr{K}^p$ if for any $s < t \in T$, $\Gamma \in \mathscr{B}$

(1.16) $$\mathsf{P}\{x_t \in \Gamma \mid x_s\} = p(s, x_s; t, \Gamma) \quad (\text{a.s. } \mathsf{P}, \ x_s \in E).$$

The conditions (1.14), (1.16) are equivalent to the condition: for all $s < t \in T$, $\Gamma \in \mathscr{B}$

(1.17) $$\mathsf{P}\{x_t \in \Gamma \mid \mathscr{N}_s\} = p(s, x_s; t, \Gamma) \quad (\text{a.s. } \mathsf{P}, \ x_s \in E).$$

(If p is a backward transition function, then $\mathscr{N}_s$ is replaced by $\mathscr{N}^s$ in (1.17)).

To state the results of [3] and the conclusions we shall draw from them, it is convenient first to introduce some general ideas.

1.6. We say that there is a *convex structure* on the measurable space Z if with each probability measure μ on Z there is associated a point z_μ "the centre of gravity of the distribution μ". A space Z together with such a structure is called a *convex measurable space.*

If $z_\mu \neq z$ for any measure μ not concentrated on z, then we say that z is an *extreme point* and write $z \in Z_e$. A convex measurable space Z is called a *simplex* if Z_e is measurable and any element z from Z is the centre of gravity of one and only one distribution μ concentrated on Z_e.

Let Z and Z' be convex measurable spaces. A one-to-one mapping φ of Z to Z' is called an *isomorphism* if φ and φ^{-1} are measurable and if $\varphi(z_\mu) = z_{\mu'}$, where $\mu'(\Gamma) = \mu[\varphi^{-1}(\Gamma)]$. It is clear that under an isomorphism extreme points go to extreme points and a space isomorphic to a simplex is itself a simplex.

Our aim is not the development of an axiomatic theory, but the analysis of a number of concrete convex measurable spaces. Each of these is representable as a family of non-negative functions $z(w)$ defined on some set W. A measurable structure on Z is subject to the requirement: for any $w \in W$ the function $F(z) = z(w)$ is measurable. We assume that the following condition is satisfied:

1.6.A. For any probability measure μ on Z the function

$$z_\mu(w) = \int_Z z(w)\,\mu(dz) \tag{1.18}$$

belongs to Z.

The convex structure on Z defined by (1.18) will be called *natural.*

Frequently we consider the measurable structure on Z generated by the system of functions $F(z) = z(w)$ ($w \in W$). We call this structure *natural.*

If Z is a simplex, then each function $z \in Z$ is representable, uniquely, in the form

$$z(w) = \int_Z \tilde{z}(w)\,\mu(d\tilde{z}) \quad (w \in W),$$

where μ is a distribution concentrated on Z_e. Therefore the formula

$$z(w) = \int_{Z_e} \tilde{z}(w)\,\mu(d\tilde{z}) \quad (w \in W) \tag{1.19}$$

establishes a one-to-one correspondence between Z and the set $\mathcal{M}(Z_e)$ of all probability measures on Z_e (the measurable structure in Z_e is induced from the measurable structure in Z).

We now return to the class $\mathcal{K}^p$, defined in **1.5.** Its elements are non-negative functions on the σ-algebra $\mathcal{N}$. We put the natural measurable and convex structures on $\mathcal{K}^p$ (**1.6.A.** is verified using (1.17)). It was established in [3] that $\mathcal{K}^p$ is a simplex.

Strictly speaking, in [3] a certain measurable set $\mathcal{K}^p_e$ in $\mathcal{K}^p$ was constructed, and it was shown that any element $\mathbf{P}$ in $\mathcal{K}^p$ is the centre of gravity of one and only one distribution concentrated on $\mathcal{K}^p_e$. To prove that $\mathcal{K}^p$ is a simplex it is also necessary to show that $\mathcal{K}^p_e$ is the set of extreme points of $\mathcal{K}^p$. From **2.2.C.** of [3] and Lemma 2.1. it follows at once that all points of $\mathcal{K}^p_e$ are extreme.[1] On the other hand, if $\mathbf{P} \,\overline{\in}\, \mathcal{K}^p_e$, then the distribution on $\mathcal{K}^p_e$, corresponding to $\mathbf{P}$ cannot be concentrated on $\mathbf{P}$. Therefore $\mathbf{P}$ is not an extreme point.

One more remark is necessary, since it was assumed in [3] that the limit

$$q^t(x) = \lim_{t \downarrow s} p(s, x;\ t, E)$$

is equal to 1 for all s, x. The fact is that on points where this is violated it is impossible to define a probability measure $\mathbf{P}_{s,x}$ in the space of trajectories for which

$$\mathbf{P}_{s,x}\{x_{t_1} \in dy_1, \ldots, x_{t_n} \in dy_n\} = p(s, x;\ t_1, dy_1) \ldots p(t_{n-1}, y_{n-1};\ t_n, dy_n).$$

---

[1] For let $\mathbf{P} \in \mathcal{K}^p_e$ be the centre of gravity of the distribution μ and let $\Omega' = \{\omega\colon \mathbf{P}^\omega = \mathbf{P}\}$ (we use the notation of § 2 of [3]). By **2.2.C.** of [3] $\mathbf{P}(\Omega') = 1$ and hence $\tilde{\mathbf{P}}(\Omega') = 0$ for μ-almost all $\mathbf{P} \in \mathcal{K}^p$. It remains to note that if $\tilde{\mathbf{P}}(\Omega') = 1$, then by **2.1.B.** of [3] for any measurable set C, $\tilde{\mathbf{P}}(C) = \tilde{\mathbf{P}}[\mathbf{P}^\omega(C)] = \tilde{\mathbf{P}}[\mathbf{P}(C)] = \mathbf{P}(C)$. Consequently μ is concentrated on $\mathbf{P}$.

However, these situations can be ignored because for any $\mathrm{P} \in \mathscr{K}^p, s \in T$

(1.20)
$$q^s(x_s) = 1 \quad (\text{a.s. } \mathrm{P},\ x_s \in E).$$

For by (1.16) when $s < t$ $p(s, x_s; t, E) = \mathrm{P}\{x_t \in E \mid x_s\} = \mathrm{P}\{\beta > t \mid x_s\}$ (a.s. $\mathrm{P}, x_s \in E$) and the right-hand side converges, as $t \downarrow s$, to $\mathrm{P}\{\beta > s \mid x_s\} = 1$ on $\{x_s \in E\}$. By (1.20) the propositions 5.2.A′ and (5.7) of [3] remain true even if we leave $\mathrm{P}_{s,x}$ undefined outside the set $\{(s, x): q^s(x) = 1\}$ (here $\mathrm{P}_{a_n}, x_{a_n}$ is, P-almost surely, defined from some n onwards). The remaining arguments of § 5 of [3] are based on (5.7) and are unchanged.

We shall construct convex measurable spaces whose elements are excessive functions and measures and show that these spaces are simplexes. An application of (1.19) to these spaces gives integral representations of excessive functions and measures.

1.7. Fix any positive function h and denote by $\mathscr{R}^{p,h}$ the set of all p-excessive measures ν for which $\{\nu, h\} = 1$. Considering the elements of $\mathscr{R}^{p,h}$ as non-negative functions on $T \times \mathscr{B}$, we introduce the natural measurable and convex structures in $\mathscr{R}^{p,h}$ (1.6.A. will be established in § 3).

We associate with the measure $\nu \in \mathscr{R}^{p,h}$ the process P_ν^h described in 1.4. It is not difficult to verify (see 2.8) that the corresponding transition function is

$$p^h(s, x; t, \Gamma) = \begin{cases} \dfrac{1}{h^s(x)} \displaystyle\int_\Gamma p(s, x; t, dy)\, h^t(y) & \text{for } 0 < h^s(x) < \infty, \\ 0 & \text{for the other } s, x, \end{cases}$$

so that we have a mapping $\mathscr{R}^{p,h}$ to $\mathscr{K}^{p^h}$. It will be proved in § 3 that this mapping is one-to-one and preserves the measurable and convex structures. Consequently $\mathscr{R}^{p,h}$ is a simplex isomorphic to $\mathscr{K}^{p^h}$. In the same § 3 it is proved that the measurable spaces $\mathscr{R}^{p,h}$ and $\mathscr{K}^p$ are standard.

We now assume that the transition function p is homogeneous and satisfies the condition:

1.7.A. For any Γ, $p(t, x, \Gamma)$ is measurable with respect to t, x (the measurable structure on T is the Borel structure).

Then $\nu_t(f)$ $(t \in T, \nu \in \mathscr{R}^{p,h})$ is measurable with respect to t, v for any non-negative measurable f (see 3.2.).

Let l be any positive measurable function on E. Denote by $\mathscr{R}_l^p$ the collection of all homogeneous p-excessive measures ν for which $\nu(l) = 1$ and introduce the natural measurable and convex structures in $\mathscr{R}_l^p$. Let us show that $\mathscr{R}_l^p$ is a simplex.

The proof goes as follows. Let ν be a p-excessive measure, h a p-excessive function, and let s be any real number. In view of the homogeneity of p the formulae $\bar{\nu}_t = \nu_{t+s}$, $\bar{h}^t = h^{t+s}$ define a new p-excessive measure $\bar{\nu}$ and a new p-excessive function $\bar{h}$. We denote these by $\theta^s\nu$ and $\theta^s h$. It is easy to see that $\{\theta^s\nu, \theta^s h\} = \{\nu, h\}$. Consider the p-excessive function

(1.21) $$h^s(x) = \frac{1}{2}\int_0^\infty e^{-|s+u|} P_u l(x)\, du$$

(it can be obtained by putting $l^t = l$, $d\lambda(t) = \frac{1}{2}e^{-|t|}dt$) in (1.5)). By (1.9) for any p-excessive measure ν

$$\{\nu, h\} = \frac{1}{2}\int_{-\infty}^{+\infty} e^{-|u|}\nu_u(l)\, du.$$

In particular, if ν is homogeneous, then $\{\nu, h\} = \nu(l)$, so that $\mathscr{R}^p_l \subseteq \mathscr{R}^{p,h}$. Put

(1.22) $$\rho_s(\nu) = \{\theta^s\nu, h\} = \frac{1}{2}\int_{-\infty}^{+\infty} e^{-|u-s|}\nu_u(l)\, du.$$

Clearly $0 < \rho_s(\nu) < \infty$ for all $\nu \in \mathscr{R}^{p,h}$ and by the measurability of $\nu_u(l)$ with respect to u and ν, $\rho_s(\nu)$ is measurable with respect to s, ν. The operators

$$\tau^s\nu = \frac{\theta^s\nu}{\rho_s(\nu)}$$

map $\mathscr{R}^{p,h}$ into itself, leaving the set $\mathscr{V}$ of extreme points invariant. It is easily seen that $(s, \nu) \to \tau^s\nu$ is a measurable map of $T \times \mathscr{R}^{p,h}$ into $\mathscr{R}^{p,h}$ and $\tau^s\tau^t = \tau^{s+t}$, $\rho_{s+t}(\nu) = \rho_s(\nu)\,\rho_t(\tau^s\nu)$ for any s, t. Let μ be a probability measure on $\mathscr{V}$ and ν its centre of gravity, that is, $\nu_t(\Gamma) = \int_{\mathscr{V}} \tilde{\nu}_t(\Gamma)\,\mu(d\tilde{\nu})$ for all $t \in T$, $\Gamma \in \mathscr{B}$. Replacing t by $t + s$ we see that

$$\theta^s\nu = \int_{\mathscr{V}} \theta^s\tilde{\nu}\mu(d\tilde{\nu}) = \int_{\mathscr{V}} \tau^s\tilde{\nu}\rho_s(\tilde{\nu})\,\mu(d\tilde{\nu}) = \int_{\mathscr{V}} \tilde{\nu}\rho_s(\tau^{-s}\tilde{\nu})\,\mu_s(d\tilde{\nu}),$$

where $\mu_s(\Gamma) = \mu\{\tau^{-s}\Gamma\}$. Clearly ν belongs to $\mathscr{R}^p_l$ if and only if $\theta^s\nu = \nu$ for all s, which is equivalent to the condition

(1.23) $$\mu\{\tau^s\Gamma\} = \int_\Gamma \mu(d\nu)\,\rho_s(\nu) \quad \text{for all} \quad s \in T,\ \Gamma \in \mathscr{B}.$$

Denote by $\mathscr{M}$ the set of all probability measures on $\mathscr{V}$ satisfying (1.23) with the natural measurable and convex structures. Associating with each measure its centre of gravity we have an isomorphic mapping of $\mathscr{M}$ onto $\mathscr{R}^p_l$. In the Appendix it is proved that $\mathscr{M}$ is a simplex. Consequently $\mathscr{R}^p_l$ is also a simplex.

1.8. Now fix a p-excessive measure ν and denote by $\mathscr{S}^{p,\nu}$ the set of all p-excessive functions h for which $\{\nu, h\} = 1$. We study the structure of $\mathscr{S}^{p,\nu}$ under the following assumptions:

1.8.A. There is a backward transition function $\hat{p}$ such that

(1.24) $$\nu_s(dx)p(s, x; t, dy) = \nu_t(dy)\,\hat{p}(t, y; s, dx).$$

1.8.B. The measure $p(s, x; t, -)$ is absolutely continuous with respect to ν_t.

1.8.C. The measure $\hat{p}(t, y; s, -)$ is absolutely continuous with respect to ν_s.

In $\mathcal{S}^{p,\nu}$ we introduce the measurable structure generated by the functions $F(h) = \nu_t\,(\chi_\Gamma h^t)$ $(t \in T, \Gamma \in \mathcal{B})$.[1] Using **1.8.B.** we shall prove the measurability of the function $\Phi(h) = h^t(x)$ $(t \in T, x \in E)$ and the validity of **1.6.A.** Consequently we can introduce a convex structure into $\mathcal{S}^{p,\nu}$. In § 4 it will be proved that $\mathcal{S}^{p,\nu}$ is a simplex. We associate with each measurable non-negative function h the measure

$$\nu_t^h(dx) = h^t(x)\nu_t(dx). \tag{1.25}$$

It follows from (1.24) that for any two functions h, r

$$\nu_s^r \mathsf{P}_t^s h^t = \nu_t^h \hat{\mathsf{P}}_s^t r^s. \tag{1.26}$$

For $r^s(x) = \chi_\Gamma(x)$ we have hence

$$\int_\Gamma \nu_s(dx)\,\mathsf{P}_t^s h^t(x) = \int_E \nu_t^h(dy)\,\hat{p}(t, y; s, \Gamma). \tag{1.26'}$$

If h is a p-excessive function, then the left-hand side does not exceed $\nu_s^h(\Gamma)$ and converges to $\nu_s^h(\Gamma)$ for $t \downarrow s$. Hence ν^h is a $\hat{p}$-excessive measure. It can be proved similarly that if r is $\hat{p}$-excessive function, then ν^r is a p-excessive measure. Comparing (1.26) and (1.3) we observe that for any p-excessive function h and $\hat{p}$-excessive function[1] r

$$\{\nu^r, h\}_p = \{\nu^h, r\}_{\hat{p}}\,. \tag{1.27}$$

Consider the function $r^t(x) = \hat{p}(t, x; t-0, E)$ (the limit on the right exists since, by **1.2.B.**, $\hat{p}(t, x; s, E)$ is monotonic in s). The equality $\hat{\mathsf{P}}_s^t r^s(x) = \hat{p}(t, x; s-0, E)$ for $t > s$ implies that r is $\hat{p}$-excessive. Putting $h = \chi_A$, $\Gamma = E$, $s \uparrow t$ in (1.26′) we have $\nu^t(A) = \nu_t^r(A)$ and by (1.27) $\{\nu, h\}_p = \{\nu^h, r\}_{\hat{p}}$. Consequently, if $h \in \mathcal{S}^{p,\nu}$, then $\nu^h \in \mathcal{R}^{\hat{p},r}$. Thus, (1.25) defines a mapping from $\mathcal{S}^{p,\nu}$ to $\mathcal{R}^{\hat{p},r}$.

To prove that the mapping is *onto* $\mathcal{R}^{\hat{p},r}$, we prove the following general lemma: if a transition function p is absolutely continuous with respect to ν, then all p- excessive measures are absolutely continuous with respect to ν. Applying this lemma to $\hat{p}$ and using **1.8.C.** we conclude that for any measure $\nu' \in \mathcal{R}^{\hat{p},r}$ there exists a density $\tilde{h}^t(x)$ relative to the measure ν_t. From (1.26′) it follows that $\tilde{h}$ has the properties:

[1] The expression $\nu(h, \Gamma)$ or $\nu(h\chi_\Gamma)$ denotes the integral of h over Γ with respect to ν (χ_Γ is the indicator of Γ, that is, the function equal to 1 on Γ and zero outside Γ).

a) $$P_t^s \tilde{h}^t \leqslant \tilde{h}^s \text{ (a.s. } \nu_s) \text{ for } s < t;$$

b) $$P_{t_n}^s \tilde{h}^{t_n} \to \tilde{h}^s \text{ (a.s. } \nu_s) \text{ if } t_n \downarrow s.$$

From these properties and **1.8.B.** we deduce the existence of the limit

$$P_t^s \tilde{h}^t(x) \uparrow h^s(x) \quad \text{for } t \downarrow s.$$

This limit (it is called the *regularizer* of the function $\tilde{h}$) is a p-excessive function, and $\tilde{h}^t = h^t$ (a.s. ν_t) for any $t \in T$, so that $\nu'_t(dx) = h^t(x)\nu_t(dx)$, or, equivalently, $\nu' = \nu^h$.

It is easily verified that under the mapping $h \to \nu^h$ distinct elements of $\mathcal{S}^{p,\nu}$ go over to distinct points of $\mathcal{R}^{\hat{p},r}$ and that this mapping preserves the measurable and convex structures. Consequently $\mathcal{S}^{p,\nu}$ is isomorphic to $\mathcal{R}^{\hat{p},r}$ and is a simplex.

1.9. Now let p be a homogeneous transition function and γ a measure on E. Denote by $\mathcal{S}_\gamma^p$ the set of all homogeneous p-excessive functions for which $\gamma(h) = 1$. We assume that **1.7.A.** holds and that for some measure ν the following conditions are satisfied:

1.9.A. There exists a homogeneous transition function $\hat{p}$ such that for any $\lambda > 0$

$$\nu(dx)\, g_\lambda(x, dy) = \nu(dy)\hat{g}_\lambda(y, dx),$$

where

$$g_\lambda(x, \Gamma) = \int_0^\infty e^{-\lambda t} p(t, x, \Gamma)\, dt, \qquad \hat{g}_\lambda(y, \Gamma) = \int_0^\infty e^{-\lambda t} \hat{p}(t, y, \Gamma)\, dt.$$

1.9.B. For any $\lambda > 0$, $x, y \in E$ the measures $g_\lambda(x, -)$, $\hat{g}_\lambda(y, -)$ are absolutely continuous with respect to ν.

1.9.C. $\gamma(dx) = l(x)\, \nu(dx)$, where $l > 0$.

Take in $\mathcal{S}_\gamma^p$ the measurable structure generated by the functions $F(h) = \nu(h, \Gamma)$ $(\Gamma \in \mathcal{B})$, and the natural convex structure on the measurable space so constructed. In § 5 it will be proved that the formula $\nu^h(dx) = h(x)\nu(dx)$ defines an isomorphic mapping of $\mathcal{S}_\gamma^p$ to $\mathcal{R}_l^{\hat{p}}$. The necessary arguments are very similar to those used in the inhomogeneous case. In addition we use the fact that conditions 1.3.α – 1.3.β are equivalent to the requirements[1]

1.9.α. $\lambda G_\lambda h(x) \leqslant h(x)$ $(x \in E)$,

1.9.β. $\lambda G_\lambda h(x) \to h(x)$ for $\lambda \to \infty$ $(\Gamma \in E)$,

and conditions 1.3.α' – 1.3.β' are equivalent to

1.9.α′. $\lambda \nu G_\lambda(\Gamma) \leqslant \nu(\Gamma)$ $(\Gamma \in \mathcal{B})$,

1.9.β′. $\lambda \nu G_\lambda(\Gamma) \to \nu(\Gamma)$ for $\lambda \to \infty$ $(\Gamma \in \mathcal{B})$.

---

[1] G_λ are the operators corresponding to the kernels $g_\lambda(x, \Gamma)$.

The equality (1.26′) is replaced by

$$\int_E \nu^h(dy)\,\hat{g}_\lambda(y,\Gamma) = \int_\Gamma \nu(dx)\,G_\lambda h(x), \tag{1.28}$$

and instead of (1.27) we use the obvious equality

$$\gamma(h) = \nu^h(l). \tag{1.29}$$

1.10. The construction of the spaces of invariant and null-excessive measures is studied in § 6. The connection between excessive functions and supermartingales is established in § 7. Finally, in the concluding § 8 we study the question for what functions p it is possible to construct ν and $\hat{p}$ to satisfy 1.8.A. – 1.8.C. It is proved that it is sufficient for p to have the following property:

1.10.A. There exist a finite measure m_t, depending measurably on t, and a non-negative function $\pi(s, x; t, y)$ measurable with respect to s, x, y such that

$$p(s, x;\ t, dy) = \pi(s, x;\ t, y)m_t(dy) \tag{1.30}$$

and for any $s < t < u$, x, $z \in E$

$$\int_E p(s, x;\ t, dy)\,\pi(t, y;\ u, z) = \pi(s, x;\ u, z) \tag{1.31}$$

(π is called a *transition density*).[1]

Choose an arbitrary strictly increasing function $\lambda(s)$ for which

$$\int_T m_s(E)\,d\lambda(s) < \infty, \tag{1.32}$$

and put

$$a_t(y) = \int_{-\infty}^{t} d\lambda(s) \int_E m_s(dx)\,\pi(s, x;\ t, y), \tag{1.33}$$

$$\pi'(s, x;\ t, y) = \begin{cases} a_s(x)\,\pi(s, x;\ t, y)\cdot\dfrac{1}{a_t(y)} & \text{for } 0 < a_t(y) < \infty, \\ 0 & \text{otherwise.}^2 \end{cases} \tag{1.34}$$

---

[1] If **1.10.A.** is satisfied for a measure m_t and if $c_t(x)$ is an arbitrary positive measurable function, then **1.10.A.** is also satisfied for $\widetilde{m}_t(dx) = c_t(x)m_t(dx)$, with $\pi(s, x; t, y)$ replaced by $\widetilde{\pi}(s, x;\ t, y) = \dfrac{\pi(s, x;\ t, y)}{c_t(y)}$. Therefore, instead of finiteness of m_t it is sufficient to require σ-finiteness.

[2] The product $\infty \cdot c$ is taken to be equal to ∞ for $c > 0$ and to zero for $c = 0$.

A p-excessive measure ν and a backward transition function $\hat{p}$ satisfying **1.8.A. – 1.8.C.** can be given by the formulae

$$(1.35) \qquad \nu_t(\Gamma) = \int_T d\lambda(s) \int_E m_s(dx)\, p(s, x; t, \Gamma),$$

$$(1.36) \qquad \hat{p}(t, y; s, dx) = m_s(dx)\, \pi'(s, x; t, y).$$

((1.35) is a particular case of (1.4) for $\gamma(ds, dx) = m_s(dx)d\lambda(s)$.)

1.11. Apart from well-known facts from measure and integration theory, we shall often use in proofs two lemmas on measurable functions and we complete this introductory section by stating them.

LEMMA A. *Suppose that the measurable structure on E is generated by a system of sets Q that contains together with any two sets their intersection. Suppose that the family $\mathcal{H}$ of non-negative functions contains the indicators of all the sets of Q and has the following properties:*

a) *if $f_1, f_2 \in \mathcal{H}$, then $c_1 f_1 + c_2 f_2 \in \mathcal{H}$ for any non-negative constants c_1, c_2;*

b) *if $f_1, f_2 \in \mathcal{H}$ and $f_1 \leqslant f_2 < \infty$, then $f_2 - f_1 \in \mathcal{H}$;*

c) *if $f_n \in \mathcal{H}$ and $f_n \uparrow f$, then $f \in \mathcal{H}$;*

d) $1 \in \mathcal{H}$.

Then $\mathcal{H}$ contains all measurable non-negative functions.

LEMMA B. *Let E_1, E_2, E_3 be measurable spaces, F a non-negative measurable function on $E_1 \times E_3$ and $\mu(x_2, \Gamma)$ a measure on E_3 with respect to Γ and a measurable function on E_2 with respect to x_2. Then the formula*

$$\Phi(x_1, x_2) = \int_{E_3} F(x_1, x_3)\, \mu(x_2, dx_3)$$

defines a measurable function on $E_1 \times E_2$.

Lemma A is easily deduced from Lemma 1.1 of the book [1]. By the latter $\mathcal{H}$ contains the indicators of all measurable sets. For any non-negative measurable function f put

$$A_{kn} = \left\{x\colon \frac{k}{2^n} \leqslant f(x) < \frac{k+1}{2^n}\right\}, \qquad f_n(x) = \sum_{k=0}^{\infty} \frac{k}{2^n} \chi_{A_{kn}}(x).$$

Clearly $f_n \uparrow f$. By virtue of a) $f_n \in \mathcal{H}$, and by virtue of c) $f \in \mathcal{H}$.

The proof of Lemma B can be found in [1] (see Lemma 1.7.).

§ 2. The construction of a Markov process from an excessive measure and function

2.1. The aim of this section is to prove the following theorem:

THEOREM 2.1. *Let p be any transition function and let the p-excessive measure ν and function h be connected by the relation*

$$(2.1) \qquad \{\nu, h\} < \infty.$$

Then there exists one and only one measure P_ν^h *in the space of trajectories* Ω *satisfying the condition: for any* $t_1 < t_2 < \ldots < t_n \in T, \Gamma_1, \Gamma_2, \ldots, \Gamma_n \in \mathscr{B}$

$$\mathsf{P}_\nu^h\{x_{t_1} \in \Gamma_1, x_{t_2} \in \Gamma_2, \ldots, x_{t_n} \in \Gamma_n\} = \int_{\Gamma_1} \ldots \int_{\Gamma_n} \nu_{t_1}(dy_1)\, p(t_1, y_1; t_2, dy_2) \ldots p(t_{n-1}, y_{n-1}; t_n, dy_n)\, h^{t_n}(y_n). \tag{2.2}$$

For P_ν^h *to be a probability measure it is necessary and sufficient that* $\{\nu, h\} = 1$. *In this case* P_ν^h *defines a Markov process with transition function*

$$p^h(s, x; t, \Gamma) = \begin{cases} \dfrac{1}{h^s(x)} \displaystyle\int_\Gamma p(s, x; t, dy)\, h^t(y) & \text{for } 0 < h^s(x) < \infty, \\ 0 & \text{for the other } s, x. \end{cases} \tag{2.3}$$

The probability distributions of the birth time α *and the death time* β *are given by*

$$\mathsf{P}_\nu^h\{\alpha < t\} = \{\nu, h\}^t, \quad \mathsf{P}_\nu^h\{\beta \leqslant t\} = \{\nu, h\}^t - \nu_t h^t, \tag{2.4}$$

where

$$\{\nu, h\}^t = \sup_{\Lambda \subseteq (-\infty, t]} \{\nu, h\}_\Lambda. \tag{2.5}$$

2.2 We first prove that (2.2) implies (2.4) – (2.5). First of all,

$$\mathsf{P}_\nu^h\{x_t \in \Gamma\} = \int_\Gamma h^t(x)\, \nu_t(dx) \quad (t \in T, \ \Gamma \in \mathscr{B}) \tag{2.6}$$

and for any $s < t \in T$, $\Gamma \in \mathscr{B}$

$$\mathsf{P}_\nu^h\{\alpha < s, x_t \in \Gamma\} = \mathsf{P}_\nu^h\{x_s \in E, x_t \in \Gamma\} = \int_E \int_\Gamma \nu_s(dx)\, p(s, x; t, dy)\, h^t(y). \tag{2.7}$$

Hence for $s < t$

$$\mathsf{P}_\nu^h\{s \leqslant \alpha < t < \beta\} = \mathsf{P}_\nu^h\{x_t \in E\} - \mathsf{P}_\nu^h\{x_s \in E, x_t \in E\} = \nu_t h^t - \nu_s P_t^s h^t. \tag{2.8}$$

Connected with each set $\Lambda = \{t_0 < t_1 < \ldots < t_n\}$ there is a random variable α_Λ defined by

$$\alpha_\Lambda = \begin{cases} t_0 & \text{for } \alpha < t_0, \\ t_k & \text{for } t_{k-1} \leqslant \alpha < t_k \quad (k = 1, 2, \ldots, n), \\ +\infty & \text{for } t_n \leqslant \alpha. \end{cases} \tag{2.9}$$

Comparing (1.3) with (2.7) – (2.8) we observe that

$$\{\nu, h\}_\Lambda = \mathsf{P}_\nu^h\{\alpha_\Lambda < \beta\}. \tag{2.10}$$

If $\Lambda \subseteq (-\infty, t]$, then $\{\alpha_\Lambda < \beta\} \subseteq \{\alpha < \alpha_\Lambda \leqslant t\} \subseteq \{\alpha < t\}$. Therefore

$$\{\nu, h\}_\Lambda \leqslant \mathsf{P}_\nu^h \{\alpha < t\}. \tag{2.11}$$

On the other hand, if Λ' is a countable dense subset of the interval $(-\infty, t]$ and $\Lambda_n \uparrow \Lambda'$ then $\{\alpha < t\} \subseteq \{\alpha_{\Lambda_n} \downarrow \alpha\}$ and therefore $\{\alpha_{\Lambda_n} < \beta\} \uparrow \{\alpha < t\}$. Consequently, $\{\nu, h\}_{\Lambda_n} \uparrow \mathsf{P}_\nu^h\{\alpha < t\}$. Hence the first formula of (2.4) follows from (2.11). The second formula follows from the first one and (2.6). In passing we have proved the useful equality

$$\{\nu, h\}^t = \lim \{\nu, h\}_{\Lambda_n}. \tag{2.12}$$

All these arguments are also applicable when $t = +\infty$. Then $\mathsf{P}_\nu^h\{\alpha < t\} = \mathsf{P}_\nu^h(\Omega)$ and $\{\nu, h\}^t = \{\nu, h\}$. Consequently, $\mathsf{P}_\nu^h\{\Omega\} = \{\nu, h\}$, so that $\mathbf{P}_\nu^h$ is a probability measure if and only if $\{\nu, h\} = 1$.

2.3. We now prove that the condition (2.2) defines $\mathbf{P}_\nu^h$ uniquely. For this it suffices, by virtue of Lemma A, to prove that for any $t_1 < t_2 < \ldots < t_n \in T$, $\Gamma_1, \ldots, \Gamma_n \in \mathscr{B}_a^b$ the probability

$$\mathsf{P}_\nu^h\{x_{t_1} \in \Gamma_1, \ldots, x_{t_n} \in \Gamma_n\} = q_{t_1 \ldots t_n}(\Gamma_1, \ldots, \Gamma_n) \tag{2.13}$$

is uniquely defined. The values $q_{t_1 \ldots t_n}$ on $\mathscr{B}_a^b$ can be calculated from the values on $\mathscr{B}^b$ by means of the recurrence formulae

$$q_{t_1}(\Gamma_1) = q_{t_1}(\Gamma_1^*) + \chi_{\Gamma_1}(a)[\{\nu, h\} - q_{t_1}(E^b)], \tag{2.14}$$

$$\begin{aligned} q_{t_1 \ldots t_n}(\Gamma_1, \ldots, \Gamma_n) &= q_{t_1 \ldots t_n}(\Gamma_1^*, \ldots, \Gamma_n^*) + \\ &+ \chi_{\Gamma_1}(a)\,[q_{t_2 \ldots t_n}(\Gamma_2, \ldots, \Gamma_n) - q_{t_1 \ldots t_n}(E^b, \Gamma_2^*, \ldots, \Gamma_n^*)], \end{aligned} \tag{2.15}$$

where $\Gamma^* = \Gamma \cap E^b$. In their turn, the values $q_{t_1 \ldots t_n}$ on $\mathscr{B}^b$ can be expressed through their values on $\mathscr{B}$ and $q_t(E^b)$ by the formulae

$$q_{t_1}(\Gamma) = q_{t_1}(\Gamma') + \chi_\Gamma(b)\,[q_{t_1}(E^b) - q_{t_1}(E)], \tag{2.16}$$

$$\begin{aligned} q_{t_1 \ldots t_n}(\Gamma_1, \ldots, \Gamma_n) &= q_{t_1 \ldots t_n}(\Gamma_1', \ldots, \Gamma_n') + \\ &+ \chi_{\Gamma_n}(b)\,[q_{t_1 \ldots t_{n-1}}(\Gamma_1, \ldots, \Gamma_{n-1}) - q_{t_1 \ldots t_n}(\Gamma_1', \ldots, \Gamma_{n-1}', E)], \end{aligned} \tag{2.17}$$

where $\Gamma' = \Gamma \cap E$. It remains to note that by (2.4)

$$q_t(E^b) = \{\nu, h\}^t. \tag{2.18}$$

2.4. Now we want to construct a measure $\mathbf{P}_\nu^h$ satisfying (2.2), starting from any pair ν, h connected by the relation (2.1.). Naturally, to this end we make use of (2.14) – (2.18). As a preliminary we have to study the function $\{\nu, h\}^t$ in (2.18).

LEMMA 2.1. *If the p-excessive measure ν and function h are connected by* (2.1), *then*:

2.4.A. *The function $\{\nu, h\}^t$ is non-negative, non-decreasing continuous on the left, and converges to $\{\nu, h\}$ as $t \to \infty$.*

2.4.B. *The function* $\rho_t = \{v, t\}^t - v_t h^t$ *is non-negative, non-decreasing and continuous on the right.*

2.4.C. *For* $s < a$ *the function*

$$(2.19) \qquad F(s, u) = \{v, h\}^u - \{v, h\}^s - v_u h^u + v_s P_u^s h^u$$

is non-negative, non-increasing and continuous on the left with respect to s, non-decreasing and continuous on the right with respect to u.

2.4.D. *For any* $\Lambda = \{t_0 < t_1 < \dots < t_n\}$

$$(2.20) \qquad \{v, h\}^{t_n} = \rho_{t_0} + \sum_{k=1}^{n} F(t_{k-1}, t_k) + \{v, h\}_\Lambda.$$

For any t

$$(2.21) \qquad F(t-0, t) = F(t, t+0) = 0.$$

PROOF. Firstly we note that the value of $\{v, h\}_\Lambda$ can only increase for expanding Λ. For if $t \notin \Lambda$ and $\Lambda' = \Lambda \cup \{t\}$, then

$$\{v, h\}_{\Lambda'} - \{v, h\}_\Lambda = \begin{cases} v_t (h^t - P_{t_0}^t h^{t_0}) & \text{for } t < t_0, \\ (v_t - v_{t_{k-1}} P_t^{t_{k-1}})(h^t - P_{t_k}^t h^{t_k}) & \text{for } t_{k-1} < t < t_k, \\ (v_t - v_{t_n} P_t^{t_n}) h^t & \text{for } t > t_n, \end{cases}$$

and the expression on the right is non-negative in view of the fact that p and h are p-excessive.

From the formula (2.5) defining $\{v, h\}^t$, it is at once clear that $\{v, h\}^t$ is non-negative, non-decreasing, and does not exceed $\{v, h\}$. Since its limit as $t \to +\infty$ is not smaller than $\{v, h\}_\Lambda$ for any Λ, this limit is equal to $\{v, h\}$. The fact that ρ_t is non-negative follows from the relation $v_t h^t = \{v, h\}_\Lambda$, where $\Lambda = \{t\}$.

If $s < u$, $\Lambda \subseteq (-\infty, s]$ and Λ' is obtained from Λ by adding the points s and u, then

$$\{v, h\}^u \geqslant \{v, h\}_{\Lambda'} \geqslant \{v, h\}_\Lambda + (v_u - v_s P_u^s) h^u.$$

Hence it is clear that $F(s, u) \geqslant 0$. From the equality

$$(2.22) \qquad F(s, u) = F(s, t) + F(t, u) + (v_t - v_s P_t^s)(h^t - P_u^t h^u) \geqslant \geqslant F(s, t) + F(t, u)$$

for $s < t < u$ it is clear that $F(s, u)$ is a non-increasing function of s and a non-decreasing function of u.

Formula (2.20) follows from (2.19). By virtue of (2.20)

$$(2.23) \qquad \sum_{k=1}^{n} F(t_{k-1}, t_k) \leqslant \{v, h\} - \{v, h\}_\Lambda.$$

Let us prove (2.21). By the definition of $\{\nu, h\}$ the right-hand side of (2.23) converges to zero for some sequence Λ_m. This property remains valid under expansion of the sets Λ_m. Therefore we can assume that $t \in \Lambda_1 \subseteq \Lambda_2 \subseteq \ldots \subseteq \Lambda_m \subseteq \ldots$, that the sum of the Λ_m is dense in T, and that the sets $\Lambda'_m = \Lambda_m \cap (-\infty, t)$ and $\Lambda''_m = \Lambda_m \cap (t, +\infty)$ are non-empty. Let s_m be the largest element of Λ'_m and u_m the smallest element of Λ''_m. By (2.23)

$$(2.24) \qquad F(s_m, t) + F(t, u_m) \leqslant \{\nu, h\} - \{\nu, h\}_{\Lambda_m}.$$

Clearly $s_m \uparrow t$, $u_m \downarrow t$. Therefore (2.21) follows from (2.24).

(2.21) and (2.22) imply that $F(t-0, u) = F(t, u)$ and $F(s, t+0) = F(s, t)$. It is clear from (2.17) and (2.15) that $\{\nu, h\}^u = \{\nu, h\}^{u-0}$. Finally, from the equality

$$\rho_u - \rho_t = F(t, u) + \nu_t(h^t - P^t_u h^u)$$

and formula (2.21) we have $\rho_{t+0} = \rho_t$.

2.5. We shall use the following well-known result (see for example, [7], Chapter III, 3).

KOLMOGOROV'S THEOREM. *Let $(X, \mathscr{B}_X)$ be a standard measurable space and suppose that for each $t_1 < \ldots < t_n$ there is a function $q_{t_1 \ldots t_n}(\Gamma_1, \ldots, \Gamma_n)$ $(\Gamma_1, \ldots, \Gamma_n \in \mathscr{B}_X)$, which is a measure relative to each Γ_i and satisfies the conditions:*

2.5.A. $q_t(X) = 1$

2.5.B. *For any $i = 1, 2, \ldots, n$ and $\Gamma_i = X$*

$$(2.25) \qquad q_{t_1 \ldots t_n}(\Gamma_1, \ldots, \Gamma_n) = q_{t_1 \ldots \hat{t}_i \ldots t_n}(\Gamma_1, \ldots, \hat{\Gamma}_i, \ldots, \Gamma_n)$$

(the hat over the arguments t_i and Γ_i means that these arguments are to be crossed out).

Then in the space Ω' of all mappings of T to X there is a probability measure $\mathbf{P}'$ such that for all $t_1 < \ldots < t_n \in T$, $\Gamma_1, \ldots \Gamma_n \in \mathscr{B}_X$

$$(2.26) \qquad \mathbf{P}'\{\omega(t_1) \in \Gamma_1, \ldots, \omega(t_n) \in \Gamma_n\} = q_{t_1 \ldots t_n}(\Gamma_1, \ldots, \Gamma_n)$$

(this measure is defined on the minimal σ-algebra $\mathscr{N}'$, containing the sets $\{\omega\colon \omega(t) \in \Gamma\}$ $(t \in T, \Gamma \in \mathscr{B}_X)$).

2.6. We shall prove that the functions $q_{t_1 \ldots t_n}$ defined by (2.2) and (2.14) – (2.18) satisfy the conditions of Kolmogorov's theorem with $X = E^b_a$, $\mathscr{B}_X = \mathscr{B}^b_a$.

It is clear that on the σ-algebra $\mathscr{B}$ these functions are measures with respect to each Γ_i, and for $\Gamma_i = E$ (2.25) is satisfied for $1 < i < n$, and for $i = 1$ and $i = n$ it is satisfied with = replaced by $\leqslant$. (This follows from **1.2.B.**, **1.3.A.** and **1.3.A'**)

Further, taking into account **2.4.B**, we can deduce from (2.16), (2.17) and (2.18), using induction on n, that the functions $q_{t_1 \ldots t_n}$ on the

σ-algebra $\mathscr{B}^b$ are also measures with respect to each Γ_i, and for $\Gamma_i = E^b$ (2.25) is satisfied for $i > 1$, but for $i = 1$ we must replace = by $\leqslant$.

Finally, using (2.14) – (2.15) and the inequality $\{\nu, h\}^t \leqslant \{\nu, h\}$ (see 2.4.A) we show by induction that the $q_{t_1 \ldots t_n}$ on $\mathscr{B}_a^b$ are measures with respect to each Γ_i and satisfy 2.5.A – 2.5.B.

2.7. Let P' be the probability measure on Ω' whose existence is asserted by Kolmogorov's Theorem. For it

$$\text{(2.27)} \qquad \mathsf{P}'\{\omega(t_1) \in \Gamma_1, \ldots, \omega(t_n) \in \Gamma_n\} = q_{t_1 \ldots t_n}(\Gamma_1, \ldots, \Gamma_n)$$
$$(t_1 < \ldots < t_n, \ \Gamma_1, \ldots, \Gamma_n \in \mathscr{B}_a^b).$$

Hence by (2.14) – (2.18) we have

$$\text{(2.28)} \qquad q_t(b) = \rho_t, \qquad q_t(a) = \{\nu, h\} - \{\nu, h\}^t$$

and for $s < t$

$$\text{(2.29)} \qquad q_{st}(E^b, a) = q_{st}(b, E^a) = 0,$$
$$\text{(2.30)} \qquad q_{st}(a, b) = F(s, t),$$
$$\text{(2.31)} \qquad q_{st}(a, E^b) = \{\nu, h\}^t - \{\nu, h\}^s,$$
$$\text{(2.32)} \qquad q_{st}(E^a, b) = \rho_t - \rho_s,$$

where ρ_t and $F(s, t)$ are defined by Lemma 2.1.

Consider the set R of all rational numbers and introduce the functions

$$\alpha(\omega) = \inf\{t: t \in R, \ \omega(t) \neq a\}, \qquad \beta(\omega) = \sup\{t: t \in R, \ \omega(t) \neq b\}.$$

It is easy to see that $\alpha \leqslant \beta$. Fix any $c \in E$ and put

$$x_t(\omega) = \begin{cases} c & \text{if } \alpha = \beta \text{ or if } \alpha < t < \beta \text{ and } \omega(t) \notin E, \\ a & \text{if } t \leqslant \alpha, \\ \omega(t) & \text{if } \alpha < t < \beta \text{ and } \omega(t) \in E, \\ b & \text{if } \beta \leqslant t. \end{cases}$$

We shall prove that

$$\text{(2.33)} \qquad \mathsf{P}'\{x_t(\omega) \neq \omega(t)\} = 0.$$

It is clear that $\{x_t(\omega) \neq \omega(t)\} \subseteq C_1 \cup C_2 \cup C_3 \cup C_4 \cup C_5$, where $C_1 = \{\alpha = \beta\}$, $C_2 = \{\alpha < t < \beta, \ \omega(t) = a\}$, $C_3 = \{\alpha < t < \beta, \ \omega(t) = b\}$, $C_4 = \{t \leqslant \alpha, \ \omega(t) \neq a\}$, $C_5 = \{t \geqslant \beta, \ \omega(t) \neq b\}$. Therefore, to prove (2.33) it is sufficient to check that $\mathsf{P}'(C_i) = 0$ $(i = 1, \ldots, 5)$.

If $s < t \in R$, then $\{s < \alpha = \beta < t\} \subseteq \{\omega(s) = a, \ \omega(t) = b\}$, and by (2.30)

$$\mathsf{P}'\{s < \alpha = \beta < t\} \leqslant q_{st}(a, b) = F(s, t).$$

Since $F(s, t)$ is continuous on the left in s and continuous on the right in t, it follows that for any $s < t$

$$\text{(2.34)} \qquad \mathsf{P}'\{s \leqslant \alpha = \beta \leqslant t\} \leqslant F(s, t).$$

For any $\Lambda = \{t_0 < t_1 < \ldots < t_n\}$

$$\mathsf{P}'\{\alpha=\beta\} \leqslant \mathsf{P}'\{\alpha=\beta<t_0\} + \sum_{k=1}^{n} \mathsf{P}'\{t_{k-1} \leqslant \alpha = \beta \leqslant t_k\} + $$
$$+\mathsf{P}'\{t_n < \alpha = \beta\} \leqslant q_{t_0}(b) + \sum_{k=1}^{n} F(t_{k-1}, t_k) + q_{t_n}(a).$$

By (2.20) and (2.28) the right-hand side does not exceed $\{\nu, h\} = \{\nu, h\}_\Lambda$, and by (2.1) $\mathbf{P}'(C_1) = 0$. The set C_2 is covered by the countable system of sets $\{\omega(s) \neq a, \ \omega(t) = a\}$ $(s \in R \cap (-\infty, t))$. According to (2.29), for $s < t$ we have $\mathbf{P}'\{\omega(s) \neq a, \ \omega(t) = a\} = q_{st}(E^b, a) = 0$. Consequently $\mathbf{P}'(C_2) = 0$. Similarly $\mathbf{P}'(C_3) = 0$. We note that $\mathbf{P}'(C_4) = \lim\limits_{s \uparrow t} \mathbf{P}' \{s < \alpha, \ \omega(t) \neq a\}$.

But for $s \in R$

$$\mathsf{P}'\{s<\alpha, \ \omega(t) \neq a\} \leqslant \mathsf{P}'\{\omega(s) = a, \ \omega(t) \neq a\} = q_{st}(a, E^b)$$

and by (2.31) and **2.4.A** $\mathbf{P}'(C_4) = 0$. Finally,

$$\mathsf{P}'(C_5) = \lim_{u \downarrow t} \mathsf{P}'\{\omega(t) \neq b, \ \beta < u\},$$
$$\mathsf{P}'\{\omega(t) \neq b, \ \ \beta < u\} \leqslant \mathsf{P}'\{\omega(t) \neq b, \ \ \omega(u) = b\} = q_{tu}(E_a, b)$$

and $\mathbf{P}'(C_5) = 0$ by (2.32) and **2.4.B**.

2.8. (2.27) and (2.33) imply that

$$\text{(2.35)} \qquad \mathsf{P}'\{x_{t_1}(\omega) \in \Gamma_1, \ \ldots, x_{t_n}(\omega) \in \Gamma_n\} = q_{t_1 \ldots t_n}(\Gamma_1, \ \ldots, \Gamma_n)$$
$$(t_1 < \cdots < t_n \in T. \ \ \Gamma_1, \ \ldots, \Gamma_n \in \mathscr{B}_a^b).$$

Consider the mapping ψ that associates with each element $\omega \in \Omega'$ the trajectory $x_t(\omega)$. It is easy to see that this is a measurable mapping of $(\Omega', \mathscr{N}')$ to $(\Omega, \mathscr{N})$. Putting $\mathbf{P}_\nu^h(A) = \mathbf{P}'\{\psi^{-1}(A)\}$ $(A \in \mathscr{N})$, we obtain a probability measure on the space of trajectories, and by (2.35) the equality (2.2) is satisfied.

Let us show that p^h is a transition function and that $\mathbf{P}_\nu^h$ belongs to $\mathscr{K}p^h$.

Obviously p^h satisfies **1.2.A**. We prove that it also satisfies **1.2.B**. Put $E_s = \{x: h^s(x) < \infty\}$, $E_s^0 = \{x: h^s(x) = 0\}$. Since $\mathbf{P}_t^s h^t(x) \leqslant h^s(x)$,

$$\text{(2.36)} \qquad \begin{cases} p(s, x; \ t, \ E \setminus E_t) = 0 & \text{for} \quad x \in E_s, \\ p(s, x; \ t, \ E \setminus E_t^0) = 0 & \text{for} \quad x \in E_s^0. \end{cases}$$

By (2.3), $p^h(t, y; u, \Gamma) = 0$ for all $t < u$ if $y \notin E_t \setminus E_t^0$. Therefore

$$\int_E p^h(s, x; \ t, dy)\, p^h(t, y; \ u, \Gamma) = \int_{E_t \setminus E_t^0} p^h(s, x; \ t, dy)\, p^h(t, y; \ u, \Gamma).$$

If $x \notin E_s \setminus E_s^0$, then the right-hand side is equal to $0 = p^h(s, x; u, \Gamma)$. If $x \in E_s \setminus E_s^0$, then by virtue of (2.3) and (2.36) it is again equal to $p^h(s, x; u, \Gamma)$.

To prove that $P_\nu^h \in \mathscr{K}^{p^h}$, we must verify for $s < t$, $\Gamma \in \mathscr{B}$ and a non-negative $\mathcal{N}_s$-measurable function ξ, that

$$P_\nu^h\{\xi,\ x_s \in E,\ \ x_t \in \Gamma\} = P_\nu^h\{\xi,\ x_s \in E,\ \ p^h(s,\ x_s;\ \ t,\ \Gamma)\}.$$

By Lemma A it suffices to check this equality for $\xi = \chi_{\Gamma_1}(x_{s_1}) \dots \chi_{\Gamma_n}(x_{s_n})$ $(s_1 < \dots < s_n \leqslant s;\ \ \Gamma_1,\ \dots,\ \Gamma_n \in \mathscr{B}_a^b)$, when it takes the form

$$(2.37)\quad q_{s_1 \dots s_n s t}(\Gamma_1, \dots, \Gamma_n,\ E,\ \Gamma) = \int_E q_{s_1 \dots s_n s}(\Gamma_1, \dots, \Gamma_n, dx)\, p^h(s, x; t, \Gamma).$$

By (2.14) – (2.18) we have (2.37) for all $\Gamma_1, \dots \Gamma_n$ from $\mathscr{B}_a^b$, if it is satisfied for $\Gamma_1, \dots, \Gamma_n \in \mathscr{B}$. We note that

$$(2.38)\qquad q_s(E_s^0) = 0,\ q_s(E \setminus E_s) = 0$$

(the latter follows from the inequality $\nu_s h^s \leqslant \{\nu,\ h\} < \infty$). From **2.5.B** and (2.38)

$$q_{s_1 \dots s_n s t}(\Gamma_1,\ \dots,\ \Gamma_n,\ E_s^0,\ \Gamma) \leqslant q_s(E_s^0) = 0,$$
$$q_{s_1 \dots s_n s t}(\Gamma_1,\ \dots,\ \Gamma_n,\ E \setminus E_s,\ \Gamma) \leqslant q_s(E \setminus E_s) = 0.$$

Therefore the left-hand side of (2.37) is equal to

$$(2.39)\qquad q_{s_1 \dots s_n s t}(\Gamma_1,\ \dots,\ \Gamma_n,\ E_s \setminus E_s^0,\ \Gamma)$$

On the other hand, by (2.3) the right-hand side of (2.37) is equal to

$$(2.40)\quad \int_{E_s \setminus E_s^0} \int_\Gamma q_{s_1 \dots s_n s}(\Gamma_1, \dots, \Gamma_n, dx) \frac{1}{h^s(x)} p(s,\ x;\ t,\ dy)\, h^t(y).$$

According to (2.2) the expressions (2.39) and (2.40) are equal. Hence (2.37) is satisfied.

§ 3 Excessive measures

3.1. In this section we study the spaces of p-excessive measures $\mathscr{R}^{p,h}$ and $\mathscr{R}_l^p$ introduced in **1.7**.

We first prove that these spaces satisfy **1.6.A**.

LEMMA 3.1. *The formula*

$$(3.1)\qquad \nu_t(\Gamma) = \int_{\mathscr{R}^{p,h}} \tilde{\nu}_t(\Gamma)\, \mu\,(d\tilde{\nu})$$

associates with each probability measure μ on $\mathscr{R}^{p,h}$ an element ν of $\mathscr{R}^{p,h}$. The formula

$$(3.2)\qquad \nu(\Gamma) = \int_{\mathscr{R}_l^p} \tilde{\nu}(\Gamma)\, \mu\,(d\tilde{\nu})$$

associates with any probability measure μ *on* $\mathscr{R}_l^p$ *a point* ν *of* $\mathscr{R}_l^p$.

PROOF. Obviously for each t, ν_t is a measure. Using Lemma A we deduce that for any non-negative $\mathscr{B}$-measurable function f, $\tilde{\nu}_t(f)$ is measurable with respect to $\tilde{\nu}$ and

$$\nu_t(f) = \int_{\mathscr{R}^{p,h}} \tilde{\nu}_t(f)\,\mu(d\tilde{\nu}). \tag{3.3}$$

In particular, for $f(x) = p(s, x; t, \Gamma)$ we have

$$\nu_s P_t^s(\Gamma) = \int_{\mathscr{R}^{p,h}} \tilde{\nu}_s P_t^s(\Gamma)\,\mu(d\tilde{\nu}).$$

Since the conditions 1.3.A′ – 1.3.B′ are satisfied for all $\tilde{\nu} \in \mathscr{R}^{p,h}$, they are satisfied for all ν. Further, it follows from (3.3) and (1.3) that

$$\{\nu, h\}_\Lambda = \int_{\mathscr{R}^{p,h}} \{\tilde{\nu}, h\}_\Lambda\,\mu(d\tilde{\nu}). \tag{3.4}$$

Let Λ' be a countable dense subset of T and let $\Lambda_n \uparrow \Lambda'$, where the Λ_n are finite. By (2.12) $\{\nu, h\}_{\Lambda_n} \uparrow \{\nu, h\}$. Therefore (3.4) implies that $\{\nu, h\} = 1$ and $\nu \in \mathscr{R}^{p,h}$.

It can be proved similarly that (3.2) defines a homogeneous p-excessive measure ν, where

$$\{\nu, h\} = \int_{\mathscr{R}_l^p} \{\tilde{\nu}, h\}\,\mu(d\tilde{\nu}).$$

To complete the proof of the lemma it is sufficient to recall (see 1.7) that if h is defined by (1.21), then $\{\nu, h\} = \nu(l)$ for any homogeneous p-excessive measure ν.

3.2. LEMMA 3.2. *Let* f *be an arbitrary non-negative measurable function on* E. *Then for each* t *the function* $\nu_t(f)$ $(\nu \in \mathscr{R}^{p,h})$ *is measurable with respect to* ν. *If* $p(s, x; t, \Gamma)$ *is measurable with respect to* t, x *for any* $s \in T$, $\Gamma \in \mathscr{B}$, *then* $\nu_t(f)$ $(\nu \in \mathscr{R}^{p,h}, t \in T)$ *is measurable with respect to* ν, t. *In particular, this is true for a homogeneous transition function satisfying* 1.7.A.

PROOF. The first assertion follows at once from Lemma A. To prove the second assertion we note that $\nu_s(\Phi_t)$ is measurable with respect to ν, t for any non-negative function $\Phi_t(x)$ that is measurable with respect to t, x. This is easily deduced by means of Lemma A and the fact that $\nu_s(\Phi_t) = \psi(t)\nu_s(\varphi)$ if $\Phi_t(x) = \psi(t)\varphi(x)$. By Lemma B the function $P_t^s f(x)$ is measurable with respect to t, x and therefore $F(s, t, \nu) = \nu_s(P_t^s f)$ is measurable with respect to ν, t. It only remains to note that

$$\nu_t(f) = \lim_{n\to\infty} \sum_{i=-\infty}^{+\infty} \chi_{\left(\frac{i-1}{n}, \frac{i}{n}\right]}(t)\, F\left(\frac{i-1}{n}, t, \nu\right).$$

3.3 THEOREM 3.1. *The correspondence* $\nu \to \mathbf{P}_\nu^h$ *defined by Theorem* 2.1. *is an isomorphism of the convex measurable spaces* $\mathscr{R}^{p,h}$ *and* $\mathscr{K}^{p^h}$.

PROOF. By Theorem 2.1. $\mathbf{P}_\nu^h \in \mathscr{K}^{p^h}$ for all $\nu \in \mathscr{R}^{p,h}$. If $\mathbf{P} = \mathbf{P}_\nu^h$, then

$$(3.5) \qquad \nu_t(\Gamma) = \int_{\Gamma} \frac{\nu_t'(dx)}{h^t(x)},$$

where

$$(3.6) \qquad \nu_t'(A) = \mathsf{P}\{x_t \in A\}, \quad \Gamma^t = \Gamma \cap \{x: 0 < h^t(x) < \infty\}.$$

Thus, no element $\mathbf{P}$ of $\mathscr{K}^{p^h}$ can have two different inverse images in $\mathscr{R}^{p,h}$. Now let $\mathbf{P}$ be any element of $\mathscr{K}^{p^h}$. Define ν by (3.5) – (3.6). Formula (1.17) implies that for any $t_1 < \ldots < t_n$, $\Gamma_1, \ldots, \Gamma_n \in \mathscr{B}$

$$(3.7) \qquad \mathsf{P}\{x_{t_1} \in \Gamma_1, \ldots, x_{t_n} \in \Gamma_n\} =$$

$$= \int_{\Gamma_1} \ldots \int_{\Gamma_n} \nu_{t_1}'(dy_1)\, p^h(t_1, y_1; t_2, dy_2) \ldots p^h(t_{n-1}, y_{n-1}; t_n, dy_n).$$

Because of (3.5) – (3.6) and (2.3) the right-hand side of (3.7) is equal to

$$\int_{\Gamma_1^t} \ldots \int_{\Gamma_n^t} \nu_{t_1}(dy_1)\, p(t_1, y_1; t_2, dy_2) \ldots p(t_{n-1}, y_{n-1}; t_n, dy_n)\, h^{t_n}(y_n).$$

The right-hand side of (2.2) is equal to this same expression by (2.36) and (2.38). Hence it follows by Lemma A that $\mathbf{P} = \mathbf{P}_\nu^h$. Note further that

$$(3.8) \qquad \nu_t(\Gamma) = \mathsf{P}\left\{\frac{\chi_\Gamma(x_t)}{h^t(x_t)}\right\}$$

and that $\nu_s \mathbf{P}_t^s(\Gamma) = \mathbf{P}\{\chi_{a<s,\, x_t \in \Gamma}/h^t(x_t)\}$. Hence it is clear that ν is p-excessive. Since $\mathbf{P}_\nu^h(\Omega) = 1$, by Theorem 2.1 $\{\nu, h\} = 1$.

We have proved that the mapping $\nu \to \mathbf{P}_\nu^h$ defines a one-to-one correspondence between $\mathscr{R}^{p,h}$ and $\mathscr{K}^{p^h}$ and that the inverse mapping is given by (3.8). The mapping (3.8) obviously preserves the convex and measurable structures. To prove that the mapping $\nu \to \mathbf{P}_\nu^h$ has the same properties it is sufficient to check that for any $\mathscr{N}$-measurable function ξ:

a) $\mathbf{P}_\nu^h(\xi)$ is measurable with respect to ν;

b) if ν is the centre of gravity of the distribution μ, then

$$(3.9) \qquad \mathsf{P}_\nu^h(\xi) = \int_{\mathscr{R}^{p,h}} \mathsf{P}_{\tilde{\nu}}^h(\xi)\, \mu(d\tilde{\nu}).$$

By Lemma A it is sufficient to prove both these assertions when $\xi = \chi_{\Gamma_1}(x_{t_1}) \ldots \chi_{\Gamma_n}(x_{t_n})$ $(t_1 < \ldots < t_n;\ \Gamma_1, \ldots, \Gamma_n \in \mathscr{B})$. But in this case

$$\mathsf{P}_\nu^h(\xi) = \nu_{t_1}(\Phi),$$

where

$$\Phi(y_1) = \chi_{\Gamma_1}(y_1) \int_{\Gamma_2} \cdots \int_{\Gamma_n} p(t_1, y_1; t_2, dy_2) \ldots p(t_{n-1}, y_{n-1}; t_n, dy_n)\, h^{t_n}(y_n),$$

so that our assertion is obvious.

3.4. We shall now prove that the measurable space $\mathscr{R}^{p,h}$ is standard.

Fix any countable dense set $\tilde{T}$ of T. We say that $\tilde{\nu}$ is a p-excessive measure on $\tilde{T}$ if $\tilde{\nu}_t$ is defined and satisfies conditions **1.3.A′ – 1.3.B′** only for $t \in \tilde{T}$. For such measures $\{\tilde{\nu}, h\}$ is defined as the supremum of $\{\tilde{\nu}, h\}_\Lambda$ over all $\Lambda \subseteq \tilde{T}$. Put $\tilde{\nu} \in \tilde{\mathscr{R}}^{p,h}$, if $\{\tilde{\nu}, h\} = 1$. A measurable structure on $\tilde{\mathscr{R}}^{p,h}$ is given by the system of functions $\tilde{\nu}_t(\Gamma)$ $(t \in \tilde{T},\ \Gamma \in \mathscr{B})$. The fact that $\mathscr{R}^{p,h}$ is standard is an obvious corollary of the two theorems:

THEOREM 3.2. *For each p-excessive measure ν put*

$$\tilde{\nu}_t(\Gamma) = \nu_t(\Gamma) \quad (t \in \tilde{T},\ \Gamma \in \mathscr{B}). \tag{3.10}$$

The mapping (3.10) *is an isomorphism of the measurable space $\mathscr{R}^{p,h}$ onto $\tilde{\mathscr{R}}^{p,h}$. The inverse isomorphism is given by the formula*[1]

$$\nu_t(\Gamma) = \lim_{s \uparrow t} \tilde{\nu}_s P_t^s(\Gamma)\ ^{1)}. \tag{3.11}$$

THEOREM 3.3 *The measurable space $\tilde{\mathscr{R}}^{p,h}$ is standard.*

3.5 PROOF OF THEOREM 3.2. If ν is a p-excessive measure, then according to **1.3.B′** $\nu_s \mathrm{P}_t^s(\Gamma) \to \nu_t(\Gamma)$ for $s \uparrow t$. If (3.10) is satisfied, then $\nu_s P_t^s(\Gamma) = \nu_s P_t^s(\Gamma)$ for $s \in \tilde{T}$. Therefore (3.10) implies (3.11). We see that ν is uniquely recoverable from its restriction $\tilde{\nu}$ on $\tilde{T}$.

Now let $\tilde{\nu}$ be an arbitrary p-excessive measure on $\tilde{T}$. By virtue of **1.3.A′**, if $s_1, s_2 \in \tilde{T}$ and $s < s_2 < t$, then

$$\tilde{\nu}_{s_1} P_t^{s_1} = \tilde{\nu}_{s_1} P_{s_2}^{s_1} P_t^{s_2} \leqslant \tilde{\nu}_{s_2} P_t^{s_2}. \tag{3.12}$$

Therefore $\nu_s P_t^s$ converges to some measure ν_t' as $s \uparrow t$. By **1.3.B′** $\nu_t' = \tilde{\nu}_t$ for $t \in \tilde{T}$.

We show that ν' is a p-excessive measure. By (3.4) $\tilde{\nu}_s P_t^s \leqslant \nu_t'$ for $s \in \tilde{T}$, $s < t$. Hence for $s < t \in T$

$$\nu_s' P_t^s = \lim_{r \uparrow s} \tilde{\nu}_r P_s^r P_t^s = \lim_{r \uparrow t} \tilde{\nu}_r P_t^r \leqslant \nu_t', \tag{3.13}$$

so that **1.3.A′** is satisfied for ν'. (3.13) implies that $\nu_s' P_t^s$ is a non-decreasing function of s. Since it coincides with $\tilde{\nu}_s P_t^s$ for $s \in \tilde{T}$, we have

$$\lim_{s \uparrow t} \nu_s' P_t^s = \lim_{s \uparrow t} \tilde{\nu}_s P_t^s = \nu_s'.$$

---

[1] We do not specify that an argument tends to a limit over $\tilde{T}$ if the function is undefined outside this set.

Thus, ν satisfies 1.3.B$'$.

Represent $\widetilde{T}$ as a union of an increasing sequence of finite sets Λ_n. According to (2.12) $\{\nu, h\} = \lim \{\nu, h\}_{\Lambda_n}$. Therefore $\{\nu, h\} = \{\widetilde{\nu}, h\}$.

We have shown that the formulae (3.10) – (3.11) establish a one-to-one correspondence between $\mathscr{R}^{p,h}$ and $\widetilde{\mathscr{R}}^{p,h}$. The mapping (3.10) is obviously measurable. To prove that (3.11) is measurable it suffices to note that $\widetilde{\nu}_s P_t^s(\Gamma) = \widetilde{\nu}_s(\varphi_t^s)$, where $\varphi_t^s(x) = p(s, x; t, \Gamma)$, and to use Lemma 3.2.

NOTE. It is not difficult to see that the correspondence established between $\mathscr{R}^{p,h}$ and $\widetilde{\mathscr{R}}^{p,h}$ also preserves the natural convex structure.

3.6. THEOREM 3.3 is proved on the same lines as Theorem 3.2 of the paper [3].

The set $\widetilde{\mathscr{R}}^{p,h}$ is contained in the space $\mathscr{M}^{\widetilde{T}}$ of all mappings of $\widetilde{T}$ into the class $\mathscr{M}$ of all measures μ on E for which $\mu(E) \leqslant 1$. We introduce, as in [3], a compact topology on $\mathscr{M}^{\widetilde{T}}$ and check that the σ-algebra of Borel sets in this topology coincides with the σ-algebra $\mathscr{F}$, generated by the functions $F(\nu) = \nu_t(\Gamma)$ $(t \in \widetilde{T}, \Gamma \in \mathscr{B})$.

It remains to verify that $\widetilde{\mathscr{R}}^{p,h} \in \mathscr{F}$. Since E is standard, there is a countable system W of functions $f \in \mathscr{B}$, separating the measures of $\mathscr{M}$. Condition 1.3.A$'$ may be written in the form

$$\widetilde{\nu}_s P_t^s f \leqslant \widetilde{\nu}_t f \quad (f \in W, \ s < t \in \widetilde{T}). \tag{3.14}$$

By 1.3.A$'$ the condition 1.3.B$'$ can be stated in the following way:

$$\lim_{s \uparrow t} \widetilde{\nu}_s P_t^s(E) = \widetilde{\nu}_t(E) \quad (t \in \widetilde{T}). \tag{3.15}$$

Finally, by virtue of (2.12), the condition $\{\widetilde{\nu}, h\} = 1$ may be written down by the formula:

$$\lim \{\widetilde{\nu}, h\}_{\Lambda_n} = 1. \tag{3.16}$$

Clearly (3.14), (3.15) and (3.16) define a set from $\mathscr{F}$.

§ 4. Excessive functions

4.1. First of all we verify that the measurable structure on $\mathscr{S}^{p,\nu}$, introduced in 1.8 satisfies 1.6.A. For this we need a general lemma.

LEMMA 4.1. *Assume that in a standard measurable space E there is given a σ-finite measure ν and a family of finite measures $p(x, -)$ depending on a parameter x from a measurable space X. Let $p(x, \Gamma)$ be measurable with respect to x for any Γ and be equal to zero if $\nu(\Gamma) = 0$. Then there is a function $\pi(x, y)$ measurable with respect to x, y such that*

$$p(x, dy) = \pi(x, y)\, \nu(dy). \tag{4.1}$$

PROOF. It is sufficient to consider the case when ν is finite. For if the sets $E_1, \ldots, E_n, \ldots$ have finite measure with respect to ν and are pairwise disjoint with union E, then applying Lemma 4.1 to the finite measures $\nu_n(\Gamma) = \nu(\Gamma \cap E_n)$ and $p_n(x, \Gamma) = p(x, \Gamma \cap E_n)$ we construct a function $\pi_n(x, y)$, measurable with respect to x, y, so that $p_n(x, dy) = \pi_n(x, y)\nu_n(dy)$ and $\pi_n(x, y) = 0$ for $y \notin E_n$; their sum is measurable and satisfies (4.1).

It is known (see for example, [6] Ch. 3, § 37, 11) that every standard measurable space is isomorphic to one of the following spaces:

a) a finite or countable space whose subsets are all measurable;

b) the half-line $(0, \infty)$ with the Borel structure.

In both cases the measurable structure of E is generated by a monotone sequence of partitions of E into disjoint sets $E_{n,1}, \ldots, E_{n,m_n}$ (each set $E_{n+1,i}$ is contained in a set $E_{n,j}$). Let

$$\pi_n(x, y) = \frac{p(x, E_{n,i})}{\nu(E_{n,i})} \quad \text{for } y \in E_{n,i}.$$

Put

$$\pi(x, y) = \lim_n \pi_n(x, y) \text{ if the limit exists.} \tag{4.2}$$

If the limit does not exist, we set $\pi(x, y) = 0$. Martingale arguments show (see, e.g., [4], Corollary to Theorem 0.2) that (4.2) holds for ν – almost all y and $\pi(x, y)$ satisfies (4.1). Since $\pi_n(x, y)$ is measurable with respect to x, y, so is $\pi(x, y)$.

REMARK. Lemma 4.1 and its proof remain valid for every space E with a measurable structure generated by a countable family of sets.

LEMMA 4.2. *Put on the space* $\mathcal{S}^{p,\nu}$ *the measurable structure generated by the functions* $F(h) = \nu_t(h^t\chi_\Gamma)$ $(t \in T,\ \Gamma \in \mathcal{B})$. *Under condition* 1.8.B *the function* $h^s(x)$ $(s \in T)$ *is measurable with respect to the pair* x *and* h *and satisfies* 1.6.A.

PROOF. By virtue of 1.8.B and Lemma 4.1 there exists a function $\pi(s, x; t, y)$, measurable with respect to x, y, such that $p(s, x; t, dy) = \pi(s, x; t, y)\,\nu_t(dy)$. According to 1.3.B $h^s(x) = \lim_{t \downarrow s} P_t^s h^t(x)$. But $P_t^s h^t(x) = \nu_t(\varphi_x^t h^t)$, where $\varphi_x^t(y) = \pi(s, x; t, y)$. Denote by $\mathcal{H}$ the set of all non-negative functions $f_x(y)$ for which $\nu_t(f_x h^t)$ is measurable with respect to x, h. Obviously $\mathcal{H}$ contains all functions of the form $f_x(y) = \chi_{\Gamma_1}(x)\chi_{\Gamma_2}(y)(\Gamma_1, \Gamma_2 \in \mathcal{B})$. By Lemma A, $\mathcal{H}$ contains all functions $f_x(y)$ that are measurable with respect to x, y, in particular, it contains the function $\varphi_x^t(y)$ for any t. Hence $h^s(x)$ is measurable with respect to x, h. Let μ be any probability measure on $\mathcal{S}^{p,\nu}$ and let

$$h^t(x) = \int_{\mathcal{S}^{p,\nu}} \tilde{h}^t(x)\,\mu(d\tilde{h}).$$

By Fubini's theorem, for any measure ν' on E

$$\nu(h^t) = \int_{\mathcal{S}^{p,\nu}} \nu'(\tilde{h}^t)\,\mu(d\tilde{h}). \tag{4.3}$$

Taking here $\nu'(\Gamma) = p(s, x; t, \Gamma)$ we have

$$P_t^s h^t = \int_{\mathcal{S}^{p,\nu}} P_t^s \tilde{h}^t \mu(d\tilde{h}),$$

and the validity of 1.3.A – 1.3.B for h follows from their validity for all $\tilde{h} \in \mathcal{S}^{p,\nu}$. From (1.3) and (4.3) it follows that

$$\{\nu, h\}_\Lambda = \int_{\mathcal{S}^{p,\nu}} \{\nu, \tilde{h}\}_\Lambda\,\mu(d\tilde{h}).$$

Using (2.12) we conclude that $\{\nu, h\} = 1$, so that $h \in \mathcal{S}^{p,\nu}$.

4.2. We prove a lemma on the limit of a monotonic sequence of measures. It will be used in this and the following sections.

LEMMA 4.3. *Let* μ_n *be a non-decreasing sequence of measures, bounded above by a* σ*-finite measure* ν. *Then the formula*

$$\mu(\Gamma) = \lim \mu_n(\Gamma) \tag{4.4}$$

defines a σ*-finite measure* μ, *and* $\mu_n(f) \uparrow \mu(f)$ *for any non-negative measurable function* f. *If all the measures* μ_n *are absolutely continuous with respect to some* σ*-finite measure* ν', *then* μ *is also absolutely continuous with respect to* ν'.

PROOF. By the Radon-Nikodym theorem we have $\mu_n(dx) = a_n(x)\nu(dx)$ for certain measurable functions $a_n(x)$. Obviously $a_n \leqslant a_{n+1} \leqslant 1$ (a.s. ν) for any n. Therefore $a_n \uparrow a \leqslant 1$ (a.s. ν). For any measurable function $f \geqslant 0$

$$\mu_n(f) = \nu(a_n f)\uparrow\nu(af). \tag{4.5}$$

In particular, $\mu_n(\Gamma) \uparrow \nu(a\chi_\Gamma)$. Hence (4.4) defines a σ-finite measure. The relation (4.5) can obviously be rewritten in the form[1] $\mu_n(f) \uparrow \mu(f)$.

The second assertion of the lemma is an obvious corollary of (4.4).

COROLLARY. *If the measure $p(s, x; t, -)$ is absolutely continuous with respect to a σ-finite measure ν_t, then all the p-excessive measures ν'_t are also absolutely continuous with respect to ν_t.*

For $\nu'_s P^s_t \uparrow \nu'_t$ for $s \uparrow t$. The equality $\nu_t(\Gamma) = 0$ implies that

$$\nu'_s P^s_t(\Gamma) = \int_E \nu'_s(dx)\, p(s, x; t, \Gamma) = 0.$$

Consequently, $\nu'_s P^s_t$ is absolutely continuous with respect to ν_t. By Lemma 4.3 ν'_t is also absolutely continuous with respect to ν_t.

4.3. We now prove two lemmas which under certain conditions allow us to construct, with respect to a function h, a p-excessive function $\widetilde{h}$ (called the regularizer of h).

LEMMA 4.4. *If $P^s_t\widetilde{h}^t(x)$ is, for any $s \in T$, $x \in E$, a non-increasing function of t for $t > s$, then for $t \downarrow s$*

$$P^s_t\widetilde{h}^t(x) \uparrow h^s(x), \tag{4.6}$$

where h is a p-excessive function.

PROOF. From the condition we have $P^s_t\widetilde{h}^t \geqslant P^s_{u'}\widetilde{h}^{u'} \geqslant P^s_v\widetilde{h}^v$ for $s < t < u' < v$ By virtue of (4.6) $P^s_{u'}\widetilde{h}^{u'} = P^s_u P^u_{u'}\widetilde{h}^{u'} \uparrow P^s_u h^u$ for $u' \uparrow u < v$. Therefore $P^s_t\widetilde{h}^t \geqslant P^s_u h^u \geqslant P^s_v\widetilde{h}^v$ for $s < t < u < v$. When $t \downarrow s$, this inequality becomes

$$h^s \geqslant P^s_u h^u \geqslant P^s_v\widetilde{h}^v. \tag{4.7}$$

Taking $u \downarrow s$ and $v \downarrow s$ we have hence

$$h^s \geqslant \overline{\lim_{u\downarrow s}}\, P^s_u h^u \geqslant \underline{\lim_{u\downarrow s}}\, P^s_u h^u \geqslant h^s. \tag{4.8}$$

Now h satisfies **1.3.A** because of (4.7) and satisfies **1.3.B** because of (4.8).

NOTE. The same arguments show with the help of Lemma 4.3 that if $\widetilde{\nu}_s P^s_t(\Gamma)$, for any $t \in T$, and $\Gamma \in \mathscr{B}$ is a non-decreasing function of s for $s < t$, then for $s \uparrow t$

$$\widetilde{\nu}_s P^s_t(\Gamma)\uparrow\nu_t(\Gamma),$$

[1] If the sequence μ_n is non-decreasing, then the same argument proves the existence of a σ-finite measure μ such that $\mu_n(f) \downarrow \mu(f)$ for any non-negative function f satisfying the condition $\nu(f) < \infty$.

where ν is a p-excessive measure.

LEMMA 4.5. *If* **1.8.B** *is satisfied and if*

$$P_t^s \widetilde{h}^t \leqslant \widetilde{h}^s \quad (n.\,н.\ \nu_s) \text{ for } s < t, \tag{4.9}$$

then (4.6) *defines a p-excessive function h. If, in addition,*

$$P_{t_n}^s \widetilde{h}^{t_n} \to \widetilde{h}^s \quad (n.\,н.\ \nu_s) \text{ for } t_n \downarrow s, \tag{4.10}$$

then $h^s = \widetilde{h}^s$ *(a.s.* ν_s*) for all s.*

PROOF. By virtue of **1.8.B**, it follows from (4.9) that for $r < s < t$

$$P_t^r \widetilde{h}^t = P_s^r P_t^s \widetilde{h}^t \leqslant P_s^r \widetilde{h}^s,$$

and the first assertion of Lemma 4.5 follows from Lemma 4.4. From (4.9) it follows that $h^s \leqslant \widetilde{h}^s$ (a.s. ν_s). On the other hand, by (4.10) $\nu_s(\widetilde{h}^s) = \lim \nu_s(P_{t_n}^s \widetilde{h}^{t_n}) = \nu_s(h^s)$. Therefore $\widetilde{h}^s = h^s$ (a.s. ν_s).

4.4. We now pass on to the study of the correspondence described in 1.8 between p-excessive functions and p-excessive measures.

LEMMA 4.6. *Under the conditions* **1.8.A** – **1.8.C** *there corresponds to each* ν' *of* $\mathcal{R}^{\hat{p},r}$ *one and only one element* $h_{\nu'}$ *of* $\mathcal{S}^{p,\nu}$ *such that*

$$\nu_t'(dx) = h_{\nu'}^t(x)\, \nu_t(dx). \tag{4.11}$$

For any t the function $h_{\nu'}^t(x)$ *is measurable with respect to* ν' *and x.*

PROOF. First of all we note that if h and $\overline{h}$ are p-excessive functions and $\nu_t'(dx) = h^t(x)\,\nu_t(dx) = \overline{h}^t(x)\,\nu_t(dx)$, then by **1.8.B** $P_t^s h^t = P_t^s \overline{h}^t$ for $s < t$ and by **1.3.B** $h^s = \overline{h}^s$ for all s. Hence to each $\nu' \in \mathcal{R}^{\hat{p},r}$ there corresponds not more than one element $h_{\nu'}$.

To prove the existence of such an element we apply the corollary to Lemma 4.3 to the transition function $\hat{p}$. By Lemma 4.1 we can find a function $\widetilde{h}_{\nu'}^t(x)$ that is measurable with respect to ν', x and such that $\nu_t'(dx) = \widetilde{h}_\nu^t(x)\nu_t(dx)$. As shown in 1.8, $\widetilde{h}_{\nu'}$ satisfies (4.9) – (4.10). By Lemma 4.5 the formula (4.6) defines a p-excessive function $h_{\nu'}$ where $\widetilde{h}_{\nu'}^s = h_{\nu'}^s$ (a.s. ν_s), so that (4.11) is satisfied. It follows from this equality (see 1.8) that $\{\nu, h\}_p = \{\nu', r\}_{\hat{p}} = 1$, hence $h_{\nu'} \in \mathcal{S}^{p,\nu}$. It remains to note that by Lemma B the function $P_t^s \widetilde{h}_{\nu'}^t(x)$, and hence also $h_{\nu'}^t(x)$, is measurable with respect to ν', x.

THEOREM 4.1. *Under the conditions* **1.8.A** – **1.8.C** *the formula*

$$\nu_t^h(dx) = h^t(x)\nu_t(dx) \tag{4.12}$$

defines an isomorphic mapping of the convex measurable space $\mathcal{S}^{p,\nu}$ *to the space* $\mathcal{R}^{\hat{p},r}$.

PROOF. In 1.8. we have seen that $\nu^h \in \mathcal{R}^{\hat{p},r}$ for all $h \in \mathcal{S}^{p,\nu}$. By Lemma 4.6 the mapping $h \to \nu^h$ is one-to-one and its inverse is the mapping

$\nu' \to h_{\nu'}$. It follows from the definition of measurability in $\mathscr{S}^{p,\nu}$ that for all $t \in T$, $\Gamma \in \mathscr{B}$ the function $\nu_t^h(\Gamma)$ is measurable with respect to h. Therefore the mapping (4.12) is measurable. Lemma 4.6 and Lemma B imply that for any $t \in T$, $\Gamma \in \mathscr{B}$ the function $\nu_t(h_{\nu'}^t \chi_\Gamma)$ is measurable with respect to ν'. Therefore the mapping $\nu' \to h_{\nu'}$ is also measurable.

If h is the centre of gravity of a distribution μ on $\mathscr{S}^{p,\nu}$, then by Fubini's theorem

$$\nu_t^h(\Gamma) = \int_{\mathscr{S}^{p,\nu}} \nu_t^{\tilde{h}}(\Gamma)\,\mu(d\tilde{h})$$

(the measurability of $h^t(x)$ with respect to h, x is used here). Hence the mapping $h \to \nu^h$ preserves the convex structure. On the other hand, if ν' is the centre of gravity of a distribution μ on $\mathscr{R}^{\hat{p},r}$, then the function

$$h^t(x) = \int_{\mathscr{R}^{\hat{p},r}} h_{\tilde{\nu}}^t(x)\,\mu(d\tilde{\nu})$$

is the density of ν_t' with respect to ν_t (this can be deduced by means of Fubini's theorem using the measurability of $h_\nu^t(x)$ with respect to ν, x). It is easy to see that $h \in \mathscr{S}^{p,\nu}$. Consequently $h = h_{\nu'}$ and the mapping $\nu' \to h_{\nu'}$ preserves the convex structures.

§ 5 Homogeneous excessive functions

5.1. In this section we assume that p is a homogeneous transition function satisfying condition **1.7.A.** We first derive some properties of the operators

$$(5.1)\qquad \begin{cases} G_\lambda f(x) = \displaystyle\int_0^\infty e^{-\lambda t} P_t f(x)\,dt, \\ (\mu G_\lambda)(\Gamma) = \displaystyle\int_0^\infty e^{-\lambda t}(\mu P_t)(\Gamma)\,dt, \end{cases}$$

then prove two lemmas, similar to Lemmas 4.4 and 4.5, and with their help check the equivalence of the conditions **1.3.**α – **1.3.**β, **1.3.**α' – **1.3.**β' and **1.9.**α – **1.9.**β, **1.9.**α' – **1.9.**β'. After this the connection between the spaces $\mathscr{S}_\gamma^p$ and $\mathscr{R}_l^{\hat{p}}$ is established in exactly the same way as that between $\mathscr{S}^{p,\nu}$ and $\mathscr{R}^{\hat{p},r}$ was established in § 4.

The semigroup property $P_s P_t = P_{s+t}$ of the operators P_t implies that

$$(5.2)\qquad G_\lambda = G_{\lambda+\delta} + \delta G_\lambda G_{\lambda+\delta},$$

$$(5.3)\qquad G_\lambda = G_{\lambda+\delta} + \delta G_{\lambda+\delta} G_\lambda \qquad (\lambda, \delta \geqslant 0).$$

We shall prove that for any measurable function $f \geqslant 0$

$$\varliminf_{\delta\to\infty} \delta G_\lambda G_{\lambda+\delta} f \geqslant G_\lambda f. \tag{5.4}$$

Put[1] $f_n = f \wedge n$. By virtue of (5.1) $G_\lambda f_n \to 0$ as $\lambda \to \infty$. Therefore (5.2) implies that $\delta G_\lambda G_{\lambda+\delta} f_n \to G_\lambda f_n$ as $\delta \to \infty$. Hence the left-hand side of (5.4) is not less than $G_\lambda f_n$ for any n. But $G_\lambda f_n \uparrow G_\lambda f$ as $n \to \infty$.

It is proved similarly that for any σ-finite measure μ

$$\varliminf_{\delta\to\infty} \mu\delta G_{\lambda+\delta} G_\lambda \geqslant \mu G_\lambda \tag{5.5}$$

(μ is approximated by finite measures $\mu_n(\Gamma) = \mu(\Gamma \cap E_n)$, where E_n is a fixed sequence of sets such that $E_n \uparrow E$, $\mu(E_n) < \infty$).

5.2. LEMMA 5.1. *If $\lambda G_\lambda \tilde{h}(x)$ is for any $x \in E$ a non-decreasing function of λ, and if as $\lambda \to \infty$*

$$\lambda G_\lambda \tilde{h}(x) \uparrow h(x), \tag{5.6}$$

then $G_\lambda h = G_\lambda \tilde{h}$ for all λ and functions h satisfying 1.9.α – 1.9.β. *If $\lambda \tilde{\nu} G_\lambda(\Gamma)$ is for any $\Gamma \in \mathscr{B}$ a non-decreasing function of λ and if the measures $\lambda \tilde{\nu} G_\lambda(\Gamma)$ are bounded above by some σ-finite measure, then the limit*

$$\lambda \tilde{\nu} G_\lambda(\Gamma) \uparrow \nu(\Gamma) \quad (\lambda \to \infty) \tag{5.7}$$

is a measure satisfying 1.9.α' – 1.9.β'. *Here $\tilde{\nu} G_\lambda = \nu G_\lambda$ for all λ.*

PROOF. Let $\delta \to \infty$. By virtue of (5.6) $(\delta + \lambda) G_\lambda G_{\lambda+\delta} \tilde{h} \uparrow G_\lambda h$ and consequently $\delta G_\lambda G_{\lambda+\delta} \tilde{h} \uparrow G_\lambda h$. By (5.4) it hence follows that $G_\lambda h \geqslant G_\lambda \tilde{h}$. On the other hand, it is clear from (5.2) that $\delta G_\lambda G_{\lambda+\delta} \tilde{h} \leqslant G_\lambda \tilde{h}$. Therefore $G_\lambda h = G_\lambda \tilde{h}$. By virtue of (5.6) h satisfies 1.9.α – 1.9.β.

The second half of the lemma is proved similarly using (5.3) and (5.5) and Lemma 4.3.

LEMMA 5.2. *If for some $\lambda > 0$, $x \in E$, the measure $g_\lambda(x, -)$ is absolutely continuous with respect to the σ-finite measure ν and if $\lambda G_\lambda \tilde{h} \leqslant \tilde{h}$ (a.s. ν), then the function h defined by formula* (5.6) *satisfies conditions* 1.9.α – 1.9.β.

PROOF. By Lemma 5.1 it is sufficient to check that $\lambda_1 G_{\lambda_1} \tilde{h} \geqslant \lambda_2 G_{\lambda_2} \tilde{h}$ for $\lambda_1 > \lambda_2 > 0$. Replacing λ by λ_2 and δ by $\lambda_1 - \lambda_2$ in (5.3) we have

$$\lambda_2 G_{\lambda_2} \tilde{h} = \lambda_2 G_{\lambda_1} \tilde{h} + \lambda_2(\lambda_1 - \lambda_2) G_{\lambda_1} G_{\lambda_2} \tilde{h}. \tag{5.8}$$

The inequality $\lambda_2 G_{\lambda_2} \tilde{h} \leqslant \tilde{h}$ (a.s. ν) implies that $\lambda_2 G_{\lambda_1} G_{\lambda_2} \tilde{h} \leqslant G_{\lambda_1} \tilde{h}$ everywhere. Therefore the right-hand side of (5.8) does not exceed $\lambda_1 G_{\lambda_1} \tilde{h}$.

5.3. THEOREM 5.1. *The conditions* 1.3.α – 1.3.β *are equivalent to* 1.9.α – 1.9.β *and the conditions* 1.3.α' – 1.3.α' *are equivalent to* 1.9.α' – 1.9.β'.

---

[1] $C_1 \wedge C_2$ denotes the smaller of the two numbers C_1, C_2.

PROOF. If we change the variable in (5.1) by means of the formula $\lambda_t = u$ we obtain

$$\lambda G_\lambda = \int_0^\infty e^{-u} P_{u/\lambda}\, du. \tag{5.9}$$

Hence it is clear that the conditions **1.9.α** – **1.9.β** follow from **1.3.α** – **1.3.β** and **1.9.α′** – **1.9.β′** follows from **1.3.α′** – **1.3.β′**.

Now assume that ν satisfies **1.9.α′** – **1.9.β′**. Note that

$$\nu' G_{\lambda'} P_t = e^{\lambda' t} \int_t^\infty e^{-s\lambda'} \nu' P_s\, ds.$$

Hence clearly

$$\mu P_t \leqslant e^{\lambda' t} \mu, \tag{5.10}$$

if $\mu = \nu' G_{\lambda'}$ for some measure ν'. Replacing λ by λ' and δ by $\lambda - \lambda'$ in (5.3) we have $\nu G_\lambda = \nu' G_{\lambda'}$ where $\nu' = \nu - (\lambda - \lambda')\nu G_\lambda$ is a measure[1] for $\lambda' \in (0, \lambda)$. According to (5.10) $\nu G_\lambda P_t \leqslant e^{\lambda' t} \nu G_\lambda$ for all $\lambda' \in (0, \lambda)$. Hence $\lambda \nu G_\lambda P_t \leqslant \lambda \nu G_\lambda$. Letting $\lambda \to \infty$ we have $\nu P_t \leqslant \nu$. Clearly νP_t is a non-increasing function of t and according to the Note to Lemma 4.4 the formula

$$\nu P_t(\Gamma) \downarrow \hat{\nu}(\Gamma) \quad \text{and} \quad t \downarrow 0 \tag{5.11}$$

defines a p-excessive measure ν. It follows from (5.9) and (5.11) that

$$\lim_{\lambda \to \infty} \nu \lambda G_\lambda(\Gamma) = \hat{\nu}(\Gamma).$$

But by **1.9.β′** the left-hand side is equal to $\nu(\Gamma)$. This means that ν satisfies **1.3.α′** – **1.3.β′**.

1.3.α – **1.3.β** are deduced from **1.9.α** – **1.9.β** similarly. The difficulty arising from the fact that $h' = h - (\lambda - \lambda')G_\lambda h$ is only defined on the set $\{G_\lambda h < \infty\}$, can be overcome as follows. Consider the truncated functions $\tilde{h}_c = h \wedge c$. Together with h they satisfy **1.9.α** and for them we have $G_\lambda \tilde{h}_c < \infty$ everywhere. Therefore

$$\lambda P_t G_\lambda \tilde{h}_c \leqslant \lambda G_\lambda \tilde{h}_c. \tag{5.12}$$

For any x and c, $\lambda G_\lambda \tilde{h}_c(x)$ is a non-decreasing function of λ (this follows, for example, from Lemma 5.1). Passing to the limit in (5.12) we have

$$P_t h_c \leqslant h_c, \tag{5.13}$$

where

$$h_c = \lim_{\lambda \to \infty} \lambda G_\lambda \tilde{h}_c.$$

---

[1] The formula written down defines $\nu'(\Gamma)$ only for sets Γ with $\nu' G_\lambda(\Gamma) < \infty$. For any measurable Γ we put $\nu'(\Gamma) = \lim \nu'(\Gamma \cap E_n)$, where E_n is a fixed sequence of sets satisfying the conditions $E_n \uparrow E$ and $\nu'(E_n) < \infty$.

It is easy to verify that $h_c \uparrow h$ as $c \to \infty$. Therefore (5.13) implies that $P_t h \leqslant h$. After this the validity of **1.3.α – 1.3.β** for h is derived in the same way as that of **1.3.α′ – 1.3.β′** was derived for ν.

5.4. We are now prepared for the study of $\mathscr{S}^p_\nu$. The following propositions hold:

5.4.A. Take the measurable structure on $\mathscr{S}^p_\nu$ generated by the functions $F(h) = \nu(h\chi_\Gamma)$ $(\Gamma \in \mathscr{B})$. Under the condition **1.9.B** the function $h(x)$ is measurable with respect to h and x and satisfied **1.6.A.**

5.4.B. If all the measures $g_\lambda(x, -)$ are absolutely continuous with respect to ν, then all homogeneous p-excessive measures are also absolutely continuous with respect to ν.

5.4.C. Under the conditions **1.9.A – 1.9.C**, to each ν' of $\mathscr{R}^{\hat{p}}_l$ there corresponds one and only one element $h_{\nu'}$ of $\mathscr{S}^p_\nu$, for which $\nu'(dx) = h_{\nu'}(x)\nu(dx)$. The function $h_{\nu'}(x)$ is measurable with respect to ν' and x.

5.4.D. Under the conditions **1.9.A – 1.9.C** the formula $\nu^h(dx) = h(x)\nu(dx)$ defines an isomorphism of the convex measurable spaces $\mathscr{S}^p_\nu$ and $\mathscr{R}^{\hat{p}}_l$.

These assertions are proved in the same way as the analogous assertion in § 4. Here the conditions **1.9.α – 1.9.β** and **1.9.α′ – 1.9.β′** are used instead of **1.3.α – 1.3.β** and **1.3.α′ – 1.3.β′**, (1.28) and (1.29) instead of (1.26) and (1.27), and finally Lemmas 5.1 and 5.2 instead of Lemmas 4.4 and 4.5.

§ 6 The faces of the simplex $\mathscr{K}^p$. Invariant and nullexcessive measures and functions. The entrance laws

6.1. A measurable subset X of a convex measurable space Z is called a *face* if the centre of gravity of a distribution μ belongs to X when and only when μ is concentrated on X. If Z is a simplex, then any face X is also a simplex, and its set of extreme points coincides with $Z_e \cap X$. Faces X and Y of a simplex Z are called complementary if X_e and Y_e are disjoint and have the sum Z_e.

We shall investigate the faces of the simplex $\mathscr{K}^p$. It is easy to see that for each measurable A the set

$$\mathscr{K}^p_A = \{\mathsf{P}: \mathsf{P} \in \mathscr{K}^p,\ \mathsf{P}(A) = 1\}$$

is a face. We shall prove that all faces are described in this way. Further, each face corresponds to some set A depending on the initial behaviour of a trajectory. The latter means that $\{A, t > \alpha\} \in \mathscr{N}_t$ for any t. The collection of such sets (it is a σ-algebra) is denoted by $\mathscr{N}_{\alpha+}$. The set $\mathscr{V} = \mathscr{K}^p_e$ of extreme points of the simplex $\mathscr{K}^p$ coincides with the set of all measures $\mathbf{P} \in \mathscr{K}^p$, for which $\mathbf{P}(B)$ is equal to 0 or 1 for all $B \in \mathscr{N}_{\alpha+}$ (see the small print on pp. 49–50). In [3] there was constructed a measure $\mathsf{P}^{\omega} \in \mathscr{K}^{p}$ depending on ω with the following properties:

a) for each measurable set A, $\mathbf{P}^{\omega}(A)$ is an $\mathscr{N}_{\alpha+}$-measurable function of ω;

b) for each $\mathsf{P} \in \mathscr{V}$ $\mathsf{P}\{\omega: \mathsf{P}^{\omega} = \mathsf{P}\} = 1$.

Let X be any face of the simplex $\mathscr{K}^{p}$. By a) the set $A = \{\omega: \mathsf{P}^{\omega} \in X\}$ belongs to $\mathscr{N}_{\alpha+}$. By b) for $\mathsf{P} \in \mathscr{V}$ the condition $\mathbf{P}(A) = 1$ is equivalent to $\mathbf{P} \in X$. Consequently $\mathscr{K}^{p}_{A} \cap \mathscr{V} = X \cap \mathscr{V}$, the faces $\mathscr{K}^{p}_{A}$ and X have a common set of extreme points, and hence $X = \mathscr{K}^{p}_{A}$.

Associate with each A the set $\mathscr{V}_{A} = \mathscr{V} \cap \mathscr{K}^{p}_{A}$ of extreme points of the face $\mathscr{K}^{p}_{A}$. Let $A_1, A_2 \in \mathscr{N}_{\alpha+}$, $\mathsf{P} \in \mathscr{V}$. Clearly $\mathbf{P}(A_1 \cap A_2) = 1$ if and only if $\mathbf{P}(A_1) = 1$ and $\mathbf{P}(A_2) = 1$. Since $\mathbf{P}(A_1)$ and $\mathbf{P}(A_2)$ must be equal to 0 or 1, the equality $\mathbf{P}(A_1 \cup A_2) = 1$ holds if and only if $\mathbf{P}(A_1) = 1$ or $\mathbf{P}(A_2) = 1$. Therefore

$$\mathscr{V}_{A_1 \cap A_2} = \mathscr{V}_{A_1} \cap \mathscr{V}_{A_2}, \quad \mathscr{V}_{A_1 \cup A_2} = \mathscr{V}_{A_1} \cup \mathscr{V}_{A_2}.$$

From these formulae it follows that the faces $\mathscr{K}^{p}_{A_1}$ and $\mathscr{K}^{p}_{A_2}$ are complementary if and only if A_2 is the complement of A_1 in Ω.

Of special interest to us are the faces $\mathscr{K}^{p}_{I}$, $\mathscr{K}^{p}_{J}$, $\mathscr{K}^{p}_{L}$, where $I = \{\alpha = -\infty\}$, $J = \{\alpha > -\infty\}$, $L = \{\alpha = 0\}$. Obviously $I, J, L \in \mathscr{N}_{\alpha+}$ and J is the complement of I. Consequently $\mathscr{K}^{p}_{I}$ and $\mathscr{K}^{p}_{J}$ are complementary faces.

6.2. Let h be a strictly positive p-excessive function. Denote by $\mathscr{R}^{p,h}_{A}$ the subset of the space $\mathscr{R}^{p,h}$ corresponding to the face $\mathscr{K}^{ph}_{A}$ under the isomorphism of the space $\mathscr{R}^{p,h}$, and $\mathscr{K}^{ph}$, studied in § 3. Based on the formulae (2.7) – (2.8), (1.3′) and the footnote to p. 69 it is not difficult to establish that

a) $\mathscr{R}^{p,h}_{I}$ is the set of all p-invariant measures ν for which $\lim_{s \to -\infty} \nu_s h^s = 1$;

b) $\mathscr{R}^{p,h}_{J}$ is the set of all p-null-excessive measures ν normalised by the condition $\{\nu, h\} = 1$;

c) $\mathscr{R}^{p,h}_{L}$ is the set of all measures ν_t for which

$$\nu_t = 0 \quad \text{for} \quad t < 0, \tag{6.1}$$

$$\nu_s P^s_t = \nu_t \quad \text{for} \quad 0 < s < t, \tag{6.2}$$

$$\lim_{s \downarrow 0} \nu_s h^s = 1. \tag{6.3}$$

Being faces of the simplex $\mathscr{R}^{p,h}$, all of these sets are simplexes.

If p is a homogeneous transition function, then the measures $\nu_t(\Gamma)$ $(t \in T, \Gamma \in B)$, obtained by the conditions (6.1) – (6.2) are called the *entry laws* for p. The set of entry laws satisfying (6.3) with the natural measurable and convex structures is obviously isomorphic to $\mathscr{R}^{p,h}_{L}$. The result that this space is a simplex was proved in [2] under a number of additional assumptions on p and h.

6.3. If the transition function p is homogeneous and h is given by (1.21), then the mappings τ^t of $\mathcal{R}^{p,h}$, constructed in 1.7 leave invariant the faces $\mathcal{R}_I^{p,h}$ and $\mathcal{R}_J^{p,h}$. Applying the argument of 1.7 to each of these faces it is not difficult to deduce that the sets $\mathcal{R}_{l,I}^p = \mathcal{R}_I^{p,h} \cap \mathcal{R}_l^p$ and $\mathcal{R}_{l,J}^p = \mathcal{R}_J^{p,h} \cap \mathcal{R}_l^p$ form a pair of complementary faces of the simplex $\mathcal{R}_l^p$. The first of these is the set of homogeneous p-invariant measures ν normalized by the condition $\nu(l) = 1$; the second is the set of homogeneous p-null-excessive measures ν for which $\nu(l) = 1$.

6.4. Denote by $\mathcal{S}_I^{p,\nu}$, $\mathcal{S}_J^{p,\nu}$ the sets that correspond to the faces $\mathcal{R}_I^{\hat{p},r}$, $\mathcal{R}_J^{\hat{p},r}$ under the isomorphism of the spaces $\mathcal{S}^{p,\nu}$ and $\mathcal{R}^{\hat{p},r}$, considered in § 4 and by $\mathcal{S}_{\gamma,I}^p$, $\mathcal{S}_{\gamma,J}^p$ the images of the faces $\mathcal{R}_{l,I}^{\hat{p}}$, $\mathcal{R}_{l,J}^{\hat{p}}$ under the isomorphism of § 5. The reader may convince himself that:

a) $\mathcal{S}_I^{p,\nu}$ is the set of p-invariant functions h for which $\lim\limits_{t\to\infty} \nu_t h^t = 1$;

b) $\mathcal{S}_J^{p,\nu}$ is the set of p-null-excessive functions h for which $\nu(h) = 1$;

c) $\mathcal{S}_{\gamma,J}^p$ is the set of homogeneous p-invariant functions h for which $\gamma(h) = 1$;

d) $\mathcal{S}_{\gamma,J}^p$ is the set of homogeneous p-null-excessive functions h for which $\gamma(h) = 1$.

§ 7 Excessive functions and supermartingales

7.1. In this short section we study the random functions obtained by substituting the trajectories of a process into the p-excessive function $h^t(x)$. To define $h^t(x_t)$ for all $t \in T$ we have to extend the function $h^t(x)$ to E_a^o. We shall show that this can be done in such a way that $h^t(x_t)$ for any $\mathsf{P} \in \mathcal{K}^p$ is a supermartingale with respect to $\mathcal{N}_t$, P, that is, so that for all $s < t$

$$\mathsf{P}\{h^t(x_t) \mid \mathcal{N}_s\} \leqslant h^s(x_s) \text{ (a.s. } \mathsf{P})$$

It turns out that

$$\lim_{t\to-\infty} \mathsf{P}h^t(x_t) = \{\nu, h\}, \tag{7.1}$$

where

$$\nu_t(\Gamma) = \mathsf{P}\{x_t \in \Gamma\}. \tag{7.2}$$

7.2. Consider the function

$$u_t(\omega) = \begin{cases} h^t(x_t) & \text{при} \quad \alpha < t < \beta, \\ 0 & \text{при} \quad t \geqslant \beta. \end{cases}$$

It is clear that for any t and c

$$(7.3)\qquad \{u_t < c,\ \alpha < t\} \in \mathcal{N}_t.$$

It follows from (1.17) that if $\mathsf{P} \in \mathcal{K}^{\mathsf{P}}$, then $\mathsf{P}\{u_t \mid \mathcal{N}_s\} = P_t^s h^t(x_s)$ (a.s. $\mathbf{P}$, $x_s \in E$) for $s < t$. It is clear that $\mathsf{P}\{u_t \mid \mathcal{N}_s\} = 0 = u_s$(a.s. $\mathbf{P}$, $\beta \leqslant s$) and from **1.3.A – 1.3.B**

$$(7.4)\qquad \mathsf{P}(u_t \mid \mathcal{N}_s) \leqslant u \quad (\text{a.s. } \mathbf{P},\ \alpha < s)$$

and

$$(7.5)\qquad \mathsf{P}\{u_{t_n} \mid \mathcal{N}_s\} \to u_s \quad (\text{a.s. } \mathbf{P},\ \alpha < s) \text{ for } t_n \downarrow s.$$

A function $\mu_t(\omega)$ defined for $t > \alpha(\omega)$ and satisfying (7.3) and (7.4). is called a $(\mathsf{P}, \mathcal{N}_t)$*-supermartingale with moment of birth* $\boldsymbol{\alpha}$. If (7.5) is also satisfied, then we say that μ_t *is a supermartingale of class C.*

In [5] it was proved that if μ_t is a non-negative $(\mathsf{P}, \mathcal{N}_t)$-supermartingale with moment of birth α, then there exists a random variable (defined only to within equivalence with respect to $\mathbf{P}$) such that $\mu_t \to u_{\alpha+}$ (a.s. $\mathbf{P}$) when t converges from the right to α over any countable dense subset of T. In the same place it was proved (see (7)) that for any finite set Λ

$$(7.6)\qquad \mathsf{P}u_{\alpha_\Lambda} \leqslant \mathsf{P}u_{\alpha+},$$

where α_Λ is a random variable defined by (2.9). It is not difficult to verify that

$$(7.7)\qquad \mathsf{P}u_{\alpha_\Lambda} = \{\nu, h\}_\Lambda,$$

where ν is given by (7.2) (see the derivative of (2.10)).

Let Λ' be a countable dense subset of the line T and let $\Lambda_n \uparrow \Lambda'$. It is clear that α_{Λ_n} converges to α_Λ from the right over Λ', and therefore $u_{\alpha_{\Lambda_n}} \to u_{\alpha+}$ (a.s. $\mathbf{P}$). By Fatou's lemma $\underline{\lim}\ \mathbf{P}u_{\alpha_{\Lambda_n}} \geqslant \mathbf{P}u_{\alpha+}$. By means of (7.6) it hence follows that $\mathbf{P}u_{\alpha_{\Lambda_n}} \to \mathbf{P}u_{\alpha+}$. Taking (7.7) and (2.12) into account we conclude that

$$(7.8)\qquad \{\nu, h\} = \mathsf{P}u_{\alpha+}.$$

7.3. In [5] it was shown that the extension

$$\bar{u}_t = \begin{cases} u_t & \text{for } t > \alpha, \\ \mathsf{P}\{u_{\alpha+} \mid \mathcal{N}_t\} & \text{for } t \leqslant \alpha \end{cases}$$

of the function u_t is a $(\mathsf{P}, \mathcal{N}_t)$-supermartingale on the whole line T. It can be foreseen that if u_t belongs to the class C, then $\bar{u}_t$ also belongs to this class.

It follows from Lemma A that every $\mathcal{N}_t$-measurable function is constant on the set $\{\alpha \geqslant t\}$. Therefore $\bar{u}_t$ does not depend on ω for $t \leqslant \alpha$. If h^t

is extended to E_a^b by putting $h^t(a) = \bar{u}_t$, $h^t(b) = 0$, then $\bar{u}_t = h^t(x_t)$ for all t and ω. Consequently $h^t(x_t)$ is a $(\mathsf{P}, \mathcal{N}_t)$-supermartingale of class C on the whole line T. The equality (7.1) follows on comparison of Theorem 1 of [5] with (7.8).

§ 8 The construction of the adjoint transition function

8.1. In conclusion we prove the result stated in **1.10.**

THEOREM 8.1. *Assume that the transition function p has property* **1.10.A** *and let $\lambda(s)$ be an arbitrary strictly increasing function for which* (1.30) *is satisfied. Define ν and $\hat{p}$ by* (1.31) – (1.34). *Then the triple p, $\hat{p}$, ν satisfies conditions* **1.8.A** – **1.8.B.**

The proof is based on the following three propositions:

8.1.A. The relations

$$\nu_t(dy) = a_t(y) m_t(dy), \tag{8.1}$$

$$\int_E \nu_s(dx)\pi(s, x; t, y) = a_t(y) \quad (s < t, \; y \in E) \tag{8.2}$$

are valid.

8.1.B. If $A_t = \{y\colon a_t(y) = \infty\}$, then $m_t(A_t) = 0$ and $p(s, x; t, A_t) = 0$ for all s, x.

8.1.C. If $a_t(y) = 0$, then $\pi(s, x; t, y) = 0$ for all $s < t$, $x \in E$. If $A_t^0 = \{y\colon a_t(y) = 0\}$, then $p(s, x; t, A_t^0) = 0$ for all s, x.

The equality (8.1) follows from (1.30), (1.33), (1.35), Fubini's theorem and Lemma B, and (8.2) follows from (1.35), (1.31), and (1.33). Further, $\nu_t(E) < \infty$ because of (1.35) and (1.32), and **8.1.B** follows from (8.1) and (1.30).

Finally, according to (1.33) the equality $a_t(y) = 0$ implies that for λ-almost all s'

$$m_{s'}\{x'\colon \pi(s', x'; t, y) > 0\} = 0. \tag{8.3}$$

If $s < t$, then (8.3) is satisified for any $s' \in (s, t)$, and by (1.30) and (1.31)

$$\pi(s, x; t, y) = \int_E \pi(s, x; s', x') m_{s'}(dx') \pi(s', x'; t, y) = 0.$$

8.2. We pass on to the proof of Theorem 8.1. Let us verify that $\hat{p}$ is a transition function. The functions $a_t(y)$ and $\pi'(s, x; t, y)$ are measurable with respect to y by Lemma B. Consequently $\hat{p}(t, y; s, \Gamma)$ is measurable with respect to y. Further, from (1.36)., (1.34), (8.1), and (8.2) we have

$$\begin{cases} \hat{p}(t, y; s, E) = \int_E \nu_s(dx)\pi(s, x; t, y)\dfrac{1}{a_t(y)} = 1 & \text{for } 0 < a_t(y) < \infty, \\ \hat{p}(t, y; s, E) = 0 & \text{for } a_t(y) = 0 \text{ and } a_t(y) = \infty. \end{cases} \tag{8.4}$$

Consequently **1.2.A** is satisfied. To prove **1.2.B** it is sufficient to verify that for any $s < t < u$, $x, z \in E$,

$$\int_E \pi'(s, x; t, y) m_t(dy) \pi'(t, y; u, z) = \pi'(s, x; u, z). \tag{8.5}$$

If $a_u(z)$ is equal to 0 or ∞, then by (1.34) both sides of this equality are zero. If $0 < a_u(z) < \infty$, then according to (1.34) and (1.30), the left-hand side of (8.5) is equal to

$$\int_{E'} a_s(x) \pi(s, x; t, y) m_t(dy) \pi(t, y; u, z) \frac{1}{a_u(z)} =$$

$$= \int_{E'} a_s(x) p(s, x; t, dy) \pi(t, y; u, z) \frac{1}{a_u(z)},$$

where $E' = \{y: 0 < a_t(y) < \infty\}$. By virtue of **8.1.B** and **8.1.C** the value of the latter integral is unchanged if we replace E' by E. According to (1.31) and (1.34) this value is $\pi'(s, x; u, z)$. We now show that **1.8.A** – **1.8.C** are satisfied for p, $\hat{p}$, ν. According to (1.30), (8.1) and (1.36)

$$\nu_s(dx) p(s, x; t, dy) = m_s(dx) a_s(x) \pi(s, x; t, y) m_t(dy),$$

$$\nu_t(dy) \hat{p}(t, y; s, dx) = a_t(y) m_t(dy) \pi'(s, x; t, y) m_s(dx).$$

To prove that these expressions are equal we note that by virtue of **8.1.C** and (1.34) $a_s(x)\pi(s, x; t, y) = a_t(y)\pi'(s, x; t, y)$ for $a_t(y) < \infty$, and by virtue of **8.1.B** $m_t\{a_t(y) = \infty\} = 0$. Therefore 1.8.A is satisfied. If $\nu_t(\Gamma) = 0$, then by virtue of (6.1) $m_t(\Gamma) = m(\Gamma \cap A_t^0)$, and (1.30) and **8.1.B** imply that $p(s, x; t, \Gamma) = p(s, x; t, \Gamma \cap A_t^0) = 0$. Consequently **1.8.B** is valid. Finally, we show that if $\nu_s(\Gamma) = 0$ then $p(t, y; s, \Gamma) = 0$, thus fulfilling **1.8.C**. In view of (8.4) it is sufficient to consider the case when $0 < a_t(y) < \infty$. In this case (1.36), (8.1) and (1.34) imply that

$$\hat{p}(t, y; s, \Gamma) = \int_\Gamma \nu_s(dx) \pi(s, x; t, y) \frac{1}{a_t(y)} = 0.$$

The manuscript was read carefully by S. E. Kuznetsov. The author is grateful to him for remarks which led to the removal of a number of inaccuracies.

Appendix

THE DECOMPOSITION OF QUASI-INVARIANT MEASURES INTO ERGODIC MEASURES

Yu. I. Kifer and S. A. Pirogov

1. In a standard space E (see 1.2.) let τ_t $(t \in T = (-\infty, +\infty))$ be a one-parameter group of measurable transformations, where $\tau(x, t) = \tau^t(x)$ is a

measurable mapping of $E \times T$ onto E (the Borel structure is put on T).

Consider the class $\mathcal{M}$ of all probability measures μ on E satisfying the condition

$$\mu(\tau^s dx) = \rho_s(x)\cdot\mu(dx), \tag{1}$$

where the functions $0 < \rho_t(x) < \infty$ on $E \times T$ is measurable and such that

$$\rho_{t+s}(x) = \rho_t(x)\cdot\rho_s(\tau^t x). \tag{2}$$

Denote the σ-algebra of measurable subsets of E by $\mathcal{B}$ and put on $\mathcal{M}$ the measurable structure generated by the functions $F(\mu) = \mu(f)$ $(f \in B)$.[1]

A measure μ is called *ergodic* if $\mu(A) = 0$ for any A in the σ-algebra of measurable sets which are invariant with respect to τ^t.

THEOREM 1. *The set $\mathcal{M}_e$ of all ergodic measures from $\mathcal{M}$ is measurable, and any measure μ from $\mathcal{M}$ can be uniquely represented in the form*

$$\mu(\Gamma) = \int_{\mathcal{M}_e} \tilde{\mu}(\Gamma)\nu(d\tilde{\mu}) \quad (\Gamma \in \mathcal{B}), \tag{3}$$

where ν is a probability measure on $\mathcal{M}$, concentrated on $\mathcal{M}_e$.

The proof of Theorem 1 is obtained by means of a general theorem on the decomposition into $\mathcal{A}$-ergodic measures, which was proved in [3] (Theorem 2.1). On the strength of this result, to prove Theorem 1 it is sufficient to construct a countable system W of functions separating measures from $\mathcal{M}$, and a kernel $\mu^x(\Gamma)$ satisfying the conditions a) $\mu^x \in \mathcal{M}$ for each $x \in E$; $\mu^x(\Gamma)$ is $\mathcal{A}$-measurable for any $\Gamma \in \mathcal{B}$; b) $\mu(\Gamma \mid \mathcal{A}) = \mu^x(\Gamma)$ (a.s. μ) for $\mu \in \mathcal{M}$, $\Gamma \in \mathcal{B}$.

2. In the construction of the μ^x a basic role is played by

THEOREM 2. *If $\mu \in M$ and $\mu(|f|) < \infty$, then the following limit exists μ-almost surely*[2]

$$\hat{f}(x) = \lim_{\substack{r\to+\infty \\ s\to-\infty}} \frac{\int_s^r f(\tau^t x)\,\rho_t(x)\,dt}{\int_s^r \rho_t(x)\,dt} = \mu(f \mid \mathcal{A}) \text{ (a.s. } \mu). \tag{4}$$

PROOF. It is sufficient to consider the case $f \geqslant 0$.

We apply theorems from [7] to the operator $Uf(x) = f(\tau^1 x)\cdot\rho_1(x)$. We put

$$F_s^r(x) = \int_s^r f(\tau^t x)\cdot\rho_t(x)\,dt,$$

$$G_s^r(x) = \int_s^r \rho_t(x)\,dt, \quad F = F_0^1, \quad G = G_0^1.$$

---

[1] As in [3], $f \in \mathcal{B}$ means that $f \geqslant 0$ and is $\mathcal{B}$-measurable.

[2] In the set where the limit in (4) does not exist we take $f = 0$.

Note that $F_n^{n+1} = U^n F$ and $F_{-\infty}^{+\infty} = \sum_n U^n F$. Similar formulae are valid for G. It is easy to see that $\mu(F) = \mu(f) < \infty$ and that G is strictly positive and $\mu(G) = 1$.

The set $C = \{x\colon G_{-\infty}^{+\infty} = \infty\}$ is called the conservative part and the set $D = E \setminus C$ is called the dissipative part of E. It is clear that $C = C' \cup C''$, where $C' = \{x\colon G_0^{+\infty} = +\infty\}$ and $C'' = \{x\colon G_{-\infty}^0 = +\infty\}$.

According to Proposition V.5.2 of [7] $F_{-\infty}^{+\infty} < \infty$ for μ-almost all $x \in D$, and consequently the limit (4) exists and

$$\hat{f}(x) = \frac{F_{-\infty}^{+\infty}(x)}{G_{-\infty}^{+\infty}(x)}. \tag{5}$$

It is obvious that $\hat{f}(\tau^t x) = \hat{f}(x)$. Let $A \in \mathcal{A}$. Rearranging the integrals and making the change of variable $y = \tau^t x$ we obtain

$$\int_A \hat{f}\, d\mu = \int_A (G_{-\infty}^{+\infty}(x))^{-1} \mu(dx) \int_{-\infty}^{\infty} f(\tau^t x)\, \rho_t(x)\, dt = \int_A f\, d\mu.$$

Consequently

$$\hat{f}(x) = \mu(f \mid \mathcal{A}) \quad (\text{a.s. } \mu,\ D) \tag{6}$$

and for $x \in D$ the relation (4) is proved.

To prove that (4) is satisfied in C we use the Chacon-Ornstein theorem ([7], 289). By this theorem the limit

$$\lim_{m\to\infty} \frac{F_0^m}{G_0^m} = \psi \quad (\text{a.s. } \mu,\ C') \tag{7}$$

exists. Here m passes through integer values, $\mathcal{J}$ denotes the set of all measurable sets for which $\tau^1 A = A$ and $\psi = \frac{\mu(F \mid \mathcal{J})}{\mu(G \mid \mathcal{J})}$. We derive from the same theorem, replacing F by UF, that

$$\lim_{m\to\infty} \frac{F_1^{m+1}}{G_0^m} = \frac{\mu(UF \mid \mathcal{J})}{\mu(G \mid \mathcal{J})} = \psi \quad (\text{a.s. } \mu,\ C'). \tag{8}$$

From (7) and (8) we have

$$\lim_{m\to\infty} \frac{F_m^{m+1}}{G_0^m} = 0 \quad (\text{a.s. } \mu,\ C'). \tag{9}$$

This relation also applies to $f = 1$ and therefore

$$\lim_{m\to\infty} \frac{G_m^{m+1}}{G_0^m} = 0 \quad (\text{a.s. } \mu,\ C'). \tag{10}$$

Since $\frac{F_0^m}{G_0^{m+1}} \leqslant \frac{F_0^t}{G_0^t} \leqslant \frac{F_0^{m+1}}{G_0^m}$ for $m \leqslant t < m+1$, by virtue of (7), (9) and (10)

$$\lim_{t \to +\infty} \frac{F_0^t}{G_0^t} = \psi \quad (\text{a.s. } \mu, C'). \tag{11}$$

It is proved similarly that

$$\lim_{s \to -\infty} \frac{F_s^0}{G_s^0} = \psi \quad (\text{a.s. } \mu, C''). \tag{12}$$

It is now not difficult to show that

$$\lim_{\substack{s \to -\infty \\ t \to +\infty}} \frac{F_s^t}{G_s^t} = \psi \quad (\text{a.s. } \mu, C). \tag{13}$$

For (11) and (12) imply that (13) is satisfied in $C' \cap C''$. On the other hand, according to proposition V.5.2 of [7], $F_0^\infty < \infty$ on $C \setminus C'$ and $F_{-\infty}^0 < \infty$ on $C \setminus C''$. Therefore (13) is satisfied on $C \setminus C'$ and $C \setminus C''$.

It follows from (13) and the definitions of the functions $\hat{f}$ and ψ that $\mu(F \mid \mathcal{I}) = \hat{f} \cdot \mu(G \mid \mathcal{I})$ (a.s. μ, C). It is easily verified that $\hat{f} \in \mathcal{A}$ and $\mathcal{A} \subset \mathcal{I}$, therefore

$$\mu(F \mid \mathcal{A}) = \hat{f} \cdot \mu(G \mid \mathcal{A}) \quad (\text{a.s. } \mu, C). \tag{14}$$

We note further that for $h \in \mathcal{A}$

$$\mu(F \cdot h) = \int_0^1 dt \int_E h(x) f(\tau^t x) \rho_t(x) \mu(dx) = \mu(f \cdot h)$$

and consequently $\mu(F \mid \mathcal{A}) = \mu(f \mid \mathcal{A})$ (a.s. μ). Similarly $\mu(G \mid \mathcal{A}) = 1$ (a.s. μ). Therefore from (6) and (4) we have $\hat{f} = \mu(f \mid \mathcal{A})$ (a.s. μ), which completes the proof of Theorem 2.

3. It is known (see [3], 3.2) that in any standard space there exists a countable system W of bounded measurable functions with the following properties:

A. Let H be a system of functions containing W and having the following properties: if $f_1, f_2 \in H$ and c_1, c_2 are numbers, then $c_1 f_1 + c_2 f_2 \in H$; if the sequence $f_n \in H$ is uniformly bounded and converges to f at each point, then $f \in H$. Then H contains all bounded measurable functions.

B. $1 \in W$ and W separates probability measures.

C. If for probability measures μ_n the sequence $\mu_n(f)$ converges for any $f \in W$, then there is a probability measure μ such that

$$\mu_n(f) \to \mu(f) \quad \text{for} \quad f \in W.$$

Now consider the set E' on which the limit (4) exists for all functions of W. By Theorem 2 $E' \in \mathcal{A}$ and $\mu(E') = 1$ for any $\mu \in \mathcal{M}$. It follows from B and C that for any $x \in E'$ there exists a unique measure μ^x such that

$f(x) = \mu^x(f)$ for $f \in W$. Hence for any bounded $h \in \mathcal{A}$, $\mu \in \mathcal{M}$ and $f \in W$

$$\mu(f \cdot h) = \mu(\mu^x(f) \cdot h). \tag{15}$$

By property **A**, (15) is satisfied for all bounded measurable f.

Since $\hat{f} \in \mathcal{A}$, we see that $\mu^x(f)$ is $\mathcal{A}$-measurable as a function of x for $f \in W$ and hence, taking property **A** into account, for any bounded measurable f. From (15) we derive that $\mu^x(f) = \mu(f \mid \mathcal{A})$ (a.s. μ).

Put $V_t f(x) = f(\tau^{-t} x)$. It is clear that for all $\mu \in \mathcal{M}$

$$\mu(V_t f \mid \mathcal{A}) = \mu(\rho_t f \mid \mathcal{A}) \text{ (a.s. } \mu).$$

Hence it follows that for μ-almost all x

$$\mu^x(V_t f) = \mu^x(\rho_t f). \tag{16}$$

Using property **A** of W it is not difficult to show that there exists a set $Q_t \in \mathcal{A}$ such that $\mu(Q_t) = 1$ for $\mu \in \mathcal{M}$ and (16) is satisfied on Q_t for all $f \in \mathcal{B}$. Denote by Q the set of $x \in E'$ for which (16) is satisfied for all $f \in \mathcal{B}$ and almost all $t \in T$. By Fubini's theorem Q is $\mathcal{A}$-measurable and $\mu(Q) = 1$.

For $x \in Q$ denote by V_x the set of those $t \in T$ for which (16) is satisfied. It is clear that V_x is a set of full Lebesgue measure on the line that is closed with respect to addition, and hence $V_x = T$.

Thus, for $x \in Q$ $\mu^x \in \mathcal{M}$. For $x \notin Q$ put μ^x equal to some measure of $\mathcal{M}$. The kernel thus constructed satisfies conditions a) and b) of **1** and this proves Theorem 1.

In conclusion we note that Theorems 1 and 2 are valid for a group of measurable transformations $\tau^n (n = 0, \pm 1, \pm 2, \ldots)$, and here the proof is simpler although it follows the same plan.

The authors are grateful to E. B. Dynkin for raising this problem and for a number of remarks contributing to its solution.

References

[1] E. B. Dynkin *Osnovaniya teorii markovskikh protsessov*, Fizmatgiz, Moscow 1959. Translation: Die Grundlagen der Theorie der Markoffschen Prozesse, Springer-Verlag, Berlin–Heidelberg–Göttingen 1962.

[2] E. B. Dynkin, Excessive measures and entry laws for Markov processes, Mat. Sb. **84** (1971), 218–253.

[3] E. B. Dynkin, Initial and final behaviour of the trajectories of Markov processes, Uspekhi Mat. Nauk **26**:4 (1971), 153–172.
= Russian Math. Surveys **26**:4 (1971).

[4] E. B. Dynkin, Markovskie protsessy, Fizmatgiz, Moscow 1963.
Translation: Markov processes, Springer–Verlag, Berlin–Göttingen–Heidelberg 1965.

[5] E. B. Dynkin, Supermartingales with random birth times, Uspekhi Mat. Nauk 26:6 (1971), 213–214.
[6] K. Kuratowski, Topologie, fourth ed., 2 vols., Polish Acad. Sci., Warsaw 1958. English translation: Topology, 2 vols., Academic Press, New York 1967–69. Russian translation: *Topologiya,* tom 1, Izdat. Mir, Moscow 1966.
[7] J. Neveu, Bases mathematiques du calcul des probabilités, Masson et Cie, Paris 1964. English translation: Mathematical foundations of the calculus of probability, Holden–Day, San Francisco 1965. Russian translation: *Matematicheskie osnovy teoriya veroyatnostei,* Izdat. Mir, Moscow 1969.

Received by the Editors, 18 October 1971

Translated by D. Newton.

REGULAR MARKOV PROCESSES

This article is concerned with the foundations of the theory of Markov processes. We introduce the concepts of a regular Markov process and the class of such processes. We show that regular processes possess a number of good properties (strong Markov character, continuity on the right of excessive functions along almost all trajectories, and so on). A class of regular Markov processes is constructed by means of an arbitrary transition function (regular re-construction of the canonical class). We also prove a uniqueness theorem.

We diverge from tradition in three respects:

a) we investigate processes on an arbitrary random time interval;

b) all definitions and results are formulated in terms of measurable structures without the use of topology (except for the topology of the real line);

c) our main objects of study are non-homogeneous processes (homogeneous ones are discussed as an important special case).

In consequence of a), the theory is highly symmetrical: there is no longer disparity between the birth time α of the process, which is usually fixed, and the death time β, which is considered random.

Principle b) does not prevent us from introducing, when necessary, various topologies in the state space (as systems of coordinates are introduced in geometry). However, it is required that the final statements should be invariant with respect to the choice of such a topology.

Finally, the main gain from c) is simplification of the theory: discarding the "burden of homogeneity" we can use constructions which, generally speaking, destroy this homogeneity.

Similar questions have been considered (for the homogeneous case) by Knight [8], Doob [2], [3] and other authors.

Contents

§ 1. Introduction

1.1. Measurable spaces. Mathematicians use the term "space" widely, understanding by it a set equipped with a definite structure (topological, linear, and so on). We shall be concerned with *measurable spaces*: the measurable structure in X is given by a certain σ-algebra $\mathscr{F}(X)$ of subsets of X (measurable subsets). Particularly important for us are spaces that are isomorphic to the Borel subsets of complete, separable metric spaces. We call them *Borel spaces.*

We shall use the following notation. If $\mathscr{F}$ is any σ-algebra in the space X and f a function on X, then the expression $f \in \mathscr{F}$ denotes that f is non-negative[1] and measurable with respect to $\mathscr{F}$.

1.2. Random processes. Suppose that to each t in a certain set T there corresponds a set X_t. To give a *process* in X_t means to associate with every t a point $x_t \in X_t$. If T is an interval of the real line, then the parameter t is interpreted as time and the function x_t as a trajectory of motion.

A *random process* is a process that depends on chance, that is, on a point ω of a certain auxiliary set Ω. Here it is assumed that Ω and X_t are measurable spaces and that for a fixed t, $x_t(\omega)$ is a measurable map of Ω into X_t. We call Ω the space of elementary events. The collection $\mathscr{F}(\Omega)$ of all measurable sets of the space Ω is often denoted by $\mathscr{M}$.

If a random process $x_t(\omega)$ is given, then to every subset $\widetilde{T}$ of T there corresponds in Ω a minimal σ-algebra $\mathscr{N}(\widetilde{T})$, containing all the sets

$$\{\omega\colon x_t(\omega) \in \Gamma\} \quad (t \in \widetilde{T},\ \Gamma \in \mathscr{F}(X_t)).$$

The measurable structure in Ω corresponding to $\mathscr{N} = \mathscr{N}(T)$ is called the *structure generated by the random process* x_t.

1.3. Filtration of a measurable space. From now on we assume that $T = [-\infty, +\infty]$ is the extended real line. We apply the term *filtration* of a measurable space Ω to a system of σ-algebras $\mathscr{M}_t$ satisfying two conditions:

1.3.A. $\mathscr{M}_t \subseteq \mathscr{M}$ for all t.

1.3.B. $\mathscr{M}_s \subseteq \mathscr{M}_t$ if $s \leqslant t$.

A filtration $\mathscr{M}_t$ is called *continuous on the right* if $\mathscr{M}_t$ for any t coincides with the intersection $\mathscr{M}_{t+}$ of all $\mathscr{M}_u$ with $u > t$.

We say that a random process $x_t(\omega)$ is adapted to a filtration $\mathscr{M}_t$ if $\{\omega\colon x_t(\omega) \in \Gamma\} \in \mathscr{M}_t$ for all $t \in T$, $\Gamma \in \mathscr{F}(X_t)$. Every random process is adapted to the filtration $\mathscr{N}_t = \mathscr{N}[-\infty, t]$. We call this the *filtration generated by the random process* x_t.

---

[1] When speaking about non-negative functions, we mean functions with values in the extended half-axis $[0, +\infty]$.

A mapping τ of the space Ω into T is called a *Markov time* with respect to the filtration $\mathscr{M}_t$ if for any $t \in T$

$$\{\omega: \tau(\omega) \leqslant t\} \in \mathscr{M}_t.$$

To every such time there corresponds a σ-algebra $\mathscr{M}_\tau$ in Ω, which is defined in the following way: $A \in \mathscr{M}_\tau$ if $\{A, \tau \leqslant t\} \in \mathscr{M}_t$ for any t. It is clear that $\mathscr{M}_\tau \subseteq \mathscr{M}$ and that τ is measurable with respect to $\mathscr{M}_\tau$.

To an arbitrary filtration $\mathscr{M}_t$ of the space Ω there corresponds a measurable structure in the space $T \times \Omega$, generated by real-valued random processes $z_t(\omega)$ that are adapted to $\mathscr{M}_t$ and continuous on the right in t for every ω. We call it the *natural measurable structure in* $T \times \Omega$ *associated with the filtration* $\mathscr{M}_t$.

1.4. Random processes on a random time interval. We suppose that for every $\omega \in \Omega$ there is given an interval $\Delta(\omega) \subseteq T$. It can be closed, open or half-open, but it must not degenerate to a point. Let a_t and b_t be two distinct points in X_t and let $E_t = X_t \setminus \{a_t, b_t\}$. We say that $x_t(\omega)$ is a random process on $\Delta(\omega)$ if the singletons $\{a_t\}$ and $\{b_t\}$ are measurable and[1]

$$\begin{aligned} x_t(\omega) &= a_t \quad \text{for } t < \Delta(\omega), \\ x_t(\omega) &\in E_t \quad \text{for } t \in \Delta(\omega), \\ x_t(\omega) &= b_t \quad \text{for } t > \Delta(\omega). \end{aligned}$$

We call E_t the *state space*, a_t and b_t *fictitious states*, x_t the *extended state space*. (The introduction of *two* fictitious states corresponds to the assumption that if a particle is not observable, then it is known whether it has not yet appeared or has already disappeared.) We denote by $\alpha(\omega)$ and $\beta(\omega)$ the ends of the interval $\Delta(\omega)$. We call the first of these the *birth time* and the second the *death time*. It is easy to see that these are Markov times. Under our assumption $\alpha(\omega) < \beta(\omega)$.

1.5. Canonical random processes. Suppose that with every $t \in T$ there is associated a measurable space E_t. A canonical random process in the spaces E_t is constructed as follows. We define a trajectory in E_t to be a function of ω associating with every $t \in T$ the point $\omega(t)$ in the extended state space $X_t = E_t \cup \{a_t, b_t\}$ such that for certain $\alpha < \beta$ the following conditions hold:

$$\begin{aligned} \omega(t) &= a_t \quad \text{for } t \leqslant \alpha, \\ \omega(t) &\in E_t \quad \text{for } \alpha < t < \beta, \\ \omega(t) &= b_t \quad \text{for } \beta \leqslant t. \end{aligned}$$

We denote by Ω the set of all trajectories and put $x_t(\omega) = \omega(t)$. The space Ω, with measurable structure generated by the x_t, serves as the space of

[1] The notation $t < \Delta$ ($t > \Delta$) means that $t < u$ ($t > u$, respectively) for all $u \in \Delta$.

elementary events. The interval $\Delta = (\alpha, \beta)$ is open.

1.6. Families of measures. Sets of probability measures given on the same measurable space will be called families of measures. The family consisting of one measure $\mathbf{P}$ is also denoted by $\mathbf{P}$. The integral of a function ξ with respect to the measure $\mathbf{P}$ is denoted by $\mathbf{P}\xi$. (If ξ is defined on a measurable subsets $\tilde{\Omega}$ of Ω, then $\mathbf{P}\xi$ is defined to be zero outside $\tilde{\Omega}$.)

Every family of measures $\mathbf{K}$ is regarded as a measurable space: the measurable structure in $\mathbf{K}$ is defined by means of the system of functions $F(\mathbf{P}) = \mathbf{P}(A)$ (A is an arbitrary measurable set).

Let $\mathbf{K}$ be a family of measures on Ω. To an arbitrary probability measure μ on $\mathbf{K}$ there corresponds the measure on Ω defined by the formula

$$\mathbf{P}_\mu(A) = \int_{\mathbf{K}} \tilde{\mathbf{P}}(A)\, \mu(d\tilde{\mathbf{P}}).$$

If $\mathbf{P}_\mu \in \mathbf{K}$ for any μ, then the set $\mathbf{K}$ is called convex. A point $\mathbf{P} \in \mathbf{K}$ is called extreme if $\mathbf{P} \neq \mathbf{P}_\mu$ for any measure μ not concentrated at $\mathbf{P}$. A measurable set $A \subset \Omega$ is called $\mathbf{K}$-*negligible* if $\mathbf{P}(A) = 0$ for all $\mathbf{P} \in \mathbf{K}$. If a property holds for all $\omega \in \tilde{\Omega}$ lying outside a $\mathbf{K}$-negligible set A, we say that this property holds $\mathbf{K}$-*almost surely on* $\tilde{\Omega}$ or briefly ($\mathbf{K}$-a.s., $\tilde{\Omega}$). The symbol $\tilde{\Omega}$ is omitted if $\tilde{\Omega} = \Omega$.

Let $\mathcal{A}$ be any subset of the σ-algebra $\mathcal{F}(\Omega)$ of all measurable sets. With every measure $\mathbf{P}$ on Ω two extensions of the system $\mathcal{A}$ can be associated. The *closure* $\mathcal{A}_{\mathbf{P}}$ is defined as the totality of all $B \in \mathcal{F}(\Omega)$ such that

$$\mathbf{P}(A \cup B \setminus A \cap B) = 0$$

for some $A \in \mathcal{A}$. The *completion* $\mathcal{A}^{\mathbf{P}}$ is given by the following condition: $B \in \mathcal{A}^{\mathbf{P}}$ if there exist $A_1, A_2 \in \mathcal{A}$ such that $A_1 \subseteq B \subseteq A_2$ and $\mathbf{P}(A_1) = \mathbf{P}(A_2)$.

The completion (closure) of $\mathcal{A}$ with respect to a family of measures $\mathbf{K}$ is defined as the intersection of $\mathcal{A}^{\mathbf{P}}$ (respectively, $\mathcal{A}_{\mathbf{P}}$) over all $\mathbf{P} \in \mathbf{K}$. The closure and completion with respect to $\mathbf{K}$ contain all $\mathbf{K}$-negligible sets. If $\mathcal{A}$ is a σ-algebra, then the closure and completion of $\mathcal{A}$ are also σ-algebras.

The closure of $\mathcal{F}(\Omega)$ always coincides with $\mathcal{F}(\Omega)$, but the completion of $\mathcal{F}(\Omega)$ is usually larger. All measures $\mathbf{P} \in \mathbf{K}$ can be uniquely extended to the completion of $\mathcal{F}(\Omega)$ with respect to $\mathbf{K}$. (The extended measures are usually denoted by the same letters.) The completion of $\mathcal{F}(\Omega)$ with respect to the family of all probability measures on Ω is called the *universal completion,* and its elements are called *universally measurable sets.*

1.7. Stochastic processes. A stochastic process is a pair $(x_t, \mathbf{P})$, where x_t is a random process and $\mathbf{P}$ is a probability measure on the space of elementary events Ω. To every family of measures $\mathbf{K}$ on Ω there corresponds a family of stochastic processes $(x_t, \mathbf{K})$. We call this family *rigid* if the σ-algebra $\mathcal{N}$ generated by x_t separates measures from $\mathbf{K}$ (that is, if the coincidence on N of two measures $\mathbf{P}_1, \mathbf{P}_2 \in \mathbf{K}$ implies their equality). In particular, the family $(x_t, \mathbf{K})$ is rigid if $\mathcal{M}$ is generated by x_t, or generated by x_t and the $\mathbf{K}$-negligible sets, or is the completion of $\mathcal{N}$ with respect to the family $\mathbf{K}$.

A set $\mathcal{X} = (x_t, \mathcal{M}_t, \mathbf{K})$, where x_t is adapted to $\mathcal{M}_t$, is called a *family of stochastic processes with filtration.* The *completion* of such a family is defined as the family $\overline{\mathcal{X}} = (x_t, \overline{\mathcal{M}}_t, \overline{\mathbf{K}})$ that is obtained in the following way: we consider the completion $\overline{\mathcal{M}}$ of the σ-algebra $\mathcal{M} = \mathcal{F}(\Omega)$ with respect to $\mathbf{K}$, we denote by $\overline{\mathbf{K}}$ the set of measures obtained as a result of extending to $\overline{\mathcal{M}}$ all measures in $\mathbf{K}$, and we form the closure $\overline{\mathcal{M}}_t$ of the σ-algebra $\mathcal{M}_t$ with respect to $\mathbf{K}$.

If $x_t(\omega) = \tilde{x}_t(\omega)$ for all $t \in T$ ($\mathbf{K}$-a.s.), then we say that the random processes x_t and $\tilde{x}_t$ are $\mathbf{K}$-*indistinguishable* or that the families of stochastic processes $(x_t, \mathbf{K})$ and $(\tilde{x}_t, \mathbf{K})$ are *indistinguishable.*

1.8. Meyer processes. Let $\mathcal{X} = (z_t, \mathcal{M}_t, \mathbf{P})$ be a real-valued stochastic process with filtration given on a time interval T. We call it a *Meyer process* if the filtration $\mathcal{M}_t$ is continuous on the right and if we can find a random process $\tilde{z}_t$ that is $\mathbf{P}$-indistinguishable from z_t and measurable with respect to the natural structure in $T \times \Omega$ associated with $\mathcal{M}_t$ (see 1.3). Let $\overline{\mathcal{X}} = (z_t, \overline{\mathcal{M}}_t, \overline{\mathbf{P}})$ be the completion of the process $\mathcal{X}$. From Meyer's results [10] (VIII, Theorems 15 and 21; IV, Theorems 49 and 52) it follows that for any constant c the formula[1]

$$\tau(\omega) = \inf \{t: z_t(\omega) \geqslant c\}$$

defines a Markov time with respect to the filtration $\overline{\mathcal{M}}_t$; for every $\varepsilon > 0$ we can find a Markov time τ_ε (with respect to $\mathcal{M}_t$ such that

$$\{\tau_\varepsilon < \infty\} = \{z_{\tau_\varepsilon} \geqslant c\}, \quad \overline{\mathbf{P}}\{\tau < \infty\} \leqslant \overline{\mathbf{P}}\{\tau_\varepsilon < \infty\} + \varepsilon.$$

1.9. Supermartingales. Let $(z_t, \mathcal{M}_t, \mathbf{P})$ be a non-negative stochastic process with filtration on a random interval $\Delta(\omega)$. We call it a *supermartingale* if for any $s < t$

$$\mathbf{P}\{z_t \mid \mathcal{M}_s\} \leqslant z_s \quad (\mathbf{P}\text{-a.s.}, \quad s \in \Delta(\omega)).$$

It is known [6] that an arbitrary supermartingale is the restriction to $\Delta(\omega)$ of a supermartingale defined on the whole line T. Moreover, if $z_t(\omega)$ is bounded above or below by a constant c for all $t \in \Delta(\omega)$, then the same bound holds for the extended supermartingale.

We assume that the filtration $\mathcal{M}_t$ is continuous on the right. Let $z_t^n(\omega)$ be a sequence of random processes on the interval $\Delta(\omega)$ with values from the half-line $[0, +\infty]$ and suppose that $z_t^n(\omega) \uparrow z_t(\omega)$ for every ω. If $(z_t^n, \mathcal{M}_t, \mathbf{P})$ are supermartingales, then $(z_t, \mathcal{M}_t, \mathbf{P})$ is also a supermartingale. If in addition the $z_t^n(\omega)$ are continuous on the right on $\Delta(\omega)$ ($\mathbf{P}$-a.s.),

---

[1] The infimum of the empty set is defined to be $+\infty$.

then $z_t(\omega)$ has the same property. The case when Δ does not depend on ω was proved by Meyer ([10], VI, Theorem 16). The proof is easily extended to the general case.

1.10. Transition functions. For each $t \in T$, let E_t be a measurable space. A function $p(t, x; u, \Gamma)$ $(t < u \in T, x \in E_t, \Gamma \in \mathscr{F}(E_u))$ is called a *transition function* if:

1.10.A. $p(t, -; u, \Gamma)$ is a measurable function on E_t.

1.10.B. $p(t, x; u, -)$ is a measure on E_u.

1.10.C. $p(t, x; u, E_u) \leqslant 1$.

1.10.D. For any $s < t < u \in T$, $x \in E_s$, $\Gamma \in \mathscr{F}(E_u)$

$$\int_{E_t} p(s, x; t, dy)\, p(t, y; u, \Gamma) = p(s, x; u, \Gamma)$$

(Kolmogorov–Chapman equation).

1.10.E. $p(t, x; u, E_u) \to 1$ as $u \downarrow t$.[1]

We say that a transition function *separates states* if the equations

$$p(t, x; u, \Gamma) = p(t, y; u, \Gamma) \quad (u > t, \Gamma \in \mathscr{F}(E_u))$$

imply that $x = y$.

It is convenient to extend the definition of $p(t, x; u, \Gamma)$ to all pairs t, u, taking it to be zero if $u \leqslant t$.

The properties 1.10.A to 1.10.E are preserved if we extend all measures $p(t, x; u, -)$ to the σ-algebra of universally measurable sets. It is natural to call this operation *completion of a transition function.*

Of special interest is the *homogeneous case,* when the state space E_t is the same for all $t \in (-\infty, +\infty)$ and for any finite s

$$p(t + s, x; u + s, \Gamma) = p(t, x; u, \Gamma).$$

The set of pairs (t, x) $(t \in T, x \in E_t)$ is denoted by E and called the *phase space.* We introduce in E a measurable structure with the help of the system of functions

$$p(t, x; u, \Gamma) \quad (u \in T, \Gamma \in \mathscr{F}(E_u)).$$

We call it the p-measurable structure.

By virtue of 1.10.A, the mapping $x \to (t, x)$ of E_t into E is measurable. From 1.10.E it follows that

$$\{(t, x): t < t_0\} = \bigcup \{(t, x): p(t, x; r, E_r) > 0\},$$

where the sum is taken over all rational $r < t_0$. Therefore the function $f(t, x) = t$ is p-measurable.

We also mention that if the transition function p separates states, then

[1] Condition 1.10.E is often not included in the definition of a transition function, and is called the requirement of normality.

it separates the points of the space $\mathscr{E}$ as well.

1.11. Markov families and Markov classes. A family of stochastic processes with filtration $\mathscr{X} = \{x_t, \mathscr{M}_t, \mathbf{K}\}$ is called a *Markov family* if for any $\mathbf{P} \in \mathbf{K}$, $t < u$, $\Gamma \in \mathscr{F}(E_u)$

$$(1.1) \qquad \mathbf{P}\{x_u \in \Gamma \mid \mathscr{M}_t\} = p(t, x_t; u, \Gamma) \qquad (\mathbf{P}\text{-a.s.}, \quad t \in \Delta(\omega)),$$

where p is a certain transition function (the so-called transition function of the Markov family). When $\mathbf{K}$ consists of one element $\mathbf{P}$, the Markov family leads to one Markov process. The term "Markov family" and "family of Markov processes" will be used as synonyms.

From (1.1) it follows that for any non-negative measurable function φ on E_u

$$(1.2) \qquad \mathbf{P}\{\varphi(x_u) \mid \mathscr{M}_t\} = p(t, x_t; u, \varphi) \quad (\mathbf{P}\text{-a.s.}, \quad t \in \Delta(\omega)),$$

where

$$(1.3) \qquad p(t, x; u, \varphi) = \int_{E_u} p(t, x; u, dy)\, \varphi(y).$$

We agree to denote by $\mathbf{K}_t$ the totality of all $\mathbf{P} \in \mathbf{K}$ satisfying the condition

$$(1.4) \qquad \mathbf{P}\{\alpha = t\} = 1.$$

For $\mathbf{P} \in \mathbf{K}_t$, $x_s = a_s$ when $s < t$, $x_s \neq a_s$ when $s > t$ ($\mathbf{P}$-a.s.) (one cannot say anything definite regarding the value of x_t).

A Markov family $\mathscr{X} = (x_t, \mathscr{M}_t, \mathbf{K})$ is called a *Markov class* if for any $t \in T$, $x \in E_t$ there exists a measure $\mathbf{P}_{t,x} \in \mathbf{K}_t$ such that for $u > t$

$$(1.5) \qquad \mathbf{P}_{t,x}\{x_u \in \Gamma\} = p(t, x; u, \Gamma).$$

By virtue of (1.4)

$$(1.6) \qquad \mathbf{P}_{t,x}\{\alpha = t\} = 1.$$

From (1.2) and (1.5) it follows that for any $\xi \in \mathscr{N}(t, \infty]$ the function $\mathbf{P}_{t,x}\xi$ is measurable in x and for any measure $\mathbf{P} \in \mathbf{K}$

$$(1.7) \qquad \mathbf{P}\{\xi \mid \mathscr{M}_t\} = \mathbf{P}_{t,x_t}\xi \qquad (\mathbf{P}\text{-a.s.}, \quad t \in \Delta).$$

Therefore

$$(1.8) \qquad \mathbf{P}\chi_\Delta(t)\xi = \mathbf{P}\mathbf{P}_{t,x_t}\xi.$$

Under the operation of completion described in 1.7, a Markov family goes over into a Markov family and a Markov class into a Markov class. Simultaneously with the completion of a Markov family one can complete the transition function p.

A Markov class $\mathscr{X} = (x_t, \mathscr{M}_t, \mathbf{K})$ in the spaces E_t is called *a canonical class* if $x_t(\omega)$ is the canonical random process in E_t, constructed in 1.5,

$\mathcal{M}_t$ is the filtration generated by it, and **K** is the totality of all probability measures in Ω satisfying (1.1). In this way the canonical class of a transition function is unambiguously defined. On the other hand, if p is an arbitrary transition function in the Borel spaces E_t, then by means of Kolmogorov's theorem we can construct measures $\mathbf{P}_{t,x}$ in the space of trajectories Ω satisfying (1.1), (1.5) and (1.6), therefore we can construct the canonical class corresponding to p.

1.12. Regular families. A Markov process $(x_t, \mathcal{M}_t, \mathbf{P})$ with transition function p is called *regular* if:

1.12.A. The state space E_t is Borel (for any t).

1.12.B. The interval $\Delta = (\alpha, \beta]$ is closed on the right and open on the left.

1.12.C. The filtration $\mathcal{M}_t$ is continuous on the right.

1.12.D. The functions $p(t, x_t; u, \Gamma)$ are continuous on the right with respect to t on the interval Δ (**P**-a.s.).

We shall study various families and, above all, classes of regular Markov processes. We call these briefly *regular families* (*regular classes*).

Let $\mathcal{K}=(x_t, \mathcal{M}_t, \mathbf{K})$ be a regular class. A state $x \in E_t$ is called *essential* if

$$\mathbf{P}_{t,x}\{x_t = x\} = 1. \tag{1.9}$$

The remaining states are called *branch points*.

In § 4 it will be proved that the set of trajectories containing branching points is **K**-negligible.

1.13. The adjoint random process. Re-construction. Let $\mathcal{K} = (x_t, \mathcal{M}_t, \mathbf{K})$ be some Markov class. The random process

$$y_t(\omega) = \mathbf{P}_{t, x_t(\omega)},$$

for which $\mathbf{K}_t$ serves as state space, is called the *adjoint random process.*

Let $\mathcal{K} = (x_t, \mathcal{M}_t, \mathbf{K})$ and $\widetilde{\mathcal{K}} = (\widetilde{x}_t, \widetilde{\mathcal{M}}_t, \mathbf{K})$ be two Markov classes with a common space of elementary events and a common set of measures **K** (but generally speaking, with different state spaces). We say that $\mathcal{K}$ and $\widetilde{\mathcal{K}}$ are *indistinguishable* if the corresponding adjoint processes are **K**-indistinguishable, that is, if

$$\mathbf{P}_{t, x_t} = \widetilde{\mathbf{P}}_{t,\widetilde{x}_t} \text{ for all } t \in \Delta = \widetilde{\Delta} \quad (\mathbf{K}\text{-a.s.}).$$

We say that $\widetilde{\mathcal{K}}$ is a re-construction of $\mathcal{K}$ if for any $\mathbf{P} \in \mathbf{K}$

$$\chi_\Delta(t) = \chi_{\widetilde{\Delta}}(t), \quad \mathbf{P}_{t, x_t} = \widetilde{\mathbf{P}}_{t,\widetilde{x}_t} \quad (\mathbf{P}\text{-a.s.}), \tag{1.10}$$

with the exception of an at most countable set $\Lambda(\mathbf{P})$ of values t.

If $\widetilde{\mathcal{K}}$ is a re-construction of $\mathcal{K}$, then the intervals Δ and $\widetilde{\Delta}$ may be different. However, they have $\mathcal{K}$-almost surely the same closure and therefore can differ only in the inclusion or exclusion of their endpoints.

Indeed, by (1.10) for any $\mathbf{P} \in \mathbf{K}$ there exists a countable everywhere dense set S such that

$$(1.11) \qquad \chi_{\tilde{\Delta}}(t) = \chi_{\Delta}(t) \text{ for all } t \in S \quad (\mathbf{P}\text{-a.s.}).$$

If the closures of $\tilde{\Delta}$ and Δ are distinct, then there exists a $t \in S$ for which $\chi_{\tilde{\Delta}}(t) \neq \chi_{\Delta}(t)$. By (1.11) the probability of this is zero.

1.14. The plan of the article. In §§ 2–5 we obtain the properties of regular processes. § 6 contains a proof that two rigid, regular re-constructions of any class $\mathscr{K}$ are indistinguishable. Finally, in §§ 7–8 we prove the existence of a regular re-construction for the canonical class corresponding to any transition function p.

§ 2. Measurable structures in the phase space

2.1. In this section we consider two types of measurable structure in the phase space E. The first of these, introduced in 1.10, is connected with the transition function. The second is defined in terms of the family of regular Mark^v processes $\mathscr{K}$.˙

We need certain lemmas.

LEMMA 2.1. *Every countable system of measurable functions separating points of a Borel space E generates a measurable structure of E.*

The proof of the lemma is based on the fact that for a one-to-one measurable mapping of one Borel space into another the image of a measurable set is measurable (see [9], § 39, 5, Theorem 2). Let $f_n (n = 1, 2, \ldots)$ be a given family of functions. Then $F(x) = \{f_1(x), f_2(x), \ldots, f_n(x), \ldots\}$ is a measurable mapping of E into the direct product T^∞ of countably many lines. The latter is a Borel space. The mapping F is one-to-one. Hence for any $\Gamma \in \mathscr{F}(E)$ the set $F(\Gamma)$ is measurable in T^∞. We investigate in E the σ-algebra generated by the functions $\{f_n\}$. It is obvious that it contains the inverse image of all measurable sets from T^∞. But for any $\Gamma \in \mathscr{F}(E)$ we have $\Gamma = F^{-1}[F(\Gamma)]$.

LEMMA 2.2. *Let $\tilde{T}$ be a countable everywhere dense subset of T, and let W_u be a countable system of functions in E_u, separating measures. If the transition function p separates states, then the same property holds for the countable system of functions*

$$(2.1) \qquad p(t, -; u, f) \quad (u \in T, f \in W_u).$$

PROOF. Let

$$p(t, x; u, f) = p(t, y; u, f)$$

for all $u \in \tilde{T}$, $f \in W_u$. Then for all $u \in \tilde{T}$

$$p(t, x; u, -) = p(t, y; u, -).$$

If $u' > t$, then we can find u in $\tilde{T}$ belonging to the interval (t, u'). By

1.10.D

$$p(t, x; u', \Gamma) = \int_{E_u} p(t, x; u, dz)\, p(u, z; u', \Gamma) =$$

$$= \int_{E_u} p(t, y; u, dz)\, p(u, z; u', \Gamma) = p(t, y; u', \Gamma).$$

For $u' \leqslant t$, both parts are equal to zero. This means that $x = y$.

LEMMA 2.3. *Let $\mathscr{K} = (x_t, \mathscr{M}_t, \mathbf{K})$ be a Markov class. If a measurable structure in Ω is generated by x_t, or by x_t and the system of all* **K**-*negligible sets, then*

a) *the sets*

$$(2.2) \qquad \{\omega: x_t(\omega) \in \Gamma\} \quad (t \in T,\ \Gamma \in \mathscr{F}(E_t)),$$

$$(2.3) \qquad \{\omega: x_t(\omega) = a_t\} \quad (t \in T)$$

separate measures from **K**.

b) *a measurable structure of* **K** *is generated by the functions* $F(\mathbf{P}) = \mathbf{P}(A)$, *where A are sets of type* (2.2) *and* (2.3).

If, moreover, $\mathbf{P}_{t,x}\{x_t = a_t\}$ *does not depend on x, then the function*

$$f(t, x) = \mathbf{P}_{t,x}\xi$$

is p-measurable for any bounded measurable function ξ.

Assertion a) *holds for any rigid class $\mathscr{K}$.*

PROOF. Let $\mathscr{H}$ consist of bounded measurable functions ξ such that

a$'$) the coincidence of two measures $\mathbf{P}_1, \mathbf{P}_2 \in \mathbf{K}$ on the sets (2.2) implies that $\mathbf{P}_1\xi = \mathbf{P}_2\xi$;

b$'$) the function $F_\xi(\mathbf{P}) = \mathbf{P}\xi$ is measurable with respect to the σ-algebra generated by the functions $F(\mathbf{P}) = \mathbf{P}(A)$, where A are sets of type (2.2) and (2.3).

It is obvious that the family $\mathscr{H}$ contains the constants and is closed with respect to linear operations, with respect to taking uniform limits, and bounded monotone limits. By a standard lemma in set theory (see, for instance, [10], Theorem 20) if $\mathscr{H}$ contains a set of functions $\mathscr{S}$, closed with respect to multiplication, then it contains all bounded functions measurable with respect to the σ-algebra generated by $\mathscr{S}$.

We denote by $\mathscr{S}$ the totality of all functions

$$(2.4) \qquad \xi(\omega) = f_1(x_{t_1}) \ldots f_n(x_{t_n}),$$

where $t_1 < \ldots < t_n \in T$, f_i is a bounded measurable function on X_{t_i} and $f_i(a_{t_i}) = 0$. by (1.8),

$$(2.5) \qquad \mathbf{P}\chi_\Delta(t_i)\xi = \mathbf{P}\Phi(x_{t_1}),$$

where Φ is the function on E_{t_1} defined by the formula

$$(2.6) \qquad \Phi(x) = \begin{cases} f_1(x)\mathbf{P}_{t_1,x} f_2(x_{t_2}) \ldots f_n(x_{t_n}) \text{ for } n > 1, \\ f_1(x) \text{ for } n = 1. \end{cases}$$

Further,

$$(2.7) \quad \mathbf{P}(1-\chi_\Delta(t_1))\,\xi = \mathbf{P}\chi_{t_1>\Delta}\xi = \\ = f_1(b_{t_1}) \ldots f_n(b_{t_n})\,[1-\mathbf{P}\{x_{t_1}\in E_{t_1}\} - \mathbf{P}\{x_{t_1}=a_{t_1}\}].$$

From (2.5) and (2.7) it follows that $\mathscr{S} \subseteq \mathscr{H}$.

Since $\mathscr{S}$ generates the same measurable structure in Ω as x_t, it follows that properties a) and b) are proved under the assumption that $\mathscr{M} = \mathscr{N}$. If $\mathscr{M}$ is generated by $\mathscr{N}$ and **K**-negligible sets, then for each $C \in \mathscr{M}$ there exists $C' \in \mathscr{N}$ such that $\mathbf{P}(C) = \mathbf{P}(C')$ for all $\mathbf{P} \in \mathbf{K}$. This means that also in this case a) and b) are true.

If $\mathscr{K}$ is any rigid class and $\mathbf{P}_1$, $\mathbf{P}_2$ are two distinct measures in **K**, then $\mathbf{P}_1(C) \neq \mathbf{P}_2(C)$ for a certain $C \in \mathscr{N}$ and, as we have proved, $\mathbf{P}_1$ and $\mathbf{P}_2$ cannot coincide on all sets (2.2) and (2.3).

We now denote by $\mathscr{H}'$ the set of bounded measurable ξ for which the function $P_{t,x}\xi$ is p-measurable. If $\mathbf{P}_{t,x}\{x_t = a_t\}$ does not depend on x, then the function $\mathbf{P}_{t,x}\, x_{t_1}\{= a_{t_1}\}$ also does not depend on x and is therefore p-measurable (see 1.6). From (2.5) and (2.7) it is clear that $\mathscr{H}'$ contains $\mathscr{S}$; hence it contains all bounded measurable functions ξ.

2.2. THEOREM 2.1. *Let p be a transition function in E_t separating states. If E_t is Borel, then the mapping $x \to (t, x)$ is an isomorphic embedding of E_t in $\mathscr{E}$.*

PROOF. By 1.10, our mapping is measurable. Distinct points of E_t go into distinct points of $\mathscr{E}$. Therefore it is sufficient to verify the fact that the images of all measurable sets are measurable.

We denote by $\mathscr{H}$ the set of all functions F on E_t such that the function

$$F(s, x) = \begin{cases} F(x) \text{ for } s = t, \\ 0 \quad \text{for } s \neq t \end{cases}$$

is p-measurable. Relying on the measurability of the mapping $(t, x) \to t$, which was proved in 1.10, it is easy to see that $\mathscr{H}$ contains all functions (2.1), and also all constants. By Lemmas 2.1 and 2.2 the functions (2.1) generate a measurable structure in E_t. As $\mathscr{H}$ is closed with respect to addition, multiplication, and taking limits, $\mathscr{H}$ contains all measurable functions in E_t. Putting $F = \chi_\Gamma$, where $\Gamma \in \mathscr{F}(E_t)$, we conclude that the image of Γ under the mapping $x \to (t, x)$ is p-measurable.

2.3. We now investigate properties of the p-measurable structure under the assumption that the transition function p corresponds to a certain regular family $\mathscr{K} = (x_t, \mathscr{M}_t, \mathbf{K})$.

A function f in $\mathscr{E}$ is called *$\mathscr{K}$-continuous* if $f(t, x_t)$ is continuous on the right on the interval (**K**-a.s.). We denote by $\mathscr{C}$ the set of all bounded p-measurable $\mathscr{K}$-continuous functions. According to 1.12.D, $\mathscr{C}$ contains all functions $f(t, x) = p(t, x; u, \Gamma)$ and therefore generates a p-measurable structure of $\mathscr{E}$. Since $\mathscr{C}$ is closed under multiplication, we obtain the following result.

LEMMA 2.4. *Let $\mathfrak{X}$ be a regular family with transition function p, and suppose that the set $\mathscr{H}$ of functions in $\mathscr{E}$ is closed under linear operations, taking uniform limits, and monotone bounded limits. If $\mathscr{H}$ contains $\mathscr{C}$, then $\mathscr{H}$ contains all bounded p-measurable functions. If, moreover, $\mathscr{H}$ contains not only non-decreasing sequences of non-negative functions but also their limits, then $\mathscr{H}$ contains all non-negative p-measurable functions.*

2.4. We derive a number of corollaries of Lemma 2.4.

LEMMA 2.5. *Let $\mathscr{X} = (x_t, \mathscr{M}_t, \mathbf{K})$ be a regular family with transition function p. Then for any p-measurable function f and any Markov time τ (with respect to $\mathscr{M}_t$) the function $f(\tau, x_\tau)$ coincides $\mathbf{K}$-almost surely with a certain $\mathscr{M}_\tau$-measurable function.*

PROOF. We denote by $\mathscr{H}$ the set of all functions f for which the assertion of the lemma holds. By Lemma 2.4 it is sufficient to show that $\mathscr{H}$ contains $\mathscr{C}$. But this assertion follows from [10] (IV, Theorems 47 and 49).

THEOREM 2.2. *Let $\mathscr{X} = (x_t, \mathscr{M}_t, \mathbf{K})$ be a regular class. If f is a bounded (or non-negative) p-measurable function on $\mathscr{E}$, then the function*

$$F(t, x; u) = \mathbf{P}_{t, x} f(u, x_u)$$

is measurable jointly[1] *with respect to (t, x) and u, and also jointly with respect to x and u.*

For F is p-measurable with respect to (t, x) and measurable with respect to x. If $f \in \mathscr{C}$, then F is continuous on the right in u and thus measurable jointly with respect to (t, x) and u, and also jointly with respect to x and u (see, for example, [7], Lemma 1.10). It remains to apply Lemma 2.4, denoting by $\mathscr{H}$ the class of all functions f for which the statement is true.

2.5. In the homogeneous case the space E_t does not depend on t, and $\mathscr{E} = T \times E$.

THEOREM 2.3. *Suppose that a homogeneous transition function p separates states and that, for every $\Gamma \in \mathscr{F}(E)$, the function $p(0, x; t, \Gamma)$ is measurable jointly with respect to t and x. Then the σ-algebra $\mathscr{F}(\mathscr{E})$ of all p-measurable sets is equal to $\mathscr{F}(T) \times \mathscr{F}(E)$.*

PROOF. For every $t \in T$, $\Gamma \in \mathscr{F}(E)$

$$p(s, x; t, \Gamma) = p(0, x; t - s, \Gamma)$$

is measurable jointly with respect to x and s; hence $\mathscr{F}(\mathscr{E}) \subseteq \mathscr{F}(T) \times \mathscr{F}(E)$.

---

[1] Let Y and Z be two measurable spaces. We say that a function $F(y, z)$ $(y \in Y, z \in Z)$ is measurable jointly with respect to y and z if it is measurable with respect to the σ-algebra $\mathscr{F}(Y) \times \mathscr{F}(Z)$.

The inclusion $\mathscr{F}(T) \subseteq \mathscr{F}(\mathscr{E})$ follows from the measurability of $f(t, x) = t$ (see 1.10). It remains to prove that $\mathscr{F}(E) \subseteq \mathscr{F}(\mathscr{E})$. Obviously, for this it is enough to construct a system $\mathscr{H}$ of measurable functions on $\mathscr{E}$ that do not depend on t and generate a measurable structure of E.

The required system is

$$(2.8) \qquad \int_t^\infty e^{-\lambda(u-t)} p(t, x; u, f)\, du,$$

where λ ranges over the non-negative rational numbers and f over a countable system W of non-negative functions separating measures on E (the existence of such a system follows from the fact that the space E is Borel). Because p is homogeneous none of the functions (2.8) depend on t. By Fubini's theorem these are measurable functions from $\mathscr{E}$ to E. Let us show that the system $\mathscr{H}$ separates points of E. If all functions in $\mathscr{H}$ coincide at the points x and y, then there exists a subset Λ of the line T having zero Lebesgue measure such that $p(0, x; u, f) = p(0, y; u, f)$ for all $u \notin \Lambda$, $f \in W$. By Lemma 2.2 it follows that $x = y$.

2.6. THEOREM 2.4. *Let $\mathscr{X} = (x_t, \mathscr{M}_t, \mathbf{K})$ be a rigid, regular class. Let $\widetilde{T}$ be a countable everywhere dense subset of T and W_t a countable system of bounded functions on E_t separating measures. Then the system of functions*

$$(2.9) \qquad \varphi^t(x_t) \quad (t \in \widetilde{T},\ \varphi^t \in W_t)$$

separates the measures in $\mathbf{K}_u$. Together with the functions $\chi_{\alpha > t}$ $(t \in \widetilde{T})$ it separates measures in $\mathbf{K}$.

PROOF. Let $\mathbf{P}_1$, $\mathbf{P}_2$ in $\mathbf{K}$ satisfy the relations

$$\mathbf{P}_1 \varphi^t(x_t) = \mathbf{P}_2 \varphi^t(x_t) \quad (t \in \widetilde{T},\ \varphi^t \in W_t).$$

Then for $t \in \widetilde{T}$

$$(2.10) \qquad \mathbf{P}_1 f^t(x_t) = \mathbf{P}_2 f^t(x_t)$$

for all measurable functions f^t in E_t. In particular, if $f^t(x) = f(t, x) \in \mathscr{C}$ then (2.10) is satisfied for all $t \in \widetilde{T}$, hence also for all t. By Lemma 2.4 it follows that (2.10) holds for all t if $f^t(x) = f(t, x)$ is any measurable function in $\mathscr{E}$. Relying on Theorem 2.1 we conclude that $\mathbf{P}_1$ and $\mathbf{P}_2$ coincide on all the sets (2.2). Further, the functions $\mathbf{P}_i\{x_t = a_t\} = \mathbf{P}_i\{\alpha > t\}$ $(i = 1, 2)$ are continuous on the right in t. Therefore, if the equation $\mathbf{P}_1\{\alpha > t\} = \mathbf{P}_2\{\alpha > t\}$ holds for $t \in \widetilde{T}$, then also for all $t \in T$. Consequently, $\mathbf{P}_1$ and $\mathbf{P}_2$ also coincide on the sets (2.3). By Lemma 2.3 $\mathbf{P}_1 = \mathbf{P}_2$. If $\mathbf{P}_1, \mathbf{P}_2 \in \mathbf{K}_u$, then the equation $\mathbf{P}_i\{\alpha > t\} = \mathbf{P}_2\{\alpha > t\}$ holds automatically, because by (1.4) $\mathbf{P}_i\{\alpha > t\} = \chi_{u > t}$.

2.7. Now let $\mathscr{X} = (x_t, \mathscr{M}_t, \mathbf{K})$ be a Markov family with transition function p. A function f on $\mathscr{E}$ is called *$\mathscr{X}$-measurable* if for any $\mathbf{P} \in \mathbf{K}$ there exist p-measurable functions f_1 and f_2 such that

a) $f_1 \leqslant f \leqslant f_2$;

b) $f_1(t, x_t) = f_2(t, x_t)$ for all $t \in \Delta$ (**P**-almost surely).

The class of $\mathscr{K}$-measurable functions contains the constants and is closed under addition, multiplication, and limit passage. From this it follows that sets whose indicator functions are $\mathscr{K}$-measurable form a σ-algebra and that the class of $\mathscr{K}$-measurable functions coincides with the class of functions measurable with respect to this σ-algebra.

LEMMA 2.6. *If* $\mathscr{K} = (x_t, \mathscr{M}_t, \mathbf{K})$ *is a regular Markov family with transition function p, then for any universally measurable* $\Gamma \subseteq E_u$ *the function* $p(t, x; u, \Gamma)$ *is* $\mathscr{K}$ *-measurable and* $\mathscr{K}$ *-continuous.*

PROOF. We consider in E_u the measure

$$\mu(A) = \int_{-\infty}^{u} e^t \mathbf{P} p(t, x_t; u, A)\, dt$$

and we introduce $\Gamma_1, \Gamma_2 \in \mathscr{F}(E_u)$ such that $\Gamma_1 \subseteq \Gamma \subseteq \Gamma_2$ and $\mu(\Gamma_2 \setminus \Gamma_1) = 0$. For brevity we write $r(t) = p(t, x_t; u, \Gamma)$, $r_i(t) = p(t, x_t; u, \Gamma_i)$ $(i = 1, 2)$, $\delta = r_2 - r_1$. Obviously,

$$\mathbf{P} \int_{-\infty}^{u} e^t \delta(t)\, dt = \mu(\Gamma_2 \setminus \Gamma_1) = 0.$$

Since $\delta(t)$ is non-negative and continuous on the right (**P**-a.s.), it follows that $\delta(t) = 0$ for all t (**P**-a.s.). It remains to note that the functions $p(t, x; u, \Gamma)$ are p-measurable and $\mathscr{K}$ -continuous.

2.8. THEOREM 2.5. *Let* $\mathscr{K} = (x_t, \mathscr{M}_t, \mathbf{P})$ *be a regular Markov process. With each function f on* $\mathscr{E}$ *we associate a real-valued random process*

$$z_t(\omega) = \begin{cases} f(t, x_t(\omega)) & \text{for } t \in \Delta(\omega), \\ 0 & \text{for } t \notin \Delta(\omega). \end{cases}$$

If f is $\mathscr{K}$ *-measurable, then* $(z_t, \mathscr{M}_t, \mathbf{P})$ *is a Meyer process.*

PROOF. The set $\mathscr{H}$ of functions f for which this assertion holds contains $\mathscr{C}$ and is closed under linear operations and limit passage. By Lemma 2.4 $\mathscr{H}$ contains all p-measurable functions. It remains to note that if f is $\mathscr{K}$ - measurable and f_1, f_2 are functions defined by conditions a)–b) of 2.7, then the processes $f_1(t, x_t)$ and $f_2(t, x_t)$ are **P**-indistinguishable.

2.9. We derive some corollaries from Theorem 2.5.

COROLLARY 1. *If f is a* $\mathscr{K}$ *-measurable function and* τ *is a Markov time with respect to the filtration* $\overline{\mathscr{M}_t}$ *then* $f(\tau, x_\tau)$ *is measurable with respect to* $\overline{\mathscr{M}}_\tau$.

For the proof it is sufficient to compare Theorem 2.5 with the results of [10] (VIII, Theorem 15 and IV, Theorem 49) and to note that if the processes z_t and z'_t are **P**-indistinguishable, then the $\overline{\mathscr{M}}_\tau$ -measurability of z_τ follows from that of z'_τ.

COROLLARY 2. *If A is a* $\mathscr{K}$ *-measurable set, then*

$$\tau_A(\omega) = \inf\{t: (t, x_t(\omega)) \in A\}$$

is a Markov time with respect to $\overline{\mathcal{M}}_t$ *and for any* $\varepsilon > 0$ *there exists a Markov time* τ_ε *(with respect to* $\overline{\mathcal{M}}_t$*) such that*

(2.11) $$\{\tau_\varepsilon \in \Delta\} = \{(\tau_\varepsilon, x_{\tau_\varepsilon}) \in A\},$$

(2.12) $$\overline{\mathbf{P}}\{\tau_A \in \Delta\} \leqslant \overline{\mathbf{P}}\{\tau_\varepsilon \in \Delta\} + \varepsilon.$$

For the proof it is sufficient to apply Theorem 2.5 and the result of Meyer stated in 1.8 with $f(t, x) = \chi_A(t, x)$ and $c = 1$.

§ 3. The strong Markov property

3.1. We shall show that regular Markov processes have the strong Markov property. A weaker form of this property will be proved in Theorem 3.1, and a stronger form (for regular classes), in Theorem 3.2.

3.2. THEOREM 3.1. *Suppose that a Markov process* $(x_t, \mathcal{M}_t, \mathbf{P})$ *with transition function* p *satisfies conditions* 1.12.B *and* 1.12.D. *Let* τ *be a Markov time with respect to* $\mathcal{M}_t$. *Then for any* $\Gamma \in \mathcal{F}(E_t)$

(3.1) $$\mathbf{P}\{x_t \in \Gamma \mid \mathcal{M}_\tau\} = p(\tau, x_\tau; t, \Gamma) \quad (\text{п. н. } \mathbf{P}, \ \tau \in \Delta_t),$$

where $\Delta_t = \Delta \cap [-\infty, t)$.

REMARK 1. Theorem 3.1 applies, in particular, to regular processes and also to their completions.

REMARK 2. From (3.1) it follows that for any non-negative measurable function f on E_t

(3.2) $$\mathbf{P}\{f(x_t) \mid \mathcal{M}_\tau\} = p(\tau, x_\tau; t, f) \qquad (\mathbf{P}\text{-a.s.}, \ \tau \in \Delta_t).$$

PROOF. Let $S = \{s_0 < s_1 < \ldots < s_n\}$ be any finite set of numbers. We put

(3.3) $$\varphi_S(t) = s_i \text{ for } s_{i-1} \leqslant t < s_i,$$

where $s_{-1} = -\infty$, $s_{n+1} = +\infty$. We consider an increasing sequence of sets S_n whose sum is everywhere dense in T, and put $\varphi_{S_n}(\tau) = \tau_n$. It is easy to see that τ_n is a Markov time. If $A \in \mathcal{M}_\tau$, then $\{A, \tau_n = s \in \Delta\} = \{A, \tau \leqslant s, \tau_n = s \in \Delta\} \in \mathcal{M}_s$ and by (1.1) for $s < t$

$$\mathbf{P}\{A, \tau_n = s \in \Delta, x_t \in \Gamma\} = \mathbf{P}\{A, \tau_n = s \in \Delta, p(s, x_s; t, \Gamma)\} = \\ = \mathbf{P}\{A, \tau_n = s \in \Delta, p(\tau_n, x_{\tau_n}; t, \Gamma)\}.$$

Summing these equations over all values s of the Markov time τ_n belonging to Δ_t we have

(3.4) $$\mathbf{P}\{A, \tau_n \in \Delta_t, x_t \in \Gamma\} = \mathbf{P}\{A, \tau_n \in \Delta_t, p(\tau_n, x_{\tau_n}; t, \Gamma)\}.$$

Since $\tau_n \downarrow \tau$, by 1.12.B and 1.12.D (3.4) implies that

$$\mathbf{P}\{A, \tau \in \Delta_t, x_t \in \Gamma\} = \mathbf{P}\{A, \tau \in \Delta_t, p(\tau, x_\tau; t, \Gamma)\}.$$

Since the functions $p(\tau, x_\tau; t, \Gamma)$ coincide **P**-almost everywhere with a $\mathcal{M}_\tau$-measurable function (see Lemma 2.5). This implies (3.1).

3.3. THEOREM 3.2. *Let* $\mathcal{K} = (x_t, \mathcal{M}_t, \mathbf{K})$ *be a Markov class satisfying conditions* 1.12.B *and* 1.12.D, *and let* τ *and* η *be Markov times with respect to* $\mathcal{M}_t$, *where* η *is measurable with respect to* $\mathcal{M}_\tau$. *Then for any p-measurable function* $f \geqslant 0$ *and* $\mathbf{P} \in \mathbf{K}$

$$(3.5) \quad \mathbf{P}\{f(\eta, x_\eta) \mid \mathcal{M}_\tau\} = F(\tau, x_\tau, \eta) \quad (\mathbf{P}\text{-a.s.}, \tau \leqslant \eta, \tau \in \Delta),$$

where

$$(3.6) \quad F(t, x, u) = \mathbf{P}_{t,x} f(u, x_u).$$

If the transition function p for the class $\mathcal{K}$ *is homogeneous then for any* $\mathbf{P} \in \mathbf{K}$, $t \geqslant 0$, *and any measurable function* $f \geqslant 0$ *in E*

$$(3.7) \quad \mathbf{P}\{f(x_{\tau+t}) \mid \mathcal{M}_\tau\} = p(t, x_\tau, f) \quad (\mathbf{P}\text{-a.s.}, \tau \in \Delta)$$

REMARK. Theorem 3.2 applies, in particular, to regular classes $\mathcal{K}$ and also to their completions. For the latter case, (3.5) extends to all $\mathcal{K}$-measurable functions $f \geqslant 0$, and (3.7) holds for all universally measurable functions $f \geqslant 0$.

PROOF. Let $\varphi_S(t)$ be defined by (3.3). We consider the sequences S_n as defined in the proof of Theorem 3.1, and put $\eta_n = \varphi_{S_n}(\eta)$.

By **1.10** $f^s(x) = f(s, x)$ is a measurable function on E_s, and by (3.2) and (1.5)

$$(3.8) \quad \mathbf{P}\{f^s(x_s) \mid \mathcal{M}_\tau\} = p(\tau, x_\tau; s, f^s) = \\ = \mathbf{P}_{\tau, x_\tau} f^s(x_s) = F(\tau, x_\tau, s) \quad (\mathbf{P}\text{-a.s.}, \tau \in \Delta_s).$$

The set $A_{ns} = \{\eta_n = s > \tau \in \Delta\}$ belongs to $\mathcal{M}_\tau$ and on it $\tau \in \Delta_s$ and $\eta_n = s$. By (3.8)

$$(3.9) \quad \mathbf{P}\{\chi_{A_{ns}} f^{\eta_n}(x_{\eta_n}) \mid \mathcal{M}_\tau\} = \\ = \chi_{A_{ns}} F(\tau, x_\tau, s) = \chi_{A_{ns}} F(\tau, x_\tau, \eta_n) \quad (\mathbf{P}\text{-a.s.})$$

Summing over all s we obtain

$$(3.10) \quad \mathbf{P}\{\chi_{\eta_n > \tau \in \Delta} f^{\eta_n}(x_{\eta_n}) \mid \mathcal{M}_\tau\} = \chi_{\eta_n > \tau \in \Delta} F(\tau, x_\tau, \eta_n) \quad (\mathbf{P}\text{-a.s.})$$

To begin with, let $f \in \mathcal{C}$. Then the function $F(t, x, u)$ is bounded and continuous on the right in u. Going to the limit in (3.10) and taking $\eta_n \downarrow \eta$ into account we get (3.5).

Lemma 2.4 applies to the set $\mathcal{H}$ of all functions f for which (3.5) holds. Thus, $\mathcal{H}$ contains all p-measurable, non-negative functions.

If the transition function p is homogeneous, then by Theorem 2.3 the measurable function f on E is a p-measurable function of (t, x). Applying (3.5)–(3.6) to $\eta = \tau + t$ and noting that $F(s, x, s + t) = p(t, x, f)$ we obtain (3.7).

3.4. For the case when $\tau = \alpha$, the strong Markov property can be expressed in the following form.

THEOREM 3.3. *Let* $\mathscr{K} = (x_t, \mathscr{M}_t, \mathbf{K})$ *be a regular class and let* $\mathscr{F}$ *be the* σ*-algebra on* Ω *generated by the random process* x_t *and by all* $\mathbf{K}$*-negligible sets. Then for all* $\mathbf{P} \in \mathbf{K}$, $\xi \in \mathscr{F}$.

$$\mathbf{P}\{\xi \mid \mathscr{M}_\alpha\} = \mathbf{P}_{\alpha\, x_\alpha}\xi \qquad (\mathbf{P}\text{-a.s.}) \tag{3.11}$$

PROOF. The set of functions ξ for which (3.11) holds is closed under linear operations and taking non-increasing limits. Therefore, to prove that (3.11) holds for all $\xi \in \mathscr{F}$ it suffices to verify that it holds for the functions

$$\xi = f_1(x_{t_1}) \dots f_n(x_{t_n}), \tag{3.12}$$

where $t_1 < \dots < t_n \in T$ and f_i is a measurable non-negative function on X_{t_i} that vanishes at a_{t_i}.

If $A \in \mathscr{M}_\alpha$ then $\{A, t_1 \in \Delta\} = \{A, \alpha \leqslant t_1\} \cap \{\beta > t_1\} \in \mathscr{M}_{t_1}$ and by (1.8)

$$\mathbf{P}\{\chi_{A,\, t_1 \in \Delta}\, \xi\} = \mathbf{P}\{\chi_{A,\, t_1 \in \Delta}\, f(x_{t_1})\},$$

where

$$f(x) = f_1(x)\, \mathbf{P}_{t_1, x} f_2(x_{t_2}) \dots f_n(x_{t_n}).$$

Hence

$$\mathbf{P}\{\xi \mid \mathscr{M}_\alpha\} = \mathbf{P}\{f(x_{t_1}) \mid \mathscr{M}_\alpha\} \qquad (\mathbf{P}\text{-a.s.},\ \ t_1 \in \Delta). \tag{3.13}$$

From Theorems 2.1 and 3.2 (for $\eta = t_1$) we have

$$\mathbf{P}\{f\cdot(x_{t_1}) \mid \mathscr{M}_\alpha\} = \mathbf{P}_{\alpha,\, x_\alpha} f(x_{t_1}) \qquad (\mathbf{P}\text{-a.s.},\ \ \alpha \leqslant t_1). \tag{3.14}$$

From (3.13) and (3.14) we have

$$\mathbf{P}\{\xi \mid \mathscr{M}_\alpha\} = \mathbf{P}_{\alpha,\, x_\alpha} f(x_{t_1}) \qquad (\mathbf{P}\text{-a.s.},\ \ t_1 \in \Delta). \tag{3.15}$$

On the other hand, by (1.7) for $s < t_1$

$$\mathbf{P}_{s,\, x}\{\xi \mid \mathscr{M}_{t_1}\} = f(x_{t_1}) \qquad (\mathbf{P}\text{-a.s.},\ \ t_1 \in \Delta). \tag{3.16}$$

If $t_1 < \alpha$, then $\xi = f(x_{t_1}) = 0$; if $\beta \leqslant t_1$, then $\xi = f(x_{t_1}) = f_1(b_{t_1}) \dots f_n(b_{t_n})$. Therefore, (3.15) and (3.16) are also satisfied for $t_1 \notin \Delta$. From (3.16) we have

$$\mathbf{P}_{s,\, x}\xi = \mathbf{P}_{s,\, x} f(x_{t_1}). \tag{3.17}$$

Comparing (3.15) and (3.17) we obtain (3.11).

§4. Essential points

4.1. We call a point (t, x) of the phase space $\mathscr{E}$ *essential* if the state $x \in E_t$ is essential, that is, if (1.9) is satisfied. The set of all essential points of $\mathscr{E}$ is called its *essential part* and is denoted by $\mathscr{E}'$. The essential part E_t' of the state space E_t is defined analogously. The aim of this section is to prove that almost all trajectories of a regular class consist only of essential points.

4.2. We shall use one general result concerning the integral representation of measures ([4], Theorem 2.1). Let $\mathbf{M}$ be a separable[1] family of measures on a σ-algebra $\mathscr{F}$ in Ω, and let $\mathscr{A}$ be a σ-algebra contained in $\mathscr{F}$. We assume that for all $\mathbf{P} \in \mathbf{M}$ and all $C \in \mathscr{F}$

$$\mathbf{P}(C \mid \mathscr{A}) = \mathbf{P}^{\omega}(C) \quad (\mathbf{P}\text{-a.s.}) \tag{4.1}$$

where $\mathbf{P}^{\omega} \in \mathbf{M}$ for every $\omega \in \Omega$ and $\mathbf{P}^{\omega}(C)$ is an $\mathscr{A}$-measurable function for every C. We denote by $\mathbf{M}_e$ the set of measures $\mathbf{P}$ in $\mathbf{M}$ such that $\mathbf{P}(A)$ is equal to 0 or 1 for all $A \in \mathscr{A}$. Then the set $\mathbf{M}_e$ is measurable in $\mathbf{M}$, is contained in the set $\{\mathbf{P}^{\omega}\}$ and any measure $\mathbf{P} \in \mathbf{M}$ can be represented in the form

$$\mathbf{P} = \int_{\mathbf{M}_e} \tilde{\mathbf{P}} \tilde{\mu}(d\tilde{\mathbf{P}}), \tag{4.2}$$

where

$$\tilde{\mu}(B) = \mathbf{P}\{\omega\colon \mathbf{P}^{\omega} \in B\}. \tag{4.3}$$

4.3. THEOREM 4.1. *Let $\mathscr{K} = (x_t, \mathscr{M}_t, \mathbf{K})$ be a class of regular Markov processes with transition function p that separates states. Then the essential part $\mathscr{E}'$ of the phase space is p-measurable, and for any $\mathbf{P} \in \mathbf{K}$, $C \in \mathscr{N}$*

$$\mathbf{P}(C) = \int_{\mathscr{E}'} \mathbf{P}_{t,x}(C)\, d\mu, \tag{4.4}$$

where

$$\mu(\Gamma) = \mathbf{P}\{(\alpha, x_{\alpha}) \in \Gamma\}. \tag{4.5}$$

PROOF. We denote by $\mathscr{F}$, $\mathscr{F}_t$ and $\mathscr{A}$ the σ-algebras in Ω generated by the system of all $\mathbf{K}$-negligible sets together with $\mathscr{N}$, $\mathscr{M}_t \cap \mathscr{N}$ and $\mathscr{M}_{\alpha} \cap \mathscr{N}$, respectively. Let $\mathbf{M}$ be the set of measures $\mathbf{P} \in \mathbf{K}$ restricted to $\mathscr{F}$. It is easy to see that $(x_t, \mathscr{F}_t, \mathbf{M})$ is a class of Markov processes. By Theorem 2.4 the family $\mathbf{M}$ is separable, and by Lemma 2.3 $\mathbf{P}_{t,x}(C)$ is p-measurable for any $C \in \mathscr{F}$.

We now apply Theorem 3.3 to $(x_t, \mathscr{F}_t, \mathbf{M})$. By this theorem, (4.1) is satisfied if we denote by $\mathbf{P}^{\omega}$ the restriction to $\mathscr{F}$ of the measure $\mathbf{P}_{\alpha, x_{\alpha}} \in \mathbf{K}$. By Lemma 2.5 it follows that $\mathbf{P}^{\omega}(C) = \mathbf{P}_{\alpha, x_{\alpha}}(C)$ is $\mathscr{A}$-measurable for any $C \in \mathscr{F}$. By 4.2 the set $\mathbf{M}_e$ is measurable in $\mathbf{M}$, is contained in the set $\{\mathbf{P}_{1,x}\}$ and any measure $\mathbf{P} \in \mathscr{M}$ can be represented in the form (4.2).

---

[1] We say that the family $\mathbf{M}$ is separable if we can find a countable system of functions separating measures in $\mathbf{M}$.

By the definition of measurable structures in $\mathscr{E}$ and $\mathbf{M}$ the formula

$$(4.6) \qquad (t, x) \to \mathbf{P}_{t,x}$$

defines a measurable mapping of $\mathscr{E}$ into $\mathbf{M}$. We shall show that $\mathscr{E}'$ coincides with the inverse image of $\mathbf{M}_e$ (consequently, $\mathscr{E}'$ is measurable) and $\mathscr{E}'$ is mapped onto $\mathbf{M}_e$ isomorphically (consequently (4.2) and (4.3) are equivalent to (4.4) and (4.5)).

If $A \in \mathscr{M}_\alpha \cap \mathscr{N}$, then by (3.11) for any $\mathbf{P} \in \mathbf{K}$

$$\mathbf{P}_{\alpha, x_\alpha}(A) = \mathbf{P}(A | \mathscr{M}_\alpha) = \chi_A \quad (\mathbf{P}\text{-a.s.})$$

In particular, for $(t, x) \in \mathscr{E}'$

$$\mathbf{P}_{t,x}(A) = \mathbf{P}_{\alpha, x_\alpha}(A) = \chi_A \quad (\mathbf{P}_{t,x}\text{-a.s.})$$

hence $\mathbf{P}_{t,x} \in \mathbf{M}_e$.

On the other hand, if $\mathbf{P}_{t,x} \in \mathbf{M}_e$, then any $\mathscr{A}$-measurable function is constant ($\mathbf{P}_{t,x}$-a.s.). In particular, this applies to $p(\alpha, x_\alpha; u, f) = \mathbf{P}^\omega f(x_u)$. However, $\mathbf{P}_{t,x}\{\alpha = t\} = 1$, hence $p(t, x_t; u, f) = \text{const}$ ($\mathbf{P}_{t,x}$-a.s.). By Lemma 2.2 it follows that for a certain $y \in E_t$

$$(4.7) \qquad x_t = y \quad (\mathbf{P}_{t,x}\text{-a.s.})$$

From (1.4), (1.8) and (4.7) we conclude that for any $u > t$ and $\Gamma \in \mathscr{F}(E_u)$

$$p(t, x; u, \Gamma) = \mathbf{P}_{t,x}\{x_u \in \Gamma\} = \mathbf{P}_{t,x}\mathbf{P}_{t,x_t}\{x_u \in \Gamma\} = \\ = \mathbf{P}_{t,x}p(t, x_t; u, \Gamma) = p(t, y; u, \Gamma)$$

hence $x = y$. Therefore (4.7) implies that $\mathbf{P}_{t,x}\{x_t = x\} = 1$, so that $(t, x) \in \mathscr{E}'$.

Since the functions $p(t, x; u, \Gamma)$ separate points of $\mathscr{E}$ (see 1.10), (4.6) establishes a one-to-one correspondence between $\mathscr{E}'$ and $\mathbf{M}_e$. We already know that this mapping is measurable. It remains to verify that the image of any p-measurable set $\Gamma \subseteq \mathscr{E}'$ is measurable in $\mathbf{M}$. But this image is described by the formula

$$(4.8) \qquad \{\mathbf{P} : \mathbf{P} \in \mathbf{M}_e, \quad \mathbf{P}[(\alpha, x_\alpha) \in \Gamma] = 1\}.$$

By Lemma 2.5 the p-measurability of $\chi_\Gamma(t, x)$ implies that of $\chi_\Gamma(\alpha, x_\alpha)$. Therefore the set (4.8) is measurable in $\mathbf{M}$.

4.4. THEOREM 4.2. *Under the assumptions of Theorem* 4.1

$$(4.9) \qquad x_t \in E'_t \textit{ for all } t \in \Delta \quad (\mathbf{K}\text{-a.s.})$$

PROOF. Putting $\Gamma = \mathscr{E}'$ in (4.5) and $C = \Omega$ in (4.4) we conclude that for any $\mathbf{P} \in \mathbf{K}$

$$\mathbf{P}\{(\alpha, x_\alpha) \in \mathscr{E}'\} = 1.$$

Hence it follows from (1.6) that for any t, x

$$\mathbf{P}_{t,x}\{(t, x_t) \in \mathscr{E}'\} = 1. \tag{4.10}$$

Since the set $A = \mathscr{E} \setminus \mathscr{E}'$ is p-measurable, by Corollary 2 of 2.9 for $\mathbf{P} \in \mathbf{K}$ and $\varepsilon > 0$ there exists a Markov time $\tau = \tau_\varepsilon$ with respect to $\overline{\mathscr{M}}_t$ for which (2.11)–(2.12) hold. We apply Theorem 3.2 to the completed process $(x_t, \overline{\mathscr{M}}_t, \overline{\mathbf{P}})$, the p-measurable function $f(t, x) = \chi_A(t, x)$ and the Markov times τ and $\eta = \tau$. We have

$$\overline{\mathbf{P}}\{f(\tau, x_\tau)|\overline{\mathscr{M}}_\tau\} = F(\tau, x_\tau, \tau) \qquad (\overline{\mathbf{P}}\text{-a.s.}, \quad \tau \in \Delta) \tag{4.11}$$

where $F(t, x, u) = \mathbf{P}_{t,x} f(u, x_u)$. But by (4.10)

$$F(t, x, t) = \mathbf{P}_{t,x} f(t, x_t) = \mathbf{P}_{t,x}\{(t, x_t) \notin \mathscr{E}'\} = 0.$$

Therefore it follows from (4.11) that

$$\overline{\mathbf{P}}\{(\tau, x_\tau) \in A\} = \overline{\mathbf{P}}F(\tau, x_\tau, \tau) = 0.$$

By (2.11)–(2.12) it follows that $\overline{\mathbf{P}}\{\tau_A \in \Delta\} \leqslant \varepsilon$. Since $\varepsilon > 0$ is arbitrary, $\overline{\mathbf{P}}\{\tau_A \in \Delta\} = 0$, which is equivalent to the assertion of the theorem.

§ 5. Excessive functions

5.1. Let $p(s, x; t, \Gamma)$ be a transition function in the spaces E_t. A non-negative function $h^t(x)$ ($t \in T$, $x \in E_t$) is called excessive with respect to p (or, more briefly, p-excessive), if:

5.1.A. For every t, $h^t(x)$ is universally measurable with respect to x.

5.1.B. For all $s < t \in T$, $x \in E_s$

$$p(s, x; t, h^t) \leqslant h^s(x).$$

5.1.C. $p(s, x; t, h^t) \to h^s(x)$ as $t \downarrow s$.

From these properties and the Kolmogorov–Chapman equation 1.10.D, it follows that $p(s, x; t, h^t)$ is a non-decreasing and right-continuous function of t.

The aim of this section is a proof of the following result.

THEOREM 5.1. *Let $\mathscr{K} = (x_t, \mathscr{M}_t, \mathbf{K})$ be any regular family with transition function p. Then all p-excessive functions are $\mathscr{K}$-measurable and $\mathscr{K}$-continuous.*

5.2. The proof of Theorem 5.1 relies on two lemmas.

LEMMA 5.1. *Let p be an arbitrary transition function. The class of all p-excessive functions can be described as the minimal class $\mathscr{H}$ of non-negative functions on $\mathscr{E}$ containing all functions*[1]

$$p(t, x; u, \Gamma) \quad (u \in T, \ \Gamma \in \overline{\mathscr{F}}(E_u)) \tag{5.1}$$

---

[1] $\overline{\mathscr{F}}(E_u)$ denotes the universal completion of the σ-algebra $\mathscr{F}(E_u)$.

and having the properties:

5.2.A. *If* $h_1, h_2 \in \mathscr{H}$, $c_1, c_2 \geqslant 0$, *then* $c_1 h_1 + c_2 h_2 \in \mathscr{H}$.

5.2.B. *If* $h_n \in \mathscr{H}$ *and* $h_n \uparrow h$, *then* $h \in \mathscr{H}$.

PROOF. From the Kolmogorov–Chapman equation the fact that the function (5.1) is p-excessive follows easily. The class $\mathscr{H}$ of all p-excessive functions obviously has the property 5.2.A. We verify that it also has the property 5.2.B. Let h_n be p-excessive and $h_t^n(x) \uparrow h_t(x)$. Obviously h satisfies 5.1.A. By Fatou's lemma h also satisfies 5.1.B. From the inequalities $p(s, x; t, h_n^t) \leqslant p(s, x; t, h^t) \leqslant h^s(x)$ it is clear that the lower and upper limits of $p(s, x; t, h^t)$ as $t \downarrow s$ are contained between $h_n^s(x)$ and $h^s(x)$. Letting $n \to \infty$, we see that h satisfies 5.1.B.

Now let $\mathscr{H}$ be any class with the properties described in the lemma. We prove at first that $\mathscr{H}$ contains all functions

$$p(t, x; u, \varphi) \quad (u \in T,\ \varphi \in \overline{\mathscr{F}}(E_u)). \tag{5.2}$$

We put

$$\varphi_n(x) = \frac{i}{2^n} \text{ for } x \in \Gamma_{in} = \left\{x: \frac{i}{2^n} < \varphi(x) \leqslant \frac{i+1}{2^n}\right\}.$$

It is clear that $\varphi_n \uparrow \varphi$ and, consequently, $p(t, x; u, \varphi_n) \uparrow p(t, x; u, \varphi)$. But

$$p(t, x; u, \varphi_n) = \sum_i \frac{i}{2^n} p(t, x; u, \Gamma_{in}).$$

By 5.2.A. and 5.2.B, (5.2) belongs to $\mathscr{H}$.

Let us prove that $\mathscr{H}$ contains every excessive function h. We consider a finite set of numbers $S = \{s_1 < s_2 < \ldots < s_m\}$ and put

$$F^t(x) = \sum_{j=1}^{m-1} p(t, x; s_j, f_j) + p(t, x; s_m, h^{s_m}), \tag{5.3}$$

where $f_j(x) = h^{s_j}(x) - p(s_j, x; s_{j+1}, h^{s_{j+1}})$. By 5.1.B $f_j \geqslant 0$ and by 5.2.A $F \in \mathscr{H}$. We put $s_0 = -\infty$. If $s_{i-1} \leqslant t < s_i$ $(i = 1, \ldots, m)$, then

$$p(t, x; s_j, f_j) = \begin{cases} p(t, x; s_j, h^{s_j}) - p(t, x; s_{j+1}, h^{s_{j+1}}) & \text{for } j \geqslant i, \\ 0 & \text{for } j < i. \end{cases}$$

Therefore

$$F^t(x) = p(t, x; s_i, h^{s_i}) \text{ for } s_{i-1} \leqslant t < s_i \quad (i = 1, 2, \ldots, m). \tag{5.4}$$

Let

$$\psi(t) = s_i \quad \text{for} \quad s_{i-1} \leqslant t < s_i \quad (i = 1, 2, \ldots, m).$$

From (5.3) and (5.4) it is clear that

$$F^t(x) = \begin{cases} p(t, x; \psi(t), h^{\psi(t)}) & \text{for } t < s_m, \\ 0 & \text{for } t \geqslant s_m. \end{cases} \tag{5.5}$$

Now let S range over an increasing sequence of sets S_n whose union is the set of all rational numbers. We denote by F_n and ψ_n the corresponding sequences of functions (5.3) and (5.4). It is easy to see that $\psi_n(t) \downarrow t$. By (5.5), 5.1.B and 5.1.C, $F_n^t(x) \uparrow h^t(x)$. From 5.2.B we conclude that $h \in \mathscr{H}$.

LEMMA 5.2. *Let* $(x_t, \mathscr{M}_t, \mathbf{P})$ *be a Markov process on the interval* $\Delta(\omega)$, *with transition function p. Then for any* $u \in T$, $f \in \mathscr{F}(E_u)$, *the triple* $(p(t, x_t; u, f), \mathscr{M}_t, \mathbf{P})$ *defines a supermartingale on* $\Delta(\omega)$. *Let* $(x_t, \overline{\mathscr{M}}_t, \overline{\mathbf{P}})$ *be the completion of the process* $(x_t, \mathscr{M}_t, \mathbf{P})$. *To each p-excessive function h there corresponds a supermartingale* $(h^t(x_t), \overline{\mathscr{M}}_t, \overline{\mathbf{P}})$ *on* $\Delta(\omega)$.

PROOF. By (1.2) and 5.1.B

$$\overline{\mathbf{P}}\{h^t(x_t) \mid \overline{\mathscr{M}}_s\} = p(s, x_s; t, h^t) \leqslant h^s(x_s) \quad (\overline{\mathbf{P}}\text{-a.s.}, \quad s \in \Delta).$$

By 1.6.A the process $p(t, x_t; u, f)$ is adapted to the filtration $\mathscr{M}_t$. From 5.1.A it follows that for any p-excessive function h the process $h^t(x_t)$ is adapted to $\overline{\mathscr{M}}_t$.

5.3. PROOF OF THEOREM 5.1. By Lemma 2.6 the class $\mathscr{H}$ of all non-negative $\mathscr{K}$-measurable, $\mathscr{K}$-continuous functions contains all functions (5.1) and satisfies 5.2.A. By Lemma 5.1 the theorem will be proved if we can show that $\mathscr{H}$ satisfies 5.2.B. If $h_n \in \mathscr{H}$ and $h_n \uparrow h$, then h, obviously, is $\mathscr{K}$-measurable. To prove its $\mathscr{K}$-continuity it is sufficient to apply the proposition at the end of 1.9 to the supermartingales $(h^t(x_t), \overline{\mathscr{M}}_t, \overline{\mathbf{P}})$, where $\overline{\mathbf{P}}$ is an arbitrary measure in $\mathbf{K}$.

5.4. We derive some corollaries from Theorem 5.1.

COROLLARY 1. *If* $(x_t\, \mathscr{M}_t, \mathbf{P})$ *is a regular Markov process with transition function p and h is any p-excessive function, then* ($\overline{\mathbf{P}}$-a.s.) $h^t(x_t)$ *has left-hand limits at all points of the interval* $(\alpha, \beta]$.

For the right-continuous non-negative supermartingale $(h^t(x_t), \overline{\mathscr{M}}_t, \overline{\mathbf{P}})$ $\overline{\mathbf{P}}$-almost surely does not have a discontinuity of the second kind and has a limit as $t \to \infty$ (if $+\infty \in \Delta$). If Δ is not random, then this is proved, for example, in [10] (VI, Theorems 3 and 6). We can pass from a random interval to a non-random one by means of the extension theorem stated in 1.9.

COROLLARY 2. *Let* $\mathscr{K}$ *be a regular class. If* $\xi \in \mathscr{N}(t, \infty]$, *then* ($\mathbf{K}$-a.s.) *the function* $\mathbf{P}_{s, x_s}\xi$ *is continuous on the right at s in the interval* $[\alpha, \beta \wedge t]$.

For in accordance with (1.8) for $s < t$

$$\mathbf{P}_{s, x}\xi\chi_{\beta > t} = \mathbf{P}_{s, x}\xi\chi_{\Delta}(t) = \mathbf{P}_{s, x}\mathbf{P}_{t, x_t}\xi = p(s, x; t, \varphi),$$

where $\varphi(x) = \mathbf{P}_{t, x}\xi$ and the right-hand side is $\mathscr{K}$-measurable by Theorem 5.1. On the other hand, $\xi\chi_{\beta \leqslant t}$ does not depend on ω, hence $\mathbf{P}_{s, x}\xi\chi_{\beta \leqslant t}$ does not depend on s and x.

§6. The uniqueness of a rigid regular re-construction

6.1. Two theorems will be proved in this section. The first one expresses in terms of $\mathscr{K}$ a random process adjoint to a regular re-construction of the class $\mathscr{K}$. The second one asserts that the rigid regular re-constructions of $\mathscr{K}$ are indistinguishable.

THEOREM 6.1. *Let* $\widetilde{\mathscr{K}} = (\widetilde{x}_t, \widetilde{\mathscr{M}}_t, \mathbf{K})$ *be a regular re-construction of the class* $\mathscr{K} = (x_t, \mathscr{M}_t, \mathbf{K})$ *and* R *a countable everywhere dense subset of* T. *Then for any* $u \in T$, $\varphi \in \mathscr{F}(E_u)$

$$\widetilde{\mathbf{P}}_{t,\widetilde{x}_t}\varphi(x_u) = \lim_{\substack{r \downarrow t \\ r \in R}} p(r, x_r; u, \varphi) \quad \textit{for all } t \in \widetilde{\Delta} \quad (\mathbf{K}\textit{-a.s.}). \tag{6.1}$$

The proof is based on the following lemma.

LEMMA 6.1. *In the notation of Theorem* 6.1 *the function*

$$\Phi^t(x) = \mathbf{P}_{t,\widetilde{x}}\varphi(x_u) \tag{6.2}$$

is excessive with respect to the transition function $\widetilde{p}$ *of* $\widetilde{\mathscr{K}}$. *For any* $t \in T$, $\mathbf{P} \in \mathbf{K}$

$$\Phi^t(\widetilde{x}_t) = \mathbf{P}\{\varphi(x_u)|\mathscr{M}_{t+}\} \qquad (\mathbf{P}\text{-a.s.}, \ \alpha \leqslant t < \beta). \tag{6.3}$$

PROOF. Applying (1.8) to the Markov process $(x_t, \mathscr{M}_t, \widetilde{\mathbf{P}}_{s,x})$ we have

$$\Phi^s(x) = \widetilde{\mathbf{P}}_{s,x}\mathbf{P}_{t,x_t}\varphi(x_u).$$

According to (1.10), for all t outside the countable set $\Lambda(\widetilde{\mathbf{P}}_{s,x})$

$$\widetilde{\mathbf{P}}_{t,\widetilde{x}_t} = \mathbf{P}_{t,x_t} \ (\widetilde{\mathbf{P}}_{s,x}\text{-a.s.}).$$

Therefore

$$\Phi^s(x) = \widetilde{\mathbf{P}}_{s,x}\widetilde{\mathbf{P}}_{t,\widetilde{x}_t}\varphi(x_u) = \widetilde{p}(s, x; t, \Phi^t) \tag{6.4}$$

for all $t \notin \Lambda(\widetilde{\mathbf{P}}_{s,x})$ in the interval (s, u). From (6.4) it follows that

$$\Phi^s(x) = \frac{1}{u-s}\int_s^u \widetilde{p}(s, x; t, \Phi^t)\,dt. \tag{6.5}$$

By Theorem 2.2 the function $\widetilde{p}(s, x; t, \Phi^t) = \widetilde{\mathbf{P}}_{s,x}\Phi^t(x_t)$ is measurable jointly with respect to x and t. Therefore Fubini's theorem is applicable, and by the Kolmogorov–Chapman equation 1.10.D, (6.5) implies that for $r < s < t$

$$\widetilde{p}(r, x; s, \Phi^s) = \frac{1}{u-s}\int_s^u \widetilde{p}(r, x; t, \Phi^t)\,dt.$$

The right-hand side does not decrease as $s \downarrow r$ and tends to $\Phi^r(x)$. Therefore Φ is $\widetilde{p}$-excessive.

By Theorem 5.1 for any $t \in T$, $\mathbf{P} \in \mathbf{K}$

(6.6) $$\Phi^t(\widetilde{x}_t) = \lim_{r \downarrow t} \Phi^r(\widetilde{x}_r) \quad (\mathbf{P}\text{-a.s.},\ t \in \widetilde{\Delta}).$$

But for $r \notin \Lambda(\mathbf{P})$ by (1.2)

(6.7) $$\Phi^r(\widetilde{x}_r) = \widetilde{\mathbf{P}}_{r,\, \widetilde{x}_r}\varphi(x_u) = \mathbf{P}_{r,\, x_r}\varphi(x_u) = \\ = \mathbf{P}\{\varphi(x_u) \mid \mathscr{M}_r\} \qquad (\mathbf{P}\text{-a.s.},\ r \in \widetilde{\Delta}).$$

(6.6) and (6.7) imply (6.3).

6.2. PROOF OF THEOREM 6.1. The existence (K-a.s.) of the limit on the right-hand side of (6.1) follows from Lemma 5.2 and a well-known property of supermartingales (see, for example, [1] (Chapter VII, Theorem 3.3). Obviously this limit is continuous on the right in t. On the other hand, from Lemma 6.1 and Theorem 5.1 it follows that the left-hand side of (6.1) is also continuous on the right in t (K-a.s.). Therefore, to prove (6.1) it suffices to verify that the equation is satisfied for any $t \in T$ and any $\mathbf{P} \in \mathbf{K}$ ($\mathbf{P}$-a.s., $t \in \widetilde{\Delta}$). This is so by (6.3) and the equations

(6.8) $$\lim_{\substack{r \downarrow t \\ r \in R}} p(r, x_r; u, \varphi) = \lim_{\substack{r \downarrow t \\ r \in R}} \mathbf{P}\{\varphi(x_u) \mid \mathscr{M}_r\} = \\ = \mathbf{P}\{\varphi(x_u) \mid \mathscr{M}_{t+}\} \quad (\mathbf{P}\text{-a.s.},\ t \in \widetilde{\Delta}).$$

(see (6.7)).

6.3. THEOREM 6.2. *Any rigid regular re-constructions $\widetilde{\mathscr{K}}$ and $\hat{\mathscr{K}}$ of the Markov class $\mathscr{K}$ are indistinguishable.*

PROOF. Let $\mathscr{K} = (x_t, \mathscr{M}_t, \mathbf{K})$, $\widetilde{\mathscr{K}} = (\widetilde{x}_t, \widetilde{\mathscr{M}}_t, \widetilde{\mathbf{K}})$, $\hat{\mathscr{K}} = (\hat{x}_t, \hat{\mathscr{M}}_t, \hat{\mathbf{K}})$. By the definition of a re-construction, $\widetilde{\mathbf{K}} = \mathbf{K} = \hat{\mathbf{K}}$. If α and β are the end-points of the interval Δ, then by 1.12.B and 1.13, $\widetilde{\Delta} = [\alpha, \beta) = \hat{\Delta}$ (K-a.s.).

We consider the system of functions (2.9) separating measures in $\mathbf{K}_u$. By (6.1)

$$\widetilde{\mathbf{P}}_{u,\, \widetilde{x}_u}\varphi^t(x_t) = \hat{\mathbf{P}}_{u,\, \hat{x}_u}\varphi^t(x_t) \text{ for all } u \in [\alpha, \beta) \quad (\mathbf{K}\text{-a.s.}).$$

Therefore (1.10) is satisfied and $\widetilde{\mathscr{K}}$ is indistinguishable from $\hat{\mathscr{K}}$.

§7. The adjoint process of a regular re-construction

7.1. The aim of the two concluding sections is to construct a regular re-construction $\widetilde{\mathscr{K}} = (\widetilde{x}_t, \widetilde{\mathscr{M}}_t, \mathbf{K})$ of a canonical class $\mathscr{K} = (x_t, \mathscr{M}_t, \mathbf{K})$ corresponding to an arbitrary transition function p.

The key to the construction is given by Theorem 6.1, which states that if a regular re-construction exists, then the corresponding adjoint random process $\mathbf{P}_t^\omega = \widetilde{\mathbf{P}}_{t,\, \widetilde{x}_t(\omega)}$ is expressed in terms of $\mathscr{K}$ by (6.1). The construction of $\mathbf{P}_t^\omega$ will be carried out in this section, and the construction of the class $\widetilde{\mathscr{K}}$ in the following section.

7.2. We fix a countable everywhere dense subset R of T and denote by $\nu(t, \varphi; u, \omega)$ the limit on the right-hand side of (6.1). By Lemma 5.2 and well-known properties of supermartingales (see, for example, [1], Chapter 7, Theorem 3.3) it follows that this limit exists for all $t \in [\alpha, \beta)$ if ω does not belong to a certain **K**-negligible set $\Omega(u, \varphi)$. We select in every space E_u $(u \in R)$ a countable family of functions W_u and denote by Ω' the sum of the $\Omega(u, \varphi)$ over all $u \in R$, $\varphi \in W_u$. It is obvious that Ω' is **K**-negligible and that

$$(7.1) \qquad \nu(t, \varphi; u, \omega) = \lim_{\substack{r \downarrow t \\ r \in R}} p(r, x_r(\omega); u, \varphi)$$

for all $\omega \notin \Omega'$, $t \geqslant \alpha(\omega)$, $u \in R$, $\varphi \in W_u$.

For a special choice of the families W_u we construct measures $\nu^t_{u,\omega}$ in the spaces E_u such that

$$(7.2) \qquad \nu(t, \varphi; u, \omega) = \nu^t_{u,\omega}(\varphi)$$

for all $\omega \notin \Omega'$, $t \geqslant \alpha$, $u \in R$, $\varphi \in W_u$. After this we construct a **K**-negligible set $\Omega_0 \supseteq \Omega'$ and measures $\mathbf{P}^\omega_t \in \mathbf{K}_t$ for which

$$(7.3) \quad \mathbf{P}^\omega_t\{x_u \in \Gamma\} = \nu^t_{u,\omega}(\Gamma) \cdot \text{ for } \omega \notin \Omega_0,\ t \in [\alpha, \beta),\ u \in R,\ \Gamma \in \mathscr{F}(E_u).$$

7.3. It is known (see [4], 3.2) that in any Borel space E a support family can be constructed, that is, a countable family of non-negative bounded functions W containing unity and satisfying the following conditions:

7.3.A. If μ_n is a sequence of finite measures on E and $\mu_n(\varphi)$ converges for any $\varphi \in W$, then we can find a finite measure μ on E such that $\mu_n(\varphi) \to \mu(\varphi)$ for all $\varphi \in W$.

7.3.B. If a set $\mathscr{H}$ of non-negative functions on E contains W and is closed under addition, multiplication by positive numbers, subtraction (leading to non-negative differences) and non-increasing limit passages, then $\mathscr{H}$ contains all functions $\varphi \in \mathscr{F}(E)$.

Let W_u be a support family in E_u. By 7.3.A the existence of measures $\nu^t_{u,\omega}$ satisfying (7.2) follows from (7.1). By (7.1) and (7.2) we have

$$(7.4) \qquad \nu^t_{u,\omega}(\varphi) = \lim_{\substack{r \downarrow t \\ r \in R}} p(r, x_r(\omega); u, \varphi)$$

for all $\omega \notin \Omega'$, $t \geqslant \alpha$, $u \in R$, $\varphi \in W_u$.

We denote by $\widetilde{\mathscr{M}}_t$ the minimal σ-algebra in Ω containing $\mathscr{M}_{t+}$ and all **K**-negligible sets. From (7.4) and (6.8) it follows that for any $u \in R$ and any $\mathbf{P} \in \mathbf{K}$

$$(7.5) \qquad \nu^t_{u,\omega}(\varphi) = \mathbf{P}\{\varphi(x_u) \mid \widetilde{\mathscr{M}}_t\} \qquad (\mathbf{P}\text{-a.s.},\ \alpha \leqslant t).$$

Since this equation holds for $\varphi \in W_u$, it follows by 7.3.B that it holds for all $\varphi \in \mathscr{F}(E_u)$.

7.4. The construction of the measure $\mathbf{P}_t^\omega$ from the system of measures $\nu_{u,\omega}^t$ is based on some lemmas.

LEMMA 7.1. *Let R^t be a countable everywhere dense subset of the half-line (t, ∞), and for every $u \in R^t$ let ν_u be a measure on E_u. Let*

$$\int_{E_u} \nu_u(dx)\, p(u, x; v, \Gamma) = \nu_v(\Gamma) \quad \textit{for all} \quad u < v \in R^t, \ \Gamma \in \mathscr{F}(E_v)$$

and

$$\lim_{u \downarrow t} \nu_u(E_u) = 1.$$

Then there exists a unique measure $\mathbf{P} \in \mathbf{K}_t$ such that

$$\mathbf{P}\{x_u \in \Gamma\} = \nu_u(\Gamma) \quad \textit{for all} \quad u \in R^t, \ \Gamma \in \mathscr{F}(E_u).$$

The proof of this lemma follows easily from Lemma 3.1 of [4].

LEMMA 7.2. *For any $u \in R$, $\varphi \in \mathscr{F}(E_u)$, $\mathbf{P} \in \mathbf{K}$, the triple $(\nu_{u,\omega}^t(\varphi), \tilde{\mathscr{M}}_t, \mathbf{P})$ defines a supermartingale on the interval $[\alpha, \infty)$ which is continuous on the right* (**P**-a.s.).

PROOF. We denote by $\mathscr{H}$ the set of all functions $\varphi \in \mathscr{F}(E_u)$, for which the assertion of the lemma holds. From (7.4)–(7.5) it follows easily that $W_u \in \mathscr{H}$. By 1.9 the set $\mathscr{H}$ is closed under non-increasing limit passages. It is obviously also closed under the other operations listed in 7.3.B. By this condition $\mathscr{H} \supseteq \mathscr{F}(E_u)$.

LEMMA 7.3. *The limit*

$$c_t(\omega) = \lim_{\substack{u \downarrow t \\ u \in R}} \nu_{u,\omega}^t(E_u) \tag{7.6}$$

exists for all $\omega \notin \Omega'$, $t \geqslant \alpha$, and for any $\mathbf{P} \in \mathbf{K}$ the triple $(c_t, \tilde{\mathscr{M}}_t, \mathbf{P})$ is a supermartingale on the interval $[\alpha, \infty)$; c_t is continuous on the right (**P**-a.s.).

PROOF. We put $\nu_u^t = \nu_{u,\omega}^t(E_u)$. By (7.6) and 1.6.D, it follows that ν_u^t does not increase in u for $u > t$. Therefore the limit (7.6) exists. By (7.5)

$$\nu_u^t = \mathbf{P}\{x_u \in E_u \mid \tilde{\mathscr{M}}_t\} = \mathbf{P}\{\beta > u \mid \tilde{\mathscr{M}}_t\} \qquad (\mathbf{P}\text{-a.s.}, \ \alpha < t). \tag{7.7}$$

Let $\varphi(t)$ be any non-decreasing function with $\varphi(t) > t$ for all t. By (7.7) for $s < t$

$$\mathbf{P}\{\nu_{\varphi(t)}^t \mid \tilde{\mathscr{M}}_s\} = \mathbf{P}\{\beta > \varphi(t) \mid \tilde{\mathscr{M}}_s\} \leqslant \mathbf{P}\{\beta > \varphi(s) \mid \tilde{\mathscr{M}}_s\} = \nu_{\varphi(s)}^s \quad (\mathbf{P}\text{-a.s.}, \ \alpha \leqslant s).$$

Moreover, $\nu_{\varphi(t)}^t$ is measurable with respect to $\tilde{\mathscr{M}}_t$. Hence $(\nu_{\varphi(t)}^t, \tilde{\mathscr{M}}_t, \mathbf{P})$ is a supermartingale.

Let $s_m \uparrow \infty$. We associate with the set $S = \{s_1, \ldots, s_m, \ldots\}$ the function

$$\varphi_S(t) = s_m \ \text{ for } s_{m-1} \leqslant t < s_m \quad (m = 1, 2, \ldots),$$

where $s_0 = -\infty$. We note that the function

$$\nu^t_{\varphi_s(t)} = \nu^t_{s_m} \text{ for } t \in [s_{m-1}, s_m)$$

is continuous on the right in t. We now consider an increasing sequence of sets S_n whose union is R. It is obvious that $\varphi_{S_n}(t) \downarrow t$. According to (7.6) $\nu^t_{\varphi_{S_n}(t)} \uparrow c_t$, and the assertion of the lemma follows from the properties of supermartingales stated in 1.9.

7.5. By (7.5) and (1.2) for any $t < u < v$, $\varphi \in W_u$, $\mathbf{P} \in \mathbf{K}$

$$\int_{E_u} \nu^t_{u,\omega}(dx)\, p(u, x; v, \varphi) = \mathbf{P}\{p(u, x_u; v, \varphi) \mid \widetilde{\mathscr{M}}_t\} =$$

$$= \mathbf{P}\{\mathbf{P}[\varphi(x_v) \mid \widetilde{\mathscr{M}}_u] \mid \widetilde{\mathscr{M}}_t\} = \mathbf{P}\{\varphi(x_v) \mid \widetilde{\mathscr{M}}_t\} = \nu^t_{v,\omega}(\varphi) \text{ (}\mathbf{P}\text{-a.s., } t \geqslant \alpha).$$

By (7.6) and (7.7)

$$c_t(\omega) = \lim_{\substack{u \downarrow t \\ u \in R}} \nu^t_{u,\omega}(E_u) = \mathbf{P}\{t < \beta \mid \mathscr{M}_{t+}\} = 1 \quad (\mathbf{P}\text{-a.s., } \alpha \leqslant t < \beta).$$

Therefore we can find a **K**-negligible set $\Omega'' \supseteq \Omega'$ such that

$$\int_{E_u} \nu^t_{u,\omega}(dx)\, p(u, x; v, \varphi) = \nu^t_{v,\omega}(\varphi) \tag{7.8}$$

for all $\omega \notin \Omega''$, $t \in R \cap (\alpha, \beta)$; $v > u \in R \cap (t, \infty)$, $\varphi \in \mathscr{F}(E_v)$:

$$c_t(\omega) = 1 \text{ for all } \omega \notin \Omega'',\ t \in R \cap (\alpha, \beta). \tag{7.9}$$

But by Lemma 7.2 both sides of (7.8) are **K**-almost surely continuous on the right in t (the left-hand side is equal to $\nu^t_{u,\omega}(f)$ for $f(x) = p(u, x; v, \varphi)$). By Lemma 7.3 a similar property holds for $c_t(\omega)$. Therefore outside a certain **K**-negligible set $\Omega_0 \supseteq \Omega''$, (7.8)–(7.9) are satisfied for all $t \in [\alpha, \beta)$, $v > u \in R \cap (t, \infty)$ and $\varphi \in W_v$. By 7.3.B it holds for any $\varphi \in \mathscr{F}(E_v)$. By Lemma 7.1 for any $\omega \notin \Omega_0$ there exist for all $t \in [\alpha, \beta)$ measures $\mathbf{P}^\omega_t$ satisfying condition (7.3).

7.6. We formulate a theorem, which summarizes and somewhat complements the results obtained in 7.1–7.5.

THEOREM 7.1. *Let* $\mathscr{K} = (x_t, \mathscr{M}_t, \mathbf{K})$ *be a canonical class of Markov processes with transition function p. Let R be a countable everywhere dense subset of T and* W_u *a support family in* E_u. *There exist a* **K**-*negligible set* Ω_0 *and measures* $\mathbf{P}^\omega_t \in \mathbf{K}_t$ *such that*

$$\mathbf{P}^\omega_t \varphi(x_u) = \lim_{\substack{r \downarrow t \\ r \in R}} p(r, x_r; u, \varphi) \tag{7.10}$$

for all $\omega \notin \Omega_0$, $t \in [\alpha, \beta)$, $u \in R$, $\varphi \in W_u$.

We denote by $\widetilde{\mathscr{M}}_t$ *the minimal σ-algebra containing* $\mathscr{M}_{t+}$ *and all* **K**-*negligible sets. For any* $\mathbf{P} \in \mathbf{K}$, $\xi \in \mathscr{N}(t, \infty]$

$$\mathbf{P}\{\xi \mid \widetilde{\mathscr{M}}_t\} = \mathbf{P}^\omega_t \xi \quad (\mathbf{P}\text{-a.s., } t \in [\alpha, \beta)). \tag{7.11}$$

For any $\xi \in \mathcal{N}[u, \infty]$ *the function* $\mathbf{P}_t^\omega \xi$ *is* **K**-*almost surely continuous on the right in t on the interval* $[\alpha, u \wedge \beta)$.

PROOF. The first assertion of the theorem follows from (7.3)–(7.4) already proved. It is sufficient to verify (7.11) for functions ξ of the form $\xi = f_1(x_{t_1}) \ldots f_n(x_{t_n})$, where $t < t_1 < \ldots < t_n$, $f_i \in \mathcal{F}(E_{t_i})$. We put $\varphi^r(x) = \mathbf{P}_{r,x}\xi$. Let $r \in R \cap (t, t_1)$. Taking into account that $\widetilde{\mathcal{M}}_t \subseteq \widetilde{\mathcal{M}}_r$ $\mathbf{P}_t^\omega \in \mathbf{K}$ and using (1.8), (7.3) and (7.5) we have

$$(7.12)\quad \mathbf{P}\{\chi_\Delta(r)\xi \mid \widetilde{\mathcal{M}}_t\} = \mathbf{P}\{\mathbf{P}(\chi_\Delta(r)\xi \mid \mathcal{M}_r) \mid \widetilde{\mathcal{M}}_t\} = \\ = \mathbf{P}\{\varphi^r(x_r) \mid \widetilde{\mathcal{M}}_t\} = \mathbf{P}_t^\omega \varphi^r(x_r) = \mathbf{P}_t^\omega \chi_\Delta(r)\xi \qquad (\mathbf{P}\text{-a.s., } t \in [\alpha, \beta)).$$

As $r \downarrow t$ $(r \in R)$, $\chi_\Delta(r) \to \chi_{[\alpha,\beta]}(t)$, and (7.12) goes over into (7.11).

Now let $\xi \in \mathcal{N}[u, \infty]$. We choose $r \in R \cap (-\infty, u)$. Since $\xi \in \mathcal{N}(r, \infty]$ and $\mathbf{P}_t^\omega \in \mathbf{K}_t$, by (1.8) and (7.3)

$$(7.13)\quad \mathbf{P}_t^\omega \chi_{\beta > r}\xi = \mathbf{P}_t^\omega \chi_\Delta(r)\xi = \mathbf{P}_t^\omega \varphi^r(x_r) = \\ = \nu_{t,\omega}^r(\varphi^r) \quad \text{for } \omega \notin \Omega_0,\ t \in [\alpha, r \wedge \beta).$$

On the other hand, $\xi \in \mathcal{N}(r, \infty]$ is constant on the set $\{\beta \leqslant r\}$ and therefore

$$(7.14)\qquad \mathbf{P}_t^\omega \chi_{\beta \leqslant r}\xi = \text{const}\, \mathbf{P}_t^\omega\{\beta \leqslant r\} = \text{const}\ [1 - \nu_{t,\omega}^r(E_r)].$$

By Lemma 7.2 the right-hand sides of (7.13) and (7.14) are continuous on the right in t for $t \geqslant \alpha$ (**K**-a.s.). Consequently $\mathbf{P}_t^\omega \xi$ is continuous on the right in t on the interval $[\alpha, r \wedge \beta)$ (**K**-a.s.). Since $[\alpha, u \wedge \beta)$ is the union of a countable number of intervals $[\alpha, r \wedge \beta)$ for $r < u$ in R, $\mathbf{P}_t^\omega \xi$ is continuous on the right on $[\alpha, u \wedge \beta)$ (**K**-a.s.).

§8. Existence of a regular re-construction

8.1. In the preceding section we have constructed measures $\mathbf{P}_t^\omega$ in terms of a canonical class $\mathcal{K}$ for all $\omega \notin \Omega_0$, $t \in [\alpha(\omega), \beta(\omega))$, where Ω_0 is a certain **K**-negligible set. We extend them to Ω_0 by the formula

$$(8.1)\qquad \mathbf{P}_t^\omega = \mathbf{P}_t \text{ for } t \in [\alpha(\omega), \beta(\omega)),$$

where $\mathbf{P}_t$ is some element of $\mathbf{K}_t$.

We put

$$(8.2)\qquad \tilde{x}_t(\omega) = \mathbf{P}_t^\omega.$$

We consider the filtration $\widetilde{\mathcal{M}}_t$ described in Theorem 7.1 and prove that the set $\widetilde{\mathcal{K}} = (\tilde{x}_t, \widetilde{\mathcal{M}}_t, \mathbf{K})$ defines a regular re-construction of $\mathcal{K}$.

8.2. For the random process $\tilde{x}_t(\omega) = \mathbf{P}_t^\omega$ the state space is $\widetilde{E}_t = \mathbf{K}_t$. We denote by $\widetilde{X}_t$ the extended state space obtained by adjoining to $\widetilde{E}_t$ two fictitious states a_t and b_t.

We need the following lemma.

LEMMA 8.1. *Let $\mathscr{A}^t$ be the minimal σ-algebra containing $\mathscr{N}(t, \infty)$ and all* **K**-*negligible sets. If F is a measurable function on $\widetilde{X}_t$, then $F[\widetilde{x}_t(\omega)]$ is measurable with respect to the σ-algebra $\widetilde{\mathscr{M}}_t \cap \mathscr{A}^t$.*

PROOF. For brevity we put $\widetilde{\mathscr{M}}_t \cap \mathscr{A}^t = \mathscr{F}$. According to Lemma 2.3 a measurable structure in **K** and hence also in $\mathbf{K}_t$ is generated by the functions

$$F(\mathbf{P}) = \mathbf{P}\{x_u \in \Gamma\} \quad (u \in T,\ \Gamma \in \mathscr{F}(X_u)).$$

Therefore to prove the lemma it suffices to convince ourselves of the $\mathscr{F}$-measurability of the functions

$$F(\widetilde{x}_t(\omega)) = F(\mathbf{P}_t^{\omega}) = \mathbf{P}_t^{\omega}\{x_u \in \Gamma\}\ (u \in T,\ \Gamma \in \mathscr{F}(X_u)). \tag{8.3}$$

Since $\mathbf{P}_t^{\omega} \in \mathbf{K}_t$, by (1.4)

$$\mathbf{P}_t^{\omega}\{x_u \in \Gamma\} = \chi_{\Gamma}(a_u) \quad \text{for } u \leqslant t,$$
$$\mathbf{P}_t^{\omega}\{x_u = a_u\} = 0 \quad \text{for } u > t.$$

It is therefore sufficient to verify the $\mathscr{F}$-measurability of the functions (8.3) for $u > t$, $\Gamma \in \mathscr{F}(E_u)$.

We note first of all that by (7.10) and (8.1) the functions $\mathbf{P}_t^{\omega}\varphi(x_u)$ are $\mathscr{F}$-measurable for $u \in R$, $\varphi \in W_u$. The $\mathscr{F}$-measurability of these functions for all $u \in R$, $\varphi \in \mathscr{F}(E_u)$ follows from 7.3.B of the support system W_u. Finally, for any $u > t$ we can find v in R belonging to the interval (t, u). By (1.4) and (1.8)

$$\mathbf{P}_t^{\omega}\varphi(x_u) = \mathbf{P}_t^{\omega}\chi_{\Delta}(u)\varphi(x_u) = \mathbf{P}_t^{\omega}p(v, x_v;\ u, \varphi).$$

Hence $\mathbf{P}_t^{\omega}\varphi(x_u)$ is $\mathscr{F}$-measurable for any $u > t$, $\varphi \in \mathscr{F}(E_u)$.

8.3. We show that the formula

$$\widetilde{p}(t, \mathbf{P};\ u, \Gamma) = \mathbf{P}\{\widetilde{x}_u \in \Gamma\}\ (\mathbf{P} \in \widetilde{E}_t,\ \Gamma \in \mathscr{F}(\widetilde{E}_u)) \tag{8.4}$$

defines a transition function in the spaces $\widetilde{E}_t = \mathbf{K}_t$.

By Lemma 8.1 for $\Gamma \in \mathscr{F}(\widetilde{E}_u)$

$$\{\widetilde{x}_u \in \Gamma\} \in \widetilde{\mathscr{M}}_u \cap \mathscr{A}^u \subseteq \mathscr{N}.$$

Therefore $\widetilde{p}$ satisfies 1.10.A. It is obvious that 1.10.B–1.10.C also hold. Let $s < t < u$ and $\mathbf{P} \in \mathbf{K}_s$. By (7.11) and (8.4)

$$\mathbf{P}\{\widetilde{x}_u \in \Gamma\} = \mathbf{P}\{\mathbf{P}[\widetilde{x}_u \in \Gamma \mid \widetilde{\mathscr{M}}_t]\} = \mathbf{P}\mathbf{P}_t^{\omega}\{\widetilde{x}_u \in \Gamma\} = \mathbf{P}\widetilde{p}(t, \mathbf{P}_t^{\omega};\ u, \Gamma) = \mathbf{P}\widetilde{p}(t, \widetilde{x}_t;\ u, \Gamma),$$

from which 1.10.D follows. Finally, as $u \downarrow t$

$$\widetilde{p}(t, \mathbf{P};\ u, \widetilde{E}_u) = \mathbf{P}\{\widetilde{x}_u \in \widetilde{E}_u\} = \mathbf{P}\{\alpha \leqslant u < \beta\} = \mathbf{P}\{u < \beta\} \to \mathbf{P}\{t < \beta) = \mathbf{P}\{\alpha < \beta\} = 1,$$

so that 1.10.E holds.

By Lemma 2.3 the transition function $\widetilde{p}$ separates states.

8.4. Let $\mathbf{P} \in \mathbf{K}$. From Lemma 8.1 and (7.11), (8.2) and (8.4) we have for any $t < u$, $\Gamma \in \mathscr{F}(\widetilde{E}_u)$

$$\mathbf{P}\{\widetilde{x}_u \in \Gamma \mid \widetilde{\mathscr{M}}_t\} = \mathbf{P}_t^{\omega}\{\widetilde{x}_u \in \Gamma\} = \widetilde{p}(t, P_t^{\omega}; u, \Gamma) = \widetilde{p}(t, \widetilde{x}_t; u, \Gamma) \ (\mathbf{P}\text{-a.s.}, \ t \in [\alpha, \beta)).$$

Therefore $(\widetilde{x}_t, \widetilde{\mathscr{M}}_t, \mathbf{P})$ is a Markov process with transition function $\widetilde{p}$.

By the definition of $\widetilde{E}_t$, to each $x \in \widetilde{E}_t$ there corresponds a measure $\widetilde{\mathbf{P}}_{t,x} \in \mathbf{K}_t$, and (8.4) implies (1.5), so that $\mathscr{K}$ is a class of Markov processes.

We show that this class is regular. According to [5], §3, $\mathbf{K}$ is a Borel space, $\mathbf{K}_t$ is a measurable set in $\mathbf{K}$ and therefore is also a Borel space. Consequently 1.12.A holds. It is obvious that 1.12.B and 1.12.C hold. Finally, Theorem 7.1 implies that 1.12.D is satisfied, since according to (8.2) and (8.4)

$$\widetilde{p}(t, \widetilde{x}_t(\omega); u, \Gamma) = \mathbf{P}_t^{\varphi}\{\widetilde{x}_u \in \Gamma\}.$$

8.5. We prove finally that $\widetilde{\mathscr{K}}$ is a re-construction of the class $\mathscr{K}$. Since $\Delta = (\alpha, \beta)$, $\widetilde{\Delta} = [\alpha, \beta)$, for any probability measure $\mathbf{P}$

$$\mathbf{P}\{\chi_{\Delta}(t) \neq \chi_{\widetilde{\Delta}}(t)\} = \mathbf{P}\{\alpha = t\}$$

can be different from zero only for a countable set of values t.

Further, the equation

$$\mathbf{P}_{t, x_t} = \widetilde{\mathbf{P}}_{t, \widetilde{x}_t} \tag{8.5}$$

is equivalent to the countable system of equations

$$\mathbf{P}_{t, x_t}\varphi(x_u) = \mathbf{P}_t^{\omega}\varphi(x_u) \quad (u \in R, \varphi \in W_u). \tag{8.6}$$

By (7.3) the equation (8.6) can be written in the form

$$p(t, x_t(\omega); u, \varphi) = \mathbf{v}_{u,\omega}^{t}(\varphi). \tag{8.7}$$

But from the theory of martingales it is known (see [1], Ch. 7, Theorem 11.2) that if $(z_t, \mathscr{F}_t, \mathbf{P})$ is a supermartingale, then

$$\lim_{\substack{r \downarrow t \\ r \in R}} z_r = z_t \quad (\mathbf{P}\text{-a.s.})$$

with the possible exception of a countable set $\Lambda(\mathbf{P})$ of values of t. Applying this result to the supermartingale $(p(t, x_t; u, \varphi), \mathscr{M}_t, \mathbf{P})$ (see Lemma 5.2) and taking (7.4) into account we conclude that if $t \notin \Lambda(\mathbf{P})$, then (8.7) holds ($\mathbf{P}$-a.s.). Consequently the system of equations (8.6) and therefore also (8.5) hold ($\mathbf{P}$-a.s.).

Thus, $\widetilde{\mathscr{K}}$ is a re-construction of $\mathscr{K}$.

8.6. We now assume that the transition function p is homogeneous and show that in this case the transition function $\widetilde{p}$ corresponding to $\widetilde{\mathscr{K}}$ is also homogeneous (under a suitable identification of the spaces $\widetilde{E}_t$).

In order to identify the spaces $\widetilde{E}_t = \mathbf{K}_t$ we consider the transformations θ_s of the space Ω defined by the formula

$$(\theta_s\omega)(t) = \omega(t+s) \quad (\alpha - s < t < \beta - s). \tag{8.8}$$

The transformations θ_s induce operators on the functions ξ and measures $\mathbf{P}$ in the space Ω acting according to the formulae

$$\begin{cases} (\theta_s\xi)(\omega) = \xi(\theta_s\omega), \\ (\mathbf{P}\theta_s)(A) = \mathbf{P}(\theta_s^{-1}A). \end{cases} \tag{8.9}$$

Here

$$(\mathbf{P}\theta_s)(\xi) = \mathbf{P}(\theta_s\xi). \tag{8.10}$$

Relying on the homogeneity of the transition function p it is not difficult to verify that $\mathbf{K}$ is invariant with respect to the operators θ_s. By (8.9) the $\mathbf{K}$-negligibility of $\theta_s^{-1}A$ follows from that of A. From the equation $\theta_s\alpha = \alpha - s$ it follows that $\theta_s\mathbf{K}_t = \mathbf{K}_{t-s}$. Moreover, θ_s is an isomorphism of the measurable spaces $\mathbf{K}_t$ and $\mathbf{K}_{t-s}$. It is this isomorphism that we use for the identification of $\mathbf{K}_t$ and $\mathbf{K}_{t-s}$. We prove that

$$\widetilde{p}(t+s,\ \mathbf{P}\theta_{-s};\ u+s,\ \Gamma\theta_{-s}) = \widetilde{p}(t,\ \mathbf{P};\ u,\ \Gamma). \tag{8.11}$$

By (8.4) this equation can be rewritten in the form

$$(\mathbf{P}\theta_{-s})\{\widetilde{x}_{u+s} \in \Gamma\theta_{-s}\} = \mathbf{P}\{\widetilde{x}_u \in \Gamma\}. \tag{8.12}$$

The left-hand side is equal to

$$\mathbf{P}\{\widetilde{x}_{u+s}(\theta_{-s}\omega) \in \Gamma\theta_{-s}\}$$

and taking (8.1) into account we can put (8.12) in the following form:

$$\mathbf{P}\{\omega: \mathbf{P}_{u+s}^{\theta_{-s}\omega}\theta_s \in \Gamma\} = \mathbf{P}\{\omega: \mathbf{P}_u^{\omega} \in \Gamma\}.$$

Therefore (8.11) will be proved if we show that outside a certain $\mathbf{K}$-negligible set

$$\mathbf{P}_{u+s}^{\theta_{-s}\omega}\theta_s = \mathbf{P}_u^{\omega} \quad \text{for all } u \in [\alpha, \beta) \tag{8.13}$$

We make use of Theorems 6.1 and 2.4. Let R be a countable everywhere dense subset of T, and let W be a supporting system of functions in the state space E. By Theorem 2.4, to verify that two measures in $\mathbf{K}_u$ (the left- and right-hand sides of (8.13) are such measures) coincide, it suffices to check that these measures coincide on the functions $\varphi(x_v)$ ($v \in R + s$, $\varphi \in W$). By Theorem 6.1 (applied to the set $R + s$) we can find a $\mathbf{K}$-negligible set Ω' such that

$$\widetilde{\mathbf{P}}_{u+s,\ \widetilde{x}_{u+s}(\omega)}\varphi(x_{v+s}) = \lim_{r\downarrow u+s,\ r\in R+s} p(r,\ x_r(\omega);\ v+s,\ \varphi) \tag{8.14}$$

for all $\omega \notin \Omega'$, $u \in [\alpha(\omega) - s, \beta(\omega) - s)$, $v \in R$, $\varphi \in W$. But $\widetilde{\mathbf{P}}_{u+s, \tilde{x}_{u+s}} = \mathbf{P}^{\omega}_{u+s}$ and by (8.10), (8.7) and (8.14)

$$(\mathbf{P}^{\theta_{-s}\omega}_{u+s}\theta_s)\,\varphi(x_v) = \widetilde{\mathbf{P}}_{u+s,\,\tilde{x}_{u+s}(\theta_{-s}\omega)}\varphi(x_{v+s}) = \lim_{r\downarrow u+s,\, r\in R+s} p(r, x_r(\theta_{-s}\omega); v+s, \varphi) =$$

$$= \lim_{r\downarrow u,\, r\in R} p(r+s, x_r(\omega); v+s, \varphi) = \lim_{r\downarrow u,\, r\in R} p(r, x_r(\omega); v, \varphi)$$

for all $\omega \notin \theta_s\Omega'$, $u \in [\alpha(\omega), \beta(\omega))$, $v \in R$, $\varphi \in W$. By Theorem 6.1 the right-hand side is equal to $\widetilde{\mathbf{P}}_{u, \tilde{x}_u(\omega)}\varphi(x_v) = \mathbf{P}^{\omega}_u\varphi(x_v)$ for all $\omega \notin \Omega''$, $u \in [\alpha, \beta)$, $v \in R$, $\varphi \in W$, where Ω'' is a **K**-negligible set. Thus (8.13) holds outside the **K**-negligible set $\theta_s\Omega' \cup \Omega''$.

References

[1] J. L. Doob, Stochastic processes, Wiley, New York 1953.
Translation: *Veroyatnostnye protsessy,* Inost. Lit., Moscow 1956.

[2] J. L. Doob, Compactification of the discrete state space of a Markov process, Z. Wahrscheinlichkeitstheorie Verw. Geb. **10** (1968), 236–251.

[3] J. L. Doob, State spaces for Markov chains, Trans. Amer. Math. Soc. **149** (1970), 279–305.

[4] E. B. Dynkin, Initial and final behaviour of trajectories of Markov processes, Uspekhi Mat. Nauk **26**:4 (1971), 153–172.
= Russian Math. Surveys **26**:4 (1971), 165–185.

[5] E. B. Dynkin, Integral representation of excessive measures and excessive functions, Uspekhi Mat. Nauk 27: 1 (1972), 43–80.
= Russian Math. Surveys **27**: 1 (1972), 43–84.

[6] E. B. Dynkin, Supermartingales with random birth times, Uspekhi Mat. Nauk **26**:6 (1971), 213–214.

[7] E. B. Dynkin, *Osnovaniya teorii markovskikh protsessov,* Fizmatgiz, Moscow 1959.
Translation: Foundations of the theory of Markov processes, Pergamon, Oxford–New York 1960.

[8] F. Knight, Markov processes on an entrance boundary, Illinois J. Math. **7** (1963), 322–336.

[9] K. Kuratowski, Topologie, fourth ed., 2 vols., Polish Acad. Sci., Warsaw 1959.
English translation: Topology, 2 vols., Academic Press, New York 1967–69.
Russian translation: *Topologiya,* tom 1, Izdat. Mir, Moscow 1966.

[10] P. A. Meyer, Probability and potentials, Blaisdell, Toronto 1966.

Translated by S. M. Rudolfer

MARKOV REPRESENTATIONS OF STOCHASTIC SYSTEMS

A great deal of research into the theory of random processes is concerned with the problem of constructing a process that has certain properties of regularity of the trajectories and has the same finite-dimensional probability distribution as a given stochastic process x_t. It is a complicated theory and one that is difficult to apply to those properties that we most need for the study of Markov processes (the strong Markov property, quasi-left-continuity, and the like.)

The problem can be usefully reformulated. In an actual experiment we do not observe the state x_t at a fixed instant t, but rather events that occupy certain time intervals. This is the motivation behind the Gel'fand-Itô theory of generalized random processes. Kolmogorov, in 1972, proposed an even more general concept of a stochastic process as a system of σ-algebras $\mathscr{F}(I)$ labelled by time intervals I. Developing this approach, we introduce the concept of a Markov representation x_t of the stochastic system $\mathscr{F}(I)$ and prove the existence of regular representations. We construct two dual regular representations (the right and the left), which we then combine into a single Markov process by two methods, the "vertical" and the "horizontal" method. We arrive at a general duality theory, which provides a natural framework for the fundamental results on entrance and exit spaces, excessive measures and functions, additive functionals, and others. The initial steps in the construction of this theory were taken in [6]. The note [5] deals with applications to additive functionals (detailed proofs are in preparation). We consider random processes defined in measurable spaces without any topology: the introduction of a reasonable topology allows of a certain arbitrariness. The relation between our definitions of regularity and more traditional properties stated in topological terms (continuity from the right, the existence of a limit from the left, etc.) are considered in the Appendix, which is written by S. E. Kuznetsov.

Contents

§1. Introduction

1.1. Stochastic systems. A σ-algebra $\mathscr{F}$ of subsets of a set Ω determines a measurable structure in Ω. We then call Ω a *measurable space* and refer to the elements of $\mathscr{F}$ as measurable sets.

Let $(\Omega, \mathscr{F})$ be a given measurable space and $T = (\alpha, \beta)$ a real interval. For each open interval $I \subseteq T$, let $\mathscr{F}(I)$ be a σ-algebra contained in $\mathscr{F}$. We call $\mathscr{F}(I)$ a *stochastic system* if the following conditions hold:

1.1.A. $\mathcal{F}(I_1) \subseteq \mathcal{F}(I_2)$ when $I_1 \subseteq I_2$.

1.1.B. If $I_n \uparrow I$, then $\mathcal{F}(I)$ is the minimal σ-algebra containing all the σ-algebras $\mathcal{F}(I_n)$.

1.1.C. Suppose that $I_n \uparrow I$ and that for each n, $\mathbf{P}_n$ is a probability measure on $\mathcal{F}(I_n)$, such that $\mathbf{P}_n(A) = \mathbf{P}_{n-1}(A)$, whenever $A \in \mathcal{F}(I_{n-1})$. Then there exists a measure $\mathbf{P}_\infty$ on $\mathcal{F}(I)$ that coincides with $\mathbf{P}_n$ on $\mathcal{F}(I_n)$ $(n = 1, 2, \ldots)$.

Let us describe an important method of constructing such systems. For each $t \in T$, let $(E_t, \mathcal{B}_t)$ be a given measurable space and let Ω be the space of all functions $\omega(t) \in E_t$ $(t \in T)$. For each $t \in T$ the formula $x_t(\omega) = \omega(t)$ defines a mapping of Ω into E_t. Let $\mathcal{F}(I)$ be the σ-algebra of subsets of Ω generated by the mappings $x_t (t \in I)$. Conditions 1.1.A and 1.1.B are obviously satisfied, and Condition 1.1.C follows from the well-known theorem of Kolmogorov on measures in infinite products, provided that E_t is a Borel space (that is, E_t is isomorphic to the Borel subsets of a complete separable metric space).

Let $\mathcal{F}(I)$ be a stochastic system. We write $\mathcal{F}_{<t} = \mathcal{F}(\alpha, t)$, $\mathcal{F}_{>t} = \mathcal{F}(t, \beta)$. For arbitrary $s \leqslant t \in T$, we define the σ-algebra $\mathcal{F}(s, t+)$ as the intersection of all $\mathcal{F}(s, u)$ with $u > t$. The σ-algebras $\mathcal{F}(s-, t)$, $\mathcal{F}(s-, t+)$ etc., are defined similarly.

A probability space is a triple $(\Omega, \mathcal{F}, \mathbf{P})$, where $(\Omega, \mathcal{F})$ is a measurable space, and $\mathbf{P}$ a probability measure on $\mathcal{F}$. We shall be interested in stochastic systems in probability spaces.

1.2. Markov representations. Transition and co-transition probabilities. By a *random process* we mean a set of measurable mappings $x_t(\omega) (t \in T)$ from a probability space $(\Omega, \mathcal{F}, \mathbf{P})$ into Borel measurable spaces E_t (*the state spaces*). We call the random process x_t a *Markov representation of the stochastic system* $\mathcal{F}(I)$, if the following hold:[1]

1.2.A. The mapping x_t is measurable with respect to $\mathcal{F}(I)$ when $t \in I$.

1.2.B. For any $t \in T$, $\xi \in \mathcal{F}_{<t}$ and $\eta \in \mathcal{F}_{>t}$

$$\mathbf{P}\{\xi\eta \mid x_t\} = \mathbf{P}\{\xi \mid x_t\}\mathbf{P}\{\eta \mid x_t\} \quad (\mathbf{P}\text{-a.s.}).$$

We denote by $\mathcal{F}_{\leqslant t}$ the minimal σ-algebra containing $\mathcal{F}_{<t}$ and all the sets $\{x_t \in \Gamma\}$ where $\Gamma \subset \mathcal{B}_t$. The notation $\mathcal{F}_{\geqslant t}$, $\mathcal{F}(s, t]$, $\mathcal{F}[s, t)$ is defined similarly. (We emphasize that all these σ-algebras depend on the choice of the representation x_t of the stochastic system $\mathcal{F}(I)$.) Then the condition 1.2.B is equivalent to each of the following conditions:

1.2.B′. For any $t \in T$ and $\eta \in \mathcal{F}_{>t}$

$$\mathbf{P}\{\eta \mid \mathcal{F}_{\leqslant t}\} = \mathbf{P}\{\eta \mid x_t\} \quad (\mathbf{P}\text{-a.s.}).$$

[1] The notation $\xi \in \mathcal{F}$ means that ξ is a non-negative $\mathcal{F}$-measurable function. By $\mathbf{P}\xi$ we denote the integral of ξ with respect to the measure $\mathbf{P}$. If $\mathbf{P}\xi < \infty$, then we say that the function ξ is $\mathbf{P}$-integrable.

1.2.B″. For any $t \in T$ and $\xi \in \mathscr{F}_{<t}$

$$\mathbf{P}\{\xi \mid \mathscr{F}_{\geq t}\} = \mathbf{P}\{\xi \mid x_t\} \quad (\mathbf{P}\text{-a.s.}).$$

Of course, not every stochastic system has a Markov representation; for one to exist it is necessary that for arbitrary $t \in T$, $\xi \in \mathscr{F}_{<t}$, $\eta \in \mathscr{F}_{>t}$

$$\mathbf{P}\{\xi\eta \mid \mathscr{F}(t-,\ t+)\} = \mathbf{P}\{\xi \mid \mathscr{F}(t-,\ t+)\}\mathbf{P}\{\eta \mid \mathscr{F}(t-,\ t+)\} \quad (\mathbf{P}\text{-a.s.}).$$

This condition is sufficient, provided that for all t the σ-algebra $\mathscr{F}(t-,\ t+)$ is generated by a measurable mapping x_t from Ω into a Borel space E_t, and by $\mathbf{P}$-negligible sets.[1] We intend to choose the Markov representation in the most expedient form, assuming that such representations do exist at all.

We have to assume even more, namely, that there is a Markov representation having transition (or co-transition) probabilities. The *transition probabilities* for x_t are a set of probability measures $\mathbf{P}_{t,x}(t \in T,\ x \in E_t)$, where the measure $\mathbf{P}_{t,x}$ is defined on the σ-algebra $\mathscr{F}_{>t}$, the function $\mathbf{P}_{t,x}\eta$ is measurable with respect to x for arbitrary $\eta \in \mathscr{F}_{>t}$ and

(1.1) $\mathbf{P}_{s,x}\xi\eta = \mathbf{P}_{s,x}\xi\mathbf{P}_{t,x_t}\eta$ for $s < t \in T$, $\xi \in \mathscr{F}(s,\ t]$, $\eta \in \mathscr{F}_{>t}$,

(1.2) $\mathbf{P}\eta = \mathbf{P}\mathbf{P}_{t,x_t}\eta$ for $t \in T$, $\eta \in \mathscr{F}_{>t}$.

From (1.2) and 1.2.B′ it follows that

(1.3) $\mathbf{P}\{\eta \mid x_t\} = \mathbf{P}_{t,x_t}\eta \quad (\mathbf{P}\text{-a.s.})$.

Therefore, under condition (1.1), the set of requirements (1.2) and 1.2.B′ is equivalent to

(1.4) $\mathbf{P}\xi\eta = \mathbf{P}\xi\mathbf{P}_{t,x_t}\eta$ for $\xi \in \mathscr{F}_{\leq t}$, $\eta \in \mathscr{F}_{>t}$.

The pair $(x_t,\ \mathbf{P}_{t,x})$, where x_t is a Markov representation of a stochastic system $\mathscr{F}(I)$ and $\mathbf{P}_{t,x}$ are transition probabilities of x_t, is called a *right Markov representation* of $\mathscr{F}(I)$. To establish that $(x_t,\ \mathbf{P}_{t,x})$ is a right Markov representation of a stochastic system $\mathscr{F}(I)$, it is sufficient to verify that the conditions 1.2.A, (1.1), and (1.4) are satisfied (and that $\mathbf{P}_{t,x}\eta$ is measurable with respect to x).

In contrast to Markov representations, we may talk of right Markov representations when a stochastic system $\mathscr{F}(I)$ is defined in a measurable space $(\Omega,\ \mathscr{F})$, but the measure $\mathbf{P}$ is not fixed. Here only the Conditions 1.2.A and (1.1) remain.

We say that the right representation $(x_t,\ \mathbf{P}_{t,x})$ *separates the states* if $\mathbf{P}_{t,x} \neq \mathbf{P}_{t,y}$ when $x \neq y$.

We say that a probability measure $\mathbf{P}'$, defined on the σ-algebra $\mathscr{F}_{>s}$, is dominated by $(x_t,\ \mathbf{P}_{t,x})$, and we write $\mathbf{P}' \in K_s(x_t,\ \mathbf{P}_{t,x})$, if

(1.5) $\mathbf{P}'\eta = \mathbf{P}'\mathbf{P}_{t,x_t}\eta$ for all $t \in (s,\ \beta)$, $\eta \in \mathscr{F}_{>t}$.

[1] We say that a set C is $\mathbf{P}$-negligible if $\mathbf{P}(C) = 0$, and $\mathbf{P}$-certain if $\mathbf{P}(C) = 1$.

By virtue of (1.1), $\mathbf{P}_{s,x} \in K_s(x_t, \mathbf{P}_{t,x})$. The condition (1.2) is equivalent to the requirement that $\mathbf{P} \in \mathbf{K}_\alpha(x_t, \mathbf{P}_{t,x})$. We say that the representation $(\bar{x}_t, \bar{\mathbf{P}}_{t,x})$ is dominated by $(x_t, \mathbf{P}_{t,x})$ if each measure $\bar{\mathbf{P}}_{t,x}$ is dominated by $(x_t, \mathbf{P}_{t,x})$, that is, if for arbitrary $s < t$, $\eta \in \mathcal{F}_{>t}$

$$(1.6) \qquad \bar{\mathbf{P}}_{s,x}\mathbf{P}_{t,x_t}\eta = \bar{\mathbf{P}}_{s,x}\eta.$$

We call two right representations $(x_t, \mathbf{P}_{t,x})$ and $(\bar{x}_t, \bar{\mathbf{P}}_{t,x})$ *equivalent* if there exist a $\mathbf{P}$-certain event Ω', measurable sets $E'_t \subseteq E_t$, $\bar{E}'_t \in \bar{E}_t$, and an isomorphism γ of E'_t onto $\bar{E}'_t$, such that $\mathbf{P}_{t,x} = \bar{\mathbf{P}}_{t,\gamma(x)}$ when $x \in E'_t$;

$$x_t(\omega) \in E'_t, \ \bar{x}_t(\omega) \in \bar{E}'_t \text{ and } \gamma[x_t(\omega)] = \bar{x}_t(\omega) \text{ when } \omega \in \Omega'.$$

Together with right representations we also consider *left* Markov representations $(x_t, \mathbf{P}^{t,x})$ of a stochastic system $\mathcal{F}$ *(I). The co-transition probabilities* $\mathbf{P}^{t,x}$ are probability measures on $\mathcal{F}_{<t}$, and (1.1)–(1.4) are replaced by

$$(1.7) \quad \mathbf{P}^{u,x}\xi\eta = \mathbf{P}^{u,x}(\mathbf{P}^{t,x_t}\xi)\eta \quad \text{for} \quad t < u \in T, \ \xi \in \mathcal{F}_{<t}, \ \eta \in \mathcal{F}[t, u);$$

$$(1.8) \quad \mathbf{P}\xi = \mathbf{P}\mathbf{P}^{t,x_t}\xi \quad \text{for} \quad t \in T, \ \xi \in \mathcal{F}_{<t};$$

$$(1.9) \quad \mathbf{P}\{\xi \mid x_t\} = \mathbf{P}^{t,x_t}\xi \quad (\mathbf{P}\text{-a.s.}) \text{ for } \quad \xi \in \mathcal{F}_{<t};$$

$$(1.10) \quad \mathbf{P}\xi\eta = \mathbf{P}(\mathbf{P}^{t,x_t}\xi)\eta \quad \text{for} \quad \xi \in \mathcal{F}_{<t}, \ \eta \in \mathcal{F}_{\geq t}.$$

We define *a two-sided Markov representation* as a set $(x_t, \mathbf{P}_{t,x}, \mathbf{P}^{t,x})$, where x_t is a Markov representation, $\mathbf{P}_{t,x}$ are its transition probabilities, and $\mathbf{P}^{t,x}$ its co-transition probabilities. The definition of equivalence extends naturally to left and two-sided Markov representations. The representation $(x_t, \mathbf{P}_{t,x}, \mathbf{P}^{t,x})$ *separates states* if from the equations $\mathbf{P}_{t,x} = \mathbf{P}_{t,y}$, $\mathbf{P}^{t,x} = \mathbf{P}^{t,y}$ it follows that $x = y$.

1.3. Transition and co-transition functions. If $(x_t, \mathbf{P}_{t,x})$ is a right Markov representation, then the function

$$(1.11) \quad p(s, x; t, \Gamma) = \mathbf{P}_{s,x}\{x_t \in \Gamma\} \quad (s < t \in T, \ x \in E_s, \ \Gamma \in \mathcal{B}_u)$$

has the following properties:

1.3.A. $p(s, -; t, \Gamma)$ is a measurable function in the space E_s.

1.3.B. $p(s, x; t, -)$ is a probability measure in the space E_t.

1.3.C. For arbitrary $s < t < u \in T$, $x \in E_s$, $\Gamma \in \mathcal{B}_u$

$$\int p(s, x; t, dy)\, p(t, y; u, \Gamma) = p(s, x; u, \Gamma)$$

(the Chapman-Kolmogorov equation).

A function satisfying the conditions 1.3.A.–1.3.C is said to be a *transition function*. We extend it to all values $s, t \in T$ by setting $p(s, x; t, \Gamma) = 0$ for $s \leq t$.

To each right Markov representation $(x_t, \mathbf{P}_{t,x})$ there corresponds a transition function (1.11). The converse holds if the stochastic system $\mathcal{F}(I)$ and the random process x_t are constructed according to the method described in §1.1; in this case the transition probabilities $\mathbf{P}_{t,x}$ are constructed from the $p(s, x; t, \Gamma)$ by means of the Kolmogorov theorem.

The co-transition function $p(t, \Gamma; u, x) = \mathbf{P}^{u,x}\{x_t \in \Gamma\}$ corresponds to the left Markov representation $(x_t, \mathbf{P}^{t,x})$. The integrals of φ with respect to the measures $p(s, x; t, -)$ and $p(t, -; u, x)$ are denoted by $p(s, x; t, \varphi)$ and $p(t, \varphi; u, x)$ respectively.

We define the *phase space*, which we denote by $\mathscr{E}$ as the union of the spaces E_t for all $t \in T$, with the measurable structure generated by the functions

$$f(t, x) = p(t, x; u, \Gamma) \qquad (u \in T, \ \Gamma \in \mathscr{B}_u).$$

The function $f(t, x) = t$ is measurable on $\mathscr{E}$, because

$$\{(t, x):\ t < u\} = \{(t, x):\ p(t, x; u, E_u) = 1\}.$$

Therefore, to each measurable subset Λ of the interval T there corresponds the measurable set $\mathscr{E}(\Lambda) = \{(t, x):\ t \in \Lambda\}$ in $\mathscr{E}$.

1.4. Regular right representations. Let $\mathscr{F}(I)$ be a stochastic system over the time interval (α, β) in the probability space $(\Omega, \mathscr{F}, \mathbf{P})$ The right Markov representation $(x_t, \mathbf{P}_{t,x})$ is said to be *regular* if it can be extended to the interval $[\alpha, \beta)$ so that the following condition holds:

1.4.A. For any $\alpha < u < \beta$ and $\eta \in \mathscr{F}_{>u}$ the function $\mathbf{P}_{t,x_t}\eta$ is $\mathbf{P}$-almost surely right-continuous in t in $[\alpha, u)$.

It is said to be *completely regular* if it satisfies the following stronger requirement:

1.4.B. For any $\alpha \leqslant s < u < \beta$. $\mathbf{P}' \in \mathbf{K}_s(x_t, \mathbf{P}_{t,x})$ and $\eta \in \mathscr{F}_{>u}$ the function $\mathbf{P}_{t,x}\eta$ is right-continuous in t in $[s, u)$, $\mathbf{P}'$-almost surely.

In contrast to 1.4.A, 1.4.B is meaningful even when no measure $\mathbf{P}$ has been fixed in $(\Omega, \mathscr{F})$.

We shall prove the following theorem in §2.

THEOREM 1.1. *Given a Markov representation $(x_t, \mathbf{P}_{t,x})$ of a stochastic system $\mathscr{F}(I)$ there exists a completely regular right representation $(x_{t+}, \mathbf{P}_{t+,x})$, which is dominated by it, separates the states, and for which the following condition holds:*

1.4.C. *For any $t \in [\alpha, \beta)$, x_t is measurable with respect to $\mathscr{F}(t, t+)$.*

Every right regular representation dominated by $(x_t, \mathbf{P}_{t,x})$ and separating the states is equivalent to $(x_{t+}, \mathbf{P}_{t+,x})$.

The representation $(x_{t+}, \mathbf{P}_{t+,x})$ constructed in Theorem 1.1. is called *the regularization* of the right representation $(x_t, \mathbf{P}_{t,x})$. The regularization of the left representation is defined similarly.

1.5. In the proof of the uniqueness assertion of Theorem 1.1. we utilize the following property of regular right representations: for any $t \in [\alpha, \beta)$ and $\eta \in \mathscr{F}_{>t}$

$$\mathbf{P}\{\eta \mid \mathscr{F}_{<t+}\} = \mathbf{P}_{t,x_t}\eta \ \ (\mathbf{P}\text{-a.s.}). \tag{1.12}$$

For completely regular processes this property can be strengthened as follows: for any $\alpha \leqslant s < t < \beta$, $\mathbf{P}' \in \mathbf{K}_s(x_t, \mathbf{P}_{t,x})$, and $\eta \in \mathscr{F}_{>t}$

(1.13) $$\mathbf{P}'\{\eta \mid \mathscr{F}(s, t+)\} = \mathbf{P}_{t, x_t}\eta \quad (\mathbf{P}\text{-a.s.}).$$

§3 begins with the derivation of the following *regularity criterion*. The right representation $(x_t, \mathbf{P}_{t,x})$ is regular if and only if for any $u \in (\alpha, \beta)$ and $\Gamma \in \mathscr{B}_u$ the function $p(t, x_t; u, \Gamma)$ is right-continuous with respect to t in $[\alpha, u)$ **P**-almost surely. It is completely regular if and only if for arbitrary $\alpha \leqslant s < u < \beta$, $\mathbf{P}' \in \mathbf{K}_s(x_t, \mathbf{P}_{t,x})$ and $\Gamma \in \mathscr{B}_u$ the function $p(t, x_t; u, \Gamma)$ is right-continuous in t in $[s, u)$, $\mathbf{P}'$-almost surely.

This criterion together with (1.12), (1.13) enables us to establish a connection between regular right representations and the concepts of a regular Markov process and a regular Markov class, introduced in [4]. Relying on the results of [4], we describe in §3 a number of important properties of regular right representations. The second half of §3 is concerned with the construction of "good" co-transition probabilities for these representations. We obtain partial results only, but they are sufficient for an exhaustive analysis, in §4, of the absolutely continuous case.

1.6. The fundamental densities of absolutely continuous Markov representations. We say that a random process x_t is *absolutely continuous* if each of its finite-dimensional distributions $m_{t_1 \ldots t_n}$ is absolutely continuous with respect to the product $m_{t_1} \times \ldots \times m_{t_n}$ of the corresponding one-dimensional distributions. If x_t is a Markov representation, the requirement that this condition is satisfied when $n = 2$ is sufficient.

Our study of absolutely continuous representations is based on the following property.

THEOREM 1.2. *If x_t is an absolutely continuous Markov representation of a stochastic system $\mathscr{F}(I)$, then the density $p(s, x; t, y)$ of the measure m_{st} with respect to $m_s \times m_t$ can be chosen in such a way that the following conditions hold*:

1.6.A. *For any $s < t < u$, $x \in E_s$, and $y \in E_u$*

$$\int p(s, x; t, z)\, m_t(dz)\, p(t, z; u, y) = p(s, x; u, y).$$

1.6.B. *For any $s < t$, $x \in E_s$ and $y \in E_t$*

$$\int m_s(dz)\, p(s, z; t, y) = \int p(s, x; t, z)\, m_t(dz) = 1.$$

By means of this density we can construct the transition and co-transition probabilities for x_t according to the formulae

(1.14) $$\mathbf{P}_{s,x}\eta = \mathbf{P}p(s, x; t, x_t)\eta \quad \text{for } s < t \in T,\ \eta \in \mathscr{F}_{>t},$$

(1.15) $$\mathbf{P}^{u,x}\xi = \mathbf{P}\xi p(t, x_t; u, x) \quad \text{for } t < u \in T,\ \xi \in \mathscr{F}_{<t}.$$

We call the density of m_{st} with respect to $m_s \times m_t$ that satisfies the conditions 1.6.A, 1.6.B the *fundamental density* of the process x_t. We say that $p(s, x; t, y)$ is the *fundamental density of the right representation*

$(x_t, \mathbf{P}_{t,x})$ if (1.14) holds, that it is the *fundamental density of the left representation* $(x_t, \mathbf{P}^{t,x})$ if (1.15) holds, and that it is the fundamental density of $(x_t, \mathbf{P}_{t,x}, \mathbf{P}^{t,x})$ if both (1.14) and (1.15) hold.

1.7. **Absolutely continuous right representations.** The transition probabilities constructed according to (1.14) have the following property: for any $s < t$, $x \in E_s$ and $\eta \in \mathscr{F}_{>t}$ it follows from $\mathbf{P}\eta = 0$ that $\mathbf{P}_{s,x}\eta = 0$. If this condition is satisfied, we say that the $\mathbf{P}_{s,x}$ are absolutely continuous and also that the right Markov representation $(x_t, \mathbf{P}_{t,x})$ is absolutely continuous.

We prove the following propositions.

1.7.A. A necessary and sufficient condition for the right representation $(x_t, \mathbf{P}_{t,x})$ to be absolutely continuous is the absolute continuity of the transition function $p(s, x; t, \Gamma) = \mathbf{P}_{s,x}\{x_t \in \Gamma\}$ (the latter means that if $m_t(\Gamma) = 0$, then $p(s, x; t, \Gamma) = 0$ for all $s < t$, $x \in E_s$).

1.7.B. A Markov representation x_t is absolutely continuous if and only if there are absolutely continuous transition probabilities $\mathbf{P}_{t,x}$ for x_t.

1.7.C. An absolutely continuous right representation has a fundamental density.

1.7.D. A right representation is absolutely continuous if it is dominated by an absolutely continuous right representation.

1.7.E. An absolutely continuous right representation is dominated by an arbitrary right representation.

1.8. **The construction of "good" co-transition probabilities.** The studies which we begin in the second half of §3 lead to a satisfactory conclusion in the absolutely continuous case. The result is the following.

THEOREM 1.3. *If x_t is measurable with respect to $\mathscr{F}(t, t+)$ and if the regular right Markov representation $(x_t, \mathbf{P}_{t,x})$ is absolutely continuous, then it has a fundamental density, jointly measurable with respect to (s, x) and (t, y), such that the co-transition probabilities defined by* (1.15) *have the following property: for arbitrary bounded $\xi \in \mathscr{F}_{<t}$ the function $\mathbf{P}^{u,x_u}\xi$ is right-continuous on $[t, \beta)$, $\mathbf{P}$-almost surely.*

(If this condition holds, we say that the left representation $(x_t, \mathbf{P}^{t,x})$ is *co-regular*).

The next theorem can easily be derived from Theorems 1.2 and 1.3.

THEOREM 1.4. *If a stochastic system $\mathscr{F}(I)$ has an absolutely continuous Markov representation x_t, then it has two-sided Markov representations $(x_{t+}, \mathbf{P}_{t+,x}, \mathbf{P}^{t+,x})$ and $(x_{t-}, \mathbf{P}_{t-,x}, \mathbf{P}^{t-,x})$ having fundamental densities $p(s+, x; t+, y)$ and $p(s-, x; t-, y)$ that are measurable (jointly in their arguments) and such that the following conditions hold*:

1.8.A. *The representations $(x_{t+}, \mathbf{P}_{t+,x})$ and $(x_{t-}, \mathbf{P}^{t-,x})$ are completely regular.*

1.8.B. *The representations $(x_{t-}, \mathbf{P}_{t-,x})$ and $(x_{t+}, \mathbf{P}^{t+,x})$ are co-regular.*

1.8.C. *If $\mathbf{P}_{t+,x} = \mathbf{P}_{t+,y}$, then $x = y$. If $\mathbf{P}^{t-,x}$ $\mathbf{P}^{t-,y}$, then $x = y$.*

1.8.D. *For arbitrary* $t \in [\alpha, \beta)$, x_{t+} *is measurable with respect to* $\mathcal{F}(t, t+)$. *For arbitrary* $t \in (\alpha, \beta)$, x_{t-} *is measurable with respect to* $\mathcal{F}(t-, t)$.

1.9. The first (vertical) central representation. We say that a representation $(x_t, \mathbf{P}_{t,x}, \mathbf{P}^{t,x})$ is *regular (respectively, completely regular)* if both the representations $(x_t, \mathbf{P}_{t,x})$ and $(x_t, \mathbf{P}^{t,x})$ are regular (respectively, completely regular).

THEOREM 1.5. *If a stochastic system* $\mathcal{F}(I)$ *has an absolutely continuous Markov representation, then it has a completely regular two-sided Markov representation* $(z_t, \mathbf{P}_{t,z}, \mathbf{P}^{t,z})$, *having a fundamental density and separating the states. It can be constructed from the representations described in Theorem* 1.4. *according to the formulae*

$$(1.16) \qquad z_t = x_{t-} \times x_{t+}, \quad \mathbf{P}_{t,\, x\times y} = \mathbf{P}_{t+,\, y}, \quad \mathbf{P}^{t,\, x\times y} = \mathbf{P}^{t-,\, x} \quad (\alpha < t < \beta),$$

and its fundamental density is given by the formula

$$(1.17) \qquad p(s,\ x \times y;\ t,\ x' \times y') = \mathbf{P}p(s+,\ y;\ q+,\ x_{q+})p(r-,\ x_{r-};\ t-,\ x'),$$

where q, r *are arbitrary points of the interval* (s, t) *with* $q < r$.

Any regular two-sided Markov representation $(\bar{z}_t, \bar{\mathbf{P}}_{t,z}\ \bar{\mathbf{P}}^{t,z})$ *of a stochastic system* $\mathcal{F}(I)$ *separating the states, is equivalent to* $(z_t, \mathbf{P}_{t,z}, \mathbf{P}^{t,z})$.

We say that the representation $(z_t, \mathbf{P}_{t,z}, \mathbf{P}^{t,z})$ constructed in Theorem 1.5 is the *first* (or *vertical*) *central representation* of $\mathcal{F}(I)$.

1.10. The second (horizontal) central representation. This representation requires the splitting of time.

We associate two points $t-$ and $t+$ with each point $t \in T = (\alpha, \beta)$ and single points $\alpha+$ and $\beta-$ with the ends α and β. We order the set V of all these points, by setting $t- < t+$ and $s\pm < t\pm$ when $s < t$. The points $t-$ and $t+$ are said to be *neighbours*. Let $v_1 < v_2 \in V$. We denote by (v_1, v_2) the set of points $v \in V$ for which $v_1 < v < v_2$ (it is empty if and only if v_1 and v_2 are neighbours). The sets $[v_1, v_2)$ and $(v_1, v_2]$ are defined similarly.

We define a mapping $v \to \hat{v}$ of V onto $[\alpha, \beta]$, by setting $\hat{v} = t$ for $v = t-$ or $t+$. Let $\mathcal{F}(I)$ be a stochastic system on the interval $T = (\alpha, \beta)$ We associate with each pair $u < v \in V$ a σ-algebra $\hat{\mathcal{F}}(u, v)$ in Ω by the following rule: if u and v are not neighbours, then $\hat{\mathcal{F}}(u, v) = \mathcal{F}(\hat{u}, \hat{v})$; if u and v are neighbours, then $\hat{\mathcal{F}}(u, v)$ is the trivial σ-algebra consisting of the two elements $\varnothing$ and Ω. Then the family $\hat{\mathcal{F}}(u, v)$ satisfies the conditions 1.1.A–1.1.C, and it is natural to call it a stochastic system with the parameter set V. The definitions of Markov representations and also of right, left and two-sided Markov representations, carry over unchanged to the situation where the parameter set is not T but V (or any other ordered set). The definition of fundamental density $p(u, x; v, y)$ also holds with one reservation: the density must be defined only for non-neighbouring pairs u and v.

We introduce a topology in V, taking as neighbourhoods of $t+$ the intervals $[t+, u)$ with $u > t+$, and as neighbourhoods of $t-$ the intervals

$(u, t-]$ with $u < t-$. We say that the two-sided representation $(x_v, \mathbf{P}_{v,x}\ \mathbf{P}^{v,x})$ is completely regular if for any $u < w \in V$ and any $\mathbf{P}$-integrable $\xi \in \hat{\mathscr{F}}_{>u}$ and $\eta \in \hat{\mathscr{F}}_{>w}$ the function $\mathbf{P}_{v,x_v}\,\eta$ is continuous in v for $v \in [u, w]$ almost surely with respect to $\mathbf{K}_{\hat{u}}(x_{t+}, \mathbf{P}_{t+,x})$ and the function $\mathbf{P}^{v,x_v}\,\xi$ is continuous in v for $v \in [u, w]$ almost surely with respect to $\mathbf{K}^{\hat{w}}(x_{t-}, \mathbf{P}^{t-,x})$.

THEOREM 1.6. *Let $\mathscr{F}(I)$ be a stochastic system on the interval T and let $\hat{\mathscr{F}}(u, v)$ be the corresponding stochastic system with the parameter set V. If $\mathscr{F}(I)$ has at least one absolutely continuous Markov representation x_t, then $\hat{\mathscr{F}}(u, v)$ has a completely regular two-sided Markov representation $(x_v, \mathbf{P}_{v,x}, \mathbf{P}^{v,x})$, which has a fundamental density, separates the states, and satisfies the following condition:*

1.10.A. *For $\mathbf{P}$-almost all ω the set of points $t \in T$ for which $\mathbf{P}_{t-,x_{t-}} \neq \mathbf{P}_{t+,x_{t+}}$ or $\mathbf{P}^{t-,x_{t-}} \neq \mathbf{P}^{t+,x_{t+}}$ is at most denumerable.*

The representation $(x_v, \mathbf{P}_{v,x}, \mathbf{P}^{v,x})$ can be constructed by combining the representations $(x_{t+}, \mathbf{P}_{t+,x}, \mathbf{P}^{t+,x})$ and $(x_{t-}, \mathbf{P}_{t-,x}, \mathbf{P}^{t-,x})$ described in Theorem 1.4. The fundamental density $p(u, x; v, y)$ is obtained from the functions $p(s+, x; t+ y)$ and $p(s-, x; t-, y)$ of Theorem 1.4. by setting

$$(1.18)\quad p(s+, x; t-, y) = \mathbf{P}p(s+, x; q+, x_{q+})p(r-, x_{r-}; t-, y),$$

$$(1.19)\quad p(s-, x; t+, y) = \mathbf{P}p(s-, x; q-, x_{q-})p(r+, x_{r+}; t+, y),$$

where q and r are arbitrary numbers in (s, t) with $q < r$.

Any regular two-sided Markov representation separating the states is equivalent to this representation.

1.11. We conclude this section by describing the apparatus we shall use in the proofs.

We make frequent use of the standard results of measure theory, such as Fubini's theorem, the Radon-Nikodym theorem, and so on. We state some less well known propositions which are also required.

1.11.A. Let $\mathscr{S}$ be a family, closed under multiplication, of non-negative functions in a space E and $\mathscr{F}$ the σ-algebra generated by this family. Suppose that $\mathscr{H}$ contains $\mathscr{S}$ and the identity and is closed under addition, under multiplication by non-negative constants, and under monotonic increasing passage to limits. Then $\mathscr{H}$ contains all functions $f \in \mathscr{F}$.

1.11.A′. Let $\mathscr{S}$ be a family, closed under multiplication, of bounded functions in a space E, and $\mathscr{F}$ the σ-algebra generated by it. Suppose that $\mathscr{H}$ contains $\mathscr{S}$ and the identity and is closed under linear operations and bounded limit passage.[1] Then $\mathscr{H}$ contains all bounded $\mathscr{F}$-measurable functions.

1.11.B. Let E_1, E_2, E_3 be measurable spaces, $F(x_1, x_3)$ $(x_1 \in E_1, x_3 \in E_3)$ a non-negative function, jointly measurable with respect to x_1, x_3, and $\mu(x_2, \Gamma)$ a measure on E_3 relative to Γ and a measurable function with respect to

[1] We say that f_n converges boundedly to f if f_n converges pointwise to f and all the functions f_n are uniformly bounded.

x_2. Then the formula

$$\Phi(x_1, x_2) = \int_{E_3} F(x_1, x_3)\, \mu(x_2, dx_3)$$

defines a function that is jointly measurable with respect to x_1 and x_2.

Propositions 1.11.A and 1.11.A′ are easily derived from Lemma 1.1 of [1] or from [9], Chapter II, Theorem 20. A proof of 1.1.B can be found in [1] (see Lemma 1.7.)

1.12. The concept of a support system is of particular importance for us. A *support system* in a measurable space $(E, \mathscr{B})$ is a denumerable family W of bounded functions $\varphi \in \mathscr{B}$, that contains the identity and has the following properties.

1.12.A. If a set $\mathscr{H}$ of non-negative functions contains W and is closed under addition, multiplication by non-negative numbers, subtraction (leading to a non-negative difference) and monotonic increasing passage to limits, then $\mathscr{H}$ contains all non-negative measurable functions.

1.12.A′. If a set $\mathscr{H}$ of bounded functions contains W and is closed with respect to addition, scalar multiplication and bounded limit passage, then $\mathscr{H}$ contains all bounded measurable functions.

1.12.B. If μ_n is a sequence of probability measures on $\mathscr{B}$ and if for each $\varphi \in W$ the limit $\mu_n(\varphi) \to l(\varphi)$ exists, then there is a unique probability measure μ on $\mathscr{B}$ such that $\mu(\varphi) = l(\varphi)$ for all $\varphi \in W$.

A support system (of course, non-unique) can be chosen in every Borel space $(E, \mathscr{B})$ (see, for example [3], §3.2; Propositions 1.11.A and 1.11.A′ are used to verify 1.12.A and 1.12.A′).

1.13. We also use the concept of a filtration.

A family of σ-algebras $\mathscr{A}_t$ $(t \in T)$ in a space Ω is said to be a *right (respectively, left) filtration* if $\mathscr{A}_s \subseteq \mathscr{A}_t$ when $s < t$ (respectively, when $s > t$). If $\mathscr{F}(I)$ is a stochastic system, then $\mathscr{F}_{<t}$ and $\mathscr{F}_{>t}$ are, respectively, right and left filtrations. Apart from these we also consider other filtrations.

A right (respectively, left) filtration $\mathscr{A}_t$ is said to be *continuous* if $\mathscr{A}_t$ for any t is equal to the intersection of all $\mathscr{A}_u$ with $u > t$ (respectively, $u < t$). A continuous right filtration can be constructed from an arbitrary right filtration $\mathscr{A}_t$ by taking the intersection $\mathscr{A}_{t+}$ of the σ-algebras $\mathscr{A}_u$ for all $u > t$. In particular, we obtain by this method a filtration $\mathscr{F}_{<t+}$ from $\mathscr{F}_{<t}$. Observe that if $\mathbf{P}$ is a probability measure whose domain of definition contains all the σ-algebras $\mathscr{A}_t$, then the $\mathbf{P}$-closures $\mathscr{A}_t^{\mathbf{P}}$ of the σ-algebras $\mathscr{A}_t$ also form a filtration; $\mathscr{A}_t^{\mathbf{P}}$ is continuous if $\mathscr{A}_t$ is.[1]

We say that a random process x_t is adapted to a filtration $\mathscr{A}_t$ if x_t for every t is measurable with respect to $\mathscr{A}_t$.

Let $\mathscr{F}(I)$ be a stochastic system in a probability space $(\Omega, \mathscr{F}, \mathbf{P})$ and $(x_t, \mathbf{P}_{t,x})$ its right Markov representation fit. We say that a right filtration

---

1) By the P closure of a σ-algebra $\mathscr{A}$ we mean the σ-algebra generated by $\mathscr{A}$ and all P-neglible sets.

$\mathcal{A}_t$ is *admissible for* $(x_t, \mathbf{P}_{t,x})$ if x_t is adapted to $\mathcal{A}_t$ and if for any $\eta \in \mathcal{F}_>$

$$\mathbf{P}\{\eta \mid \mathcal{A}_t\} = \mathbf{P}_{t,x_t}\eta \quad (\mathbf{P}\text{-a.s.}). \tag{1.20}$$

By (1.4.), the filtration $\mathcal{F}_{\leqslant t}$ is always admissible. According to (1.12), for a regular representation $(x_t, \mathbf{P}_{t,x})$ the filtration $\mathcal{F}_{<t+}$ is also admissible.

Let $\mathcal{A}_t$ be a right filtration of Ω. A *Markov time* (with respect to $\mathcal{A}_t$) is a mapping τ of Ω into an interval T, extended to include the point ∞, such that $\{\omega: \tau(\omega) \leqslant t\}$ for every $t \in T$ belongs to $\mathcal{A}_t$. Here $\mathcal{A}_\tau$ denotes the class of all sets A for which $A \cap \{\tau \leqslant t\} \in \mathcal{A}_t$ for any $t \in T$. (This is a σ-algebra in Ω.)

A Markov time τ is said to be *predictable* if there exists a sequence of Markov times τ_n such that $\tau_n \uparrow \tau$ and $\tau_1 < \tau_2 \ldots < \tau_n \ldots$ for the set $\{\tau < \infty\}$. If τ is a predictable Markov time, then $\mathcal{A}_{\tau-}$ denotes the minimal σ-algebra, which contains all the algebras $\mathcal{A}_{\tau_n}$. (It is independent of the choice of the sequence τ_n.)

1.14. Let $\mathcal{A}_t$ $(t \in T)$ be a right filtration, ξ_t a non-negative random process adapted to it, and $\mathbf{P}$ a probability measure whose domain of definition contains all the σ-algebras $\mathcal{A}_t$. The triple $(\xi_t, \mathcal{A}_t, \mathbf{P})$ is called a *supermartingale* if for any $s < t \in T$

$$\mathbf{P}\{\xi_t \mid \mathcal{A}_s\} \leqslant \xi_s \quad (\mathbf{P}\text{-a.s.}).$$

It is called a *martingale* if equality holds.

We need the following properties of non-negative supermartingales.

1.14.A. Let $(\xi_t, \mathcal{A}_t, \mathbf{P})$ be a non-negative supermartingale and R a denumerable everywhere dense subset of $T = (\alpha, \beta)$. Then there is a $\mathbf{P}$-certain set Ω' such that for $\omega \in \Omega'$ the function $\xi_r(\omega)$ has right limits in R at all points $t \in [\alpha, \beta)$ and left limits in R at all points $t \in (\alpha, \beta]$.

1.14.B. Let $\mathcal{A}_t$ be a continuous right filtration and ξ_t^n a non-decreasing sequence of non-negative random processes such that $\xi_t^n \uparrow \xi_t$. If $(\xi_t^n, \mathcal{A}_t, \mathbf{P})$ is a supermartingale, then so is $(\xi_t, \mathcal{A}_t, \mathbf{P})$ Furthermore, if the functions ξ_t^n are right-continuous $\mathbf{P}$-almost surely, then ξ_t has the same property.

Proofs of 1.14.A and 1.14.B can be found in [11], Corollary 2.2 and [9], Chapter VI, Theorem 16, respectively.

The following proposition can be derived from 1.14.A (see, for example, [7], Chapter VII, Theorem 4.3).

1.14.C. Let $\mathcal{F}_n \subseteq \mathcal{F}$, be an increasing sequence of σ-algebras in a probability space $(\Omega, \mathcal{F}, \mathbf{P})$ and $\mathcal{F}_\infty$ the minimal σ-algebra containing all the $\mathcal{F}_n$. If $\xi \in \mathcal{F}$ and $\mathbf{P}\,\xi < \infty$, then

$$\lim \mathbf{P}\{\xi \mid \mathcal{F}_n\} = \mathbf{P}\{\xi \mid \mathcal{F}_\infty\} \quad (\mathbf{P}\text{-a.s.}).$$

§ 2. Regularization of a right Markov representation

2.1. We begin with a stochastic system $\mathscr{F}(I)$ in a measurable space $(\Omega, \mathscr{F})$ with parameter set $T = (\alpha, \beta)$. Let $(x_t, \mathbf{P}_{t,x})$ be any right Markov representation of $\mathscr{F}(I)$. Its regularization $(x_{t+}, \mathbf{P}_{t+,x})$ proceeds in three stages. Firstly, in §§2.1.–2.3, we construct the state space E_{t+} as a certain class of probability measures on the σ-algebra $\mathscr{F}_{>t}$. Then, in §§2.4 and 2.5, we construct a random process $x_{t+} = \Pi_{t,\omega}$ with values in E_{t+}. Finally, in §2.6, we prove that the set $(x_{t+}, \mathbf{P}_{t+,x})$ satisfies all the requirements of Theorem 1.1, where $\mathbf{P}_{t+,x}$ is the measure corresponding to the point $x \in E_{t+}$, according to the definitions of E_{t+}. §2.7 deals with the uniqueness question.

We consider the class $\mathbf{K}_s = \mathbf{K}_s(x_t, \mathbf{P}_{t,x})$, introduced in §1.2, of all probability measures $\mathbf{P}$ on $\mathscr{F}_{>s}$ dominated by $(x_t, \mathbf{P}_{t,x})$. If $\mathbf{P} \in \mathbf{K}_s$, then

$$(2.1) \qquad \mathbf{P}\xi\eta = \mathbf{P}\xi\mathbf{P}_{t,x_t}\eta \quad \text{for all} \quad s < t,\ \xi \in \mathscr{F}(s,t],\ \eta \in \mathscr{F}_{>t}.$$

For by 1.1.B it suffices to verify this for $\xi \in \mathscr{F}(r,t]$, $r \in (s,t)$. But according to (1.5), for such ξ the left-hand side is equal to $\mathbf{P}\mathbf{P}_{r,x_r}\xi\eta$ and the right-hand side to $\mathbf{P}\mathbf{P}_{r,x_r}\xi\,\mathbf{P}_{t,x_t}\eta$, and (2.1) follows from (1.1). The condition (2.1) can be written as

$$(2.2) \qquad \mathbf{P}\{\eta \mid \mathscr{F}(s,t]\} = \mathbf{P}_{t,x_t}\eta \quad (\mathbf{P}\text{-a.s.}) \text{ for } s < t,\ \eta \in \mathscr{F}_{>t}.$$

We equip $\mathbf{K}_s$ with the measurable structure generated by the functions $F(\mathbf{P}) = \mathbf{P}\eta$ $(\eta \in \mathscr{F}_{>s})$. The measurable space so obtained serves as a state space E_{t+} for the process x_{t+}. Our next task is to study the structure of this space.

2.2. Throughout the remainder of this paper, we denote by R an *arbitrary denumerable everywhere dense subset of* (α,β), and by $R_{<s}$, $R_{>s}$, R_{st} *its intersections with the intervals* $(-\infty, s)$, $(s, +\infty)$, (s,t). We denote by W_u an *arbitrary support system in the state space* E_u.

LEMMA 2.1. *Let* $s < v$. *If* $\mathbf{P}'$ *and* $\mathbf{P}''$ *are measures in* $\mathbf{K}_s$ *and*

$$(2.3) \qquad \mathbf{P}'\varphi(x_u) = \mathbf{P}''\varphi(x_u) \quad \textit{for all} \quad u \in R_{sv},\ \varphi \in W_u,$$

then $\mathbf{P}' = \mathbf{P}''$.

PROOF. By 1.11.A, since (2.3) holds for all $\varphi \in W_u$ it holds for all $\varphi \in \mathscr{B}_u$. For $u \in R_{sv}$, $\eta \in \mathscr{F}_{>u}$ and $\varphi(x) = \mathbf{P}_{u,x}\eta$, from the definition of $\mathbf{K}_s$ we have $\mathbf{P}'\eta = \mathbf{P}'\varphi(x_u)$, $\mathbf{P}''\eta = \mathbf{P}''\varphi(x_u)$ and hence $\mathbf{P}'\eta = \mathbf{P}''\eta$. It remains to observe that, by 1.1.B, if two measures coincide on all σ-algebras $\mathscr{F}_{>u}$ for $u \in R_{sv}$, then they coincide on $\mathscr{F}_{>s}$.

2.3. We denote by $\mathscr{M}_t$ the set of all probability measures on the space E_t with the measurable structure generated by the functions $f(\nu) = \nu(\Gamma)$ $(\Gamma \in \mathscr{B}_t)$. Let $\mathscr{M}_{sv}$ be the product of the spaces $\mathscr{M}_t$ for all $t \in R_{sv}$. Then it is known (see, for example, [3], §3.2) that $\mathscr{M}_t$ is a Borel space. Hence $\mathscr{M}_{sv}$ is also a Borel space.

LEMMA 2.2. *We fix $s < v \in T$ and associate with each $\mathbf{P} \in \mathbf{K}_s$ the family of measures*

$$m_t(\Gamma) = \mathbf{P}\{x_t \in \Gamma\} \qquad (t \in R_{sv}). \tag{2.4}$$

Then the formula (2.4) *defines an isomorphic mapping of $\mathbf{K}_s$ onto the subset $\mathscr{L}_{sv}$ of $\mathscr{M}_{sv}$ that is characterized by the conditions*

$$m_{u_2}(\Gamma) = \int\limits_{E_{u_1}} m_{u_1}(dx)\, p(u_1, x; u_2, \Gamma) \quad \textit{for all} \quad u_1 < u_2 \in R_{s,v}, \quad \Gamma \in \mathscr{B}_{u_2}. \tag{2.5}$$

$\mathscr{L}_{sv}$ is measurable, so that $\mathbf{K}_s$ is a Borel space.

PROOF. We note that (2.5) is equivalent to the conditions

$$m_{u_2}(\varphi) = \int\limits_{E_{u_1}} m_{u_1}(dx)\, p(u_1, x; u_2, \varphi) \quad \text{for all} \quad u_1 < u_2 \in R_{s,v}, \quad \varphi \in W_{u_2}. \tag{2.6}$$

(To derive (2.5) from (2.6) we have to use 1.12.A.) Both sides of (2.6) are measurable functions on $\mathscr{M}_{sv}$. Therefore, the set $\mathscr{L}_{sv}$ is a measurable subset of $\mathscr{M}_{sv}$.

We now show that the set of one-dimensional distributions m_u corresponding to a measure $\mathbf{P} \in \mathbf{K}_s$ satisfies (2.6). Indeed, the left-hand side of (2.6) is equal to $\mathbf{P}\varphi(x_{u_2})$, and the right-hand side to $\mathbf{P}\mathbf{P}_{u_1, x_{u_1}}\varphi(x_{u_2})$, so that (2.6) follows from (2.1). Thus, (2.4) defines a mapping from $\mathbf{K}_s$ into $\mathscr{L}_{sv}$. By Lemma 2.1 distinct elements have distinct images.

We now show that to every $m \in \mathscr{L}_{sv}$ there corresponds a $\mathbf{P} \in \mathbf{K}_s$ that is connected with m by (2.4). We choose a sequence $r_n \downarrow s (r_n \in R_{sv})$ and consider on $\mathscr{F}_{>r_n}$ the measure $\mathbf{P}_n(A) = \int\limits_{E_{r_n}} m_{r_n}(dx)\, \mathbf{P}_{r_n, x}(A)$. By (2.6), these measures are compatible, and by 1.1.C there is a measure $\mathbf{P}$ on $\mathscr{F}_{>s}$ that coincides with $\mathbf{P}_r$ on $\mathscr{F}_{>r_n}$. We claim that $\mathbf{P} \in \mathbf{K}_s$. For every $t > s$ there exists an $r_n \in (s, t)$. By (1.1), $\mathbf{P}_{r_n, x}\, \eta = \mathbf{P}_{r_n, x}\, \mathbf{P}_{t, x_t}\, \eta$ for all $\eta \in \mathscr{F}_{>t}$, and $x \in E_{r_n}$, hence, $\mathbf{P}_n \eta = \mathbf{P}_n \mathbf{P}_{t, x_t} \eta$. But $\mathbf{P} = \mathbf{P}_n$ on $\mathscr{F}_{>t}$, therefore, $\mathbf{P}\eta = \mathbf{P}\mathbf{P}_{t, x_t}\eta$. It remains to verify that (2.4) holds. For any $u \in R_{sv}$ we choose an $r_n \in (s, u)$. Then for $\Gamma \in \mathscr{B}_u$ by (2.5),

$$m_u(\Gamma) = \int m_{r_n}(dx)\, p(r_n, x; u, \Gamma) = \mathbf{P}_n\{x_u \in \Gamma\} = \mathbf{P}\{x_u \in \Gamma\}.$$

We have shown that (2.4) defines a bijective mapping of $\mathbf{K}_s$ onto $\mathscr{L}_{sv}$. It is clearly measurable. To show that the inverse mapping is likewise measurable, it suffices to check that the functions $F(\mathbf{P}) = \mathbf{P}\eta$ go over into measurable functions on $\mathscr{L}_{sv}$. But by (2.4), these functions go into $f(m) = m_t(\varphi)$, where $\varphi(x) = \mathbf{P}_{t, x}\eta$.

2.4. We proceed to construct the random process $\Pi_{t,\omega}$ in the state space $\mathbf{K}_t$.

THEOREM 2.1. *Let $\mathcal{F}(I)$ be a stochastic system in a measurable space $(\Omega, \mathcal{F})$ on the time interval $T = (\alpha, \beta)$, and let $(x_t, \mathbf{P}_{t,x})$ be a right Markov representation of it. Then there are measures $\Pi_{t,\omega} \in \mathbf{K}_t (t \in T, \omega \in \Omega)$ having the following properties.*

2.4.A. *For any $\eta \in \mathcal{F}_{>t}$ the function $\Pi_{t,\omega}\eta$ is measurable with respect to $\mathcal{F}(t, t+)$, and for $s \leqslant t$ and $\mathbf{P} \in \mathbf{K}_s$*

$$\mathbf{P}\{\eta \mid \mathcal{F}(s, t+)\} = \Pi_{t,\omega}\eta \quad (\mathbf{P}\text{-a.s.}). \tag{2.7}$$

2.4.C. *Let $s < u$, $\mathbf{P} \in \mathbf{K}_s$, $\eta \in \mathcal{F}_{>u}$ and R any denumerable everywhere dense subset of T. There is a* **P**-*negligible set Ω' such that for all $\omega \in \Omega'$, $t \in [s, u)$*

$$\Pi_{t,\omega}\eta = \lim_{r \downarrow t,\, r \in R} \mathbf{P}_{r, x_r}\eta. \tag{2.8}$$

The way to prove the theorem is shown by Lemma 2.2: to construct the measure $\Pi_{t,\omega}$ it is sufficient to construct a set of measures $m_u = m^t_{u,\omega}$ in $\mathcal{L}_{tv}$. By comparing (2.4) and (2.8), we see that if the required measures $m^t_{u,\omega}$ exist, then for any $u \in R_{sv}$, $\mathbf{P} \in \mathbf{K}_s$ and $\varphi \in W_u$ we must have

$$m^t_{u,\omega}(\varphi) = \lim_{\substack{r \downarrow t \\ r \in R}} p(r, x_r; u, \varphi) \text{ for all } t \in [s, u] \quad (\mathbf{P}\text{-a.s.}). \tag{2.9}$$

This equation serves us as a guiding light. First of all, we must verify that the limit on the right-hand side exists. To do this we use martingale theory.

We set $p(r, x_r; u, \varphi) = \xi_r$, $\mathcal{F}(s, r] = \mathcal{A}_r$. By (2.2), for $s < r < u$ and $\mathbf{P} \in \mathbf{K}_s$

$$\xi_r = \mathbf{P}\{\varphi(x_u) \mid \mathcal{A}_r\} \quad (\mathbf{P}\text{-a.s.}).$$

Thus, $(\xi_r, \mathcal{A}_r, \mathbf{P})$ is a martingale on (s, u). By 1.14.A, except in a **P**-negligible set $D(s, u, \varphi) \in \mathcal{F}(s, u)$ for all $t \in [s, u)$, the limits of $\xi_r(\omega)$ exist as r tends to t through R. The set $D(s, u, \varphi)$ does not depend on choice of **P** in $\mathbf{K}_s$. We form the union of the $D(s, u, \varphi)$ for all $u \in R_{sv}$, $\varphi \in W_u$ and denote its complement by C_{sv}. Clearly, $C_{sv} \in \mathcal{F}(s, v)$, $\mathbf{P}(C_{sv}) = 1$ and for $\omega \in C_{sv}$ the limit on the right-hand side of (2.9) exists for all $u \in R_{sv}$ and $\varphi \in W_u$, $t \in [s, u)$. By 1.12.B, there are measures $m^t_{u,\omega}$ on $\mathcal{B}_u$ such that (2.9) holds for all $\omega \in C_{sv}$, $u \in R_{sv}$, $\varphi \in W_u$.

2.5. We now fix $u \in R_{sv}$, $\mathbf{P} \in \mathbf{K}_s$ and prove that for arbitrary $\varphi \in \mathcal{B}_u$.

2.5.A. $m^t_{u,\omega}(\varphi) = \mathbf{P}\{\varphi(x_u) \mid \mathcal{F}(s, t+)\}$ (**P**-a.s.) for $t \in [s, u)$.

2.5.B. Outside a certain **P**-negligible set, the function $m^t_{u,\omega}(\varphi)$ is continuous from the right in t on $[s, u)$.

We denote by $\mathcal{H}$ the set of all $\varphi \in \mathcal{B}_u$, for which these assertions hold. By (2.9) and (2.2), $W_u \subseteq \mathcal{H}$. Hence, it suffices to confirm that $\mathcal{H}$ is invariant under the operations listed in 1.12.A. Only for 2.5.B and the monotonic ascending limit passage is this not obvious. We observe that if $\varphi \in \mathcal{H}$, then the system $(m^t_u(\varphi), \mathcal{F}(s, t+), \mathbf{P})$ defines a supermartingale on $[s, u)$ that is continuous from the right (this is clear from 2.5.A and 2.5.B). It remains to refer to 1.14.B.

We write $\omega \in \Omega_{tv}$ if $\omega \in C_{tv}$ and the set $m_u = m^t_{u,\omega} (u \in R_{tv})$ defines a point of $\mathscr{L}_{tv}$, that is, if

$$\int m^t_{u_1, \omega}(dx)\, p(u_1, x; u_2, \varphi) = m^t_{u_2, \omega}(\varphi) \tag{2.10}$$

$$\text{for all } u_1 < u_2 \in R_{tv}, \quad \varphi \in W_{u_2}.$$

Clearly, $\Omega_{tv} \in \mathscr{F}(t, v)$. By Lemma 2.2., to each point $\omega \in \Omega_{tv}$ there corresponds a unique measure $\Pi_{t,\omega}$ in $\mathbf{K}_t$ such that

$$\Pi_{t,\omega}\varphi(x_u) = m^t_{u,\omega}(\varphi) \text{ for } u \in R_{tv}, \ \varphi \in \mathscr{B}_u. \tag{2.11}$$

This measure clearly does not depend on the choice of $v \in (t, \beta)$, so that the measures $\Pi_{t,\omega}$ are defined on the union Ω_{t+} of all the $\Omega_{tv} (v \in (t, \beta))$. This union belongs to the σ-algebra $\mathscr{F}(t, t+)$. Outside Ω_{t+} we define $\Pi_{t,\omega}$ to be an arbitrary fixed measure in $\mathbf{K}_t$.

Let $\mathbf{P} \in \mathbf{K}_s$. By (2.2) and 2.5.A, for any $u_2 > u_1 > t \geqslant s$ and $\varphi \in W_{u_2}$

$$\int m^t_{u_1,\omega}(dx)\, p(u_1, x; u_2, \varphi) = \mathbf{P}\{p(u_1, x_{u_1}; u_2, \varphi) \mid \mathscr{F}(s, t+)\} =$$

$$= \mathbf{P}\{\mathbf{P}\{\varphi(x_{u_2}) \mid \mathscr{F}(s, u_1)\} \mid \mathscr{F}(s, t+)\} = \mathbf{P}\{\varphi(x_{u_2}) \mid \mathscr{F}(s, t+)\}$$

$$= m^t_{u_2,\omega}(\varphi) \quad (\mathbf{P}\text{-a.s.}),$$

so that (2.10) is satisfies $\mathbf{P}$-almost surely. But according to 2.5.B, both sides of this equation are continuous from the right $\mathbf{P}$-almost surely. Therefore, the intersection of the Ω_{tv} for all $v > t \geqslant s$ is $\mathbf{P}$-certain. A fortiori, the intersection of the Ω_{t+} for all $t \geqslant s$ is $\mathbf{P}$-certain.

We prove now that the measures $\Pi_{t,\omega}$ satisfy 2.4.A. By (2.9) and (2.11),

$$\Pi_{t,\omega}\varphi(x_u) = \lim_{\substack{r \downarrow t \\ r \in R}} p(r, x_r; u, \varphi) \text{ for } u \in R_{tv}, \ \varphi \in W_u, \ \omega \in \Omega_{tv}. \tag{2.12}$$

Moreover, since $\Pi_{t,\omega} \in \mathbf{K}_t$, by (2.1), for $t < u$ and $\eta \in \mathscr{F}_{>u}$ we have

$$\Pi_{t,\omega}\eta = \Pi_{t,\omega}\mathbf{P}_{u,x_u}\eta. \tag{2.13}$$

We write $\eta \in \mathscr{H}$, if $\Pi_{t,\omega}\eta$ is measurable with respect to $\mathscr{F}(t, v)$. From (2.12) it is clear that $\mathscr{H}$ contains all the functions $\varphi(x_u)$ $(u \in R_{tv}, \varphi \in W_u)$. By 2.2, this is also the case for all $\varphi \in \mathscr{B}_u$. From (2.13) it follows that $\mathscr{H} \supseteq \mathscr{F}_{>u}$ for any $u \in R_{t,v}$. By 1.1.B, $\mathscr{H} \supset \mathscr{F}_{>t}$. Thus, for any $\eta \in \mathscr{F}_{>t}, \Pi_{t,\omega}\eta$ is measurable with respect to $\mathscr{F}(t, v)$. Since v is arbitrary, $\Pi_{t,\omega}\eta$ is measurable with respect to $\mathscr{F}(t, t+)$. By comparing (2.11) and 2.5.A we conclude that (2.7) holds if $\eta = \varphi(x_u)$, $u \in R_{tv}$, $\varphi \in \mathscr{B}_u$. Relying on (2.13), we extend this formula to $\eta \in \mathscr{F}_{>u}$, and by using 1.1.B, to $\eta \in \mathscr{F}_{>t}$.

It remains to prove 2.4.B. Now (2.8) is satisfied $\mathbf{P}$-almost surely for fixed t, because by (2.7) and (2.2), both sides of (2.8) are equal to $\mathbf{P}\{\eta \mid \mathscr{F}(s, t+)\}$. The right-hand side of (2.8) is right-continuous on $[s, u)$.

Therefore, it is sufficient to show that the left hand-side is right-continuous P-almost surely. This follows from 2.5.B., (2.11), and (2.13).

2.6. We proceed to the proof of Theorem 1.1. We consider the random process $x_{t+}(\omega) = \Pi_{t,\omega} (t \in [\alpha, \beta))$ in the Borel space $E_{t+} = \mathbf{K}_t$ and denote by $\mathbf{P}_{t+}$ the measure on $\mathcal{F}_{>t}$ corresponding to $x \in E_{t+}$ in accordance with the definition of E_{t+}. Clearly,

(2.14) $$\mathbf{P}_{t+, x_{t+}} = \Pi_{t, \omega}.$$

We prove that $(x_{t+}, \mathbf{P}_{t+,x}$ is a right Markov representation of the stochastic system $\mathcal{F}(I)$, satisfying the conditions of Theorem 1.1.

By 2.4.A, the mapping $x_{t+}: \Omega \to E_{t+}$ is measurable with respect to $\mathcal{F}(t, t+)$ and for any $\mathbf{P}' \in \mathbf{K}_s$

(2.15) $\mathbf{P}'\{\eta \mid \mathcal{F}(s, t+)\} = \mathbf{P}_{t+, x_{t+}(\omega)}\eta$ ($\mathbf{P}'$-a.s.) for $\eta \in \mathcal{F}_{>t}$

In particular, setting $\mathbf{P}' = \mathbf{P}_{s+,x}$, we see that

$$\mathbf{P}_{s+,x}\xi\eta = \mathbf{P}_{s+,x}\xi\mathbf{P}_{t+,x_{t+}}\eta \text{ for } \xi \in \mathcal{F}(s, t+), \eta \in \mathcal{F}_{>t},$$

so that $(x_{t+}, \mathbf{P}_{t+,x})$ satisfies (1.1). On the other hand, applying (2.15) to $\mathbf{P}' = \mathbf{P}_{s,x}$, we have

(2.16) $$\mathbf{P}_{s,x}\eta = \mathbf{P}_{s,x}\mathbf{P}_{t+,x_{t+}}\eta \text{ for } \eta \in \mathcal{F}_{>t}.$$

We verify that $(x_{t+}, \mathbf{P}_{t+,x})$ satisfies (1.4). It is sufficient to prove that the condition is satisfied for bounded ξ and η. Let $t < r < u$, $\xi \in \mathcal{F}_{\leqslant t}$, $\eta \in \mathcal{F}_{>u}$ (the σ-algebra $\mathcal{F}_{\leqslant t}$ is constructed in accordance with the representation x_t). Since $(x_t, \mathbf{P}_{t,x})$ satisfies (1.4), and since $\mathcal{F}_{\leqslant t} \subset \mathcal{F}_{<r}$ and $\mathcal{F}_{>u} \subseteq \mathcal{F}_{>r}$, we have $\mathbf{P}\xi\eta = \mathbf{P}\xi\mathbf{P}_{r,x_r}\eta$. Passing to the limit as $r \downarrow t$, and taking 2.4.B and (2.14) into account, we conclude that (1.4) holds for all $\eta \in \mathcal{F}_{>u}$, $t < u$. By 1.1.B, (1.14) holds for all $\eta \in \mathcal{F}_{>t}$.

From 2.4.B and (2.14) it follows that the representation $(x_{t+}, \mathbf{P}_{t+,x})$ is completely regular. Clearly, it separates states. Since $\mathbf{P}_{t+,x} \in \mathbf{K}_t$, it is dominated by $(x_t, \mathbf{P}_{t,x})$. Finally, 1.4.C has already been established above.

2.7. Before proving the last part of Theorem 1.1, we verify that each regular right representation $(x_t, \mathbf{P}_{t,x})$ satisfies (1.12). ((1.13) can be verified similarly.) For any $t < r < u$ and any bounded functions $\xi \in \mathcal{F}_{<t+}$ and $\eta \in \mathcal{F}_{>u}$ by (1.4),

$$\mathbf{P}\xi\eta = \mathbf{P}\xi\mathbf{P}_{r,x_r}\eta.$$

Passing to the limit as $r \downarrow t$, we have

(2.17) $$\mathbf{P}\xi\eta = \mathbf{P}\xi\mathbf{P}_{t,x_t}\eta.$$

By 1.1.B, since this equation is valid for all $u > t$ and $\eta \in \mathcal{F}_{>u}$ it holds for all $\eta \in \mathcal{F}_{>t}$. Since $\mathbf{P}_{t,x_t}\eta$ is measurable with respect to $\mathcal{F}_{<t+}$, (1.12) now follows from (2.17).

The final statement of Theorem 1.1 follows immediately from the following lemma.

LEMMA 2.3. *If two regular right representations* $(\tilde{x}_t, \tilde{\mathbf{P}}_{t,x})$ *and* $(\bar{x}_t, \bar{\mathbf{P}}_{t,x})$

are dominated by some representation $(x_t, \mathbf{P}_{t,x})$ *then*

$$(2.18) \qquad \widetilde{\mathbf{P}}_{t,\widetilde{x}_t} = \overline{\mathbf{P}}_{t,\overline{x}_t} \text{ for all } t \in [\alpha, \beta) \text{ (P-a.s.).}$$

If, in addition, $(\widetilde{x}_t, \widetilde{\mathbf{P}}_{t,x})$ *and* $(\overline{x}_t, \overline{\mathbf{P}}_{t,x})$ *separate states, then they are equivalent.*

PROOF. From (1.12) it follows that for any $t < u, \eta \in \mathscr{F}_{>u}$

$$\widetilde{\mathbf{P}}_{t,\widetilde{x}_t}\eta = \overline{\mathbf{P}}_{t,\overline{x}_t}\eta \quad \text{(P-a.s.)} .$$

Therefore, by 1.4.A, on a P-certain set Ω',

$$(2.19) \qquad \widetilde{\mathbf{P}}_{t,\widetilde{x}_t}\varphi(x_u) = \overline{\mathbf{P}}_{t,\overline{x}_t}\varphi(x_u) \quad \text{for all} \quad t \in [\alpha, \beta),\ u \in R_{>t},\ \varphi \in W_u.$$

But the measures $\widetilde{\mathbf{P}}_{t,\widetilde{x}_t}$ and $\overline{\mathbf{P}}_{t,\overline{x}_t}$ belong to $\mathbf{K}_t = \mathbf{K}_t\ (x_t, \mathbf{P}_{t,x})$, and by Lemma 2.1, it follows from (2.19) that $\widetilde{\mathbf{P}}_{t,\widetilde{x}_t} = \overline{\mathbf{P}}_{t,\overline{x}_t}$ on Ω' for all $t \in [\alpha, \beta)$, so that (2.18) holds.

The formula $x \to \widetilde{\mathbf{P}}_{t,x}$ defines a measurable mapping from the state space $\widetilde{E}_t$ of $\widetilde{x}_t$ into $\mathbf{K}_t$. It is bijective if $(\widetilde{x}_t, \widetilde{\mathbf{P}}_{t,x})$ separates states. It is known (see [9], §39.5, Theorem 2) that under a bijective measurable mapping from one Borel space to another the images of measurable sets are measurable. Therefore, the image $\widetilde{Q}_t$ of $\widetilde{E}_t$ in $\mathbf{K}_t$ under the mapping $\widetilde{x} \to \widetilde{\mathbf{P}}_{t,x}$ is measurable, and this mapping is an isomorphism of $\widetilde{E}_t$ onto $\widetilde{Q}_t$. Similarly, the mapping $x \to \overline{\mathbf{P}}_{t,x}$ is an isomorphism from $\overline{E}_t$ onto a measurable subset $\overline{Q}_t$ of $\mathbf{K}_t$. We set $\mathbf{K}_t' = \widetilde{Q}_t \cap \overline{Q}_t$ and denote by $\widetilde{E}_t'$ and $\overline{E}_t'$ the inverse images of this set in $\widetilde{E}_t$ and $\overline{E}_t$. To each $x \in \widetilde{E}_t$ there corresponds a unique $\gamma(x) \in \overline{E}_t'$ such that $\widetilde{\mathbf{P}}_{t,x} = \overline{\mathbf{P}}_{t,\gamma(x)}$, and in this way we establish an isomorphism from $\widetilde{E}_t'$ to $\overline{E}_t'$. By (2.18), for $\omega \in \Omega'$ and any $t \in [\alpha, \beta)$ we have $\widetilde{x}_t(\omega) \in \widetilde{E}_t'$, $\overline{x}_t(\omega) \in \overline{E}_t'$ and $\gamma[\widetilde{x}_t(\omega)] = \overline{x}_t(\omega)$.

§3. Regular right representations

3.1. In §1.5 a criterion for regularity and for complete regularity were formulated in terms of transition functions. The necessity is obvious. Let us prove the sufficiency. The cases of regularity and complete regularity are treated in exactly parallel fashion, hence we give the details only for the former. We denote by $\mathscr{H}$ the set of all functions φ on E_u for which $p(t, x_t; u, \varphi)$ is continuous from the right in t on the interval $[\alpha, u)$, P-almost surely. By hypothesis, $\mathscr{H}$ contains indicators of all sets $\Gamma \in \mathscr{B}_u$. Clearly, $\mathscr{H}$ is invariant under linear operations. By 1.14.B, $\mathscr{H}$ is also invariant under monotonic increasing limit passage. Therefore, according to 1.11.A, $\mathscr{H}$ contains all functions $\varphi \in \mathscr{B}_u$. Next, if $\eta \in \mathscr{F}_{>u}$, then

$$(3.1) \qquad \mathbf{P}_{t,x_t}\eta = p(t, x_t; u, \varphi),$$

where $\varphi(x) = \mathbf{P}_{u,x}\eta$. Hence $\mathbf{P}_{t,x_t}\eta$ is right-continuous in $[\alpha, u)$, P-almost surely.

We note two further properties of regular right representations.

3.1.A. If $\eta \in \mathcal{F}(t, t+)$, then there is a function $\varphi \in \mathcal{B}_t$ such that

$$\eta = \varphi(x_t) \quad (\mathbf{P}\text{-a.s.}) . \tag{3.2}$$

For (1.12) is applicable to η, and the right-hand side is equal to η, the left-hand side to $\varphi(x_t)$ for $\varphi(x) = \mathbf{P}_{t,x}\eta$.

3.1.B. If $\eta \in \mathcal{F}_{>u}$, then for **P**-almost all ω the function $\mathbf{P}_{t, x_t(\omega)}\eta$ has left limits at each point of the interval $[\alpha, u)$.

For it is clear from (1.12) that the triple $(\mathbf{P}_{t, x_t}\eta, \mathcal{F}_{<t+}, \mathbf{P})$ defines a martingale on $[\alpha, u)$. By 1.4.A, this is continuous from the right. Hence the assertion follows from 1.14.A.

3.2. Let $\mathcal{A}_t$ be a (right or left) filtration of Ω, and **P** a probability measure whose domain contains all the σ-algebras $\mathcal{A}_t$. A real-valued random process $\eta_t(\omega)$ that is adapted to $\mathcal{A}_t$ and **P**-almost surely right-continuous in t is said to be *strictly measurable* if $\mathcal{A}_t$ is a right filtration and *strictly reconstructable* if $\mathcal{A}_t$ is a left filtration. A process $\eta_t(\omega)$ that is **P**-almost surely left-continuous in t and adapted to the right filtration $\mathcal{A}_t$ is called *strictly predictable*. In $T \times \Omega$ we consider the σ-algebra generated by all the strictly measurable processes. We call sets and functions that are measurable with respect to this σ-algebra *well-measurable*. Similarly, we define *reconstructible* and *predictable* sets and functions as those that are measurable relative to the structure generated by the strictly reconstructible and strictly predictable processes.

Under the hypothesis that the measure **P** is complete, that the right filtration $\mathcal{A}_t$ is continuous, and that each σ-algebra $\mathcal{A}_t$ contains all **P**-negligible sets, a well-measurable structure has a number of remarkable properties.

3.2.A. If A is a well-measurable set, the formula $\tau(\omega) = \inf\{t: (t, \omega) \in A\}$ defines a Markov time.[1] (It is called the debut of A).

3.2.B. If the function $\varphi(t, \omega)$ is well-measurable, then for any Markov time τ the function $\varphi[\tau(\omega), \omega]$ is measurable with respect to $\mathcal{A}_\tau$.

3.2.C. For two well-measurable functions φ_1 and φ_2 to be indistinguishable, it is necessary and sufficient that $\mathbf{P}\{\varphi_1(\tau, \omega) \neq \varphi_2(\tau, \omega)\} = 0$ for every Markov time τ.

(Two functions φ_1 and φ_2 are said to be indistinguishable if **P**-almost surely $\varphi_1(t) = \varphi_2(t)$ for all t.)

3.2.D. For a bounded well-measurable function $\varphi(t, \omega)$ to be right-continuous in t **P**-almost surely, it is necessary and sufficient that $\mathbf{P}\varphi(\tau_n) \to \mathbf{P}\varphi(\lim \tau_n)$ for any non-increasing sequence of Markov times τ_n.

---

[1] If Λ is empty, we set $\inf \Lambda = +\infty$.

3.2.E. For a predictable function $\varphi(t, \omega)$ to be left-continuous **P**-almost surely, it is necessary and sufficient that $\varphi(\tau_n) \to \varphi(\lim \tau_n)$ **P**-almost surely for any non-decreasing sequence of predictable Markov times τ_n.

Proofs of these properties can be found in [9] (Ch. VIII, Theorems 15 and 21; Ch. IV, Theorems 49 and 52) and [10] (Ch. IV, Theorem 28.II and Theorem 24). (Properties 3.2.A and 3.2.B are proved under the requirement of progressive measurability, which is weaker than well-measurability.)

3.3. We recall that with each transition function there is associated a measurable structure in the phase space $\mathscr{E}$ (see §1.3). It is generated by the functions

$$f(t, x) = p(t, x; u, \Gamma) \quad (u \in T, \ \Gamma \in \mathscr{B}_u). \tag{3.3}$$

To each right Markov representation there corresponds a transition function and hence, a measurable structure in $\mathscr{E}$.

The application of Propositions 3.2.A–3.2.D to Markov representations is based on the following lemma.

LEMMA 3.1. *Let* $(x_t, \mathbf{P}_{t,x})$ *be a regular right Markov representation. If* f *is a measurable function on* $\mathscr{E}$, *then* $f(t,x_t)$ *is well-measurable with respect to* $\mathscr{F}_{<t+}$.

PROOF. The functions f for which the assertion holds form a set that is invariant under linear operations, multiplication, and limit passage. The functions (3.3) belong to this set, since the corresponding process $f(t, x_t)$ is continuous from the right **P**-almost surely and compatible with $\mathscr{F}_{<t+}$.

3.4. We now derive some properties of the measurable structure in $\mathscr{E}$, which we need later.

3.4.A. If $\eta \in \mathscr{F}_{>u}$, then $\mathbf{P}_{t,x}\eta$ is a measurable function on $\mathscr{E}_{\leq u} = \mathscr{E}[\alpha, u]$.

The proof proceeds on the same plan as the derivation of the regularity criterion in §3.1.

3.4.B. The measurable structure in $\mathscr{E}$ is generated by the denumerable family of functions

$$f(t, x) = p(t, x; u, \varphi) \qquad (u \in R, \ \varphi \in W_u). \tag{3.4}$$

For let $\mathscr{S}$ be the σ-algebra in $\mathscr{E}$ generated by the functions (3.4). Relying on 1.12.A, we can see that all the functions $p(t, x; u, \varphi)$ $(u \in R, \varphi \in \mathscr{B}_u)$ are measurable with respect to $\mathscr{S}$. For any $u \in T$ and any $\Gamma \in \mathscr{B}_u$ by 1.3.C,

$$p(t, x; u, \Gamma) = \lim_{r \uparrow u, \, r \in R} p(t, x; r, \varphi_r),$$

where $\varphi_r(y) = p(r, y; u, \Gamma)$, from which it follows that the functions (3.3) are measurable with respect to $\mathscr{S}$.

3.4.C. Suppose that the right representation $(x_t, \mathbf{P}_{t,x})$ is regular. Then there is a **P**-certain set $\hat{\Omega}$ such that $F(t, x_t(\omega))$ is a Borel function of t for any measurable function F on $\mathscr{E}$.

For let us write $\omega \in \hat{\Omega}$ when $F(t, x_t(\omega))$ is right-continuous in t for all functions (3.4). Clearly $\mathbf{P}(\hat{\Omega}) = 1$. Thus, the set $\mathscr{H}$ of functions F for which 3.4.C is valid contains the functions (3.4) and the identity and is closed under addition, multiplication, and limit passage. It remains to use 1.1.A.

3.4.D. Let $(x_t, \mathbf{P}_{t,x})$ be regular. If f is a measurable function on $\mathscr{E}_{\geq t} = \mathscr{E}[t, \beta)$, then $f(t, x_t)$ is measurable with respect to the **P**-closure $\mathscr{F}^{\mathbf{P}}_{>t}$ of $\mathscr{F}_{>t}$. Moreover, it is reconstructable with respect to the filtration $\mathscr{F}^{\mathbf{P}}_{>t}$.

From the relation

$$p(t, x_t;\ u, \Gamma) = \lim_{r \downarrow t} p(r, x_r;\ u, \Gamma) \quad (\mathbf{P}\text{-a.s.})$$

it follows that the assertion holds for all the functions (3.3). The further arguments are the same as in the proof of 3.4.C.

3.5. Let $\mathscr{A}_t$ be an admissible filtration for $(x_t, \mathbf{P}_{t,x})$. We assume that to every $u \in T$ there corresponds a random variable $\eta_u \in \mathscr{F}_{>u}$, and we set

$$F^u(t, x) = \mathbf{P}_{t,x}\eta_u \qquad (t < u). \tag{3.5}$$

Since the filtration $\mathscr{A}_t$ is admissible, by (1.20) for $t < u$

$$\mathbf{P}\{\eta_u \mid \mathscr{A}_t\} = F^u(t, x_t)\ (\mathbf{P}\text{-a.s.}). \tag{3.6}$$

The representation $(x_t, \mathbf{P}_{t,x})$ is said to be *strong Markov* if the corresponding equation

$$\mathbf{P}\{\eta_u \mid \mathscr{A}_\tau\} = F^u(\tau, x_\tau) \quad (\mathbf{P}\text{-a.s., } \tau < u) \tag{3.7}$$

holds for any Markov time τ (with respect to $\mathscr{A}_t$).

Every right regular representation $(x_t, \mathbf{P}_{t,x})$ is strong Markov. This follows at once from Theorem 3.1. of [4].[1]

We need a corollary to (3.7). Let σ be an A_τ-measurable random variable with values in T, having at most denumerably many values. Multiplying (3.7) by the indicator function of the set $\sigma = u > \tau$ and summing over u we obtain

$$\mathbf{P}\{\eta_\sigma \mid A_\tau\} = F^\sigma(\tau, x_\tau) \quad (\mathbf{P} - \text{a.s., } \tau < \sigma). \tag{3.8}$$

If η_u is right-continuous on $[s, \infty)$ $P_{s,x}$-a.s. for all s, x and if the representation $(x_t, \mathbf{P}_{t,x})$ is completely regular, then (3.8) also holds when σ has arbitrarily (not necessarily denumerably) many values. This can be shown by a limit argument or it can be derived from Theorem 3.2 of [4].

3.6. A set Γ in $\mathscr{E}$ is said to be *inaccessible for* $(x_t, \mathbf{P})$ if $x_t \notin \Gamma$ for all t, $\mathbf{P}$-a.s.

Let $(x_t, \mathbf{P}_{t,x})$ be a fixed completely regular right Markov representation.

---

1) Strictly speaking, (3.7) is proved in [4] only for $\eta_u = \chi_\Gamma(x_u)$, but the proof also holds for arbitrary η_u.

We call a set $\Gamma \subseteq \mathscr{E}$ *polar* if for any s with $\alpha \leqslant s < \beta$ and any $\mathbf{P}' \in \mathbf{K}_s(x_t, \mathbf{P}_{t,x})$ $\mathbf{P}'\{x_t \notin \Gamma$ for all $t \in (s, \beta)\} = 1$. Finally, we call a point $x \in E_t$ *essential* if $\mathbf{P}_{t,x}\{x_t = x\} = 1$.

From results in [4], §4 it follows that the set of all inessential points is a) measurable in $\mathscr{E}$ and a Borel space; b) polar; and c) inaccessible. Therefore, from any arbitrary completely regular representation we can obtain a completely regular representation without inessential points (by simply discarding the latter).

3.7. Let $(x_t, \mathbf{P}_{t,x})$ be a right Markov representation and $(x_{t+}, \mathbf{P}_{t+,x})$ its regularization, as described in §2. The state space coincides with the class of measures $\mathbf{K}_t = \mathbf{K}_t(x_t, \mathbf{P}_{t,x})$. It is not difficult to show that a measure $\mathbf{P}' \in \mathbf{K}_t$ is an essential point if and only if $\mathbf{P}'(A) = 0$ or 1 for all $A \in \mathscr{F}(t, t+)$. In the terminology of [3], this means that the set of essential points can be identified with the entrance space for $(x_t, \mathbf{P}_{t,x})$. Thus, the regularization $(x_{t+}, \mathbf{P}_{t+\,x})$ of a right Markov representation can be regarded as defined in the entrance space for $(x_t, \mathbf{P}_{t,x})$.

3.8. Let $p(s, x; t, \Gamma)$ be a transition function in spaces E_t. A non-negative function $h^t(x)$ $(t \in T, x \in E_t)$ is said to be *excessive* if a) for any t it is measurable in x; b) $p(s, x; t, h^t) \leqslant h^s(x)$ for all $s < t$; c) $p(s, x; t, h^t) \to h^s(x)$ as $t \downarrow s$.

From [4], §5 it follows that if $p(s, x; t, \Gamma)$ is a transition function of a regular right representation $(x_t, \mathbf{P}_{t,x})$, then every excessive function h has the following property: $h^t(x_t)$ is $\mathbf{P}$-almost surely right-continuous in t on $[\alpha, \beta)$. If the representation $(x_t, \mathbf{P}_{t,x})$ is completely regular, then for any $s \in [\alpha, \beta)$ and $\mathbf{P}' \in \mathbf{K}_s(x_t, \mathbf{P}_{t,x})$, the function $h^t(x_t)$ is $\mathbf{P}'$-almost surely continuous from the right in t on $[s, \beta)$.

3.9. We fix a regular right Markov representation $(x_t, \mathbf{P}_{t,x})$ and set $\mathscr{A}_t = \mathscr{F}^{\mathbf{P}}_{<t+}$.

We say that the two functions f_1 and f_2 on $\mathscr{E}$ are *equal quasi-everywhere* or *quasi-equivalent* if they are equal outside some inaccessible set. By 3.2.C and Lemma 3.1, this is the case if and only if $\mathbf{P}\{f_1(\tau, x_\tau) \neq f_2(\tau, x_\tau)\} = 0$ for any Markov time τ (with respect to $\mathscr{A}_t$) (assuming that both functions are measurable.)

Let $\eta_t(\omega)$ be a non-negative function on $T \times \Omega$, jointly measurable with respect to t and ω (in T we consider the Borel measurable structure). We define the *projection of* η *on* $\mathscr{E}$ to be a measurable function on $\mathscr{E}$ satisfying the following condition: for any Markov time τ

$$(3.9) \qquad \mathbf{P}\{\eta_\tau \mid \mathscr{A}_\tau\} = f(\tau, x_\tau) \quad (\mathbf{P}\text{-a.s.}, \tau < \infty)$$

If the projection exists, it is unique to within quasi-equivalence.

THEOREM 3.1. *The projection on* $\mathscr{E}$ *exists for any bounded reconstructable function* η. *If* η *is strictly reconstructable, then its projection is given by the formula*

(3.10) $$f(t, x) = \lim_{r\downarrow t,\, r\in R} \mathbf{P}_{t,x}\eta_r \quad \textit{quasi-everywhere.}$$

and $f(t, x_t)$ is **P**-*almost surely continuous from the right in t.*

PROOF. The family of functions η having projections on $\mathscr{E}$, is closed under linear operations and monotonic increasing passage to a limit. (If $\eta^n \uparrow \eta$ and f^n is the projection of η^n, then outside a certain inaccessible set f^n is non-decreasing and its limit is the projection of η). By 1.11.A, it is sufficient to prove the theorem for strictly reconstructable functions.

Thus, let η be strictly reconstructable. We set $(t, x) \in B$ if the limit on the right-hand side of (3.10) exists. We denote the limit by $f(t, x)$ and set $f(t, x) = +\infty$ outside B. We now claim that f is the projection of η on $\mathscr{E}$.

To show that f is measurable it is sufficient to establish the measurability of the lower limit f_1 and the upper limit f_2 of $F^r(t, x) = \mathbf{P}_{t,x}\eta_r$ as $r \downarrow t$, $r \in R$. We observe that

$$f_1(t, x) = \lim_{n\to\infty} \inf_{r\in R} F^r_n(t, x),$$

where $F^r_n(t, x) = F^r(t, x)$ where $t < r < t + \frac{1}{n}$ and $+\infty$ elsewhere. Since F^r and t are measurable functions on $\mathscr{E}$, so is F^r_n, and hence also f_1.

The measurability of f_2 is proved similarly.

We choose a version $\xi(\omega)$ of the conditional mathematical expectation $\mathbf{P}\{\eta_\tau \mid A_\tau\}$ and let $C = \{\tau < \infty,\ \xi \neq f(\tau, x_\tau)\}$. If $\omega \in C$, then for some positive $\varepsilon(\omega) > 0$ there is in any interval $\left(\tau, \tau + \frac{1}{n}\right)$ a point $r \in R$ such that

(3.11) $$|\xi - F^r(\tau, x_\tau)| > \varepsilon(\omega).$$

We enumerate the points of R and denote by σ_n the first point belonging to $\left(\tau, \tau + \frac{1}{n}\right)$ for which (3.11) holds (if there is no such point, we set $\sigma_n = +\infty$). It is easily seen that the function σ_n is measurable with respect to $\mathscr{A}_\tau$. On C, $\sigma_n > \tau$, $\sigma_n \to \tau$ and $|\xi - F^{\sigma_n}(\tau, x_\tau)| > \varepsilon(\omega)$ for all n. On the other hand, by (3.8),

$$F^{\sigma_n}(\tau, x_\tau) = \mathbf{P}\{\eta_{\sigma_n} \mid \mathscr{A}_\tau\} \to \mathbf{P}\{\eta_\tau \mid \mathscr{A}_\tau\} \quad (\mathbf{P}\text{-a.s., } C)$$

Hence $\mathbf{P}(C) = 0$ and f satisfies (3.9).

Since η is bounded, it follows from (3.9) that $\mathbf{P}\{f(\tau, x_\tau) = \infty\} = 0$ for any Markov time τ. Since f is measurable, the set $f = \infty$ is inaccessible. But this set is the complement of B. Therefore, (3.10) holds.

From (3.9) it follows that $\mathbf{P}\eta_\tau = \mathbf{P}f(\tau, x_\tau)$, and from 3.2.D and the right continuity of η_t it follows that $f(t, x_t)$ is continuous from the right.

3.10. We extend Theorem 3.1. to functions whose values are probability measures.

We denote by $\mathscr{M}(Y)$ the set of all probability measures on a measurable Borel space Y. A function $\eta_{t,\omega}$ with values in $\mathscr{M}(Y)$ is said to be *reconstructable* (respectively, *strictly reconstructable*) if for any non-negative measurable function φ on Y the real-valued function $\eta_{t,\omega}(\varphi)$ is *reconstructable* (respectively, *strictly reconstructable*).

THEOREM 3.2. *If $\eta_{t,\omega}$ is a strictly reconstructable function with values in $\mathcal{M}(Y)$ then there exists a function $p_{t,x}$ with values in $\mathcal{M}(Y)$ such that $p_{t,x}(\varphi)$ is the projection of $\eta_{t,\omega}(\varphi)$ on $\mathscr{E}$ for any non-negative measurable function φ on Y.*

PROOF. We consider a support system W in the space Y and set $\nu_r^{t,x}(\Gamma) = \mathbf{P}_{t,x}\eta_r(\Gamma)$. By (3.10), the limit

$$\lim_{r\downarrow t,\, r\in R} \nu_r^{t,x}(\varphi) \tag{3.12}$$

exists for all $\varphi \in W$ if (t, x) is outside some inaccessible set C. By 1.12.B, to each $(t, x) \notin C$ there corresponds a probability measure $p_{t,x}$ on Y such that the limit (3.12) is equal to $p_{t,x}(\varphi)$ for all $\varphi \in W$. For $(t, x) \in C$ we set $p_{t,x} = \nu$, where ν is an arbitrary fixed probability measure on Y. By Theorem 3.1., $p_{t,x}(\varphi)$ for $\varphi \in W$ is the projection of $\eta_{t,\omega}(\varphi)$ on $\mathscr{E}$. But (3.9) remains valid under linear operations and monotonic increasing passage to a limit, and by 1.12.A, the theorem holds for all non-negative measurable φ.

3.11. We now apply the general results on projections to the study of Markov processes.

THEOREM 3.3. *Let $(x_t, \mathbf{P}_{t,x})$ be a regular right representation and $y(\omega)$ a measurable mapping of $(\Omega, \mathscr{F}_{<s+})$ into a Borel space $(Y, \mathscr{B}_Y)$. Then there is a function $p_{t,x}$ defined on $\mathscr{E}_{\geq s}$, taking values in $\mathcal{M}(Y)$ and such that for any bounded function $\varphi \in \mathscr{B}_Y$ the following conditions hold:*

3.11.A. *$p_{t,x}(\varphi)$ is measurable with respect to (t, x).*

3.11.B. *$p_{t,x_t}(\varphi)$ is* **P***-almost surely right-continuous in t on $[s, \beta)$.*

3.11.C. *For $t \geqslant s$,*

$$\mathbf{P}\{\varphi(y) \mid \mathscr{F}_{>t}\} = p_{t,x_t}(\varphi) \quad (\mathbf{P}\text{-a.s.}) \tag{3.13}$$

These properties define $p_{t,x}$ uniquely to within quasi-equivalence.

We call $p_{t,x}$ the *indicatrix of $y(\omega)$*.

PROOF. 1°. Let W be a support system in Y. If p^1 and p^2 are two indicatrices of the mapping y, then, by 3.11.C, **P**-almost surely $p^1_{t,x_t}(\varphi) = p^2_{t,x_t}(\varphi)$ for all $t \in R$ and $\varphi \in W$; by 3.11.B, equality holds for all $t \in [s, \beta)$ and $\varphi \in W$, outside some **P**-negligible set C. By 1.12.A, it follows that outside C it holds for all $t \in [s, \beta)$ and all $\varphi \in \mathscr{B}_Y$. Hence $p^1_{t,x} = p^2_{t,x}$ quasi-everywhere. This proves the uniqueness of the indicatrix. We now proceed to prove the existence.

2°. Since E_t and Y are Borel spaces, there is a function $\pi_{t,x}$ with values in $\mathcal{M}(Y)$ such that $\pi_{t,x}(\varphi)$ for $\varphi \in \mathscr{B}_Y$ is measurable in x and

$$\mathbf{P}\{\varphi(y) \mid x_t\} = \pi_{t,x_t}(\varphi) \quad (\mathbf{P}\text{-a.s.}). \tag{3.14}$$

Since $\varphi[y(\omega)] \in \mathscr{F}_{<t}$ for $s < t$, according to (1.10) it follows from (3.14) that

$$\mathbf{P}\{\varphi(y) \mid \mathscr{F}_{\geqslant t}\} = \pi_{t,x_t}(\varphi) \quad (\mathbf{P}\text{-a.s.}), \tag{3.15}$$

so that $(\pi_{t,x_t}(\varphi), \mathscr{F}_{\geqslant t}, \mathbf{P})$ is a martingale on the interval (s, β). According to 1.14.A, there is a **P**-negligible set C such that for $\omega \notin C$ the limit

$$\lim_{r\downarrow t,\, r\in R} \pi_{r,x_r(\omega)}(\varphi)$$

exists for all $t \in [s, \beta)$ and $\varphi \in W$. By 1.12.B, for $\omega \notin C$ and $t \in [s, \beta)$ there is a measure $\eta_{t,\omega} \in \mathscr{M}(Y)$ such that

$$\eta_{t,\omega}(\varphi) = \lim_{r\downarrow t,\, r\in R} \pi_{r,x_r(\omega)}(\varphi). \tag{3.16}$$

3°. We denote by $\mathscr{H}$ the set of all bounded functions φ for which (3.16) holds **P**-almost surely, and prove that $\mathscr{H}$ contains all bounded $\mathscr{B}$-measurable functions. By 2°, $\mathscr{H}$ contains W. Clearly, $\mathscr{H}$ is closed under addition and scalar multiplication. By 1.12.A′, it suffices to verify that $\mathscr{H}$ is closed under bounded convergence.

Thus, suppose that functions φ_n in $\mathscr{H}$ *converge boundedly* to φ; we claim that $\varphi \in \mathscr{H}$. Clearly $\eta_{t,\omega}(\varphi_n) \to \eta_{t,\omega}(\varphi)$. Therefore, it is sufficient to verify that for some sequence n_k

$$\sup_{r\in R_{\geqslant s}} |\pi_{r,x_r}(\varphi) - \pi_{r,x_r}(\varphi_{n_k})| \to 0 \quad (\mathbf{P}\text{-a.s.}). \tag{3.17}$$

For brevity we write $z_n(r) = \pi_{r,x_r}(\varphi - \varphi_n)$. Using the well-known Kolmogorov martingale inequality, for any $c > 0$

$$\mathbf{P}\{\sup_{r\in R_{\geqslant s}} |z_n(r)| > c\} \leqslant c^{-1} \sup_{r\in R} \mathbf{P}|z_n(r)|.$$

By (3.15), the right-hand side is equal to $c^{-1}\mathbf{P}\,|\varphi(y) - \varphi_n(y)|$ and tends to zero as $n \to \infty$. We choose the n_k so that $\mathbf{P}\,|\varphi(y) - \varphi_{n_k}(y)| < 1/k^3$. Taking $c = 1/k$ we observe that, with probability not less than $1 - 1/k^2$, $\sup |z_n(r)| \leqslant 1/k$. (3.17) now follows from the Borel-Cantelli lemma.

4°. By (3.16), $\eta_{t,\omega}$ is a strictly reconstructable function on $[s, \beta) \times \Omega$ with values in $\mathscr{M}(Y)$. We denote by $p_{t,x}$ its projection on $\mathscr{E}_{\geqslant s}$ as defined in Theorem 3.2. Then 3.11.A is satisfied by the definition of a projection. 3.11.B follows from Theorem 3.1, since by (3.16), η is strictly reconstructable.

It remains to establish (3.13). The function $\eta_{t,\omega}(\varphi)$ is measurable with respect to $\mathscr{A}_t = \mathscr{F}^{\mathbf{P}}_{<t+}$. Applying (3.9) to $\tau = t$, we have

$$p_{t,x_t}(\varphi) = \mathbf{P}\{\eta_{t,\omega}(\varphi) \mid \mathscr{A}_t\} = \eta_{t,\omega}(\varphi) \quad (\mathbf{P}\text{-a.s.}).$$

But from (3.15), (3.16) and 1.14.C it follows that

$$\eta_{t,\omega}(\varphi) = \lim_{r\downarrow t,\, r\in R} \mathbf{P}\{\varphi(y) \mid \mathscr{F}_{\geqslant r}\} = \mathbf{P}\{\varphi(y) \mid \mathscr{F}_{>t}\} \quad (\mathbf{P}\text{-a.s.}).$$

3.12. LEMMA 3.2. *Let $(x_t, \mathbf{P}_{t,x})$ be a regular right representation, $r_1 \leqslant r_2 \in T$, and $y^i(\omega)$ measurable mappings from $(\Omega, \mathscr{F}_{<r_i+})$ into a Borel space $(Y^i, \mathscr{B}^i)$ that are connected by the relations*

$$\mathbf{P}\{y^1(\omega) \in \Gamma \mid \mathscr{A}\} = p(\Gamma, y^2(\omega)) \quad (\mathbf{P}\text{-a.s.}), \tag{3.18}$$

where $\mathcal{A}$ is a σ-algebra containing $\mathcal{F}_{>r_2}$, p is a probability measure with respect to $\Gamma \in \mathcal{B}_1$ and a $\mathcal{B}^2$-measurable function with respect to y^2. Then the indicatrices p^i of the mappings y^i are connected by the relations

$$p^1_{t,x}(\varphi) = \int p(\varphi, z) p^2_{t,x}(dz) \quad \textit{quasi-everywhere on } \mathcal{E}_{\geqslant r_2},$$

and the exceptional inaccessible set is measurable and does not depend on $\varphi \in \mathcal{B}^1$.

PROOF. It is sufficient to verify that the right-hand side is an indicatrix of y_1. It can be written in the form $p^2_{t,x}(\tilde{\varphi})$, where $\tilde{\varphi}(z) = p(\varphi, z)$. Now 3.11.A and 3.11.B are obvious from this. It remains to prove 3.11.C, but by (3.18) and (3.13), for $t > r_2$ we have

$$\mathbf{P}\{\varphi(y^1) \mid \mathcal{F}_{>t}\} = \mathbf{P}[\mathbf{P}\{\varphi(y^1) \mid \mathcal{A}\} \mid \mathcal{F}_{>t}] = \mathbf{P}[\tilde{\varphi}(y^2) \mid \mathcal{F}_{>t}] = p^2_{t,x}(\tilde{\varphi}) \ (\mathbf{P}\text{-a.s.}).$$

§4. Absolutely continuous Markov representations

4.1. To prove Theorem 1.2. we need a number of auxiliary propositions.

Let x_t be an absolutely continuous Markov representation of a stochastic system $\mathcal{F}(I)$. We consider a density $\bar{p}(s, x; t, y)$ of measure m_{st} with respect to $m_s \times m_t$ and set

$$\bar{p}(s, x; t, B) = \int_B \bar{p}(s, x; t, y)\, m_t(dy), \tag{4.1}$$

$$\bar{p}(s, A; t, y) = \int_A m_s(dx)\, \bar{p}(s, x; t, y). \tag{4.2}$$

4.1.A. For any $s < t < u$ and $f \in \mathcal{B}_t$

$$\mathbf{P}\{f(x_t) \mid \mathcal{F}_{\leqslant s}\} = \bar{p}(s, x_s; t, f) \quad (\mathbf{P}\text{-a.s.}), \tag{4.3}$$

$$\mathbf{P}\{f(x_t) \mid \mathcal{F}_{\geqslant u}\} = \bar{p}(t, f; u, x_u) \quad (\mathbf{P}\text{-a.s.}). \tag{4.4}$$

To establish (4.3) it suffices to note that for any $\varphi \in \mathcal{B}_s$

$$\mathbf{P}\varphi(x_s)\, \bar{p}(s, x_s; t, f) = \int\int m_s(dx)\, \varphi(x)\, \bar{p}(s, x; t, y)\, f(y)\, m_t(dy) = \mathbf{P}\varphi(x_s)\, f(x_t),$$

and to use 1.2.B′. (4.4.) is proved similarly.

4.1.B. For any $s < t < u$

$$\bar{p}(s, x; t, E_t) = 1 \quad \text{for } m_s\text{-almost all } x. \tag{4.5}$$

$$\bar{p}(t, E_t; u, y) = 1 \quad \text{for } m_u\text{-almost all } y. \tag{4.6}$$

This follows at once from 4.1.A.

4.1.C. For any $s < t < u$, $\varphi \in \mathcal{B}_s$, and $\psi \in \mathcal{B}_u$

$$\int \bar{p}(s, x; t, dy)\, \bar{p}(t, y; u, \psi) = \bar{p}(s, x; u, \psi) \tag{4.7}$$

for m_s-almost all x, and

$$\int \bar{p}(s, \varphi; t, y)\, \bar{p}(t, dy; u, z) = \bar{p}(s, \varphi; u, z) \tag{4.8}$$

for m_u-almost all z.

Let us prove (4.7), for example. By (4.3),

$$\bar{p}(s, x_s; u, \psi) = \mathbf{P}\{\psi(x_u) \mid \mathcal{F}_{\leqslant s}\} = \mathbf{P}\{\mathbf{P}[\psi(x_u) \mid \mathcal{F}_{\leqslant t}] \mid \mathcal{F}_{\leqslant s}\} = \\ = \mathbf{P}\{\bar{p}(t, x_t; u, \psi) \mid \mathcal{F}_{\leqslant s}\} = \bar{p}(s, x_s; t, f) \text{ (P-a.s.)},$$

where $f(y) = p(t, y; u, \psi)$. This is equivalent to (4.7).

LEMMA 4.1. *We can construct a transition function $p(s, x; t, B)$ and a cotransition function $p(s, A; t, y)$ such that for any $s < t$, $A \in \mathcal{B}_s$, and $B \in \mathcal{B}_t$*

$$\text{(4.9)} \qquad \mathbf{P}\{x_s \in A, x_t \in B\} = \int_A m_s(dx)\, p(s, x; t, B) = \int_B p(s, A; t, y)\, m_t(dy)$$

and that the measures $p(s, x; t, -)$, $p(t, -; u, y)$ are absolutely continuous with respect to m_t.

PROOF. We write $x \in E_s'$ if (4.5) holds for all $t \in R_{>s}$ and (4.7) for all $t < u \in R_{>s}$ and $\psi \in W_u$. By means of 1.12.A (4.7) can be extended to all $\psi \in \mathcal{B}_u$. By 4.1.B and 4.1.C, we have $m_s(E_s \setminus E_s') = 0$.

We denote the left-hand side of (4.7) by $p(s, x; t, u, \psi)$. Let $r_1 < r_2 \in R_{st}$. Setting $t = r_1$, $u = r_2$, $\psi(z) = \bar{p}(r_2, z; t, B)$ in (4.7), we conclude that for $x \in E_s'$ the expression

$$\text{(4.10)} \qquad \iint \bar{p}(s, x; r_1, dy)\, \bar{p}(r_1, y; r_2, dz)\, \bar{p}(r_2, z; t, B)$$

is equal to $p(s, x; r_2; t, B)$. On the other hand, for m_{r_1}-almost all y

$$\int \bar{p}(r_1, y; r_2, dz)\, \bar{p}(r_2, z; t, B) = \bar{p}(r_1, y; t, B),$$

therefore (4.10) is equal to $\bar{p}(s, x; r_1; t, B)$. Hence, for $x \in E_s'$, $p(s, x; r, t, B)$ does not depend on the choice of $r \in R_{st}$. We denote it by $p(s, x; t, B)$. For $x \in E_s \setminus E_s'$ we set $p(s, x; t, B) = p(s, a_s; t, B)$, where a_s is a fixed point of E_s'. Since $p(s, x; t, B) = \bar{p}(s, x; t, B)$ for m_s-almost all x, the first equation of (4.9) is satisfied. It is easy to see that $p(s, x; t, B)$ satisfies 1.3.A–1.3.C and is absolutely continuous with respect to m_t.

LEMMA 4.2. *Let $p(s, x; t, B)$ be a transition function satisfying the conditions of Lemma* 4.1. *Then we can choose a density $p(s, x; t, y)$ of the measure $p(s, x; t, -)$ relative to m_t that is jointly measurable with respect to (s, x) and y, and satisfies* 1.6.A.–1.6.B.

PROOF. By the definition of the measurable structure in $\mathscr{E}$ (see §3.3), the function $p(s, x; t, B)$ is measurable with respect to (s,x) for each B. Therefore, the measure $p(s, x; t, -)$ has density $\bar{p}(s, x; t, y)$ relative to m_t, which is jointly measurable with respect to the pair (s, x) and y (see, for example [2], §0.15). Clearly, $\bar{p}(s, x; t, y)$ is also a density for m_{st} with respect to $m_s \times m_t$, so that Propositions 4.1.A–4.1.C are applicable to it.

Let $\hat{E}_u$ be the subset of E_u such that (4.6) holds for all $t \in R_{<u}$ and

(4.8) for all $s<t\in R_{<u}$ and $\varphi\in W_s$. By 4.1.B–4.1.C, $m_u(E_u\setminus\hat{E}_u)=0$. We write

$$(4.11)\qquad p(s, x; r; t, y)=\int p(s, x; r, dz)\bar{p}(r, z; t, y)=$$
$$=\int \bar{p}(s, x; r, z)\bar{p}(r, dz, t, y).$$

From (4.8) it follows that for $z\in\hat{E}_u$ and $r_1<r_2\in R_{<u}$

$$p(r_1, y; r_2; u, z)=\bar{p}(r_1, y; u, z) \text{ for } m_{r_1}\text{-almost all } y.$$

Integrating with respect to $p(s, x; r_1, -)$ and using 1.3.C, we have $p(s, x; r_2; u, z) = p(s, x; r_1; u, z)$. Hence, for arbitrary $s<t$, $x\in E_s$, and $z\in\hat{E}_u$ the function (4.11) does not depend on the choice of $r\in R_{su}$. We denote it by $p(s, x; u, z)$. For $z\in E_u\setminus\hat{E}_u$ we set $p(s, x; u, z)=p(s, x; u, b_u)$, where b_u is a fixed point of $\hat{E}_u$. By the definition of $p(s, x; t, y)$, for $z\in\hat{E}_u$

$$(4.12)\qquad \int p(s, x; t, y)\,m_t(dy)\,p(t, y; u, z)=$$
$$=\iiint p(s, x; r_1, dz_1)\bar{p}(r_1, z_1; t, y)\,m_t(dy)\,p(t, y; r_2, dz_2)\bar{p}(r_2, z_2; u, z)=$$
$$=\iiint p(s, x; r_1, dz_1)\,p(r_1, z_1; t, dy)\,p(t, y; r_2, dz_2)\bar{p}(r_2, z_2; u, z)=$$
$$=\int p(s, x; r_2, dz_2)\bar{p}(r_2, z_2; u, z)=p(s, x; u, z)$$

(here $r_1\in R_{st}$, $r_2\in R_{tu}$, and we have used 1.3.C). Moreover,

$$(4.13)\qquad \int_B p(s, x; t, y)\,m_t(dy)=\int\int_B p(s, x; r, dz)\bar{p}(r, z; t, y)\,m_t(dy)=$$
$$=\int p(s, x; r, dz)\,p(r, z; t, B)=p(s, x; t, B).$$

By (4.6), for $y\in\hat{E}_t$

$$(4.14)\qquad \int m_s(dz)\,p(s, z; t, y)=$$
$$=\iint m_s(dz)\bar{p}(s, z; r, v)\bar{p}(r, dv; t, y)=\int \bar{p}(r, dv; t, y)=1.$$

Clearly, 1.6.A follows from (4.12) and 1.6.B from (4.14). By (4.13), $p(s, x; t, y)$ is a density for $p(s, x, t, -)$ with respect to m_t. Finally, from (4.11) it is clear that $p(s, x; t, y)$ is jointly measurable in (s, x) and y.

4.2. The first part of Theorem 1.2., asserting the existence of a fundamental density follows at once from Lemmas 4.1 and 4.2. We claim that (1.14) defines transition probabilities for x_t.

For any $s<t$ and $x\in E_s$ we consider the measure $\mathbf{P}^t_{s,x}$ on $\mathcal{F}_{>t}$ defined by

$$\mathbf{P}^t_{s,x}(C) = \mathbf{P}p(s, x; t, x_t)\chi_C.$$

Suppose that $t < u$, $\varphi \in \mathscr{B}_t$, $\eta \in \mathscr{F}_{>u}$ and that $p(t, y; u, x_u)$ is measurable jointly in y and ω. Applying Fubini's theorem and using (4.4) we have

$$\mathbf{P}\varphi(x_t)\mathbf{P}^u_{t,x_t}\eta = \int \varphi(y)\, m_t(dy)\, \mathbf{P}^u_{t,y}\eta = \mathbf{P}\int \varphi(y)\, m_t(dy)\, p(t, y; u, x_u)\,\eta =$$
$$= \mathbf{P}p(t, \varphi; u, x_u)\,\eta = \mathbf{P}\{\mathbf{P}[\varphi(x_t) \mid \mathscr{F}_{\geq u}]\,\eta\} = \mathbf{P}\varphi(x_t)\,\eta.$$

Hence it follows that for $t < u$ and $\eta \in \mathscr{F}_{>u}$

$$\mathbf{P}\{\eta \mid x_t\} = \mathbf{P}^u_{t,x_t}\eta \quad (\mathbf{P}\text{-a.s.}) \tag{4.15}$$

Now let $s < t < u$ and $\eta \in \mathscr{F}_{>u}$. By 1.6.A and (4.15),

$$\mathbf{P}^u_{s,x}\eta = \mathbf{P}p(s, x; u, x_u)\,\eta = \tag{4.16}$$
$$= \mathbf{P}\int p(s, x; t, z)\, m_t(dz)\, p(t, z; u, x_u)\,\eta =$$
$$= \int p(s, x; t, z)\, m_t(dz)\, \mathbf{P}^u_{t,z}\eta = \mathbf{P}p(s, x; t, x_t)\,\mathbf{P}^u_{t,x_t}\eta =$$
$$= \mathbf{P}p(s, x; t, x_t)\,\mathbf{P}\{\eta \mid x_t\} = \mathbf{P}p(s, x; t, x_t)\,\eta = \mathbf{P}^t_{s,x}\eta.$$

We consider an arbitrary sequence $t_n \downarrow s$. By (4.16), $\mathbf{P}^{t_n}_{s,x} = \mathbf{P}^{t_{n+1}}_{s,x}$ on $\mathscr{F}_{>t_n}$. Therefore, by 1.1.C, there exists a measure $\mathbf{P}_{s,x}$ on $\mathscr{F}_{>s}$ that coincides with $\mathbf{P}^{t_n}_{s,x}$ on $\mathscr{F}_{>t_n}$. For any $t > s$ there is a $t_n \in (s, t)$, and by (4.16), $\mathbf{P}_{s,x} = \mathbf{P}^{t_n}_{s,x} = \mathbf{P}^t_{s,x}$ on $\mathscr{F}_{>t}$. Hence (1.14) is satisfied. From (1.14) it follows that $\mathbf{P}_{s,x}\eta$ is a measurable function of x.

According to (4.15), for $t < u$ and $\eta \in \mathscr{F}_{>u}$

$$\mathbf{P}\{\eta \mid x_t\} = \mathbf{P}_{t,x_t}\eta \quad (\mathbf{P}\text{-a.s.}),$$

therefore (1.2) is satisfied. As x_t satisfies 1.2.B′, (1.4) also follows from this.

Let $s < r < t$, $\xi \in \mathscr{F}(r, t]$ and $\eta \in \mathscr{F}_{>t}$. By (1.12) and (1.4),

$$\mathbf{P}_{s,x}\xi\eta = \mathbf{P}p(s, x; r, x_r)\xi\eta = \mathbf{P}p(s, x; r, x_r)\xi\mathbf{P}_{t,x_t}\eta = \mathbf{P}_{s,x}\xi\mathbf{P}_{t,x_t}\eta.$$

Hence (1.1) holds for $\xi \in \mathscr{F}(r, t]$ and $r \in (s, t)$. Using 1.1.C, we conclude that (1.1) holds for all $\xi \in \mathscr{F}(s, t]$. Thus, the $\mathbf{P}_{s,x}$ are the transition probabilities for x_t.

4.3. We now prove the results stated in § 1.7.

PROOF of 1.7.A. The absolute continuity of the transition function $p(s, x; t, \Gamma)$ clearly follows from that of $(x_t, \mathbf{P}_{t,x})$. On the other hand, if $\eta \in \mathscr{F}_{>t}$, then, with $\varphi(x) = \mathbf{P}_{t,x}\eta$ we have, by (1.1) and (1.2),

$$\mathbf{P}\eta = \mathbf{P}\varphi(x_t) = m_t(\varphi), \quad \mathbf{P}_{s,x}\eta = \mathbf{P}_{s,x}\varphi(x_t) = p(s, x; t, \varphi).$$

Therefore, the absolute continuity of $p(s, x; t, \Gamma)$ implies that of $(x_t, \mathbf{P}_{t,x})$.

PROOF of 1.7.B. If $(x_t, \mathbf{P}_{t,x})$ is absolutely continuous, then according to 1.7.A, $p(s, x; t, dy) = p(s, x; t, y)m_t(dy)$. The density $p(s,x; t, y)$ can be chosen to be jointly measurable in x and y. For any $A \in \mathscr{B}_s$ and $B \in \mathscr{B}_t$ we have

$$m_{st}(A \times B) = \mathbf{P}\{x_s \in A,\ x_t \in B\} = \mathbf{P}\chi_A(x_s)\mathbf{P}_{s,x_s}\{x_t \in B\} =$$
$$= \int_A \int_B m_s(dx)\, p(s, x; t, y)\, m_t(dy).$$

Therefore, m_{st} is absolutely continuous with respect to $m_s \times m_t$. This proves one half of 1.7.B. The other half follows from Theorem 1.2.

PROOF of 1.7.C. According to Lemma 4.2, we can choose a fundamental density $p(s, x; t, y)$ so that

$$\mathbf{P}_{s,x}\{x_t \in \Gamma\} = \int_\Gamma p(s, x; t, y)\, m_t(dy) = \mathbf{P}p(s, x; t, x_t)\,\chi_\Gamma(x_t).$$

By 1.11.A, it follows from this that for any $\varphi \in \mathscr{B}_t$

$$\mathbf{P}_{s,x}\varphi(x_t) = \mathbf{P}p(s, x; t, x_t)\varphi(x_t).$$

Setting $\varphi(x) = \mathbf{P}_{t,x}\eta$ and bearing (1.1) and (1.2) in mind, we arrive at (1.14).

1.7.D. is clear.

PROOF of 1.7.E. If $(x_t, \mathbf{P}_{t,x})$ is absolutely continuous and if $p(s, x; t, y)$ is a fundamental density for it, then for any right representation $(\bar{x}_t, \bar{\mathbf{P}}_{t,x})$, for $s < r < t$ and $\eta \in \mathscr{F}_{>t}$,

$$\mathbf{P}_{s,x}\eta = \mathbf{P}p(s, x; r, x_r)\eta = \mathbf{P}p(s, x; r, x_r)\bar{\mathbf{P}}_{t,\bar{x}_t}\eta = \mathbf{P}_{s,x}\bar{\mathbf{P}}_{t,\bar{x}_t}\eta.$$

4.4. THE PROOF OF THEOREM 1.3. is divided into several steps.

1°. Let $\bar{p}(s, x; t, y)$ be a fundamental density for the representation $(x_t, \mathbf{P}_{t,x})$ (its existence is guaranteed by 1.7.C). Then

$$\mathbf{P}_{s,x}\eta = \mathbf{P}\bar{p}(s, x; t, x_t)\eta \quad (s < t,\ \eta \in \mathscr{F}_{>t}). \tag{4.17}$$

By (4.4), for $s_1 < s_2$

$$\mathbf{P}\{x_{s_1} \in \Gamma \mid \mathscr{F}_{\geq s_2}\} = \bar{p}(s_1, \Gamma; s_2, x_{s_2}) \ (\mathbf{P}\text{-a.s.}). \tag{4.18}$$

We denote by $p^s_{t,y}$ the indicatrix of the mapping $x_s(\omega)$. By Lemma 3.2, outside some measurable inaccessible set (depending on s_1 and s_2)

$$\int \bar{p}(s_1, \varphi; s_2, z)\, p^{s_2}_{t,y}(dz) = p^{s_1}_{t,y}(\varphi) \text{ on } \mathscr{E}_{\geq s_2} \tag{4.19}$$

for all $\varphi \in \mathscr{B}_{s_1}$. Hence there is a measurable inaccessible set C outside which (4.19) holds for all $s_1 < s_2 \in R$ and $\varphi \in \mathscr{B}_{s_1}$. Now we set

$$p(s, x; r; t, y) = \int \bar{p}(s, x; r, z)\, p^r_{t,y}(dz). \tag{4.20}$$

From (4.19) it follows that for $r_1 < r_2 \in R_{st}$

$$p(s, x; r_2; t, y) = \int \bar{p}(s, x; r_1, z)\, m_{r_1}(dz)\, p(r_1, z; r_2; t, y). \tag{4.21}$$

By (4.2) and (4.19), for $(t, y) \notin C$

$$p^{s_1}_{t,y}(dx) = m_{s_1}(dx)p(s_1, x; r_2; t, y), \tag{4.22}$$

and by (4.21), (4.22) and (4.20),

$$p(s, x; r_2; t, y) = \int \overline{p}(s, x; r_1, z)\, p^{r_1}_{t,y}(dz) = p(s, x; r_1; t, y).$$

Consequently, if $(t, y) \notin C$, the value of $p(s, x; r; t, y)$ does not depend on choice of r in R_{st}, so that we can omit the argument r and rewrite (4.20) in the form

$$(4.23) \qquad p(s, x; t, y) = \int \overline{p}(s, x; r, z)\, p^{r}_{t,y}(dz) \qquad ((t, y) \notin C).$$

For $(t, y) \in C$ we set

$$(4.24) \qquad p(s, x; t, y) = p(s, x; t, x_t(\omega_0)),$$

where ω_0 is an arbitrary point of Ω such that $(t, x_t(\omega_0)) \notin C$ for all t, and $f(t, x_t(\omega_0))$ is measurable in t for any function f that is measurable on $\mathscr{C}$. (By 3.4.C, almost all ω_0 have these properties).

2°. To prove that the function defined by (4.23) and (4.24) is jointly measurable in (s, x) and (t, y) it is sufficient to verify that

a) the function $F_1(s, x; t, y) = \sup_{r \in R} p(s, x; r; t, y)$ is jointly measurable in (s, x) and (t, y) (here we assume that $p(s, x; r; t, y) = 0$ for $r \notin (s, t)$;

b) the function $F_2(s, x; t) = p(s, x; t, x_t(\omega_0))$ is jointly measurable in (s, x) and t.

Now a) follows from the joint measurability of $\overline{p}(s, x; r, z)$ in (s, x) and z (see Lemma 4.2), from the measurability of $p^r_{t,y}(\Gamma)$ with respect to (t, y) (see 3.11.A), and from 1.11.B. To prove b), we observe that $F_2(s, x, t) = F_1(s, x; t, x_t(\omega_0))$. But if $F(s, x; t, y)$ is any function that is jointly measurable in (s, x) and (t, y), then $F(s, x; t, x_t(\omega_0))$ is jointly measurable in (s, x) and t. For a function F that can be expressed as a product $G_1(s, x)G_2(t, y)$ this follows from the choice of ω_0. Since the property in question is conserved under linear operations and limit passage, it extends to any measurable F.

3°. Since $x_t \in \mathscr{F}(t, t+)$ is measurable, we have $\mathscr{F}_{\geqslant t} = \mathscr{F}_{>t}$ and from (3.15) and (4.18) we obtain

$$p^r_{t, x_t}(\varphi) = \mathbf{P}\{\varphi(x_t) \mid \mathscr{F}_{\geqslant t}\} = \overline{p}(r, \varphi; t, x_t) \quad (\mathbf{P}\text{-a.s.}).$$

Consequently, $p^r_{t,y}(\varphi) = \overline{p}(r, \varphi; t, y)$ for m_t-almost all y. But by (4.23), $p(s, x; t, y) = p^r_{t,y}(\varphi)$ for $\varphi(z) = \overline{p}(s, x; r, z)$, and bearing in mind 1.6.A, we have

$$(4.25) \qquad p(s, x; t, y) = \overline{p}(r, \varphi; t, y) = \overline{p}(s, x; t, y) \text{ for } m_t\text{-almost all } y.$$

From (4.17) and (4.25) it follows that $p(s, x; t, y)$ satisfies (1.14).

4°. We now verify that $p(s, x; t, y)$ satisfies 1.6.A and 1.6.B. Since $\overline{p}$ satisfies these conditions, by (4.23)

$$(4.26)\qquad \int \bar{p}(s,\ x;\ t,\ y)\, m_t(dy)\, p(t,\ y;\ u,\ z) =$$
$$= \int\int \bar{p}(s,\ x;\ t,\ y)\, m_t(dy)\, \bar{p}(t,\ y;\ r,\ v)\, p^r_{u,z}(dv) =$$
$$= \int \bar{p}(s,\ x;\ r,\ v)\, p^r_{u,z}(dv) = p(s,\ x;\ u,\ z) \quad \text{for } (u,\ z) \notin C,$$

$$(4.27)\qquad \int m_s(dx)\, p(s,\ x;\ t,\ y) = \int\int m_s(dx)\, \bar{p}(s,\ x;\ r,\ z)\, p^r_{t,y}(dz) =$$
$$= \int p^r_{t,y}(dz) = 1 \quad \text{for } (t,\ y) \notin C.$$

From (4.26) and (4.25) it follows that outside C 1.6.A holds and 1.6.B holds, by (4.27) and (1.14). The extension of these properties to $(t,\ y) \in C$ is obvious.

5°. By Theorem 1.2, (1.15) defines co-transition probabilities for x_t. Let $\xi \in \mathscr{F}_{<t}$ be bounded. We now prove that the function $\mathbf{P}^{u,x_u}\,\xi$ is $\mathbf{P}$-almost surely continuous from the right on $(t,\ \beta)$. To prove this, it is sufficient to verify that for any $r \in R_{>t}$ the function $\mathbf{P}^{u,x_u}\,\xi$ is $\mathbf{P}$-almost surely continuous from the right on $(r,\ \beta)$. We choose $s \in (t,\ r)$ and set $\varphi(z) = \mathbf{P}^{s,z}\xi$. By (1.7), (1.15), and (4.23), for $u > r$ and $(u,\ z) \notin C$ we have

$$\mathbf{P}^{u,z}\xi = \mathbf{P}^{u,z}\varphi(x_s) = \mathbf{P}\varphi(x_s)\, p(s,\ x_s;\ u,\ z) =$$
$$= \int m_s(dx)\, \varphi(x)\, \bar{p}(s,\ x;\ r,\ y)\, p^r_{u,z}(dy) = p^r_{u,z}(f),$$

where

$$f(y) = \int m_s(dx)\, \varphi(x)\, \bar{p}(s,\ x;\ r,\ y) = \bar{p}(s,\ \varphi;\ r,\ y).$$

Therefore, the required property follows from 3.11.B.

6°. It remains to show that, in the notation of 5°,

$$\lim_{u \downarrow t} \mathbf{P}^{u,x_u}\xi = \mathbf{P}^{t,x_t}\xi \quad (\mathbf{P}\text{-a.s.}).$$

By (1.10), for $u \geqslant t$

$$\mathbf{P}\{\xi \mid \mathscr{F}_{\geqslant u}\} = \mathbf{P}^{u,x_u}\xi \ (\mathbf{P}\text{-a.s.}).$$

Hence, $(\mathbf{P}^{u,x_u}\xi,\ \mathscr{F}_{\geqslant u},\ \mathbf{P})$ is a martingale on $(t,\ \beta)$, and by 1.14.A and 5°, the limit in question exists. By 1.14.C, it is equal $\mathbf{P}$-almost surely to $\mathbf{P}\{\xi \mid \mathscr{F}_{>t}\}$. It remains for us to observe that

$$\mathbf{P}\{\xi \mid \mathscr{F}_{>t}\} = \mathbf{P}\{\xi \mid \mathscr{F}_{\geqslant t}\} = \mathbf{P}^{t,x_t}\xi \quad (\mathbf{P}\text{-a.s.}).$$

4.5. The proof of Theorem 1.4 takes up a few lines. By Theorem 1.2, from the existence of an absolutely continuous Markov representation x_t it follows that there are an absolutely continuous right representation $(x_t,\ \mathbf{P}_{t,x})$ and a left representation $(x_t,\ \mathbf{P}^{t,x})$. Their regularizations

$(x_{t+}, \mathbf{P}_{t+,x})$ and $(x_{t-}, \mathbf{P}^{t-,x})$ satisfy the conditions 1.8.A, 1.8.B, and 1.8.D. By 1.7.D, they are absolutely continuous. The remaining statements of Theorem 1.4. follow from Theorem 1.3.

4.6. We divide the proof of Theorem 1.5 into several steps.

1°. We show that (1.16) defines a completely regular, two-sided Markov representation of the system $\mathscr{F}(I)$.

Since 1.2.A holds for x_{t-} and x_{t+}, it also holds for z_t. Next, since x_{t-} is measurable with respect to $\mathscr{F}_{<t}$ (see 1.8.D), the σ-algebras $\mathscr{F}(s, t]$ and $\mathscr{F}_{\leqslant t}$ for the process z_t coincide with the σ-algebras $\mathscr{F}(s, t]$ and $\mathscr{F}_{\leqslant t}$ for x_{t+}. Therefore, the validity of (1.1) and (1.4) for $(z_t, \mathbf{P}_{t,z})$ follows from their validity for $(x_{t+}, \mathbf{P}_{t+,x})$. Thus, $(z_t, \mathbf{P}_{t,z})$ is a right Markov representation of $\mathscr{F}(I)$. Since $(x_{t+}, \mathbf{P}_{t+,x})$ is completely regular and $\mathbf{P}_{t,z_t} = \mathbf{P}_{t+,x_{t+}}$, the representation $(z_t, \mathbf{P}_{t,z})$ is also completely regular.

Similarly it can be verified that $(z_t, \mathbf{P}^{t,z})$ is a completely regular left representation of $\mathscr{F}(I)$.

2°. We prove that (1.17) defines a fundamental density for $(z_t, \mathbf{P}_{t,z}, \mathbf{P}^{t,z})$.

In the first place, by (1.14) and (1.15),

$$(4.28)\qquad p(s, x \times y;\ t, x' \times y') = \mathbf{P}_{s+,y}\, p(r-, x_{r-};\ t-, x') = \mathbf{P}^{t-,x'} p(s+, y;\ q+, x_{q+}).$$

From this it is clear that the expression does not depend on r nor on q.

We now check that (4.28) is connected with $\mathbf{P}_{t,z}$ by (1.14).

We denote the one-dimensional distributions of the processes z_t, x_{t-}, and x_{t+} by m_t, m_{t-}, and m_{t+}, respectively, and set

$$p(t-, x;\ t+, B) = \mathbf{P}_{t-,x}\{x_{t+} \in B\}.$$

Let $\xi \in \mathscr{F}_{<t}$, $\eta \in \mathscr{F}_{>t}$, $\varphi \in \mathscr{B}_{t-}$. Since $(x_{t-}, \mathbf{P}_{t-,x}, \mathbf{P}^{t-,x})$ is a two-sided Markov representation of $\mathscr{F}(I)$, by (1.14) and (1.10)

$$(4.29)\quad \mathbf{P}\varphi(x_{t-})\xi\eta = \mathbf{P}\varphi(x_{t-})\xi\, \mathbf{P}_{t-,x_{t-}}\eta = \mathbf{P}\varphi(x_{t-})(\mathbf{P}_{t-,x_{t-}}\eta)\mathbf{P}^{t-,x_{t-}}\xi.$$

In particular,

$$\mathbf{P}\{x_{t-} \in A,\ x_{t+} \in B\} = \mathbf{P}\chi_A(x_{t-})\,\mathbf{P}_{t-,x_{t-}}\{x_{t+} \in B\} = \int_A m_{t-}(dx)\, p(t-, x;\ t+, B),$$

which can be rewritten more compactly as

$$(4.30)\qquad m_t(dx, dy) = m_{t-}(dx)p(t-, x;\ t+, dy).$$

Relying on (1.16), (1.14), and (4.29) we deduce that for $s < q < t$ and $z = x \times y$

$$\mathbf{P}_{s,z}\{z_t \in A \times B\} = \mathbf{P}p(s+, y;\ q+, x_{q+})\,\chi_A(x_{t-})\,\chi_B(x_{t+}) = \int_A m_{t-}(dx)\,\mathbf{P}^{t,-x} p(s+, y;\ q+, x_{q+})\, p(t-, x;\ t+, B).$$

By (4.30) and (4.28), the right-hand side is equal to

$$\int_{A\times B} m_t(dx', dy')\, p(s, x\times y;\ t,\ x'\times y') = \mathbf{P}p(s,\ z;\ t,\ z_t)\,\chi_{A\times B}(z_t).$$

By means of 1.11.A we derive from this, that for any non-negative measurable function f

$$\mathbf{P}_{s,z}f(z_t) = \mathbf{P}p(s,\ z;\ t,\ z_t)f(z_t).$$

Setting $f(z) = \mathbf{P}_{t,z}\eta$ and bearing (1.1) and (1.4) in mind we arrive at (1.14). (1.15) can be verified similarly. 1.6.B follows from (1.14) and (1.15).

Let us verify 1.6.A. By (4.28), for $s < q < t < r < u$

$$(4.31)\qquad \int p(s,\ z;\ t,\ z')\, m_t(dz')\, p(t,\ z',\ u,\ z'') =$$

$$= \mathbf{P}(\mathbf{P}^{t,z_t}p(s+,\ y;\ q+,\ x_{q+}))\,(\mathbf{P}_{t,z_t}p(r-,\ x_{r-};\ u-,\ x'')).$$

Applying (1.4) and (1.10) to z_t, $\mathbf{P}_{t,z}$, and $\mathbf{P}^{t,z}$, we conclude that the right-hand side of (4.31) is equal to

$$\mathbf{P}p(s+,\ y;\ q+,\ x_{q+})p(r-,\ x_{r-};\ u-,\ x'') = p(s,\ x\times y;\ u,\ x''\times y'').$$

By 1.8.C the representation $(z_t, \mathbf{P}_{t,z}, \mathbf{P}^{t,z})$ separates the states.

3°. We now prove the final statement of the theorem.

By 1.7.E, the representation $(z_t, \mathbf{P}_{t,z})$ is dominated by $(\bar{z}_t, \bar{\mathbf{P}}_{t,z})$, so that Lemma 2.3 can be applied to these two representations. Therefore, $\mathbf{P}_{t,z} = \bar{\mathbf{P}}_{t,\bar{z}_t}$ for all $t \in (\alpha, \beta)$ (**P**-a.s.). By symmetry arguments, $\mathbf{P}^{t,z_t} = \bar{\mathbf{P}}^{t,\bar{z}_t}$ for all $t \in (\alpha, \beta)$ (**P**-a.s.). The measures $\mathbf{P}_{t,z}$ and $\bar{\mathbf{P}}_{t,z}$ belong to the class $\mathbf{K}_t = \mathbf{K}_t(\bar{z}_t, \bar{\mathbf{P}}_{t,z})$ and the measures $\mathbf{P}^{t,z}$ and $\bar{\mathbf{P}}^{t,z}$ to the corresponding class $\mathbf{K}^t$ associated with the left representation $(\bar{z}_t, \bar{\mathbf{P}}^{t,z})$. We map the state spaces of the processes z_t and $\bar{z}_t$ into $\mathbf{K}_t \times \mathbf{K}^t$ by means of the formulae $z \to (\mathbf{P}_{t,z}, \mathbf{P}^{t,z})$ and $z \to (\bar{\mathbf{P}}_{t,z}, \bar{\mathbf{P}}^{t,z})$. The proof of the equivalence of $(z_t, \mathbf{P}_{t,z}, \mathbf{P}^{t,z})$ and $(\bar{z}_t, \bar{\mathbf{P}}_{t,z}, \bar{\mathbf{P}}^{t,z})$ is completed in the same way as that of Lemma 2.2.

4.7. To prove Theorem 1.6 we need several lemmas.

LEMMA 4.3. *Let $t < u \in T$ and let η be a bounded $\mathscr{F}_{>u}$-measurable function. We set* $\varphi(t, \omega) = \mathbf{P}_{t+,\, x_{t+}(\omega)}\eta$, $\psi(t, \omega) = \mathbf{P}_{t-,\, x_{t-}(\omega)}\eta$. *Then*

$$(4.32)\qquad \lim_{r\uparrow t}\varphi(r, \omega) = \psi(t, \omega)\ \textit{for all}\ t \in (\alpha, u]\ (\mathbf{P}\text{-a.s.}),$$

$$(4.33)\qquad \lim_{r\downarrow t}\psi(r, \omega) = \varphi(t, \omega)\ \textit{for all}\ t \in [\alpha, u)\ (\mathbf{P}\text{-a.s.})$$

PROOF. By the regularity of $(x_t, \mathbf{P}_{t+,x})$ and by Proposition 3.1.B, $\varphi(t, \omega)$ is for **P**-almost all ω right-continuous and has left-hand limits on $[\alpha, u]$. By the co-regularity of $(x_{t-}, \mathbf{P}_{t-,x})$, the function $\psi(t, \omega)$ is **P**-almost surely left continuous and has right-hand limits on the same interval (the latter is proved by considering the martingale $(\psi(t,\omega), \mathscr{F}_{<t}, \mathbf{P})$). We observe that by (1.12), $\mathbf{P}\{\eta \mid \mathscr{F}_{<t+}\} = \varphi(t, \omega)$ (**P**-a.s.), and applying (1.4) to the

right representation $(x_{t-}, \mathbf{P}_{t-,x})$, we have $\mathbf{P}\{\eta \mid \mathscr{F}_{<t}\} = \psi(t, \omega)$ (**P**-a.s.). Hence, for every t,

$$(4.34) \quad \lim_{r \uparrow t} \varphi(r, \omega) = \lim_{r \uparrow t} \mathbf{P}\{\eta \mid \mathscr{F}_{<r+}\} = \mathbf{P}\{\eta \mid \mathscr{F}_{<t}\} = \psi(t, \omega) \quad (\mathbf{P}\text{-a.s.}).$$

Since both sides of (4.34) are almost surely left-continuous, (4.32) follows from (4.34), and (4.33) is proved similarly.

LEMMA 4.4. *Let $\mathscr{A}_t = \mathscr{F}^{\mathbf{P}}_{<t+}$. For any predictable Markov time τ and any $\eta \in \mathscr{F}_{>u}$*

$$(4.35) \qquad \mathbf{P}\{\eta \mid \mathscr{A}_{\tau-}\} = \mathbf{P}_{\tau-, x_{\tau-}}\eta \quad (\mathbf{P}\text{-a.s., } \tau \leqslant u).$$

If η_t is a non-negative recoverable function, then

$$(4.36) \qquad \mathbf{P}\{\eta_\tau \mid \mathscr{A}_{\tau-}\} = \psi^\tau(\tau, x_{\tau-}) \quad (\mathbf{P}\text{-a.s.}),$$

where $\psi^u(t, x) = \mathbf{P}_{t-,x}\eta_u$.

PROOF. Suppose first that η is bounded. Applying (3.7) to the representation $(x_{t+}, \mathbf{P}_{t+,x})$ and using 1.14.C, we have

$$\mathbf{P}\{\eta \mid \mathscr{A}_{\tau-}\} = \lim \mathbf{P}\{\eta \mid \mathscr{A}_{\tau_n}\} = \lim \mathbf{P}_{\tau_n+, x_{\tau_n+}}\eta.$$

(4.35) now follows, by Lemma 4.3. (4.35) can be extended to unbounded functions η by means of 1.11.A. Finally, (4.36) is derived from (4.35) just as (3.8) from (3.7).

LEMMA 4.5.[1] *If the functions $\xi \in \mathscr{F}_{<u}$ and $\eta \in \mathscr{F}_{>u}$ are* **P**-*integrable, then for* **P**-*almost all ω the function $\mathbf{P}^{t+,x_{t+}}\xi$ is right continuous in $[u, \beta)$, while the function $\mathbf{P}_{t-,x_{t-}}\eta$ is left continuous in $(\alpha, u]$.*

PROOF. We denote by $\psi(t, \omega)$ the function that is equal to $\mathbf{P}_{t-,x_{t-}}\eta$ for $t \leqslant u$ and vanishes for $t > u$. According to 1.8.B, if η is bounded, it is strictly predictable. Relying on 1.11.A, we conclude from this that it is predictable for any $\eta \in \mathscr{F}_{>u}$. According to 3.2.E, the left continuity of the function $\mathbf{P}_{t-,x_{t-}}\eta$ will follow if we verify that

$$(4.37) \qquad \psi(\tau_n, \omega) \to \psi(\tau, \omega) \quad (\mathbf{P}\text{-a.s.})$$

for any non-decreasing sequence of predictable Markov times $\tau_n \to \tau$. According to (4.35),

$$(4.38) \qquad \mathbf{P}\{\eta \mid \mathscr{A}_{\tau_n-}\} = \psi(\tau_n, \omega) \quad (\mathbf{P}\text{-a.s., } \tau_n \leqslant u).$$

But the minimal σ-algebra containing all the $\mathscr{A}_{\tau_n-}$ is $\mathscr{A}_{\tau-}$ and by 1.14.C (4.37) follows from (4.38).

The second statement of Lemma 4.5 is proved similarly.

COROLLARY. *Lemma* 4.3 *holds for all* **P**-*integrable* $\eta \in \mathscr{F}_{>u}$.

For the boundedness of η was required only to use the left continuity of $\psi(t, \omega)$ and right continuity of $\varphi(t, \omega)$ (**P**-a.s.). But according to Lemma 4.5, this continuity follows from the **P**-integrability of η.

---

[1] Communicated to the author by S. E. Kuznetsov.

4.8. We now prove Theorem 1.6. We consider a triple $(x_v, \mathbf{P}_{v,x}, \mathbf{P}^{v,x})$ obtained by combining the representations $(x_{t+}, \mathbf{P}_{t+,x}, \mathbf{P}^{t+,x})$ and $(x_{t-}, \mathbf{P}_{t-,x}, \mathbf{P}^{t-,x})$, constructed in Theorem 1.4.

1°. We verify that $(x_v, \mathbf{P}_{v,x})$ is a right Markov representation for $\hat{\mathscr{F}}(v_1, v_2)$. Clearly, x_v is measurable with respect to $\hat{\mathscr{F}}(v_1, v_2)$ for $v_1 < v < v_2$. We now show that

$$\mathbf{P}_{u,x}\xi\eta = \mathbf{P}_{u,x}\xi\mathbf{P}_{v,x_v}\eta \quad \text{for all} \quad u < v \in V,\ \xi \in \hat{\mathscr{F}}(u, v],\ \eta \in \hat{\mathscr{F}}_{>v}, \tag{4.39}$$

$$\mathbf{P}\xi\eta = \mathbf{P}\xi\mathbf{P}_{v,x_v}\eta \quad \text{for all} \quad v \in V,\ \xi \in \hat{\mathscr{F}}_{\leqslant v},\ \eta \in \hat{\mathscr{F}}_{>v}. \tag{4.40}$$

We note that

$$\hat{\mathscr{F}}_{\leqslant v} = \mathscr{F}_{<t},\quad \hat{\mathscr{F}}(u, v] = \mathscr{F}(\hat{u}, t) \quad \text{for} \quad v = t-,$$

$$\hat{\mathscr{F}}_{\leqslant v} \subseteq \mathscr{F}_{<t+},\quad \hat{\mathscr{F}}(u, v] \subseteq \mathscr{F}(\hat{u}, t+) \quad \text{for} \quad v = t+.$$

Since $(x_{t-}, \mathbf{P}_{t-,x})$ satisfies (1.4) and $(x_{t+}, \mathbf{P}_{t+,x})$ satisfies (1.12), (4.40) holds for any v. Next, being absolutely continuous, the two representations $(x_{t-}, \mathbf{P}_{t-,x})$ and $(x_{t+}, \mathbf{P}_{t+,x})$ dominate each other. Therefore, $\mathbf{P}_{u,x} \in \mathbf{K}_{\hat{u}}(x_{t+}, \mathbf{P}_{t+,x})$, and (4.39) follows from (1.13) for $v = t+$. For $v = t-$ the equation is satisfied because $\mathbf{P}_{u,x} \in \mathbf{K}_{\hat{u}}(x_{t-}, \mathbf{P}_{t-,x})$.

Let $\eta \in \hat{\mathscr{F}}_{>u}$ and let $A_t = \{\omega\colon \mathbf{P}_{v,x_v}\eta$ is continuous in v on $[t+, u]\}$. From the corollary to Lemma 4.5 it follows that $\mathbf{P}\,A_t = 1$ for $t \in [\alpha, \hat{u})$. By (1.14) and (1.5) it follows from this that $\mathbf{P}'\,A_t = 1$ for $s \in [\alpha, t)$ and any $\mathbf{P}' \in \mathbf{K}_s(x_{t+}, \mathbf{P}_{t+,x})$. For $s = t$ the corresponding result follows from 1.8.A.

Combining the properties of $(x_v, \mathbf{P}_{v,x})$ with corresponding properties of $(x_v, \mathbf{P}^{v,x})$ (which are valid by symmetry), we conclude that $(x_v, \mathbf{P}_{v,x}, \mathbf{P}^{v,x})$ is a regular two-sided Markov representation of the stochastic system $\mathscr{F}(I)$. By 1.8.C, this representation separates states.

2°. We now prove 1.10.A. The measures $\mathbf{P}_{t-,x}$ and $\mathbf{P}_{t+,y}$ belong to $\mathbf{K}_t(x_{t+}, \mathbf{P}_{t+,x})$. By Lemma 2.1, we need only verify that for $r \in R$, $\varphi \in W_r$ and for $\mathbf{P}$-almost all ω the set

$$\{t\colon\ t \in (\alpha, r),\ \mathbf{P}_{t-,x_{t-}}\varphi(x_r) \neq \mathbf{P}_{t+,x_{t+}}\varphi(x_r)\}.$$

is at most denumerable. But the points of this set are discontinuity points of $\varphi(t, \omega) = \mathbf{P}_{t+,x_{t+}}\varphi(x_r)$. Since the latter has almost surely no discontinuities of the second kind, the number of discontinuity points is at most denumerable.

3°. We now verify that the function $p(u, x; v, y)$ defined in the statement of the theorem is a fundamental density for $(x_v, \mathbf{P}_{v,x}, \mathbf{P}^{v,x})$.

From (1.18), (1.19) and (1.14)–(1.15) we have

$$p(s+, x; t-, y) = \mathbf{P}_{s+,x}p(r-, x_{r-}; t-, y) = \mathbf{P}^{t-,y}p(s+, x; q+, x_{q+}), \tag{4.41}$$

$$(4.42) \qquad p(s-, x; t+, y) = \mathbf{P}_{s-, x} p(r+, x_{r+}; t+, y) = \\ = \mathbf{P}^{t+, y} p(s-, x; q-, x_{q-})$$

(cf. (4.28). From this it follows that the expressions (4.41) and (4.42) do not depend on r and q.

We claim that for any non-neighbouring $u < v \in V$, $\eta \in \mathscr{F}_{>v}$

$$(4.43) \qquad \mathbf{P}_{u, x}\eta = \mathbf{P}p(u, x; v, x_v)\eta.$$

Let $u = s\pm$, $v = t\pm$. For the combinations of signs $++$ and $--$ (4.39) reduces to (1.14). In the case of $+-$, by (4.41), (1.10) and (1.11) the right-hand side of (4.43) is equal to

$$\mathbf{P}\eta\mathbf{P}^{t-, x_{t-}} p(s+, x; q+, x_{q+}) = \mathbf{P}\eta p(s+, x; q+, x_{q+}) = \mathbf{P}_{s+, x}\eta.$$

Finally, in the case of $-+$, by (4.42), (1.10) and (1.14) it is equal to

$$\mathbf{P}\eta\mathbf{P}^{t+, x_{t+}} p(s-, x; q-, x_{q-}) = \mathbf{P}\eta p(s-, x; q-, x_{q-}) = \mathbf{P}_{s-, x}\eta.$$

Thus, (4.43) is proved. It can be shown similarly that for any non-neighbouring $u < v \in V$ and $\xi \in \hat{\mathscr{F}}_{<u}$

$$(4.44) \qquad \mathbf{P}^{v, x}\xi = \mathbf{P}\xi p(u, x_u; v, x).$$

From (4.43) and (4.44) it follows that $p(u, x; v, x_v)$ satisfies the conditions of 1.6.B and is a density for m_{uv} with respect to $m_u \times m_v$. It remains to verify that for any non-neighbouring $u < v < w$

$$(4.45) \qquad \int p(u, x; v, z)\, m_v(dz)\, p(v, z; w, y) = p(u, x; w, y)$$

Let $u = s\pm$, $v = r\pm$, $w = t\pm$. We know already that the equation holds for the combinations of signs $+++$ and $---$. Let $s < \rho < r < \sigma < t$. Using (4.41) we note that, in the case $+--$ the left-hand side of (4.45) is equal to

$$\int \mathbf{P}_{s+, x} p(\rho-, x_{\rho-}; r-, z)\, m_{r-}(dz)\, p(r-, z; t-, y) = \\ = \mathbf{P}_{s+, x} p(\rho-, x_{\rho-}; t-, y) = p(s+, x; t-, y),$$

and in the case $++-$ it is equal to

$$\int p(s+, x; r+, z)\, m_{r+}(dz)\, \mathbf{P}^{t-, y} p(r+, z; \sigma+, x_{\sigma+}) = \\ = \mathbf{P}^{t-, y} p(s+, x; \sigma+, x_{\sigma+}) = p(s+, x; t-, y).$$

The cases $-++$ and $--+$ are dealt with similarly by means of (4.42). Finally, in the case $+-+$, using (4.41) and (4.42) we observe that the left-hand side of (4.45) is equal to

$$\int \mathbf{P}^{r-, z} p(s+, x; \rho+, x_{\rho+})\, m_{r-}(dz)\, \mathbf{P}_{r, -z} p(\sigma+, x_{\sigma+}; t+, y) = \\ = \mathbf{P}\,[\mathbf{P}^{r-, x_{r-}} p(s+, x; \rho+, x_{\rho+})]\,[\mathbf{P}_{r-, x_{r-}} p(\sigma+, x_{\sigma+}; t+, y)] = \\ = \mathbf{P}\mathbf{P}\,[p(s+, x; \rho+, x_{\rho+}) \mid x_{r-}]\, \mathbf{P}\,[p(\sigma+, x_{\sigma+}; t+, y) \mid x_{r-}].$$

Since x_{t-} satisfies 1.2.A, this equation is equivalent to

$$\mathbf{P}p(s+, x; \rho+, x_{\rho+})p(\sigma+, x_\sigma; t+, y) =$$
$$= \mathbf{P}_{s+, x}p(\sigma+, x_{\sigma+}; t+, y) = p(s+, x; t+, y)$$

(here (1.14) and the property 1.6.A of $p(s+, x; t+, y)$) are used). The case $-+-$ is treated similarly.

4°. Finally, let $(\bar{x}_v, \bar{\mathbf{P}}_{v,x}, \bar{\mathbf{P}}^{v,x})$ be any regular two-sided representation of a stochastic system $\hat{\mathscr{F}}(I)$ separating states. Then $(\bar{x}_t, \bar{\mathbf{P}}_{t,x})$ is a right representation of the system $\mathscr{F}(I)$ and by 1.7.E, it dominates the right representations $(x_{t+}, \mathbf{P}_{t+,x})$ and $(x_{t-}, \mathbf{P}_{t-,x})$. By Lemma 2.3 $\mathbf{P}_{t+,x_{t+}} = \bar{\mathbf{P}}_{t+,\bar{x}_{t+}}$ for all $t \in [\alpha, \beta)$ (P-a.s.). By the regularity of $(x_v, \mathbf{P}_{v,x}, \mathbf{P}^{v,x})$ and $(\bar{x}_v, \bar{\mathbf{P}}_{v,x}, \bar{\mathbf{P}}^{v,x})$ it follows from this that $\mathbf{P}_{v,x_v} = \bar{\mathbf{P}}_{v,\bar{x}_v}$ for all $v \in [\alpha+, \beta-)$ (P-a.s.). By similar arguments, $\mathbf{P}^{v,x_v} = \bar{\mathbf{P}}^{v,\bar{x}_v}$ for all $v \in [\alpha_+, \beta-)$ (P-a.s.). The measures $\mathbf{P}_{v,x}$ and $\bar{\mathbf{P}}_{v,x}$ belong to the class $\mathbf{K}_v = \mathbf{K}_{\hat{v}}(\bar{x}_{t+}, \bar{\mathbf{P}}_{t+,x})$, and the measures $\mathbf{P}^{v,x}$ and $\bar{\mathbf{P}}^{v,x}$ to the corresponding class $\mathbf{K}^v$ associated with the left representation $(\bar{x}_{t-}, \bar{\mathbf{P}}^{t-,x})$.

We map the state spaces of the processes x_v and $\bar{x}_v$ into $\mathbf{K}_v \times \mathbf{K}^v$ by the formulae

$$x \to (\mathbf{P}_{v,x}, \mathbf{P}^{v,x}), \quad x \to (\bar{\mathbf{P}}_{v,x}, \bar{\mathbf{P}}^{v,x}).$$

The proof that $(x_v, \mathbf{P}_{v,x}, \mathbf{P}^{v,x})$ and $(\bar{x}_v, \bar{\mathbf{P}}_{v,x}, \bar{\mathbf{P}}^{v,x})$ are equivalent is completed in the same way as that of Lemma 2.3.

Appendix

REGULAR TOPOLOGICAL REPRESENTATIONS

S. E. Kuznetsov

1. We shall show that if a stochastic system $(\mathscr{F}(I), \mathbf{P})$ has an absolutely continuous Markov representation, then we can construct for it in a good topological space a pair of regular Markov representations $(x_t, \mathbf{P})$ and $(y_t, \mathbf{P})$ with trajectories having no discontinuities of the second kind. Moreover, x_t is almost surely continuous from the right and y_t from the left, $x_{t-} = y_t$ and $y_{t+} = x_t$. If we reverse the time in the process y_t, then we obtain a pair of dual Markov processes of the kind that is usually treated in potential theory.

2. Let $(\mathscr{F}(I), \mathbf{P})$ be a stochastic system. Let x_t and y_t be two two-sided Markov representations with a common state space E_t and common transition and co-transition probabilities $\mathbf{P}_{t,x}$ and $\mathbf{P}^{t,x}$ (but, generally speaking, distinct transition and co-transition functions!). We call the pair (x_t, y_t) a regular topological representation if the following conditions are satisfied.

2.A. $\mathscr{E} = \bigcup_t E_t$ is a Borel subset of a complete separable compact metric space and the function $f(t, x) = t$ is continuous.

2.B. The trajectories of the process x_t are **P**-a.s. continuous from the right and have left-hand limits, and the trajectories of y_t are **P**-a.s. left-continuous and have right-hand limits. Moreover, $\lim_{u \downarrow t} y_u = x_t$, $\lim_{s \uparrow t} x_s = y_t$, (**P**-a.s.) for all t.

2.C. The representations $(x_t, \mathbf{P}_{t,x})$ and $(y_t, \mathbf{P}^{t,x})$ are completely regular, and $(x_t, \mathbf{P}^{t,x})$ and $(y_t, \mathbf{P}_{t,x})$ are co-regular.

3. THEOREM. *Suppose that a stochastic system* $(\mathscr{F}(I), \mathbf{P})$ *has an absolutely continuous Markov representation* $\bar{x}_t$. *Then* $\mathscr{F}(I)$ *also has a regular topological representation.*

We construct a topological representation using the representations $(x_{t+}, \mathbf{P}_{t+,x}, \mathbf{P}^{t+,x})$ and $(x_{t-}, \mathbf{P}_{t-,x}, \mathbf{P}^{t-,x})$ constructed in Theorem 1.4. Let $x \in E_{t-}$, $y \in E_{t+}$. We write $(x, y) \in D_t$ if

$$\mathbf{P}_{t-,x} = \mathbf{P}_{t+,y}, \quad \mathbf{P}^{t-,x} = \mathbf{P}^{t+,y}. \tag{1}$$

Let R be a denumerable everywhere dense subset of T, $R_{>t} = R \cap (t, \beta)$, and $R_{<t} = R \cap (\alpha, t)$. By Lemma 2.1 the equations (1) are equivalent to the denumerable system of equations

$$\begin{cases} \mathbf{P}_{t-,x}\varphi(\bar{x}_u) = \mathbf{P}_{t+,y}\varphi(\bar{x}_u), & \varphi \in W_u, \ u \in R_{>t}, \\ \mathbf{P}^{t-,x}\psi(\bar{x}_s) = \mathbf{P}^{t+,y}\psi(\bar{x}_s), & \psi \in W_s, \ s \in R_{<t}. \end{cases} \tag{2}$$

(Here W_u denotes a support system of functions in $\bar{E}_u$, the state space of $\bar{x}_u$). Since the functions in (2) are measurable, D_t is a Borel subset of the product space $Z_t = E_{t-} \times E_{t+}$. Similarly, $\mathscr{D} = \bigcup_t D_t$ is a Borel subset of $\mathscr{Z} = \bigcup_t Z_t$.

We write $D_{t-} = \mathrm{pr}_{E_{t-}} D_t$, $D_{t+} = \mathrm{pr}_{E_{t+}} D_t$. Since the representations $(x_{t+}, \mathbf{P}_{t,x})$ and $(x_{t-}, \mathbf{P}^{t-,x})$ separate states, every point of D_{t-} and D_{t+} has a unique inverse image under projection. Hence, D_{t-} and D_{t+} are one-to-one measurable images of a Borel set, and by [8], §39.V are also Borel sets. Hence

$$E_t = D_t \cup (E_{t-} \setminus D_{t-}) \cup (E_{t+} \setminus D_{t+})$$

is a Borel space.[1] By similar arguments

$$\mathscr{E} = \bigcup_t E_t = \mathscr{D} \cup (\mathscr{E}_- \setminus \mathscr{D}_-) \cup (\mathscr{E}_+ \setminus \mathscr{D}_+)$$

is also a Borel space, where $\mathscr{E}_+ = \bigcup_t E_{t+}$, $\mathscr{E}_- = \bigcup_t E_{t-}$, $\mathscr{D}_+ = \bigcup_t D_{t+}$ and $\mathscr{D}_- = \bigcup_t D_{t-}$.

The spaces E_{t-} and E_{t+} have natural embeddings in E_t (here D_{t-} and D_{t+} are identified). The transition and co-transition probabilities $\mathbf{P}_{t,x}$ and $\mathbf{P}^{t,x}$ can be transferred to E_t without ambiguity. We set $x_t = x_{t+}$ and $y_t = x_{t-}$. Then 2C is satisfied.

We set $\varphi(t) = 0$ for $t \leqslant 0$, $\varphi(t) = t$ for $0 < t < 1$, and $\varphi(t) = 1$ for $t \geqslant 1$. We take it that $\mathbf{P}_{t,x}f(\bar{x}_u) = 0$ for $u \leqslant t$ and $\mathbf{P}^{t,x}f(\bar{x}_s) = 0$ for $s \leqslant t$.

---

[1] If X and Y are two disjoint measurable spaces, then $\Gamma \subset X \cup Y$ is taken to be measurable if $\Gamma \cap X$ and $\Gamma \cap Y$ are.

We take as a denumerable coordinate system in $\mathscr{E}$ the function of the form

$$
(3) \quad \begin{cases} F(t, x) = \varphi(u-t)\,\mathbf{P}_{t,x} f(\bar{x}_u), & u \in R,\ f \in W_u, \\ F(t, x) = \varphi(t-s)\,\mathbf{P}^{t,x} f(\bar{x}_s), & s \in R,\ f \in W_s. \end{cases}
$$

We may assume that all the functions $f \in W_u$ are bounded by 1. Then the coordinate system separates states, and no coordinate exceeds 1. By means of this coordinate system $\mathscr{E}$ is mapped injectively into the Hilbert cube H, and according to [8], §39. V its image in H is a Borel set. Now 2B is a corollary to Theorem 1.6(the regularity of the horizontal representation).

4. We show now that not only the transition and the co-transition probabilities $\mathbf{P}_{t,x}$ and $\mathbf{P}^{t,x}$, but also the fundamental density $p(s, x; t, y)$ transfer in a natural way to E_t. It suffices to prove that

$$
(4) \quad \begin{cases} p(t-, x; u\pm, z) = p(t+, y; u\pm, z) & \text{for } t < u, \quad (x, y) \in D_t, \\ p(s\pm, x; t-, y) = p(s\pm, x; t+, z) & \text{for } s < t, \quad (y, z) \in D_t. \end{cases}
$$

We prove, for example, that $p(t-, x; u-, z) = p(t+, y; u-, z)$ for $t < u$ and $(x, y) \in D_t$. By (1.18) and (1.14), for $t < q < r < u$

$$
(5) \quad p(t+, y; u-, z) = \mathbf{P}p(t+, y; q+, x_{q+})p(r-, x_{r-}; u-, z) = \\ = \mathbf{P}_{t+,y}p(r-, x_{r-}; u-, z).
$$

On the other hand, from 1.6.A and (1.14) it follows easily that

$$
(6) \quad p(t-, x; u-, z) = \mathbf{P}p(t-, x; q-, x_{q-})p(r-, x_{r-}; u-, z) = \\ = \mathbf{P}_{t-,x}p(r-, x_{r-}; u-, z).
$$

For $(x, y) \in D_t$, the right-hand sides of (5) and (6) coincide by definition. The remaining equalities are proved similarly.

Using 1.6.A, (1.18) and (1.19), we can easily verify that the fundamental density $p(s, x; t, y)$ satisfies the Chapman-Kolmogorov equations

$$
\int_{E_t} p(s, x; t, y)\, m_{t-}(dy)\, p(t, y; u, z) = p(s, x; u, z),
$$

$$
\int_{E_t} p(s, x; t, y)\, m_{t+}(dy)\, p(t, y; u, z) = p(s, x; u, z),
$$

where $m_{t-}(dy) = \mathbf{P}\{x_{t-} \in dy\}$, $m_{t+}(dy) = \mathbf{P}\{x_{t+} \in dy\}$. Therefore, $p(s, x; t, y)$ is really a fundamental density in the sense of §1.6.

5. We mention one important case. We assume that for every t and any $s \in R_{<t}$, $u \in R_{>t}$, $\varphi \in W_{u+}$ and $\psi \in W_{s-}$ for **P**-almost all ω the functions $p(t+, x_{t+}; u+, \varphi)$ and $p(s-, \psi; t-, x_{t-})$ are continuous at t. Then by Theorem 1.6 and Lemma 2.1., for every t

$$
(7) \quad \mathbf{P}_{t-, x_{t-}} = \mathbf{P}_{t+, x_{t+}}, \quad \mathbf{P}^{t-, x_{t-}} = \mathbf{P}^{t+, x_{t+}} \quad (\mathbf{P}\text{-a.s.}).
$$

By definition of D_t and E_t, $x_{t-} = x_{t+}$ (**P**-a.s.) in E_t. Thus, the processes x_{t-} and x_{t+} have identical one-dimensional distributions $m_t = m_{t-} = m_{t+}$.

The transition and cotransition functions can be expressed in terms of the fundamental density and the one-dimensional distributions of the process by the formulae

$$p(s, x; t, dy) = p(s, x; t, y)m_t(dy),$$
$$p(s, dx; t, y) = m_s(dx)p(s, x; t, y).$$

Therefore, when (7) holds, the processes x_{t-} and x_{t+} have common transition and cotransition functions that are dual, that is,

$$m_s(dx)p(s, x; t, dy) = p(s, dx; t, y)m_t(dy).$$

References

[1] E. B. Dynkin, *Osnovaniya teorii markovskikh protesssov,* Fitzmagiz., Moscow 1959. Translation: Foundations of the theory of Markov processes, Pergamon Press, Oxford-New York 1960.

[2] E. B. Dynkin, *Markovskie protsessy,* Fizmatgiz, Moscow 1953. Translation: Markov processes, Springer-Verlag, Berlin-Göttingen-Heidelberg 1965.

[3] E. B. Dynkin, Initial and final behaviour of trajectories of Markov processes, Uspekhi Mat. Nauk **26**:4 (1971), 153–172. = Russian Math. Surveys **26**:4 (1971), 165–185.

[4] E. B. Dynkin, Regular Markov processes, Uspekhi Mat. Nauk **28**:2 (1973), 35–64. = Russian Math. Surveys **28**:2 (1973), 33–64.

[5] E. B. Dynkin, Additive functionals of Markov processes and their spectral measures, Dokl. Akad. Nauk SSSR **214** (1974), 1214–1244. = Soviet Math. Dokl. **15** (1974), 330–334.

[6] E. B. Dynkin and S. E. Kuznetsov, Defining functions of Markov processes and corresponding dual regular classes, Dokl. Akad. Nauk SSSR **214** (1974), 25–28. = Soviet Math. Dokl. **15** (1974), 20–23.

[7] J. L. Doob, Stochastic processes, Wiley, New York 1953. Translation: *Veroyatnostnye protsessy,* Izdat. Inost. Lit., Moscow 1956.

[8] K. Kuratowski, Topologie, 4th ed., 2 vols, Polish Acad. Sci., Warsaw 1958. Translation: Topology, 2 vols, Academic Press, New York 1967-69. Russian translation: *Topologiya*, vol. 1, Mir, Moscow 1966.

[9] P.-A. Meyer, Probability and potentials, Blaisdell, Toronto 1966. Translation: *Veroyatnost' i potentsialy*, Mir, Moscow 1973.

[10] C. Dellacherie, Capacités et processus stochastiques, Springer-Verlag, Berlin-Heidelberg-New York 1972.

[11] J. L. Snell, Applications of martingale system theorems, Trans. Amer. Math. Soc. **73** (1952), 293–312.

Received by the Editors, 26 March 1974

Translated by G. and R. L. Hudson

Reprinted from E.B. Dynkin, "Sufficient Statistics and Extreme Points" *Annals of Probability*, volume 6, pages 705–730.

SPECIAL INVITED PAPER

SUFFICIENT STATISTICS AND EXTREME POINTS[1]

A convex set M is called a simplex if there exists a subset M_e of M such that every $P \in M$ is the barycentre of one and only one probability measure μ concentrated on M_e. Elements of M_e are called extreme points of M. To prove that a set of functions or measures is a simplex, usually the Choquet theorem on extreme points of convex sets in linear topological spaces is cited. We prove a simpler theorem which is more convenient for many applications. Instead of topological considerations, this theorem makes use of the concept of sufficient statistics.

1. Introduction.

1.1. If M is a simplex in a finite-dimensional linear space, the set M of extreme points is finite, and to say that $\bar{P}$ is a barycentre of a probability measure μ concentrated on M_e means that

$$\bar{P} = \sum_{P \in M_e} \mu(P)P ,$$

where $\mu(P) \geqq 0$ for all $P \in M_e$ and $\sum_{P \in M_e} \mu(P) = 1$. The concept of a barycentre can be naturally extended to probability measures on spaces of functions and measures. Simplexes in such spaces play an important role in various fields of mathematics. Here are some examples:

1.1.A. The set of all probability measures invariant with respect to a measurable transformation T of a measurable space $(\Omega, \mathscr{F})$. (Extreme points are ergodic measures.)

1.1.B. The set of all Gibbs states specified by a given family of conditional distributions.

1.1.C. The set of all symmetric probability measures on a product space (with infinite number of factors). Extreme points are product measures.

1.1.D. The set of all Markov processes with a given transition function.

1.1.E. The set of all stationary probability distributions for a given stationary transition function.

1.1.F. The class of all normed excessive functions associated with a given transition function. A particular case is the class of all positive superharmonic functions h in a domain D of a Euclidean space normed by the condition $h(c) = 1$,

---

Received August 3, 1977.

[1] Research supported by NSF Grant No. MCS 77-03543.

AMS 1970 *subject classifications.* Primary 60-02; Secondary 60J50, 60K35, 82A25, 28A65.

Key words and phrases. Extreme points, sufficient statistics, Gibbs states, ergodic decomposition of an invariant measure, symmetric measures, entrance and exit laws, excessive measures and functions.

where c is a fixed point of D. This class is associated with the Brownian motion in D.

These classes were treated by many authors from different points of view. We mention here the works of Krylov and Bogolubov [13] (related to the class 1.1.A); Dobrushin [2] (class 1.1.B); de Finetti [9], [10]; and Hewitt–Savage [11] (1.1.C); Martin [15]; Doob [3]; and Hunt [12] (1.1.F).

In the present paper, all these classes of measures and functions and some others will be investigated by constructing suitable sufficient statistics.

1.2. The role of a special type of sufficient statistics (we call them H-sufficient) is revealed by Theorem 3.1. This theorem was first published in 1971 ([4], Section 2) in a slightly different form and without explicitly mentioning sufficient statistics. The theorem was applied to the class of all Markov processes with a given transition function (class 1.1.D) in [4] and to excessive measures and excessive functions (1.1.F) in [5].

We start with general definitions of a barycentre, extreme points, etc., in Section 2. Relations between H-sufficient statistics and decomposition into extreme points are investigated in Section 3. The main method of constructing H-sufficient statistics is a special kind of passage to the limit which is studied in Section 4. The rest of the paper is devoted to various applications. In particular, Sections 9—12 contain an improved version of the results on Markov processes published in [4] and [5]. The presentation is self-contained, but we refer to [5] for some technical details.

2. Convex measurable spaces.

2.1. Let $(M, \mathscr{B}_M)$ be an arbitrary measurable space. We say that a *convex structure* is introduced into M if a point P_μ, the barycentre of μ, is associated with each probability measure μ on $\mathscr{B}_M$. A space $(M, \mathscr{B}_M)$ provided with such a structure will be called a *convex measurable* space.

We say that P is an *extreme point* of M, and write $P \in M_e$, if P is not a barycentre of any measure μ except the measure concentrated on P. A convex measurable space M is called a *simplex* if M_e is measurable and each $P \in M$ is a barycentre of one and only one probability measure μ concentrated on M_e.

Let $(M, \mathscr{B}_M)$ and $(M', \mathscr{B}_{M'})$ be convex measurable spaces and let T be a mapping of M into M'. We say that *T preserves the convex structure* if T is measurable and transforms the barycentre of a measure μ into the barycentre of the measure

$$\mu'(\Gamma) = \mu(T^{-1}\Gamma), \quad \Gamma \in \mathscr{B}_{M'}.$$

We say that T is an *isomorphism* if it is invertible and T and T^{-1} preserve convex structure.

An axiomatic theory of convex measurable spaces can be developed but our task is rather an analysis of concrete spaces.

2.2. Let M be a collection of positive functions on an arbitrary set Z. (By a positive function we mean a function with values in an extended real half-line $[0, +\infty]$.) Let $\mathscr{B}_M$ be an arbitrary σ-algebra in M with the property:

2.2.A. For each $z \in Z$, the function $F_z(\varphi) = \varphi(z)$ is $\mathscr{B}_M$-measurable.

Let μ be a probability measure on $\mathscr{B}_M$. We define a *barycentre* φ_μ of μ by the formula

$$\varphi_\mu(z) = \int_M \varphi(z)\mu(d\varphi) . \tag{2.1}$$

If M contains the barycentres of all probability measures, it is a convex measurable space.

A measurable structure in M is called *natural* if it is determined by the minimal σ-algebra $\mathscr{B}_M$ with the property 2.2.A. Unless otherwise stipulated we consider in M the natural measurable structure, and we always consider in M the convex structure defined by formula (2.1).

Formula (2.1) makes sense also for finite nonprobabilistic measures μ. In this case, we call φ_μ a *generalized barycentre* of μ. If M contains all generalized barycentres, we say that M is a *convex cone.*

2.3. Now let M be a set of probability measures on a measurable space $(\Omega, \mathscr{F})$. The set M can be considered also as a class of positive functions on $\mathscr{F}$, and we can apply all the definitions of Subsection 2.2.

If M is a simplex, the formula

$$\bar{P}(A) = \int_{M_e} P(A)\mu(dP) \tag{2.2}$$

establishes a one-to-one correspondence between M and the set of all probability measures on M_e.

We consider one example. Let $M(\mathscr{F})$ be the class of all probability measures on a σ-algebra $\mathscr{F}$. It is easy to check, step by step, that:

(i) $M(\mathscr{F})$ is convex.

(ii) Measures $Q^\omega(A) = 1_A(\omega)$, $A \in \mathscr{F}$ are extreme points of $M(\mathscr{F})$.

(iii) Each $P \in M(\mathscr{F})$ is a barycentre of a measure μ defined by formula

$$\mu(\Gamma) = P\{\omega : Q^\omega \in \Gamma\} . \tag{2.3}$$

This measure is concentrated on the set $M_e(\mathscr{F})$ of extreme points of $M(\mathscr{F})$.

(iv) If P is an extreme point, then $P = Q^\omega$ for some ω.

(v) If μ is a measure concentrated on $M_e(\mathscr{F})$ and P is a barycentre of μ, then μ and P satisfy (2.3).

(vi) $M(\mathscr{F})$ is a simplex.

We prove all these statements in a much more general situation in Section 3.

2.4. We shall use the following abbreviations. If f is a function and $\mathscr{F}$ is a σ-algebra, then the expression $f \in \mathscr{F}$ means that f is $\mathscr{F}$-measurable and bounded. An expression Pf (or $P(f)$) means an integral of f with respect to a measure P.

Let M be a class of probability measures on $(\Omega, \mathscr{F})$. A set A is called *M-null* if $A \in \mathscr{F}$ and $P(A) = 0$ for all $P \in M$. We say that $A, B \in \mathscr{F}$ are *P-equivalent* if

$1_A = 1_B$ a.s. P. Two σ-algebras $\mathscr{A}$, $\mathscr{B} \subset \mathscr{F}$ are *M-equivalent* if, for each $P \in M$, every $A \in \mathscr{A}$ is P-equivalent to a $B \in \mathscr{B}$ and vice versa.

3. *H*-sufficiency and the decomposition into extreme points.

3.1. Let M be an arbitrary class of probability measures on a measurable space $(\Omega, \mathscr{F})$. We say that M is *separable* if $\mathscr{F}$ contains a countable family $\mathscr{A}$ separating the measures in M (which means that for each pair of different elements P_1, P_2 of M there exists $A \in \mathscr{A}$ such that $P_1(A) \neq P_2(A)$). The class $M(\mathscr{F})$ is separable if $\mathscr{F}$ is *countably generated* (i.e., generated by a countable family of sets).

A σ-algebra $\mathscr{F}^0 \subset \mathscr{F}$ is called *sufficient* for M if all measures $P \in M$ have a common conditional distribution relative to $\mathscr{F}^0$; in other words, if for each $\omega \in \Omega$ there exists a probability measure Q^ω on $\mathscr{F}$ such that, for each A, $Q^\omega(A)$ is $\mathscr{F}$-measurable and

$$P(A \mid \mathscr{F}^0) = Q^\omega(A) \quad \text{a.s.} \quad P \qquad \text{for all} \quad P \in M . \tag{3.1}$$

A sufficient σ-algebra will be called *H-sufficient* if, in addition,

$$Q^\omega \in M \quad \text{a.s.} \quad M \tag{3.2}$$

(which means that $P(Q^\omega \in M) = 1$ for all $P \in M$).

If $\mathscr{F}^1$ is M-equivalent to $\mathscr{F}^0$ and if $\mathscr{F}^0$ is sufficient (H-sufficient) for M, then so is $\mathscr{F}^1$.

THEOREM 3.1. *Let $\mathscr{F}^0$ be an H-sufficient σ-algebra for a separable class M. Then the set M_e of extreme points of M is measurable and each $P \in M$ is a barycentre of one and only one probability measure μ_P concentrated on M_e. If M is convex, it is a simplex.*

Let Q^ω be measures satisfying (3.1) *and* (3.2). *Then M_e is a subset of a set $\{Q^\omega\}$ and the measure μ_P is given by formula*

$$\mu_P(\Gamma) = P\{\omega : Q^\omega \in \Gamma\} . \tag{3.3}$$

A measure $P \in M$ belongs to M_e if and only if

$$P\{\omega : Q^\omega = P\} = 1 . \tag{3.4}$$

PROOF. 1°. We start with the following elementary observation: If P is any probability measure on a σ-algebra $\mathscr{F}$ and if $\mathscr{F}^0$ is any subalgebra of $\mathscr{F}$, then the conditions (i), (ii), (iii) are equivalent:

(i) P is trivial on $\mathscr{F}^0$.

(ii) Each $\mathscr{F}^0$-measurable function Z is constant a.s. P.

(iii) $P\{P(A \mid \mathscr{F}^0) \neq P(A)\} = 0$ for each $A \in \mathscr{F}$.

2°. Denote by M_0 the set of all measures $P \in M$ which are trivial on $\mathscr{F}^0$. According to 1°, M_0 can be described by the condition (iii). Taking into account (3.1), we rewrite (iii) in the form

$$P\{Q^\omega(A) \neq P(A)\} = 0 \qquad \text{for all} \quad A \in \mathscr{F} . \tag{3.5}$$

Let $\mathscr{A}$ be a countable family of sets separating measures of M. Obviously (3.5) implies that

$$P\{Q^\omega(A) \neq P(A) \text{ for all } A \in \mathscr{A}\} = 0 . \tag{3.6}$$

Since P and Q^ω belong to M, (3.6) implies that

$$P\{Q^\omega \neq P\} = 0 . \tag{3.7}$$

It is clear that (3.5) follows from (3.7); hence each of the conditions (3.5), (3.6) and (3.7) characterizes the set M_0. The condition (3.5) can be rewritten also in the following form:

$$f_A(P) = 0 \qquad \text{for all} \quad A \in \mathscr{A} , \tag{3.8}$$

where

$$f_A(P) = \int_\Omega Q^\omega(A)^2 P(d\omega) - P(A)^2 = \int [Q^\omega(A) - P(A)]^2 P(d\omega) . \tag{3.9}$$

Evidently, f_A is $\mathscr{B}_M$-measurable. Therefore $M_0 \in \mathscr{B}_M$. It follows from (3.7) that for each $P \in M_0$ there exists $\omega \in \Omega$ such that $P = Q^\omega$.

3°. Now we prove that

$$Q^\omega \in M_0 \quad \text{a.s.} \quad M . \tag{3.10}$$

It follows from (3.1) that $Q^\omega Y = P(Y|\mathscr{F}^0)$ a.s. P. Setting $Y_A = Q^\omega(A)^2$, we conclude from (3.9) that

$$f_A(Q^\omega) = Q^\omega Y_A - Y_A = P(Y_A|\mathscr{F}^0) - Y_A ,$$

and hence

$$Pf_A(Q^\omega) = 0 . \tag{3.11}$$

But it is clear from (3.9) that $f_A \geqq 0$. Therefore (3.11) implies that $f_A(Q^\omega) = 0$ a.s. P. We see that, for almost all ω, the measure Q^ω satisfies the condition (3.8) which implies that $Q^\omega \in M_0$.

4°. Let a measure μ_P be defined by formula (3.3). Then the formula

$$\int_M F(\tilde{P})\mu_P(d\tilde{P}) = \int_\Omega F(Q^\omega)P(d\omega) \tag{3.12}$$

holds for indicator functions $F = 1_\Gamma$, $\Gamma \in \mathscr{B}_M$. Standard arguments show that (3.12) is true for all bounded $\mathscr{B}_M$-measurable functions F. For $F(\tilde{P}) = \tilde{P}(A)$, $A \in \mathscr{F}$ the right side of (3.12) is equal to $P(A)$. Thus P is a barycentre of μ_P. According to 3°, μ_P is concentrated on M_0.

5°. Now let $P \in M$ be a barycentre of a measure μ concentrated on M_0. For every $\Gamma \subset M_0$, $\Gamma \in \mathscr{B}_M$

$$P\{Q^\omega \in \Gamma\} = \int_{M_0} \tilde{P}(Q^\omega \in \Gamma)\mu(d\tilde{P}) . \tag{3.13}$$

The left side is equal to $\mu_P(\Gamma)$. By (3.7) $\tilde{P}(Q^\omega \in \Gamma) = 1_\Gamma(\tilde{P})$ for $\tilde{P} \in M_0$. Therefore the right side of (3.13) is equal to

$$\int_{M_0} 1_\Gamma(\tilde{P})\mu(d\tilde{P}) = \mu(\Gamma) .$$

Hence $\mu_P = \mu$.

6°. Let $P \in M_e$. According to 4°, P is a barycentre of μ_P. Therefore μ_P is concentrated on P which means that $P\{Q^\omega \neq P\} = 0$, and $P \in M_0$ by 2°.

7°. Now let $\bar{P} \in M_0$ be a barycentre of a measure μ on M. According to 2°, $\bar{P}\{Q^\omega \neq \bar{P}\} = 0$. Hence μ is concentrated on the set $M' = \{P\colon P(Q^\omega \neq \bar{P}) = 0\}$. But if $P \in M'$, $C \in \mathscr{F}$, then $P(C) = PP(C \mid \mathscr{F}^0) = \int Q^\omega(C)P(d\omega) = \bar{P}(C)$. Therefore $M' = \{\bar{P}\}$ and μ is concentrated on $\bar{P}$. This proves that $\bar{P} \in M_e$.

3.2.

THEOREM 3.2. *Let a separable class M have an H-sufficient σ-algebra and let $\mathscr{F}^1$ be the class of all sets $A \in \mathscr{F}$ with the following property*:

$$P(A) = 0 \quad \text{or} \quad P(A) = 1 \quad \text{for all} \quad P \in M_e\,. \tag{3.14}$$

Then a σ-algebra $\mathscr{F}^0$ is H-sufficient for M if and only if it is M-equivalent to $\mathscr{F}^1$.

PROOF. We need only to prove that each H-sufficient σ-algebra $\mathscr{F}^0$ is M-equivalent to $\mathscr{F}^1$. By Theorem 3.1, $\mathscr{F}^0 \subset \mathscr{F}^1$. Therefore it is sufficient to construct, for every fixed $P \in M$, $A \in \mathscr{F}^1$, a set $B \in \mathscr{F}^0$ which is P-equivalent to A. A function $Q^\omega(A)$ is P-equivalent to a $\mathscr{F}^0$-measurable function f. Sets $B = \{\omega : f(\omega) = 1\}$ and $C = \{\omega : f(\omega) = 0\}$ belong to $\mathscr{F}^0$, and

$$1_B + 1_C = 1 \quad \text{a.s.} \quad P\,,$$

$$P(BA) = P1_B Q^\omega(A) = P(B)\,, \qquad P(CA) = P1_C Q^\omega(A) = 0\,.$$

Hence $1_A = 1_A 1_B = 1_B$ a.s. P. Our theorem is proved.

Now suppose that a class M is a simplex and let $\mathscr{F}^1$ be defined by (3.14). It is clear that

$$P(A \mid \mathscr{F}^1) = P(A) \qquad \text{for each} \quad P \in M_e\,. \tag{3.15}$$

Therefore $\mathscr{F}^1$ is H-sufficient for M_e (and consequently for M) if and only if a measurable mapping $\omega \to Q^\omega$ of $(\Omega, \mathscr{F}^1)$ into M_e exists such that $P(Q^\omega = P) = 1$ for all $P \in M_e$. In this case, every two measures of M_e are singular on $\mathscr{F}^1$ with respect to each other. If M_e is at most countable, this condition is not only necessary but also sufficient: It implies the existence of decomposition of Ω into the sets $\Omega_P \in \mathscr{F}^1$, $P \in M_e$ with the property $P(\Omega_P) = 1$, and the mapping Q^ω can be defined by formula $Q^\omega = P$ for $\omega \in \Omega_P$.

3.3. We discuss now the concept of H-sufficiency from a slightly different, more algebraic point of view.

A real-valued function $Q^\omega(A) = Q(\omega, A)$, $\omega \in \Omega$, $A \in \mathscr{F}$ is called a *Markov kernel* if, for each $\omega \in \Omega$, $Q(\omega, \cdot)$ is a probability measure and, for each $A \in \mathscr{F}$, $Q(\cdot, A)$ is an $\mathscr{F}$-measurable function. A linear operator on the space of bounded $\mathscr{F}$-measurable functions and a linear operator on the space $M(\mathscr{F})$ of all probability measures are associated with every Markov kernel Q. We denote them by the same letter and call them *Markov operators*. They are defined by the formulas

$$Qf(\omega) = \int Q(\omega, d\omega')f(\omega') = Q^\omega(f)\,, \tag{3.16}$$

and

$$(PQ)(A) = \int P(d\omega)Q(\omega, A). \tag{3.17}$$

We shall consider the first operator also on unbounded functions f (in this case Qf is defined only on a part of Ω). The second operator can be extended too: the formula (3.17) makes sense not only for $P \in M(\mathscr{F})$ but also for $P \in M(\mathscr{F}')$ if $\mathscr{F}'$ is a σ-algebra with the property that $Qf \in \mathscr{F}'$ for all $f \in \mathscr{F}$. Two Markov operators Q and Q' are called *M-equivalent* if $Qf = Q'f$ a.s. M for all $f \in \mathscr{F}$.

We say that a set $A \in \mathscr{F}$ is *Q-invariant* if $Q1_A = 1_A$.

LEMMA 3.1. *If all sets of a σ-algebra $\mathscr{F}^0$ are Q-invariant, then*

$$Q(gf) = gQf \tag{3.18}$$

for each $f \in \mathscr{F}$, $g \in \mathscr{F}^0$ and

$$P\{f \mid \mathscr{F}^0\} = P\{Qf \mid \mathscr{F}^0\} \quad \text{a.s.} \quad P \tag{3.19}$$

for every Q-invariant measure P and every $f \in \mathscr{F}$.

PROOF. It suffices to check (3.18) for $f = 1_A$, $g = 1_B$ where $A \in \mathscr{F}$, $B \in \mathscr{F}^0$. In this case

$$Q(gf) - gQf = (1 - g)Q(gf) - gQ[(1 - g)f]$$

and

$$0 \leqq (1 - g)Q(gf) \leqq (1 - g)Qg = 0,$$
$$0 \leqq gQ[(1 - g)f] \leqq g(1 - g) = 0.$$

Formula (3.19) follows immediately from (3.18).

A Markov operator Q is called a *sufficient statistic* for M if there exists a σ-algebra $\mathscr{F}^0 \subset \mathscr{F}$ such that

$$P(f \mid \mathscr{F}^0) = Qf \quad \text{a.s.} \quad P \tag{3.20}$$

for all $P \in M$ and all $f \in \mathscr{F}$. If, in addition, (3.2) holds, we say that Q is *H-sufficient* for M. Obviously (3.20) is equivalent to (3.1).

If Q is a sufficient (or an H-sufficient) statistic for M, then so are all operators M-equivalent to Q.

THEOREM 3.3. *If a convex separable class M has an H-sufficient statistic, then there exists an H-sufficient statistic Q, such that*

$$Q(fQg) = QfQg \qquad \text{for all} \quad f, g \in \mathscr{F} \tag{3.21}$$

and M coincides with the class of all Q-invariant measures.

Every Markov operator Q with the property (3.21) is H-sufficient for the class M of all Q-invariant measures. The corresponding H-sufficient σ-algebra $\mathscr{F}^0$ can be defined as the collection of all Q-invariant sets. A mapping $P \to P^0$, where P^0 is the restriction of the probability measure P to $\mathscr{F}^0$, is an isomorphism of M onto $M(\mathscr{F}^0)$. The inverse mapping is given by the formula $P = P^0Q$.

PROOF. 1°. Let $\tilde{Q}^\omega$ be an H-sufficient statistic for M. By Theorem 3.1 $\Omega_1 = \{\omega : Q^\omega \bar{\in} M_e\}$ is an M-null set. Hence an operator

$$\begin{aligned} Qf &= \tilde{Q}f \quad \text{on} \quad \Omega_1^c, \\ &= \tilde{Q}f(\omega^*) \quad \text{on} \quad \Omega_1, \end{aligned}$$

where ω^* is a fixed point of Ω_1^c, is H-sufficient for M too.

By (3.4), for all $\omega_1 \in \Omega_1^c$, $Q^{\omega_1}\{\omega : Q^\omega = Q^{\omega_1}\} = 1$, and

$$\begin{aligned} Q(fQg)(\omega_1) &= \int f(\omega)Q^\omega(g)Q^{\omega_1}(d\omega) = \int f(\omega)Q^{\omega_1}(g)Q^{\omega_1}(d\omega) \\ &= Q^{\omega_1}(g)Q^{\omega_1}(f) = Qf(\omega_1)Qg(\omega_1) . \end{aligned}$$

2°. It follows from (3.20) that $PQf = Pf$ for all $P \in M$, $f \in \mathscr{F}$. Therefore all $P \in M$ are Q-invariant. On the other hand, if P is Q-invariant, then

$$P(A) = \int P(d\omega)Q^\omega(A)$$

and $P \in M$ since $Q^\omega \in M$ for all $\omega \in \Omega$.

3°. Let Q be a Markov operator with the property (3.21) and let $\mathscr{F}^0$ be the totality of all Q-invariant sets. It is easy to see that $\mathscr{F}^0$ is a σ-algebra. By Lemma 3.1, all functions $f \in \mathscr{F}^0$ are Q-invariant. To prove the converse, we denote by H the class of all measurable transformations Φ of the real line such that $\Phi(f)$ is Q-invariant for every Q-invariant f. The class H contains linear functions and is closed under addition and monotone convergence. By virtue of (3.21), it is closed also under multiplication. Therefore it contains all bounded Borel functions, in particular, functions $\Phi(u) = 1_{u<c}$ for all constant c. Hence for each Q-invariant f, the sets $\{\omega : f(\omega) > c\}$ belong to $\mathscr{F}^0$, and f is $\mathscr{F}^0$-measurable.

4°. Setting $f = 1$ in (3.21), we see that $Q^2 = Q$. Hence $Qf \in \mathscr{F}^0$ for all $f \in \mathscr{F}$. The identity (3.21) implies that $Q(gf) = gQf$ for $f \in \mathscr{F}$, $g \in \mathscr{F}^0$. Hence, for each Q-invariant measure P,

$$P(gQf) = PQ(gf) = P(gf) ,$$

and (3.20) is satisfied; Q is a sufficient statistic for the class M of all Q-invariant measures and $\mathscr{F}^0$ is the corresponding sufficient σ-algebra. On the other hand, the identity $Q^2 = Q$ implies that $Q^\omega \in M$ of all ω, and Q is H-sufficient.

Since $Qf \in \mathscr{F}^0$ for all $f \in \mathscr{F}$, an equality $PQ = P$ implies that $P^0Q = P$ where P^0 is a restriction of $P \in M$ to $\mathscr{F}^0$. Obviously $P^0Q \in M$ for every $P^0 \in M(\mathscr{F}^0)$. Therefore we have a one-to-one correspondence between M and $M(\mathscr{F}^0)$. It is easy to check that this correspondence is an isomorphism in the sense of Subsection 2.1.

3.4. We shall prove that under certain circumstances sufficiency implies the H-sufficiency.

A family of Markov operators V_t satisfying the condition $V_sV_t = V_{s+t}$ for all s, t is called a one-parameter semigroup if t takes values on the positive real half-line, and it is called a one-parameter group if t takes values on the real line.

We say that V_t is measurable if, for each $f \in \mathscr{F}$, the function $V_t f(\omega)$ is measurable with respect to the pair t, ω (the measurable structure on Ω is given by $\mathscr{F}$ and on the real line by the σ-algebra $\mathscr{B}$ of all Borel sets).

THEOREM 3.4. *Let $\mathscr{V}$ be a finite or countable family of Markov operators or a measurable one-parameter semigroup or group in* $(\Omega, \mathscr{F})$ *and let $\mathscr{F}$ be countably generated. Suppose that $\mathscr{F}^0 \subset \mathscr{F}$ is sufficient for the class M of all $\mathscr{V}$-invariant measures and* (3.19) *holds for all $P \in M$, $Q \in \mathscr{V}$. Then $\mathscr{F}^0$ is H-sufficient for M.*

PROOF. Consider a Markov operator Q satisfying condition (3.20). To prove (3.2), we need only to check that for each $P \in M$ and each $f \in \mathscr{F}$

$$QVf = Qf \quad \text{for all} \quad V \in \mathscr{V} \quad \text{a.s.} \quad P. \tag{3.22}$$

(Indeed (3.22) implies that, for almost all ω, all measures $Q^\omega V$, $V \in \mathscr{V}$ coincide with Q^ω on a countable family of sets separating measures of $M(\mathscr{F})$ and therefore coincide everywhere.)

It follows from (3.19) that

$$QVf = Qf \quad \text{a.s.} \quad P \qquad \text{for every} \quad V \in \mathscr{V}. \tag{3.23}$$

If $\mathscr{V}$ is at most countable, then (3.23) implies (3.22) and our theorem is proved.

In the case of a Markov semigroup or a group, we consider the set $A = \{(t, \omega) : Q^\omega V_t = Q^\omega\}$. It follows from (3.23) that for each t, $P\{\omega : (t, \omega) \in A\} = 1$. The set A belongs to $\mathscr{B} \times \mathscr{F}$. By Fubini's theorem there exists a set Ω_1 such that $P(\Omega_1) = 1$ and, if $\omega \in \Omega_1$, then, for almost all t, $(t, \omega) \in A$, that is, $Q^\omega V_t = Q^\omega$. Taking into account that $V_s V_t = V_{s+t}$ for all s, t, we easily prove that $Q^\omega V_t = Q^\omega$ for all $\omega \in \Omega_1$ and all t.

REMARK. Theorem 3.4 and its proof are valid for a group $\mathscr{G}$ of Markov operators if there exists a σ-algebra $\mathscr{B}_\mathscr{G}$ in $\mathscr{G}$ and a σ-finite measure λ on $\mathscr{B}_\mathscr{G}$ such that: (i) $Vf(\omega)$ is $\mathscr{B}_\mathscr{G} \times \mathscr{F}$-measurable for each $f \in \mathscr{F}$; (ii) $\lambda(VB) = \lambda(B)$ for each $B \in \mathscr{B}_\mathscr{G}$ and each $V \in \mathscr{G}$.

3.5. Let $\mathscr{F}_1$ and $\mathscr{F}_2$ be sufficient σ-algebras for a class M and let Q_1 and Q_2 be correspondent sufficient statistics. It is easy to see that $Q_1 Q_2 = Q_2 Q_1$ a.s. M for all $f \in \mathscr{F}$ if and only if $\mathscr{F}_1$ and $\mathscr{F}_2$ are conditionally independent given $\mathscr{F}^0 = \mathscr{F}_1 \cap \mathscr{F}_2$. In this case $\mathscr{F}^0$ is a sufficient σ-algebra for M and $Q_{12} = Q_1 Q_2$ and $Q_{21} = Q_2 Q_1$ are corresponding sufficient statistics.

Now let M be a convex class and let Q_1 and Q_2 be H-sufficient. The set $\Omega_0 = \{\omega' : Q_2^{\omega'} \bar{\in} M\}$ is M-null and therefore

$$Q_{12}^\omega(A) = \int_{\Omega_0^c} Q_1^\omega(d\omega') Q_2^{\omega'}(A) .$$

Hence $Q_{12}^\omega \in M$ if $Q_1^\omega \in M$, and Q_{12} is H-sufficient for M.

4. Asymptotic sufficiency.

4.1. We say that a sequence of Markov operators Q_n converges M-almost surely to a Markov operator Q and write $Q_n \to Q$ a.s. M if for each $P \in M$ and each $f \in \mathscr{F}$

$$Qf = \lim Q_n f \quad \text{a.s.} \quad P. \tag{4.1}$$

A sequence Q_n is called an *asymptotically sufficient* statistic for M if there exists a sufficient statistic Q such that $Q_n \to Q$ a.s. M. If Q is H-sufficient, we say that Q_n is *asymptotically H-sufficient*.

To prove that a sequence Q_n is asymptotically sufficient, we use a concept of a support system.

4.2. A countable family W of bounded measurable functions in a measurable space $(\Omega, \mathscr{F})$ is called a *support system* if the following two conditions are satisfied:

4.2.A. If μ_n is a sequence of probability measures on $\mathscr{F}$ and if $\lim \int f\, d\mu_n = l(f)$ exists for each $f \in W$, then there is a probability measure μ such that $l(f) = \int f\, d\mu$ for all $f \in W$.

4.2.B. If a class H of real-valued functions contains W and is closed under addition, multiplication by constants, and bounded convergence, then H contains all bounded measurable functions. (We say that f_n converges boundedly to f if f_n converges pointwise to f and all the functions f_n are uniformly bounded.)

A measurable space $(\Omega, \mathscr{F})$ will be called a *B-space* if there exists a support system in $(\Omega, \mathscr{F})$. The unit interval $I = [0, 1]$ with the Borel measurable structure is an example of a B-space: a support system is formed by functions $1, x, x^2, \cdots, x^n, \cdots$.

A measurable space $(\Omega, \mathscr{F})$ is called a *Borel space* if it is isomorphic to a Borel subset of a complete separable metric space. It is well known (see, e.g., [7] or [19]) that all uncountable Borel spaces are isomorphic. By this fact it is easy to prove that all Borel spaces are B-spaces.

It follows from 4.2.B that a support system generates σ-algebra $\mathscr{F}$. Therefore, for any B-space $(\Omega, \mathscr{F})$, the σ-algebra $\mathscr{F}$ is countably generated and $M(\mathscr{F})$ is separable.

4.3.

LEMMA 4.1. *Let $(\Omega, \mathscr{F})$ be a B-space and let*

$$P\{f \mid \mathscr{F}^0\} = \lim_{n\to\infty} Q_n f \quad \text{a.s.} \quad P \tag{4.2}$$

for every $P \in M$ and all $f \in \mathscr{F}$. Then Q_n is asymptotically sufficient for M and $\mathscr{F}^0$ is a sufficient σ-algebra for M.

PROOF. Put $\omega \in \Omega'$ if $\lim Q_n{}^\omega(f) = l^\omega(f)$ exists for all elements f of a support system W. If $\omega \in \Omega'$, then, by 4.1.A, there exists a probability measure Q^ω such that $Q^\omega(f) = l^\omega(f)$ for all $f \in W$. It follows from (4.2) that $P(\Omega') = 1$ and that

$$P\{f \mid \mathscr{F}^0\} = Q^\omega(f) = Qf(\omega) \quad \text{a.s.} \quad P \tag{4.3}$$

for all $f \in W$, $P \in M$. By 4.1.B, (4.3) holds for all $f \in \mathscr{F}$. Therefore Q is sufficient for M. It follows from (4.2) and (4.3) that $Q_n \to Q$ a.s. M.

4.4. It follows from Theorem 3.1 that, if Q_n is an asymptotically H-sufficient statistic for M, then

$$(4.4) \qquad Pf = \lim Q_n f \quad \text{a.s.} \quad P$$

for $P \in M_e$ and $f \in \mathscr{F}$. Formula (4.4) is valid also for all unbounded functions f for which (4.2) is true. In most applications, (4.2) and (4.4) hold for all P-integrable f.

Let us fix an arbitrary countable family W of bounded $\mathscr{F}$-measurable functions and define a convergence of measures by the condition that $P_n \to P$ if $P_n(f) \to P(f)$ for all $f \in W$. The formula (4.4) implies that each P in M_e is the limit of Q_n^ω for some ω.

4.5.

LEMMA 4.2. *Let M be a class of probability measures on a B-space $(\Omega, \mathscr{F})$. If $\mathscr{F}_1$, $\mathscr{F}_2$ are sufficient σ-algebras for M, then $\mathscr{F}^0 = \bar{\mathscr{F}}_1 \cap \bar{\mathscr{F}}_2$ is also sufficient for M. (Here $\bar{\mathscr{F}}_i$ is the minimal σ-algebra which contains $\mathscr{F}_i$ and all M-null sets.) If V_i is a sufficient statistic corresponding to $\mathscr{F}_i$ $(i = 1, 2)$, then formulas*

$$(4.5) \qquad Q_1 = V_1, \qquad Q_{2k} = V_2 Q_{2k-1}, \qquad Q_{2k+1} = V_1 Q_{2k} \qquad \text{for} \quad k = 1, 2, \cdots$$

define an asymptotically sufficient statistic corresponding to $\mathscr{F}^0$.

PROOF. According to Lemma 4.1 it is sufficient to check formula (4.2). This formula follows from one result of Burkholder ([1], Theorem 4).

COROLLARY. *If $\mathscr{F}_i$ $i = 1, 2, \cdots$ are sufficient for M, then $\mathscr{F}^0 = \bigcap \bar{\mathscr{F}}_n$ is also sufficient for M.*

PROOF. By Lemma 4.2 all σ-algebras $\mathscr{A}_n = \bar{\mathscr{F}}_1 \cap \cdots \cap \bar{\mathscr{F}}_n$, $n = 1, 2, \cdots$ are sufficient for M. Let Q_n be corresponding sufficient statistics. Then

$$\lim Q_n f = \lim P\{f \mid \mathscr{A}_n\} = P\{f \mid \mathscr{F}^0\} \quad \text{a.s.} \quad P$$

for all $P \in M$ and all $f \in \mathscr{F}$.

5. Gibbs states.

5.1. Let L be a directed set, i.e., a partially ordered set with the property that for each two elements Λ_1, Λ_2 of L, there exists $\Lambda \in L$ such that $\Lambda > \Lambda_1$ and $\Lambda > \Lambda_2$. We consider two directed families indexed by L: a family of σ-algebras $\mathscr{F}_\Lambda \subset \mathscr{F}$ and a family of Markov operators Π_Λ in $(\Omega, \mathscr{F})$.

Following H. Föllmer, we say that $(\mathscr{F}_\Lambda, \Pi_\Lambda)$ is a *specification* in $(\Omega, \mathscr{F})$ if:

5.1.A. $\mathscr{F}_{\tilde{\Lambda}} \subset \mathscr{F}_\Lambda$ for $\tilde{\Lambda} > \Lambda$.

5.1.B. $\Pi_{\tilde{\Lambda}} \Pi_\Lambda = \Pi_{\tilde{\Lambda}}$ if $\tilde{\Lambda} > \Lambda$.

5.1.C. $\Pi_\Lambda f \in \mathscr{F}_\Lambda$ for $f \in \mathscr{F}$.

5.1.D. $\Pi_\Lambda f = f$ for $f \in \mathscr{F}_\Lambda$.

Concrete examples of specifications will be discussed in Sections 8 and 9.

A probability measure P on $(\Omega, \mathscr{F})$ is called a *Gibbs state specified by*

$\Pi = (\mathscr{F}_\Lambda, \Pi_\Lambda)$ if

$$P\{f \mid \mathscr{F}\} = \Pi_\Lambda f \quad \text{a.s.} \quad P \tag{5.1}$$

for each $f \in \mathscr{F}$ and $\Lambda \in L$.

We assume that the directed set Λ contains a *cofinal sequence* Λ_n, i.e., a sequence with the property that for every $\Lambda \in L$ there exists $\Lambda_n > \Lambda$.

Evidently, each Gibbs state P is a Π_Λ-invariant measure for all $\Lambda \in L$. On the other hand, if a probability measure P is invariant relative to the family $\{\Pi_\Lambda\}$, then, by 1.5.D, 1.5.C, and Lemma 3.1,

$$P\{f \mid \mathscr{F}_\Lambda\} = P\{\Pi_\Lambda f \mid \mathscr{F}_\Lambda\} = \Pi_\Lambda f \quad \text{a.s.} \quad P$$

and P is a Gibbs state. Now let P be invariant with respect to operators Π_{Λ_n} corresponding to a cofinal sequence Λ_n. For each $\Lambda \in L$ there exists a $\Lambda_n > \Lambda$ and, by 5.1.B, $\Pi_{\Lambda_n} \Pi_\Lambda = \Pi_{\Lambda_n}$. Therefore

$$P = P\Pi_{\Lambda_n} = P\Pi_{\Lambda_n}\Pi_\Lambda = P\Pi_\Lambda .$$

We see that the class $G(\Pi)$ of all Gibbs states specified by Π coincides with the class of all probability measures which are invariant with respect to a countable family Π_{Λ_n}.

5.2. We define the *tail σ-algebra* $\mathscr{F}^0$ as the intersection of all $\mathscr{F}_\Lambda$.

THEOREM 5.1. *Let $\Pi = (\mathscr{F}_\Lambda, \Pi_\Lambda)$ be a specification in a B-space $(\Omega, \mathscr{F})$. Then the tail σ-algebra $\mathscr{F}^0$ is H-sufficient for the class $G(\Pi)$ and, to each cofinal sequence Λ_n, there corresponds an asymptotically H-sufficient statistic Π_{Λ_n}.*

PROOF. It is clear that $\mathscr{F}_{\Lambda_n} \downarrow \mathscr{F}^0$. Therefore

$$\lim P\{f \mid \mathscr{F}_{\Lambda_n}\} = P\{f \mid \mathscr{F}^0\} \quad \text{a.s.} \quad P \tag{5.2}$$

for each probability measure P and each P-integrable f. If $P \in G(\Pi)$ then (5.2) implies (5.1). By Lemma 4.1, Π_{Λ_n} is an asymptotically sufficient statistic and $\mathscr{F}^0$ is a sufficient σ-algebra for $G(\Pi)$. Since

$$P\{\Pi_{\Lambda_n} f \mid \mathscr{F}^0\} = P\{P\{f \mid \mathscr{F}_{\Lambda_n}\} \mid \mathscr{F}^0\} = P\{f \mid \mathscr{F}^0\} \quad \text{a.s.} \quad P$$

for $P \in G(\Pi)$, $f \in \mathscr{F}$, the σ-algebra $\mathscr{F}^0$ is H-sufficient for $G(\Pi)$ by Theorem 3.4.

6. Shifts.

6.1. To each measurable transformation T of a space $(\Omega, \mathscr{F})$, there corresponds a Markov operator which transforms functions according to the formula

$$Tf(\omega) = f(T\omega)$$

and measures according to the formula

$$(PT)(A) = \int P(d\omega) 1_A(T\omega) = P(T^{-1}A) .$$

Markov operators of this kind will be called *shifts*.

LEMMA 6.1. *If T is a shift of a B-space $(\Omega, \mathscr{F})$, then*

$$Q_n = n^{-1} \sum_{k=0}^{n-1} T^k \tag{6.1}$$

is an asymptotically H-sufficient statistic for the class M of all T-invariant measures and the σ-algebra $\mathscr{F}^0$ *of all T-invariant sets is the corresponding H-sufficient σ-algebra.*

PROOF. By Birkhoff's ergodic theorem (see, e.g., [18], V-6), the relation (4.2) is satisfied, and Q_n is asymptotically sufficient for M by Lemma 4.1. H-sufficiency follows from Theorem 3.4 and Lemma 3.1.

COROLLARY. *Suppose that a shift T of a B-space* $(\Omega, \mathscr{F})$ *transforms into itself a class* $M \subset M(\mathscr{F})$ *and a σ-algebra* $\mathscr{F}^0 \subset \mathscr{F}$. *If* $\mathscr{F}^0$ *is H-sufficient for M, then the collection* $\mathscr{F}_T{}^0$ *of all T-invariant sets of* $\mathscr{F}^0$ *is H-sufficient for the class* M_T *of all T-invariant measures* $P \in M$.

This follows from Lemma 6.1 and 3.5 because the limit Q of operators (6.1) commutates with the conditioning with respect to $\mathscr{F}^0$.

THEOREM 6.1. *Let G be a finite or countable group of shifts of a B-space* $(\Omega, \mathscr{F})$. *The σ-algebra* $\mathscr{F}^0$ *of all G-invariant sets is H-sufficient for the class M of all G-invariant measures.*

PROOF. Denote by $\mathscr{A}_T$ the minimal σ-algebra containing all M-null sets and all T-invariant sets $A \in \mathscr{F}$. By Lemma 6.1, $\mathscr{A}_T$ is sufficient for M. By Lemma 4.2, an intersection $\mathscr{A}$ of $\mathscr{A}_T$ over all $T \in G$ is sufficient for M. Obviously, $\mathscr{F}^0 \subset \mathscr{A}$. On the other hand, if $A \in \mathscr{A}$, then $T1_A = 1_A$ a.s. M for each $T \in G$. The union B of $T^{-1}(A)$ over all $T \in G$ is G-invariant and $1_A = 1_B$ a.s. M. Hence $\mathscr{A}$ and $\mathscr{F}^0$ are M-equivalent and $\mathscr{F}^0$ is sufficient for M. H-sufficiency of $\mathscr{F}^0$ follows from Theorem 3.4 and Lemma 3.1.

REMARK. Theorem 6.1 holds for important classes of uncountable groups G. Suppose that there exists a countable subgroup G^1 of group G with the property that $\mathscr{F}^0$ is M-equivalent to the σ-algebra $\mathscr{F}^1$ of all G^1-invariant sets. As we know, $\mathscr{F}^1$ is sufficient for the class of all G^1-invariant measures. Hence $\mathscr{F}^1$ is sufficient for M, and $\mathscr{F}^0$ is sufficient for M too. By the remark at the end of 3.4, $\mathscr{F}^0$ is H-sufficient for M if G satisfies conditions (i), (ii).

Now let G be a locally compact group. Then condition (ii) is satisfied for Borel σ-algebra $\mathscr{B}_G$ and Haar measure λ. Condition (i) implies that, for each $P \in M$ and every square integrable f, $T \to Tf$ is a continuous mapping of G into $L^2(\Omega, P)$ (see, e.g., [17], Section 29). Using this fact, it is easy to prove that, if G has a countable everywhere dense subgroup G^1, then $\mathscr{F}^1$ is M-equivalent to $\mathscr{F}^0$ and $\mathscr{F}^0$ is H-sufficient for M.

The rôle of σ-algebra $\mathscr{F}^0$ for decomposition of invariant measures into extreme elements was discovered independently by Farrell [8] and Varadarajan [21]. The fact that $\mathscr{F}^0$ is a sufficient σ-algebra for M is proved in [8] also for a certain class of abelian semigroups.

6.2. We consider now a slightly wider class of operators than shifts.

THEOREM 6.2. *Let T be an invertible transformation of a B-space* $(\Omega, \mathscr{F})$, *let*

T and T^{-1} be measurable, and let $Y(\omega)$ be a strictly positive $\mathscr{F}$-measurable function. Let $Uf(\omega) = Y(\omega)f(T\omega)$. Then

$$V_n f = \frac{\sum_{-n}^{n} U^k f}{\sum_{-n}^{n} U^k 1}$$

is an asymptotically H-sufficient statistic for the class M of all U-invariant probability measures. The corresponding H-sufficient σ-algebra $\mathscr{F}^0$ consists of all T-invariant sets.

PROOF. We prove that

$$P\{f \mid \mathscr{F}^0\} = \lim_{n\to\infty} V_n f \quad \text{a.s.} \quad P \tag{6.2}$$

for every $P \in M$ and every P-integrable f.

Put $\gamma^{-1} = \sum U^k 1$, $\varphi = U^k f$, summing over all integers k. Let $\Omega_0 = \{\omega : \gamma = 0\}$, $\Omega_1 = \{\omega : \gamma > 0\}$. By the Chacon-Ornstein theorem (see, e.g., [18], V, 6.4) (6.2) holds on Ω_0 and, in order to prove that it holds on Ω_1, we need only check that

$$P\{f \mid \mathscr{F}^0\} = \varphi\gamma \quad \text{a.s.} \quad P \qquad \text{on} \quad \Omega_1 . \tag{6.3}$$

The obvious relations $U\varphi = \varphi$, $U\gamma^{-1} = \gamma^{-1}$ imply that $T\varphi = \varphi Y^{-1}$, $T\gamma = \gamma Y$, and $T(\varphi\gamma) = \varphi\gamma$. Hence $\varphi\gamma$ is $\mathscr{F}^0$-measurable. On the other hand, $(Uf)g = U(fT^{-1}g)$ and therefore

$$P(gUf) = P(fT^{-1}g) \tag{6.4}$$

for all $P \in M$ and all positive $\mathscr{F}$-measurable f, g. It follows from this that

$$P(g\gamma U^k f) = P(gfT^{-k}\gamma)$$

for $g \in \mathscr{F}^0$ and $k = 0, \pm 1, \cdots$. Hence

$$P(g\gamma\varphi) = P(gf\alpha), \tag{6.5}$$

where $\alpha = \sum T^{-k}\gamma$. Since α is $\mathscr{F}^0$-measurable, (6.5) implies that

$$\alpha P\{f \mid \mathscr{F}^0\} = \gamma\varphi . \tag{6.6}$$

Now α does not depend on f. Taking $f = 1$, we see that $\alpha = 1$, and (6.6) goes into (6.3). By Lemma 4.1, $\mathscr{F}^0$ is sufficient for M. Formula (6.4) implies (3.19), and $\mathscr{F}^0$ is H-sufficient for M by Theorem 3.4.

REMARK. Suppose that T_t is a one-parameter group of shifts and $U_t f(\omega) = Y_t f(T_t\omega)$ where $Y_{s+t} = Y_s T_s Y_t$. Then Theorem 6.2 holds with

$$V_t = \frac{\int_{-t}^{t} U_s\, ds}{\int_{-t}^{+t} U_s 1\, ds}$$

instead of V_n. This result was first proved by Yu. I. Kifer and S. A. Pirogov in an appendix to [5].

6.3.

THEOREM 6.3. *Let a class M of positive functions be a B-space and a simplex and let $k\varphi \,\bar{\in}\, M$ if $\varphi \in M$ and $k \neq 1$. Suppose that T is an automorphism of a* *cone* $M^* = \{k\varphi : \varphi \in M, k > 0\}$ *(which means that* T *and* T^{-1} *preserve generalized barycentres). Then the set* M_T *of all points* $\varphi \in M$ *such that* $T\varphi = \varphi$ *is also a simplex. This statement is true also for a one-parameter group of transformations* T_t.

PROOF. For each $\varphi \in M^*$ there exists one and only one positive number $k(\varphi)$ such that $\varphi/k(\varphi) \in M$. Put $Y(\varphi) = k(T\varphi)$ and $\bar{T}\varphi = T\varphi/Y(\varphi)$. Obviously $\bar{T}$ is an invertible transformation of the measurable space (M_e, B_{M_e}), and $\bar{T}$ and $\bar{T}^{-1}$ are measurable. Each $\varphi \in M$ can be uniquely represented in the form

$$\varphi = \int_{M_e} \tilde{\varphi}\mu(d\tilde{\varphi}) . \tag{6.7}$$

Hence

$$T\varphi = \int_{M_e} T\tilde{\varphi}\mu(d\tilde{\varphi}) = \int_{M_e} Y(\tilde{\varphi})\bar{T}\tilde{\varphi}\mu(d\tilde{\varphi}) = \int_{M_e} \psi\mu_1(d\psi)$$

where

$$\mu_1(d\psi) = Y(\bar{T}^{-1}\psi)\mu(\bar{T}^{-1}\, d\psi) .$$

It is clear that $T\varphi = \varphi$ if and only if $\mu_1 = \mu$ which is equivalent to the relation $\mu U = \mu$ where $UF(\psi) = Y(\psi)F(\bar{T}\psi)$. By Theorem 6.2, the class of all U-invariant probability measures is a simplex, and formula (6.12) establishes an isomorphism of this class and M_T.

7. Symmetric measures.

7.1. In the rest of the paper we investigate various classes of measures on product spaces. We start with the necessary notations.

Let there be given an arbitrary set S and a set E_s associated with each s of S. We call a *configuration* and denote by x_S a collection of $x_s \in E_s$, $s \in S$. The *product space* E_S is the set of all configurations x_S. A space E_Λ of configurations x_Λ over Λ corresponds to each subset Λ of the set S.

Now let a σ-algebra $\mathscr{B}_s$ in E_s be fixed for each $s \in S$. We denote by $\mathscr{B}_\Lambda$ the minimal σ-algebra in E_Λ which contains sets $\{x_\Lambda : x_s \in \Gamma\}$ for all $s \in \Lambda$, $\Gamma \in \mathscr{B}_s$. To each probability measure P on $(E_S, \mathscr{B}_S)$ and each $\Lambda \subset S$, there corresponds a probability measure P_Λ on $(E_\Lambda, \mathscr{B}_\Lambda)$ defined by the formula

$$P_\Lambda(A) = P\{x_S : x_\Lambda \in A\} , \quad A \in \mathscr{B}_\Lambda . \tag{7.1}$$

A collection of measures P_Λ for all finite $\Lambda \subset S$ is called a *system of finite-dimensional distributions*. If $(E_s, \mathscr{B}_s)$ are Borel spaces, then (7.1) establishes a one-to-one correspondence between all probability measures P on $(E_S, \mathscr{B}_S)$ and all consistent systems of finite-dimensional distributions (Kolmogorov's theorem). In particular, to each family of probability measures p_s, $s \in S$, there corresponds a product measure P for which all finite-dimensional distributions P_Λ are the products of p_s, $s \in \Lambda$.

A system of random variables on the probability space $(E_S, \mathscr{B}_S, P)$ is given by the formula

$$X_s(\omega) = x_s \qquad \text{for} \quad \omega = x_S , \quad s \in S .$$

These random variables are independent if and only if P is a product measure.

Sets $\{x_S : x_\Lambda \in A\}$, Λ finite, $A \in \mathscr{B}_\Lambda$, are called *cylinders*. Two measures on $\mathscr{B}_S$ are identical if they coincide on all cylinders.

7.2. In this section we assume that $(E_s, \mathscr{B}_s) = (E, \mathscr{B})$ does not depend on s and we write E^S for E_S and $\mathscr{B}^S$ for $\mathscr{B}_S$. Any transformation g of S induces a transformation $x_S' = T_g x_S$ of the space E^S given by the formula $x_s' = x_{gs}$. Put $g \in G$ if g is invertible and $gs \neq s$ only for a finite number of s. Measures, measurable sets, and functions invariant with respect to the family of operators T_g, $g \in G$, will be called *symmetric*.

Let Γ be a finite subset of S. Denote by G^Γ the totality of all $g \in G$ such that $gs = s$ outside G. Denote by $\mathscr{F}^\Gamma$ the class of all elements of $\mathscr{B}^S$-invariant relative to T_g, $g \in G^\Gamma$. Let V_Γ be an arithmetic mean of operators T_g, $g \in G^\Gamma$. It is easy to see that

$$P\{f \mid \mathscr{F}^\Gamma\} = V_\Gamma f \quad \text{a.s.} \quad P \tag{7.2}$$

for each symmetric measure P and each P-integrable f.

THEOREM 7.1. *Let S be a countable set, $(E, \mathscr{B})$ a Borel space and M the class of all symmetric measures on $(E^S, \mathscr{B}^S)$. Then:*

(a) *M is a simplex;*

(b) *a measure P is an extreme point of M if and only if P is a product of identical probability measures $p_s = p$, $s \in S$ (in other words, if X_s, $s \in S$ are identically distributed independent random variables);*

(c) *the class $\mathscr{F}^0$ of all symmetric sets is an H-sufficient σ-algebra for M and V_{Γ_n} is the corresponding asymptotically H-sufficient statistic if $\Gamma_n \uparrow S$.*

PROOF. The fact that $\mathscr{F}^0$ is H-sufficient for M follows immediately from Theorem 6.1. If $\Gamma_n \uparrow S$, then $\mathscr{F}^{\Gamma_n} \downarrow \mathscr{F}^0$ and (7.2) implies that

$$P\{f \mid \mathscr{F}^0\} = \lim V_{\Gamma_n} f \quad \text{a.s.} \quad P$$

for all $P \in M$ and all $f \in \mathscr{F}$. By Lemma 4.1, V_{Γ_n} is an asymptotically sufficient statistic for M. The statement (c) is proved. By Theorem 3.1, (c) implies (a).

It remains to prove (b). Let $S = \{0, 1, \cdots, n, \cdots\}$ and $\Gamma_n = \{0, 1, \cdots, n-1\}$. By virtue of (c) and (4.4)

$$Pf = \lim V_{\Gamma_n} f \quad \text{a.s.} \quad P \tag{7.3}$$

for $P \in M$, and $f \in \mathscr{B}^S$. In order to prove that P is a product measure, it suffices to check that, for all m and all $A \in \mathscr{B}^{\Gamma_m}$, $B \in \mathscr{B}$,

$$P\{x_{\Gamma_m} \in A, x_m \in B\} = P\{x_{\Gamma_m} \in A\} P\{x_m \in B\}\,. \tag{7.4}$$

It follows from (7.3) that

$$P\{1_A(x_{\Gamma_m}) P 1_B(x_m)\} = \lim_{n\to\infty} P\{1_A(x_{\Gamma_m}) V_{\Gamma_n} 1_B(x_m)\}\,. \tag{7.5}$$

Evidently $V_{\Gamma_n} 1_B(x_m) = n^{-1} \sum_{k=1}^{n} 1_B(x_k)$ for $n \geqq m$. Since

$$P\{x_{\Gamma_m} \in A, x_k \in B\} = P\{x_{\Gamma_m} \in A, x_m \in B\}$$

for all $k \geqq m$, (7.5) implies (7.4).

Now we prove that each product measure $\bar{P} \in M$ belongs to M_e. Since $\bar{P}$ is a barycentre of a probability measure μ concentrated on M_e, we have

$$\bar{P}[\varphi(x_0)\varphi(x_1)] = \int_{M_e} P[\varphi(x_0)\varphi(x_1)]\mu(dP) \tag{7.6}$$

for every $\varphi \in \mathscr{B}$. Here $\bar{P}$ and P are symmetric product measures and therefore (7.6) is equivalent to

$$[\bar{P}\varphi(x_0)]^2 = \int_{M_e} P\varphi(x_0)^2\mu(dP) ,$$

which implies that

$$\int_{M_e} [P\varphi(x_0) - \bar{P}\varphi(x_0)]^2\mu(dP) = 0 . \tag{7.7}$$

It follows from (7.7) that $\mu\{P: P \in M_e, P = \bar{P}\} = 1$. Thus $\bar{P} \in M_e$.

7.3. The statements (a) and (b) of Theorem 7.1 are true for uncountable S too. Indeed, if Λ is a countable subset of S, then the measure P_Λ, introduced by (7.1), characterizes a symmetric measure P uniquely because it defines all finite-dimensional distributions of P. Hence the mapping $P \to P_\Lambda$ is a one-to-one mapping of M onto the set of all symmetric measures on $(E^\Lambda, \mathscr{B}^\Lambda)$. This mapping preserves the convex structure, and P is a product measure if and only if P_Λ is a product measure also.

The statement (c) has to be modified as follows. Let Λ be an arbitrary countable subset of S. Denote by $\mathscr{F}^0$ the collection of all sets of the form $A \times E^{S\setminus\Lambda}$ where A is a symmetric subset of E^Λ. Then $\mathscr{F}^0$ is an H-sufficient σ-algebra for M.

8. Stochastic fields.

8.1. Let $(E_S, \mathscr{B}_S)$ be a product of spaces $(E_s, \mathscr{B}_s)$, $s \in S$, and let L be a collection of subsets of S ordered by inclusion. Denote by $\mathscr{F}_\Lambda$ a σ-algebra in E_S generated by random variables X_s, $s \in S\setminus\Lambda$. Assume that, for each $\Lambda \in L$, a measure $p_\Lambda(\cdot \mid x_{S\setminus\Lambda})$ is given on $(E_\Lambda, \mathscr{B}_\Lambda)$ which depends on $x_{S\setminus\Lambda}$, and put

$$\Pi_\Lambda f(x_S) = \int_{E_{S\setminus\Lambda}} f(x_{S\setminus\Lambda} y_\Lambda) p_\Lambda(dy_\Lambda \mid x_{S\setminus\Lambda}) . \tag{8.1}$$

(We denote by $x_{S\setminus\Lambda} y_\Lambda$ a configuration which coincides with y_Λ over Λ and with $x_{S\setminus\Lambda}$ over $S\setminus\Lambda$.) If $\Pi = (\mathscr{F}_\Lambda, \Pi_\Lambda)$ is a specification (i.e., if 5.1.A—5.1.D are satisfied), we say that p is a *specifying function.*

We say that (X_s, P) is a *stochastic field specified by p* if

$$P\{X_\Lambda \in A \mid X_{S\setminus\Lambda}\} = p_\Lambda(A \mid X_{S\setminus\Lambda}) \quad \text{a.s.} \quad P \tag{8.2}$$

for each $\Lambda \in L$ and each $A \in \mathscr{B}_\Lambda$. Obviously (8.2) is equivalent to (5.1). Hence Theorem 5.1 can be applied to the set of all stochastic fields specified by p.

8.2. Let S be a countable set and let L be the collection of all finite subsets of S. To each real-valued function $U(\Gamma, x_\Gamma)$, $\Gamma \in L$, $x_\Gamma \in E_\Gamma$, there corresponds a specifying function

$$p_\Lambda(C \mid x_{S\setminus\Lambda}) = Z^{-1} \int_C [\exp \textstyle\sum U(\Gamma, x_\Gamma)] \prod_{s \in \Lambda} \lambda_s(dx_s) , \tag{8.3}$$

where Γ runs over all finite subsets of S such that $\Gamma \cap \Lambda \neq \emptyset$, λ_s is a measure on $(E_s, \mathscr{B}_s)$ and Z is independent of x_Λ and can be calculated from the condition $p_\Lambda(E_\Lambda \mid x_{S\setminus\Lambda}) = 1$. (The only restriction on U is convergence of series in (8.3).)

Now suppose that S is a graph. A specifying function p is called Markov if $p_\Lambda(\cdot \mid x_{S\setminus\Lambda})$ depends only on $x_{\partial\Lambda}$ where $\partial\Lambda$ is a collection of all points of $S\setminus\Lambda$ which have neighbors in Λ. A function (8.3) is Markov if and only if the inequality $U(\Gamma, x_\Gamma) \neq 0$ implies that each two points s_1, s_2 of Γ are neighbors.

Proofs of all statements of subsection 8.2 can be found, for example, in [20].

8.3. Now suppose $(E_s, \mathscr{B}_s) = (E, \mathscr{B})$ does not depend on s. Let L be the collection of all finite subsets of a countable set S. Consider the family $\{\mathscr{F}^\Gamma, \Gamma \in L\}$ of σ-algebras in E^S which has been defined in 7.2. Obviously $\mathscr{F}^\Gamma \supset \mathscr{F}^{\tilde{\Gamma}}$ if $\Gamma \subset \tilde{\Gamma}$. Suppose that, for each $\Gamma \in L$, a measure $p_\Gamma(\cdot \mid x_S)$ on $(E^\Gamma, \mathscr{B}^\Gamma)$ is given depending on x_S and such that operators

$$\Pi_\Gamma f(x_S) = \int_{E^S} f(x_{S\setminus\Gamma}) p_\Gamma(dy_\Lambda \mid x_S)$$

satisfy conditions 5.1.B—5.1.D. Theorem 5.1 can be applied to the class of all probability measures P satisfying the condition

$$P\{X_\Gamma \in A \mid \mathscr{F}^\Gamma\} = p_\Gamma\{A \mid X_S\} \quad \text{a.s.} \quad P \qquad \text{for all} \quad \Gamma \in L \quad \text{and all} \quad A \in \mathscr{B}^\Gamma .$$

The tail σ-algebra $\mathscr{F}^0 = \bigcap \mathscr{F}^\Gamma$ coincides with a collection of all symmetric subsets of E^S.

9. Markov processes with a given transition function.

9.1. A stochastic field (X_s, P), $s \in S$, is called a *stochastic process* if S is a subset of a real line. The case when S is an interval is the most important.

We denote by $\mathscr{F}_{\leq s}$ the σ-algebra in E_S generated by X_t, $t \leq s$, $t \in S$. The notations $\mathscr{F}_{<s}$, $\mathscr{F}_{>s}$, $\mathscr{F}_{\geq s}$ have an analogous meaning.

A real-valued function $p(s, x; t, \Gamma)$, $s < t \in S$, $x \in E_s$, $\Gamma \in \mathscr{B}_t$, is called a *Markov transition function* if $p(s, x; t, \cdot)$ is a probability measure, $p(s, \cdot; t, \Gamma)$ is a $\mathscr{B}_s$-measurable function, and

$$p(s, x; u, \Gamma) = \int_{E_t} p(s, x; t, dy) p(t, y; u, \Gamma) \tag{9.1}$$

for all $s < t < u \in S$, $x \in E_s$, $\Gamma \in \mathscr{B}$.

Starting from a transition function p, we define a specification $\Pi(p)$ in the following way. We consider a family of finite-dimensional distributions

$$\begin{aligned} &p(s_1, dx_1, \cdots, s_n, dx_n) \\ &\quad = p(s, x; s_1, dx_1) p(s_1, x_1; s_2, dx_2) \cdots p(s_{n-1}, x_{n-1}; s_n, dx_n), \\ &\qquad s_1 < s_2 < \cdots < s_n \in \Lambda_s = S \cap (s, +\infty) \end{aligned} \tag{9.2}$$

and denote by $P_{s,x}$ the corresponding probability measure on $\mathscr{F}_{>s}$. We define L as the totality of all sets Λ_s, $s \in S$, and put

$$\mathscr{F}^{\Lambda_s} = \mathscr{F}_{\leq s}, \qquad \Pi_{\Lambda_s} f(x_S) = \int f(x_{S\setminus\Lambda_s} y_{\Lambda_s}) P_{s,x_s}(dy_{\Lambda_s}) = P_{s,x_s} f(x_{S\setminus\Lambda_s} X_{\Lambda_s}) . \tag{9.3}$$

Then $\Pi(p) = (F_\Lambda, \Pi_\Lambda)$ is a specification. (Formula (9.3) is a particular case of (8.1) with $p_{\Lambda_s}(A \mid x_{S\setminus\Lambda_s}) = P_{s,x_s}\{x_{\Lambda_s} \in A\}$.) We will use an abbreviation Π_s for the operator Π_{Λ_s}.

We say that (X_s, P) is a *Markov process with a transition function p* and we write $P \in M(p)$ if

$$(9.4) \qquad P\{A \mid \mathscr{F}_{\leq s}\} = P_{s,X_s}(A) \quad \text{a.s.} \quad P$$

for each $s \in S$, $A \in \mathscr{F}_{>s}$. Obviously $M(p)$ coincides with the set of all Gibbs states specified by $\Pi(p)$.

9.2. THEOREM 9.1. *Let $r = \inf S$. If $r \in S$, then Π_r is an H-sufficient statistic for $M(p)$ and the corresponding H-sufficient σ-algebra is generated by X_r. If $r \bar{\in} S$, then an intersection $\mathscr{F}^0$ of all σ-algebras $\mathscr{F}_{<s}$, $s \in S$ is an H-sufficient σ-algebra for $M(p)$ and, to each sequence $s_n \downarrow r$, $s_n \in S$ these corresponds an asymptotically H-sufficient statistic Π_{s_n}.*

PROOF. In the case $r \in S$, we need only check that

$$(9.5) \qquad P\{f \mid X_r\} = \Pi_r f \quad \text{a.s.} \quad P$$

if $P \in M(p)$ and $f \in \mathscr{B}_s$. It is sufficient to prove this only for functions of the form $f(x_S) = \varphi(x_r)\psi(x_{\Lambda_r})$. But for such f the left side of (9.5) is equal to

$$\varphi(X_r)P\{\psi(X_{\Lambda_r}) \mid \mathscr{F}_{\leq r}\} = \varphi(X_r)P_{r,X_r}\psi(X_{\Lambda_r})$$

and, by the definition of Π_r, the right side of (9.5) is the same.

Suppose now that $r \bar{\in} S$. Then to each $s_n \downarrow r$, $s_n \in S$ there corresponds a cofinal sequence Λ_{s_n}, and, if (E_S, B_S) is a B-space, we can apply Theorem 5.1. This is the case if S is countable.

If S is not countable, we consider a countable subset $\Lambda = \{s_n\}$ of S where $s_n \downarrow r$ and we use the same trick as in 7.3 replacing each measure P by P_Λ.

Since $\mathscr{F}_{\leq s_m} \downarrow \mathscr{F}^0$, we have

$$(9.6) \qquad P\{f \mid \mathscr{F}^0\} = \lim_{m\to\infty} \Pi_{s_m} f \quad \text{a.s.} \quad P$$

for each $P \in M(p)$ and each P-integrable f, and our theorem will be proved if we construct a Markov operator Q with the properties

$$(9.7) \qquad \begin{aligned} P\{f \mid \mathscr{F}^0\} &= Qf \quad \text{a.s.} \quad P, \\ Q^\omega &\in M(p) \quad \text{a.s.} \quad P, \end{aligned}$$

for each $P \in M(p)$. It follows from (9.4) that, for $P \in M(p)$,

$$P\{f \mid \mathscr{F}^0\} = P\{P\{f \mid \mathscr{F}_{\leq s}\} \mid \mathscr{F}^0\} = P\{P_{s,X_s} f \mid \mathscr{F}^0\} \quad \text{a.s.} \quad P.$$

Therefore if

$$(9.8) \qquad P\{\varphi(X_{s_n}) \mid \mathscr{F}^0\} = Q\varphi(X_{s_n}) \quad \text{a.s.} \quad P \in M(p)$$

for all $n = 1, 2, \cdots$ and all $\varphi \in \mathscr{B}_{s_n}$, then (9.7) is true for all n and all $f \in \mathscr{F}$, and hence, it is true for all $f \in \mathscr{F}$.

Denote by $\tilde{M}(p)$ the class of all Markov processes on $(E_\Lambda, \mathscr{B}_\Lambda)$ with the transition function p and by $\tilde{\mathscr{F}}^0$ the corresponding tail σ-algebra. It follows from (9.5) and an analogous formula for $\tilde{\mathscr{F}}^0$ that

$$P\{\varphi(x_{s_n}) \mid \tilde{\mathscr{F}}^0\} = P\{\varphi(x_{s_n}) \mid \mathscr{F}^0\} \quad \text{a.s.} \quad P \in M(p)\,. \tag{9.9}$$

The mapping $P \to P_\Lambda$ is a one-to-one mapping of $M(p)$ onto $\tilde{M}(p)$. As we know, there exists a Markov operator $\tilde{Q}$ in (E_Λ, B_Λ) such that $\tilde{Q}^\omega \in \tilde{M}(p)$ a.s. $\tilde{M}(p)$ and

$$P\{\varphi(x_{s_n}) \mid \tilde{\mathscr{F}}^0\} = \tilde{Q}\varphi(x_{s_n}) \quad \text{a.s.} \quad P \tag{9.10}$$

for each $P \in \tilde{M}(p)$ and each $\varphi \in \mathscr{B}_{s_n}$. Denote by Q^ω a measure of class $M(p)$ which corresponds to $\tilde{Q}^\omega$. It follows from (9.9) and (9.10) that Q satisfies (9.8).

10. Entrance and exit laws.

10.1. Let $p(s, x; t, \Gamma)$ be a Markov transition function. Put

$$P_t^s h^t(x) = \textstyle\int_{E_t} p(s, x; t, dy) h^t(y)\,,$$
$$(\nu_t P_t^s)(\Gamma) = \textstyle\int_{E_t} \nu_s(dx) p(s, x; t, \Gamma)\,.$$

(Here h^t is a $\mathscr{B}_t$-measurable function with the values in the extended half-line $[0, +\infty]$; ν_t is a measure on $\mathscr{B}_t$.)

We say that ν is an *entrance law* if $\nu_s P_t^s = \nu_t$ for all $s < t \in S$ and we say that h is an *exit law* if $P_t^s h^t = h^s$ for all $s < t \in S$.

If ν is an entrance law and h is an exit law, then the value of $\nu_t(h_t)$ does not depend on t and we denote it by $\{\nu, h\}$. If $\{\nu, h\} = 1$, then the formula

$$\begin{aligned} &p(t_1, dx_1, \cdots, t_n, dx_n) \\ &\quad = \nu_{t_1}(dx_1) p(t_1, x_1; t_2, dx_2) \cdots p(t_{n-1}, x_{n-1}; t_n, dx_n) h^{t_n}(x_n)\,, \\ &\qquad\qquad t_1 < \cdots < t_n \in S \end{aligned} \tag{10.1}$$

defines a family of consistent finite-dimensional distributions, and we denote by P_ν^h the corresponding probability measure on $(E_S, \mathscr{B}_S)$.

Let $\infty > h^s(x) > 0$ for all s, x.

Let R_p^h be the class of all entrance laws ν normed by the condition $\{\nu, h\} = 1$ with natural measurable and convex structures. Let $M(p^h)$ be the class of all Markov processes with the transition function

$$p^h(s, x; t, dy) = h^s(x)^{-1} p(s, x; t, dy) h^t(y)\,. \tag{10.2}$$

It is easy to see that $\nu \to P_\nu^h$ is an isomorphism of convex measurable spaces R_ν^h and $M(p^h)$. According to Theorem 9.1 and 3.1, the space R_ν^h is a simplex. If ν is an extreme point of R_p^h, and if $P_\nu^h[\varphi^t(X_t)] < \infty$, then by 4.4, for every sequence $s_n \downarrow r_1$,

$$P_\nu^h[\varphi^t(X_t)] = \lim P^h_{s_n, X_{s_n}} \varphi^t(X_t) \quad \text{a.s.} \quad P_\nu^h\,. \tag{10.3}$$

Formula (10.3) implies that for each ν_t-integrable f

$$\nu_t(f) = \lim h^{s_n}(X_{s_n})^{-1} \textstyle\int_{E_{s_n}} p(s_n, X_{s_n}; t, dy) f(y) \quad \text{a.s.} \quad P_\nu^h\,. \tag{10.4}$$

10.2. Now we investigate a class S_ν^p of all exit laws normed by the condition $\{\nu, h\} = 1$ under the following additional assumption:

10.2.A. If $\nu_t(\Gamma) = 0$, then $p(s, x; t, \Gamma) = 0$ for all $s \in t$, $x \in E_s$.

We proved in [6] (see Lemma 4.2) that a density $\rho(s, x; t, y) = p(s, x; t, dy)/\nu_t(dy)$ can be selected in such a way that

$$\int_{E_t} \rho(s, x; t, y)\nu_t(dy)\rho(t, y; u, z) = \rho(s, x; u, z) \tag{10.5}$$

for all $s < t < u$, $x \in E_s$, $z \in E_u$ and

$$\int \rho(s, x; t, y)\nu_s(dx) = 1 \qquad \text{for all} \quad s, \quad x. \tag{10.6}$$

The formula

$$\hat{p}(s, dx; t, y) = \nu_s(dx)\rho(s, x; t, y) \tag{10.7}$$

defines a backward transition function. Starting from $\hat{p}$, we define probability measures $P^{u,z}$ on $\mathscr{F}_{<u}$ exactly in the same way as measures $P_{s,x}$ were defined with the help of forward transition function p. We say that (X_s, P) is a Markov process with a backward transition function $\hat{p}$ if

$$P\{A \mid \mathscr{F}_{\geq u}\} = P^{u,X_u}(A) \quad \text{a.s.} \quad P \qquad \text{for} \quad A \in \mathscr{F}_{<u}\,.$$

We consider a measurable structure in S_ν^p generated by functions $F(h) = \nu_s(\varphi h^s)$, $s \in S$, $\varphi \in \mathscr{B}_s$. It was proved in [4] (Lemma 4.2) that $h^s(x)$ is measurable with respect to the pair h, x, and hence the condition 2.2.A is satisfied. It is easy to check that the mapping $h \to P_\nu^h$ is an isomorphism of S_ν^p onto the class $M(\hat{p})$ of all Markov processes with the backward transition function $\hat{p}$ defined by (10.7). Now we use Theorem 3.1 and propositions dual to Theorem 9.1 and to formula (4.4), and we conclude that S^p is a simplex and that

$$\nu_t(h^t\varphi^t) = \lim P^{u_n,X_{u_n}}\varphi^t(x_t) = \lim \int \varphi^t(x)\hat{p}(t, dx; u_n, X_{u_n}) \quad \text{a.s.} \quad P_\nu^h \tag{10.8}$$

if h is an extreme point of S_ν^p, if $u_n \uparrow r_2$ and if $P_\nu^h|\varphi^t(X_t)| < \infty$.

It follows from (10.8) and (10.7) that

$$\int h^t(x)\varphi^t(x)\,\nu_t(dx) = \lim \int \rho(t, x; u_n, X_{u_n})\varphi^t(x)\nu_t(dx) \quad \text{a.s.} \quad P_\nu^h$$

if $h^t\varphi^t$ is ν_t-integrable. Applying the last formula to

$$\begin{aligned} \varphi^t(x) &= \rho(s, x; t, y) && \text{for} \quad t > s\,, \\ &= 0 && \text{for} \quad t \leq s\,, \end{aligned}$$

we see that, if h is extreme and if $h^s(x) < \infty$, then

$$h^s(x) = \lim \rho(s, x; u_n, X_{u_n}) \quad \text{a.s.} \quad P_\nu^h\,.$$

Remark. S. E. Kuznecov [14] has proved that the assumption 10.2.A is not only sufficient but also necessary for the class S_ν^p to be a simplex.

11. Excessive measures and excessive functions.

11.1. In this section, the results of Section 10 will be extended to wider classes of measures and functions associated with a transition function p. Let

S coincide with the set of all real numbers. An *excessive function* h and an *excessive measure* ν are defined, respectively, by conditions

$$P_t^s h^t \leqq h^s, \qquad P_t^s h^t \uparrow h^s \quad \text{as} \quad t \downarrow s$$

and

$$\nu_s P_t^s \leqq \nu_s, \qquad \nu_s P_t^s \uparrow \nu_t \quad \text{as} \quad s \uparrow t .$$

It is convenient to replace in the definition of a Markov transition function the condition $p(s, x; t, E_t) = 1$ by a weaker condition $p(s, x; t, E_t) \leqq 1$. An immediate gain is that an extended class is invariant with respect to transformation $p \to p^h$ defined by formula (10.2) for each strictly positive finite excessive function h.

Let ν be an excessive measure and h be an excessive function. We put $\{\nu, h\} = +\infty$ if $\nu_t(h^t) = +\infty$ for some t. If $\nu_t(h^t) < \infty$ for all t, we define $\{\nu, h\}$ as a supremum of sums

$$\nu_{t_1}(h^{t_1}) + \sum_{k=2}^{m} [\nu_{t_k}(h^{t_k}) - \nu_{t_{k-1}}(P_{t_k}^{t_{k-1}} h^{t_k})]$$

over all finite subsets $t_1 < t_2 < \cdots < t_m$ of S. (This is consistent with the definition given in Section 10 for the case of an entrance law ν and an exit law h.)

The crucial point is the construction of a probability measure P_ν^h corresponding to a triple p, ν, h such that $\{\nu, h\} = 1$. As in Section 10, we start from formula (10.1). However P_ν^h will be defined not on $(E_S, \mathscr{B}_S)$ but on a different space (Ω, F). In order to construct this space, we add to E_s two extra points a_s and b_s and denote by $\bar{\mathscr{B}}_s$ a σ-algebra in $\bar{E}_s = E_s \cup a_s \cup b_s$ generated by $\mathscr{B}_s$ and the one-point sets $\{a_s\}$ and $\{b_s\}$. We define Ω as a subset of the product space $(\bar{E}_S, \bar{\mathscr{B}}_S) = \Pi_{s \in S} (\bar{E}_s, \bar{\mathscr{B}}_s)$, namely, x_S of $\bar{E}_S$ belongs to Ω if there exist two real numbers $\alpha < \beta$ such that

$$x_s = a_s \quad \text{for} \quad s \leqq \alpha , \qquad x_s \in E_s \quad \text{for} \quad s \in (\alpha, \beta) , \qquad x_s = b_s \quad \text{for} \quad s \geqq \beta .$$

The random variables $\alpha(\omega)$ and $\beta(\omega)$ are called *the birth time* and *the death time*. To each $s \in S$ there corresponds a function X_s on Ω defined by the formula

$$X_s(\omega) = x_s \qquad \text{for} \quad \omega = x_S ,$$

and we denote by $\mathscr{F}$ the σ-algebra in Ω generated by X_s, $s \in S$.

We proved in [5] that, if $\{\nu, h\} = 1$, then there exists one and only one probability measure P_ν^h on (Ω, F) such that, for every $t_1 < \cdots < t_n \in S$, $\Gamma_1 \in \mathscr{B}_{t_1}, \cdots,$ $\Gamma_n \in \mathscr{B}_{t_n}$,

$$P_\nu^h\{\alpha < t_1, X_{t_1} \in \Gamma_1, \cdots, X_{t_n} \in \Gamma_n, \beta > t_n\} = p(t_1, \Gamma_1, \cdots, t_n, \Gamma_n) ,$$

where the right side is defined by formula (10.1).

To each $s \in S$, $x \in E_s$ there corresponds an excessive measure

$$\begin{aligned} \nu_t^{s,x}(\Gamma) &= p(s, x, t, \Gamma) \qquad &\text{for} \quad t > s , \\ &= 0 \qquad &\text{for} \quad t \leqq s . \end{aligned}$$

We say that a probability measure P on $\mathscr{F}$ defines *a Markov process* (X_s, P)

with a transition function p^k if, for all $s \in S$ and $A \in \mathscr{F}_{>s}$,

$$P\{A \mid \mathscr{F}_{\leq s}\} = P^k_{s,x_s}(A) \quad \text{a.s. } P \qquad \text{on } \{\omega : \alpha < s < \beta\} .$$

Let $\Lambda = \{t_1 < \cdots < t_m\}$ be a finite subset of S and let $t_0 = -\infty$, $t_{m+1} = +\infty$. Put

$$\alpha_\Lambda = t_{k+1} \quad \text{if} \quad t_k \leqq \alpha < t_{k+1} , \qquad \beta_\Lambda = t_k \quad \text{if} \quad t_k < \beta \leqq t_{k+1} ,$$
$$k = 0, 1, \cdots, m ;$$

$$\Pi_\Lambda f(\omega) = \Pi_{(\alpha_\Lambda(\omega), +\infty)} f(\omega) .$$

Theorem 9.1 can be extended to processes with random birth and death times as follows.

THEOREM 11.1. *Put* $A \in \mathscr{F}^0$ *if* $\{A, \alpha \leqq s\} \in \mathscr{F}_{\leq s}$ *for all* $s \in S$. *Then* $\mathscr{F}^0$ *is an H-sufficient σ-algebra for* $M(p^k)$. *To each increasing sequence of finite sets* Λ_n *with a union everywhere dense in S, there corresponds an asymptotically H-sufficient statistic* Π_{Λ_n}.

Now all the results of Section 10 can be easily carried over to excessive measure and functions. We have to replace s_n by α_{Λ_n} in (10.4) and u_n by β_{Λ_n} in (10.6).

11.2. We proved in [5] that the space of all p-excessive measures is a Borel space (the main point is that each p-excessive measure is defined uniquely by the values ν_t for rational t). Therefore all simplexes $M(p)$, $R_p{}^k$, $S_\nu{}^p$ investigated in Sections 9, 10 and 11 are Borel spaces.

12. Stationary transition functions.

12.1. We suppose now that a Markov transition function $p(s, x; t, \Gamma)$ is *stationary* which means that: (i) S is a subgroup of the additive group of real numbers; (ii) all spaces $(E_s, \mathscr{B}_s) = (E, \mathscr{B})$ are identical; (iii) $p(s, x; t, \Gamma) = p(t - s, x, \Gamma)$ depends only on the difference $t - s$. We shall consider only two possibilities: S is the group of all integers (the discrete case) and S is the group of all real numbers (the continuous case). In the second case we assume that $p(t, x, \Gamma)$ is measurable with respect to the pair t, x.

We denote by θ_t a shift in $(E^S, \mathscr{B}^S)$ which corresponds to the transformation $s \to s + t$ of S. A Markov process (X_s, P) is called *stationary* if P is invariant with respect to the group θ_t, $t \in S$.

THEOREM 12.1. *Let* $\mathscr{F}^0 = \bigcap \mathscr{F}_{\leq s}$ *be the tail σ-algebra and let* $\mathscr{F}_\theta{}^0$ *be a collection of all* $A \in \mathscr{F}^0$ *which are invariant with respect to the group* θ_t. *Then* $\mathscr{F}_\theta{}^0$ *is an H-sufficient σ-algebra for the class* $M_\theta(P)$ *of all stationary Markov processes with a transition function p.*

PROOF. In the discrete case we can apply the corollary to Lemma 6.1 to $T = \theta_1$, the class $M = M(p)$, and σ-algebra $\mathscr{F}^0$ (which are invariant with respect to T). In the continuous case $(E^S, \mathscr{B}^S)$ is not a B-space. This obstacle can be overcome in the same way as in the proof of Theorem 9.1 but we will not go into details.

12.2. A probability measure ν is called a *stationary distribution* for p if $\nu P_t = \nu$ for all $t \in S$. (Here $P_t = P_t^0$ are the Markov operators associated with the transition function p.) Each stationary distribution defines an entrance law $\nu_t = \nu$, $t \in S$. The corresponding Markov process P_ν^1 belongs to the class $M_\theta(p)$ investigated in Theorem 12.1. In this way we establish an isomorphism between the class N of all stationary distributions and $M_\theta(p)$. Hence N is a simplex.

12.3. An excessive measure ν_t is called *stationary* if ν_t does not depend on t. Obviously a measure ν on $(E, \mathscr{B})$ is a stationary excessive measure if and only if $\nu P_t \leqq \nu$ for all $t \in S$ and $\nu P_t \uparrow \nu$ as $t \downarrow 0$. In a similar way, we introduce the concept of a stationary excessive function.

THEOREM 12.2. *Let l be a strictly positive measurable function on $(E, \mathscr{B})$. A class of all stationary excessive measures ν normed by the condition $\nu(l) = 1$ is a simplex.*

PROOF. Since p is stationary, the formula $(T_t\nu)_s = \nu_{s+t}$ defines for each t a transformation of the set of all p-excessive measures. Obviously, ν is stationary if and only if it is invariant with respect to the group T_t.

Consider a p-excessive function

$$h^s(x) = \tfrac{1}{2} \int_0^\infty e^{-|s+u|} P_u l(x)\, du .$$

A simple calculation shows that, for each excessive measure ν,

$$\{\nu, h\} = \tfrac{1}{2} \int_{-\infty}^{+\infty} e^{-|u|} \nu_u(l)\, du \tag{12.1}$$

which implies that

$$\{\nu, h\} = \nu(l) \quad \text{if} \quad \nu \quad \text{is stationary,} \tag{12.2}$$

and

$$\{T_s\nu, h\} \leqq e^{|s|}\{\nu, h\} . \tag{12.3}$$

Denote by M^* the set of all excessive measures ν satisfying the condition $\{\nu, h\} < \infty$ and by M the set of $\nu \in M^*$ for which $\{\nu, h\} = 1$. According to Section 11, M is a Borel space and a simplex. It follows from (12.3) that M^* is invariant with respect to T_t, and Theorem 12.2 follows from Theorem 6.3.

12.4.

THEOREM 12.3. *Suppose a stationary transition function $p(t, x, \Gamma)$ is absolutely continuous with respect to a measure γ for each t and x. Then the set of all stationary excessive functions h normed by the condition $\gamma(h) = 1$ is a simplex.*

PROOF. We consider transformations $(T_t h)^s = h^{t+s}$ of the set of all excessive functions. The formula

$$\nu_t(\Gamma) = \tfrac{1}{2} \int_0^\infty e^{-|t-u|} (\gamma P_u)(\Gamma)\, du$$

defines an excessive measure, and we have

$$\{\nu, h\} = \tfrac{1}{2} \int_{-\infty}^{+\infty} e^{-|u|} \gamma(h^u)\, du$$

for every excessive function h. In particular $\{\nu, h\} = \gamma(h)$ for a stationary h. To complete the proof, we apply Theorem 6.3 in the same way as in the proof of Theorem 12.2.

REMARK. Put

$$g_\lambda(x, \Gamma) = \int_0^\infty e^{-\lambda t} p(t, x, \Gamma)\, dt\,, \quad \lambda \geqq 0\,.$$

If a measure

$$\eta_\lambda(\Gamma) = \int_E \gamma(dx) g_\lambda(x, \Gamma)$$

is σ-finite for some λ, then Theorem 12.2 remains true if $p(t, x, \Gamma)$ is absolutely continuous with respect to η_λ (Kuznecov [14], Theorem 3).

REFERENCES

[1] BURKHOLDER, D. L. (1961). Sufficiency in the undominated case. *Ann. Math. Statist.* **32** 1191-1200.

[2] DOBRUSHIN, R. L. (1968). Description of a random field by means of conditional probabilities and the conditions governing its regularity. *Theor. Probability Appl.* **13** 197-224.

[3] DOOB, J. L. (1959). Discrete potential theory and boundaries. *J. Math. and Mech.* **8** 433-458.

[4] DYNKIN, E. B. (1971). Initial and final behavior of trajectories of Markov processes. *Uspehi Mat. Nauk* **26** 4 153-172. (English translation: *Russian Math. Surveys* **26** 4 165-185.)

[5] DYNKIN, E. B. (1972). Integral representation of excessive measures and excessive functions. *Uspehi Mat. Nauk* **27** 1 43-80. (English translation: *Russian Math. Surveys* **27** 1 43-84.)

[6] DYNKIN, E. B. (1975). Markov representations of stochastic systems. *Uspehi Mat. Nauk* **30** 1 61-99. (English translation: *Russian Math. Surveys* **30** 1 65-104.)

[7] DYNKIN, E. B. and JUSKEVIČ, A. A. (1975). *Controlled Markov Processes and Their Applications.* Nauka, Moscow. (English translation will be published by Springer-Verlag.)

[8] FARRELL, R. H. (1962). Representation of invariant measures. *Illinois J. Math.* **6** 3 447-467.

[9] DE FINETTI, B. (1931). Fuzione caratteristica di un fenomeno aleatorio. *Atti Accad. Naz. Lincei, Mem. Cl. Sci. Fis. Mat. Natur.* **4** 251-299.

[10] DE FINETTI, B. (1937). La prévision: ses lois logiques, ses sources subjectives. *Ann. Inst. H. Poincaré* **7** 1-68.

[11] HEWITT, E. and SAVAGE, L. J. (1955). Symmetric measures on cartesian products. *Trans. Amer. Math. Soc.* **80** 2 470-501.

[12] HUNT, G. A. (1960). Markov chains and Martin boundaries. *Illinois J. Math.* **4** 313-340.

[13] KRYLOV, N. and BOGOLIOUBOV, N. (1937). La théorie générale de la mesure dans son application à l'étude des systèmes de la mécanique non linéaires. *Ann. of Math.* **38** 65-113.

[14] KUZNECOV, S. E. (1974). On decomposition of excessive functions. *Dokl. Akad. Nauk SSSR* **214** 276-278.

[15] MARTIN, R. S. (1941). Minimal positive harmonic functions. *Trans. Amer. Math. Soc.* **49** 137-172.

[16] MEYER, P. A. (1966). *Probability and Potential.* Blaisdell, Waltham, Mass.

[17] NAIMARK, M. A. (1964). *Normed Rings.* P. Noordhoff N. V., Groningen.

[18] NEVEU, J. (1964). Bases mathématiques du calcul des probabilités. Masson et Cie, Paris.

[19] PARTHASARATHY, K. R. (1967). *Probability Measures on Metric Spaces.* New York, London.

[20] PRESTON, C. I. (1974). Gibbs states on countable sets. Cambridge Univ. Press.

[21] VARADARAJAN, V. S. (1963). Groups of automorphisms of Borel spaces. *Trans. Amer. Math. Soc.* **109** 2 191–220.

DEPARTMENT OF MATHEMATICS
CORNELL UNIVERSITY
ITHACA, N.Y. 14853

Reprinted from E.B. Dynkin, "Minimal Excessive Measures and Functions," *Transactions of the AMS*, volume 258, pages 217–244, by permission of the American Mathematical Society.

MINIMAL EXCESSIVE MEASURES AND FUNCTIONS[1]

ABSTRACT. Let H be a class of measures or functions. An element h of H is minimal if the relation $h = h_1 + h_2$, $h_1, h_2 \in H$ implies that h_1, h_2 are proportional to h. We give a limit procedure for computing minimal excessive measures for an arbitrary Markov semigroup T_t in a standard Borel space E. Analogous results for excessive functions are obtained assuming that an excessive measure γ on E exists such that $T_t f = 0$ if $f = 0$ γ-a.e. In the Appendix, we prove that each excessive element can be decomposed into minimal elements and that such a decomposition is unique.

1. Introduction.

1.1. In 1941 R. S. Martin [13] published a paper where positive harmonic functions in a domain D of a Euclidean space were investigated. Let H stand for the class of all such functions subject to condition $f(a) < \infty$ where a is a fixed point of D. Martin has proved that:

(a) each element of H can be decomposed in a unique way into minimal elements normalized by the condition $f(a) = 1$;

(b) if the Green function of the Laplacian in D is known, then all minimal elements can be computed by a certain limit process.

J. L. Doob [2] has discovered that the Martin decomposition of harmonic functions is closely related to the behaviour of Brownian paths at the first exit time from D. G. A. Hunt [9] has shown that, using these relations, it is possible to get Martin's results by probabilistic considerations. Actually only discrete Markov chains were treated in [1] and [5], however, the methods are applicable to Brownian motion as well.

In [10] Hunt has studied Markov processes with a continuous time parameter on a separable locally compact space and he has proved results of Martin type under certain regularity conditions for the transition functions. The regularity conditions were relaxed by H. Kunita and T. Watanabe [11] and by the author [3], [4]. Now we are able to eliminate them completely and to develop a theory applicable to arbitrary Markov processes in standard Borel spaces. In particular, the theory is easy to apply to general diffusion processes without any restrictions on diffusion

Received by the editors March 14, 1979 and, in revised form, April 30, 1979. Presented in an invited address at the annual meeting of the Institute of Mathematical Statistics, August 13–16, 1979 in Washington, D. C.

AMS (MOS) subject classifications (1970). Primary 60J50; Secondary 60J45, 28A65.

Key words and phrases. Martin boundary, excessive measures and functions, entrance and exit laws, decomposition into minimal (extreme) elements.

[1]Research supported by NSF Grant No. MCS 77-03543.

and drift coefficients. (In the case of Brownian motion we get in this way a new proof of Martin's results.)

The decomposition of excessive measures and functions into minimal elements was studied in [6] and [7]. In the present paper we concentrate on computation of minimal elements. In the Appendix, we give a new proof of existence and uniqueness of the decomposition into minimal elements. This proof is based on constructing sufficient statistics for certain classes of Markov processes.

1.2. We say that a function f is positive if it takes values from the extended real half-line $[0, +\infty]$. We write $f \in \mathfrak{B}$ if f is a positive function measurable with respect to a σ-algebra $\mathfrak{B}$. If m is a measure on $\mathfrak{B}$, we write $f \in L^1(m)$ if f is $\mathfrak{B}$-measurable and m-integrable, and we write $f \in L^1_+(m)$ if in addition $f > 0$ m-a.e. We denote by $m(f)$ the integral of f with respect to m.

2. Discussion of results.

2.1. Let $(E, \mathfrak{B})$ be a measurable space. A function $p_t(x, B)$, $t > 0$, $x \in E$, $B \in \mathfrak{B}$ is called a *stationary transition function* if it is $\mathfrak{B}$-measurable in x, is a measure with respect to B and if

$$p_t(x, E) \leqslant 1 \quad \text{for all } s, x; \tag{2.1}$$

$$\int_E p_s(x, dy)p_t(y, B) = p_{s+t}(x, B) \quad \text{for all } 0 < s < t,\ x \in E,\ B \in \mathfrak{B}. \tag{2.2}$$

If (2.2) but not necessarily (2.1) holds, we call p a *generalized stationary transition function*.

We put

$$mT_t(B) = \int_E m(dx)p_t(x, B) \tag{2.3}$$

and

$$T_t h(x) = \int_E p_t(x, dy)h(y). \tag{2.4}$$

These formulae are meaningful for all measures m on $\mathfrak{B}$ and all functions $h \in \mathfrak{B}$. We say that m is an excessive measure if it is σ-finite and if, for each $B \in \mathfrak{B}$,

$$mT_t(B) \uparrow m(B) \quad \text{as } t \downarrow 0. \tag{2.5}$$

A function $h \in \mathfrak{B}$ is called *excessive*[2] if it is finite a.s. with respect to all measures $p_t(x, \cdot)$ and if, for every $x \in E$,

$$T_t h(x) \uparrow h(x) \quad \text{as } t \downarrow 0. \tag{2.5a}$$

2.2. Throughout this paper we assume that:

2.2.A. *$(E, \mathfrak{B})$ is a standard Borel space.*

2.2.B. For each $B \in \mathfrak{B}$, $p_t(x, B)$ is a $\mathfrak{B}_R \times \mathfrak{B}$-measurable function ($\mathfrak{B}_R$ denotes the σ-algebra of all Borel subsets of the real line R).

[2]Hunt's definition of excessive functions requires measurability with respect to the completion of $\mathfrak{B}$ relative to an arbitrary probability measure. This looks less restrictive than $\mathfrak{B}$-measurability. However, under the assumption 2.2.B, both conditions are equivalent for functions h with the property (2.5.a).

In all propositions on excessive functions we assume in addition that:

2.2.C. All the measures $p_t(x, \cdot)$ are absolutely continuous with respect to a σ-finite measure γ.

The role of this condition is revealed by the following lemma.

LEMMA 2.1. *Under condition 2.2.C, a Radon-Nikodým derivative*

$$\rho_t(x, y) = p_t(x, dy)/\gamma(dy) \tag{2.6}$$

can be chosen to be measurable in x, y and to satisfy the relation

$$\int_E \rho_s(x, y)\, \gamma(dy)\rho_t(y, z) = \rho_{s+t}(x, z) \tag{2.7}$$

for all $x, z \in E$, $s, t > 0$. If the measure γ is excessive, then we can assume in addition, that

$$\int \gamma(dx)\rho_t(x, y) \leqslant 1 \tag{2.8}$$

for all t, y.

We say that γ is a *reference measure* if $\gamma(B) = 0$ if and only if $p_t(x, B) = 0$ for all t and x. If a measure γ satisfies condition 2.2.C, then

$$\gamma_1(B) = \int_0^\infty dt\, e^{-t} \int_E \gamma(dx) p_t(x, B) \tag{2.9}$$

is a reference measure. Obviously all reference measures are equivalent, and each excessive measure γ satisfying 2.2.C is a reference measure.

2.3. We fix a stationary transition function p and we denote by M the set of all excessive measures and by H the set of all excessive functions.

All minimal elements of M can be obtained by passage to a limit from the *Green measure* g_x, $x \in E$, and the *truncated Green measure* g_x^u, $x \in E$, $u > 0$, which are defined by the following formulae

$$g_x(B) = \int_0^\infty p_t(x, B)\, dt, \tag{2.10}$$

$$g_x^u(B) = \int_0^u p_t(x, B)\, dt. \tag{2.11}$$

We say that an element m of M is *conservative* and we write $m \in M_c$ if $g_x(l) = \infty$ a.s. m for all strictly positive measurable l. We say that $m \in M$ is *dissipative* and we write $m \in M_d$ if $g_x(l) < \infty$ a.s. m for all m-integrable positive l.

LEMMA 2.2. *Each minimal element m of M belongs either to M_c or to M_d. If $m \in M_c$ then $mT_t = m$ for all $t > 0$.*

THEOREM 2.1. *Let a minimal element m of M belong to M_c. If $\varphi, \psi \in L^1(m)$, $m(\psi) \neq 0$, then*

$$m(\varphi)/m(\psi) = \lim_{u\to\infty} (g_x^u(\varphi)/g_x^u(\psi)) \tag{2.12}$$

for m-almost all x.

Theorem 2.1 is true for all generalized stationary transition functions.

THEOREM 2.2. *Let a minimal element m of M belong to M_d. There exists a probability measure P on the space E^∞ of all sequences $x_1, x_2, \ldots, x_k, \ldots \in E$ such that, if $\varphi, \psi \in L^1(m)$, $m(\psi) \neq 0$, then*

$$m(\varphi)/m(\psi) = \lim(g_{x_k}(\varphi)/g_{x_k}(\psi)) \tag{2.13}$$

for P-almost all sequences $\{x_k\}$.

2.4. The following implications of Theorems 2.1–2.2 rather than the theorems themselves are useful for practical computation of minimal elements.

COROLLARY. *Let S be a countable family of positive $\mathfrak{B}$-measurable functions. We write $m = S\text{-}\lim m_k$ if $m_k(\varphi) \to m(\varphi)$ for all m-integrable functions $\varphi \in S$. Let m be a minimal element of M. If m is conservative, then*

$$m = S\text{-}\lim_{u\to\infty} c(u)g_x^u \quad \textit{for some } x \in E.$$

If m is dissipative, then

$$m = S\text{-}\lim c_k g_{x_k} \quad \textit{for some } x_1, x_2, \ldots \in E.$$

Here c are constants (which can be expressed by formulae

$$c(u) = m(\psi)/g_x^u(\psi), \qquad c_k = m(\psi)/g_{x_k}(\psi)$$

for an arbitrary function $\psi \in L^1_+(m)$).

2.5. Now let γ be a reference measure and let

$$g^y(x) = \int_0^\infty \rho_t(x, y)\, dt, \tag{2.14}$$

$$g_u^y(x) = \int_0^u \rho_t(x, y)\, dt \tag{2.15}$$

where ρ is described in Lemma 2.1. We call g^y the *Green function* and g_u^y the *truncated Green function.*

Put $\gamma^h(dx) = h(x)\gamma(dx)$, $\gamma^y(dx) = g^y(x)\gamma(dx)$. An element h of H is called *conservative* if $\gamma^y(\varphi) = \infty$ γ^h-a.e. for each strictly positive φ, and it is called *dissipative*, if $\gamma^y(\varphi) < \infty$ γ^h-a.e. for each γ^h-integrable positive φ. These definitions are independent of the choice of reference measure. The set of all conservative elements of H is denoted by H_c and the set of all dissipative elements by H_d.

LEMMA 2.3. *Suppose that condition 2.2.C holds for an excessive measure γ. Then each minimal element h of H belongs either to H_c or to H_d. If $h \in H_c$ then $T_t h = h$ for all $t > 0$.*

THEOREM 2.3. *Suppose that h is a conservative minimal element of H. If h is integrable relative to measures ξ and η and if $\eta(h) \neq 0$, then*

$$\xi(h)/\eta(h) = \lim_{u\to\infty} (\xi(g_u^y)/\eta(g_u^y)) \tag{2.16}$$

for γ^h-almost all y.

THEOREM 2.4. *Let h be a dissipative minimal element of H. Then there exists a probability measure P on the space E^∞ such that, if h is integrable with respect to ξ and η and if $\eta(h) \neq 0$, then*

$$\xi(h)/\eta(h) = \lim(\xi(g^{y_k})/\eta(g^{y_k})) \tag{2.17}$$

for almost all sequences $\{y_k\}$.

COROLLARY. *Let S be a countable family of measures on $(E, \mathcal{B})$. We write $f = S\text{-}\lim f_k$ if $\xi(f_k) \to \xi(f)$ for all $\xi \in S$ such that f is ξ-integrable. Let h be a minimal element of H. If $h \in H_c$, then*

$$h = S\text{-}\lim c(u)g_u^y \quad \textit{for some } y \in E.$$

If $h \in H_d$, then

$$h = S\text{-}\lim c_k g^{y_k} \quad \textit{for some } y_1, y_2, \ldots \in E.$$

(Constants c can be calculated by the formulae

$$c(u) = \eta(h)/\eta(g_u^y), \qquad c_k = \eta(h)/\eta(g^{y_k}) \tag{2.18}$$

where η is an arbitrary measure such that $\eta(h) < \infty$.)

REMARK. For $\xi(B) = 1_B(x)$, formulae (2.16) and (2.17) can be rewritten in the following form

$$h(x) = \lim_{u\to\infty} c(u)g_u^y(x), \tag{2.16a}$$

$$h(x) = \lim c_k g^{y_k}(x) \tag{2.17a}$$

where c are given by formulae (2.18). Let ν be a σ-finite measure such that $\nu\{h = \infty\} = 0$. By Fubini's theorem, for γ^h-almost all y, formula (2.16a) is true for ν-almost all x. Analogously, for P-almost all sequences y_k, formula (2.17a) holds for ν-almost all x.

2.6. Theorem 2.4 implies immediately Martin's results on minimal positive harmonic functions. The condition 2.2.C is satisfied for Lebesgue measure γ. The density $\rho_t(x, y)$ is symmetric and

$$\int_D g^y(x)\gamma(dx) = \int_D \int_0^\infty \gamma(dx)\rho_t(x, y)\,dt = \int_0^\infty \rho_t(y, D)\,dt = E_y\beta$$

where β is the first exit time of Brownian motion from the domain D. For a bounded domain D, the right side is finite and therefore all elements of H are dissipative. Suppose that h is a minimal element and that $h(a) = 1$. According to the remark at the end of Subsection 2.5, there exists a sequence $y_k \in E$ such that

$$h(x) = \lim(g^{y_k}(x)/g^{y_k}(a)) \quad \text{for } \gamma\text{-almost all } x \in D. \tag{2.19}$$

Take a convergent subsequence and, changing notations, denote it y_k again. If $\lim y_k = y \in D$, then $h(x) = g^y(x)/g^y(a)$ γ-a.e., hence everywhere in D (because both functions are superharmonic). If $y \overline{\in} D$, then the limit in the right side of (2.19) is a harmonic function (g^y is a harmonic in $D \setminus \{y\}$ and Harnack's inequality implies uniform convergence on each compact subset of D). The sequence y_k

corresponds to a point of the Martin boundary and the function defined by (2.19) is the minimal harmonic function associated with this point.[3]

We see how small the part of this picture which depends on the analytic properties of classical harmonic functions is.

2.7. Although an explicit description of the measure P in Theorems 2.2 and 2.4 is not important for computing minimal elements, it is instructive from the point of view of stochastic processes.

Consider a decreasing sequence t_n, $n = 0, 1, \ldots$, and a sequence of σ-finite measures ν_{t_n} subject to the condition

$$\nu_{t_n} T_{t_{n-1}-t_n} = \nu_{t_{n-1}}, \qquad n = 1, 2, \ldots . \tag{2.20}$$

Let l be a positive function and $\nu_{t_0}(l) = 1$. Formulae

$$m_{t_0}(dx_0) = \nu_{t_0}(dx_0) l(x_0), \tag{2.21}$$

$$m_{t_n t_{n-1} \cdots t_0}(dx_n, dx_{n-1}, \ldots, dx_0)$$
$$= \nu_{t_n}(dx_n) p_{t_{n-1}-t_n}(x_n, x_{n-1}) \cdots p_{t_0-t_1}(x_1, dx_0) l(x_0)$$

define a compatible family of finite-dimensional distributions, and by Kolmogorov's theorem, there exists a sequence X_{t_n} of random variables taking values in E such that $m_{t_n t_{n-1} \cdots t_0}$ is the probability distribution of $X_{t_n}, X_{t_{n-1}}, \ldots, X_{t_0}$. In other words, there exists a measure P in E^∞ such that $m_{t_n \ldots t_0}(\Gamma_n \times \cdots \times \Gamma_0)$ is the measure of the cylinder with the base $\Gamma_n \times \cdots \times \Gamma_0$.

Now each minimal element m of M is either *invariant*, i.e., $mT_t = m$ for all t, or $mT_t \downarrow 0$ as $t \to \infty$ (in the second case we call m *null-excessive*).

If m is invariant, then (2.20) is satisfied for $\nu_{t_0} = \nu_{t_1} = \cdots = \nu_{t_n} = \cdots = m$. For $m \in M_d$, Theorem 2.2 holds for every measure P corresponding to a sequence $\nu_{t_n} = m$, $t_n \downarrow -\infty$.

If m is null-excessive, then it can be represented in the form

$$m = \int_0^\infty \nu_t \, dt \tag{2.22}$$

where

$$\nu_s T_{t-s} = \nu_t \quad \text{for all } 0 < s < t.$$

Theorem 2.2 holds for every measure P corresponding to a sequence ν_{t_n}, $t_n \downarrow 0$.

The random variables X_{t_n} form a Markov process. It is natural to interpret the set of minimal elements of M (with proportional elements identified) as the *entrance space* for this process. This space consists of two parts: *the entrance space at time* 0 corresponding to the null-excessive elements and *the entrance space at* $-\infty$ corresponding to the invariant elements.

---

[3]The points corresponding to the minimal harmonic functions form only a part of the Martin boundary. The fundamental results on minimal positive harmonic functions described in Subsection 1.1 have been obtained originally by using a representation of harmonic functions as integrals over all the boundary. The subsequent development has shown that not the entire boundary but only its minimal part is of real importance.

Using (2.21) for a fixed t_0 and variable $t_1, t_2, \ldots,$ it is possible to define a Markov process X_t for all $t < t_0$. The statement of Theorem 2.2 remains true if t tends to 0 or $-\infty$ over any countable subset of $(-\infty, t_0)$. It can take all real values if the paths of X_t have certain regularity properties (as in the case of Brownian motion).

2.8. To construct a measure P mentioned in Theorem 2.4, we consider an increasing sequence t_n, $n = 0, 1, \ldots,$ a sequence of positive measurble functions φ^{t_n} subject to conditions

$$T_{t_n - t_{n-1}} \varphi^{t_n} = \varphi^{t_{n-1}}, \qquad n = 1, 2, \ldots,$$

and a measure ν such that $\nu(\varphi^{t_0}) = 1$. Let P be a measure on E^∞ corresponding to the finite-dimensional distributions

$$m_{t_0}(dx_0) = \nu(dx_0)\varphi^{t_0}(x_0),$$

$$m_{t_0 t_1 \cdots t_n}(dx_0, dx_1, \ldots, dx_n)$$
$$= \nu(dx_0) p_{t_1 - t_0}(x_0, dx_1) \cdots p_{t_n - t_{n-1}}(x_{n-1}, dx_n)\varphi^{t_n}(x_n).$$

If a minimal element h of H is *invariant* (i.e., $T_t h = h$ for all t), then Theorem 2.4 holds for a measure P corresponding to $\varphi^{t_n} = h$, $t_n \uparrow +\infty$. If h is *null-excessive* (i.e., $T_t h \downarrow 0$ as $t \to \infty$), then

$$h = \int_{-\infty}^{0} \varphi^t \, dt \tag{2.23}$$

where

$$T_{t-s}\varphi^t = \varphi^s \quad \text{for all } s < t < 0,$$

and we can use any measure P corresponding to φ^{t_n} and $t_n \uparrow 0$.

The set of minimal elements of H, with the identification of proportional elements, can be interpreted as the exit space at time 0 for null-excessive elements, and at time $+\infty$ for invariant elements. (Time 0 can be replaced with any other finite time s_0.)

From a probabilistic point of view, it is more natural to consider a stochastic process with random birth time α and death time β and to interpret elements of the entrance and exit spaces as possible birth and death places (cf. Theorems 7.2 and 7.4).

3. Conservative minimal elements.

3.1. A function $v(x, B)$, $x \in E$, $B \in \mathcal{B}$, is called a *kernel* if it is a $\mathcal{B}$-measurable function of x for each $B \in \mathcal{B}$, and is a measure relative to B for each $x \in E$. A kernel $v(x, B)$ defines a transformation of measures

$$mV(B) = \int_E m(dx) v(x, B)$$

and a transformation of positive measurable functions

$$V\varphi(x) = \int_E v(x, dy)\varphi(y).$$

Two kernels v and v^* are *dual* relative to a measure m if

$$\int \varphi(x) V\psi(x) m(dx) = \int \psi(x) V^*\varphi(x) m(dx)$$

for all φ and ψ.

To each V there correspond the *Green operator*

$$G\varphi(x) = \sum_{k=0}^{\infty} V^k\varphi(x) \tag{3.1}$$

and the *truncated Green operator*

$$G_n\varphi(x) = \sum_{k=0}^{n-1} V^k\varphi(x). \tag{3.2}$$

The following proposition is one form of the Chacon-Ornstein ergodic theorem.

THEOREM 3.1. *Let v be a kernel on $(E, \mathfrak{B})$ and let a σ-finite measure m on $\mathfrak{B}$ satisfy the condition*

$$mV(B) \leq m(B) \quad \textit{for all } B \in \mathfrak{B}. \tag{3.3}$$

Then $m = m_c + m_d$ where

3.1.A. *The measures m_c and m_d are singular with respect to each other,*

3.1.B. *$G\varphi = 0$ or $+\infty$ m_c-a.e. for each $\varphi \in \mathfrak{B}$,*

3.1.C. *$G\varphi < \infty$ m_d-a.e. for each $\varphi \in L^1_+(m)$.*

These properties define the measures m_c and m_d uniquely. For each $l \in L^1_+(m)$, we have $m_c(B) = m(B \cap E_c)$, $m_d(B) = m(B \cap E_d)$ where $E_c = \{x: Gl(x) = \infty\}$, $E_d = \{x: Gl(x) < \infty\}$.

The measures m_c and m_d are called the conservative and dissipative parts of m relative to V.

The following statments hold.

3.1.D. *If $\varphi = 0$ m_d-a.e., then $V\varphi = 0$ m_d-a.e. and $m(V\varphi) = m(\varphi)$.*

3.1.E. *If $V1 \leq 1$ m-a.e., then the equality $\psi = 0$ m_c-a.e. implies the equality $V\psi = 0$ m_c-a.e.*

3.1.F. *Suppose that $m_d = 0$. Put $B \in \mathfrak{B}^V_m$ if $B \in \mathfrak{B}$ and if*

$$\int_B \varphi dm = \int_B V\varphi dm \quad \textit{for all } \varphi \in \mathfrak{B}.$$

The class $\mathfrak{B}^V_m$ is a σ-algebra in $\mathfrak{B}$ and

$$\lim(G_n\varphi(x)/G_n\psi(x)) = m^\psi\left(\frac{\varphi}{\psi}|\mathfrak{B}^V_m\right) \tag{3.4}$$

for m-almost all x if $\varphi \in L^1(m)$, $\psi \in L^1_+(m)$. (Here $m^\psi(dx) = \psi(x)m(dx)$.)

All these statements, except 3.1.E, are proved, e.g., in [14] (§§V.5 and V.6).

Let us prove 3.1.E. If $V1 \leq 1$, m-a.e., then $m(V^*\varphi) = m(\psi V1) \leq m(\psi)$ for all $\psi \in \mathfrak{B}$, and V^* satisfies condition (3.3). Let G^* be the corresponding Green operator. Take $l \in L^1_+(m)$. As we know, m_c and m_d are the restrictions of m to $E_c = \{Gl = \infty\}$ and $E_d = \{Gl < \infty\}$. Because of 3.1.C, there exists a function $\hat{l} \in L^1(m)$ such that $\hat{l} = 0$ on E_c, $\hat{l} > 0$ on E_d and $m(\hat{l}Gl) = m_d(\hat{l}Gl) < \infty$. We have $m(lG^*\hat{l}) < \infty$, hence $m\{G^*\hat{l} = \infty\} = 0$. Let $\hat{m}_c$ and $\hat{m}_d$ be the conservative and dissipative parts of m relative to V^*. By 3.1.B $\hat{m}_c\{0 < G^*\hat{l} < \infty\} = 0$ and $\hat{m}_c\{0 < G^*\hat{l}\} = 0$. However $G^*\hat{l} \geq \hat{l} > 0$ on E_d. Hence $\hat{m}_c(E_d) = 0$ and, for all $B \in \mathfrak{B}$, $m_d(B) = m(B \cap E_d) = \hat{m}_d(B \cap E_d) \leq \hat{m}_d(B)$. Because the roles of V and V^* are symmetric, we also have $\hat{m}_d \leq m_d$. Hence $\hat{m}_d = m_d$, $\hat{m}_c = m_c$.

Now let $\varphi = 0$ m_d-a.e., $\psi = 0$ m_c-a.e. Then, by 3.1.D, $V^*\varphi = 0$, $\hat{m}_d = m_d$-a.e. and $m_c(\varphi V\psi) = m(\psi V^*\varphi) = 0$. Hence $V\psi = 0$ m_c-a.e.

REMARK. Using the relation $G_{n+1}\varphi = G_n V\varphi + \varphi$, it is easy to prove that the limit (3.4) coincides m_c-a.e. with the limits

$$\lim_{n\to\infty} (G_{n+1}\varphi(x)/G_n\psi(x)) = \lim_{n\to\infty} (G_n\varphi(x)/G_{n+1}\psi(x)). \tag{3.5}$$

3.2. Now we apply Theorem 3.1 to investigating the class M of all excessive measures associated with a stationary transition function $p_t(x, B)$. Let T_t be operators defined by formulas (2.3) and (2.4). Consider the Green measure g_x and the truncated Green measure g_x^u introduced by (2.10) and (2.11).

We put

$$E_0 = \{x: p_t(x, B) = 0 \text{ for all } t, B\}$$

and we notice that, if l is strictly positive, then

$$g_x(l) > 0 \quad \text{on } E \setminus E_0. \tag{3.6}$$

Indeed, if $g_x(l) = 0$, then $p_t(x, E) = 0$ for almost all t, and $x \in E_0$ because of (2.2).

LEMMA 3.1. *Fix a strictly positive function $l \in \mathfrak{B}$ and consider, for each measure m on $\mathfrak{B}$, its restrictions m_c and m_d to the sets*

$$E_c = \{x: g_x(l) = \infty\}, \qquad E_d = \{x: g_x(l) < \infty\}.$$

If $m \in M$ and $m(l) < \infty$, then $m_c \in M_c$ and $m_d \in M_d$. Moreover m_c is invariant with respect to operators T_t.

PROOF. 1°. Since (3.3) holds for $V = T_t$, we have a decomposition $m = m_c^t + m_d^t$ where m_c^t and m_d^t satisfy conditions 3.1.A, B, C. For each $\varphi \in \mathfrak{B}$

$$g_x(\varphi) = G\tilde{\varphi}(x) \quad \text{where } \tilde{\varphi}(x) = g_x^t(\varphi). \tag{3.7}$$

It follows from 3.1.B and (3.7) that $m_d^t(E_c) = 0$, $m_c^t(E_d \setminus E_0) = 0$. Besides $GT_t l(x) = l(x)$ on E_0 and, since $0 < l < \infty$ m-a.e., we have $m_c^t(E_0) = 0$ by 3.1.B. Let B_c, B_d stand for the intersections of $B \in \mathfrak{B}$ with E_c and E_d. We have $m_c(B) = m(B_c) = m_c^t(B_c) = m_c^t(B)$. Thus $m_c = m_c^t$. Analogously $m_d = m_d^t$.

2°. Let φ_c, φ_d be the restrictions of $\varphi \in \mathfrak{B}$ to E_c, E_d. By 3.1.D and 3.1.E, we have

$$m_c(T_t\varphi) = m_c(T_t\varphi_c) = m(T_t\varphi_c) = m(\varphi_c) = m_c(\varphi),$$

$$m_d(T_t\varphi) = m_d(T_t\varphi_d) = m(T_t\varphi_d)\uparrow m(\varphi_d) = m_d(\varphi) \quad \text{as } t\downarrow 0.$$

Hence m_c and m_d belong to M and m_c is invariant with respect to T_t.

3°. It follows from 3.1.B, C, (3.6) and (3.7), that $g_x(\varphi) = \infty$, $m_c = m_c^t$-a.e. if $\varphi > 0$, and $g_x(\varphi) < \infty$, $m_d = m_d^t$-a.e. if $\varphi \geq 0$, $m(\varphi) < \infty$. Hence $m_c \in M_c$, $m_d \in M_d$.

3.3. Evidently Lemma 3.1 implies Lemma 2.2. Theorem 2.1 follows from the following result.

THEOREM 3.2. *Let* $m \in M_c$. *Put* $B \in \mathfrak{B}_m^T$ *if* $B \in \mathfrak{B}$ *and*

$$\int_B \varphi dm = \int_B T_t \varphi dm \quad \text{for all } t > 0, \varphi \in \mathfrak{B}. \tag{3.8}$$

The class $\mathfrak{B}_m^T$ *is a* σ-*algebra in* E. *If* $\varphi \in L^1(m)$, $\psi \in L_+^1(m)$, *then*

$$\lim_{u \to \infty} (g_x^u(\varphi)/g_x^u(\psi)) = m^{\psi}\left(\frac{\varphi}{\psi} | \mathfrak{B}_m^T\right) \tag{3.9}$$

for m-*almost all* x. *If* m *is a minimal element of* M, *then the right side of* (3.9) *is equal to* $m(\varphi)/m(\psi)$.

PROOF. We apply part 3.1.F of Theorem 3.1 to $V = T_t$. Notice that G_n in formula (3.4) and g_x^u in (2.11) are connected by the relation

$$G_n\tilde{\varphi}(x) \leqslant g_x^u(\varphi) \leqslant G_{n+1}\tilde{\varphi}(x) \quad \text{for } nt \leqslant u < (n+1)t \tag{3.10}$$

where $\tilde{\varphi}(x) = g_x^t(\varphi)$. We can write (3.10) in the following form

$$G_n\tilde{\varphi}/c_x(u) \leqslant g_x^u(\varphi)/c_x(u) \leqslant G_{n+1}\tilde{\varphi}/c_x(u) \tag{3.11}$$

for $nt \leqslant u < (n+1)t$ where $c_x(u) = G_n\psi(x)$ for $nt \leqslant u < (n+1)t$. Since $m(\tilde{\varphi}) = tm(\varphi) < \infty$, it follows from (3.4), (3.5) and (3.11) that

$$\lim_{u \to \infty} (g_x^u(\varphi)/c_x(u)) = m^{\psi}\left(\frac{\tilde{\varphi}}{\psi} | \mathfrak{B}_m^V\right), \quad m\text{-a.e.} \tag{3.12}$$

It is easy to check that the right side of (3.12) equals $tm^{\psi}(\varphi/\psi|\mathfrak{B}_m^V)$. Since this expression is equal to t for $\varphi = \psi$, we have from (3.12)

$$\lim_{u \to \infty} (g_x^u(\varphi)/g_x^u(\psi)) = m^{\psi}\left(\frac{\varphi}{\psi} | \mathfrak{B}_m^V\right), \quad m\text{-a.e.} \tag{3.13}$$

Denote by F the left side of (3.13) (on the set where the limit does not exist, we replace it by lim sup). It follows from (3.13) that $F \in \mathfrak{B}_m^V$ for all $V = T_t$. Hence $F \in \mathfrak{B}_m^T$ and (3.13) implies (3.9).

It is easy to see that if $m \in M_c$, then its restriction m_B to any set $B \in \mathfrak{B}_m^T$ belongs to M. If m is minimal then m_B is proportional to m, hence $m(B) = 0$ or $m(E \setminus B) = 0$. Therefore each $\mathfrak{B}_m^T$-measurable function is constant a.s. m. The last statement of Theorem 3.2 follows easily from this observation.

REMARK. In Subsections 3.2, 3.3, only the proof of Lemma 3.1 makes use of 3.1.E and therefore depends on the part (2.1) of the definition of a stationary transition function. The rest is valid for generalized transition functions as well.

3.4. PROOF OF LEMMA 2.1. We start from any version $\bar{p}_t(x, y)$ of the Radon-Nikodým derivative $p_t(x, dy)/\gamma(dy)$ measurable in x, y and we put

$$\begin{aligned} \rho_s^r(x, z) &= \int \bar{p}_{s-r}(x, y)\gamma(dy)\bar{p}_r(y, z) \\ &= \int p_{s-r}(x, dy)\bar{p}_r(y, z), \qquad 0 < r < s. \end{aligned}$$

We set $z \in E'$ if, for all rational $s > r > 0$,

$$\rho_s^r(x, z) = \bar{\rho}_s(x, z) \quad \text{for } \gamma\text{-almost all } x. \tag{3.14}$$

It is easy to check that

$$\int_{B_1} \int_{B_2} \gamma(dx)\rho_s^r(x, z)\gamma(dz) = \int_{B_1} \int_{B_2} \gamma(dx)\bar{\rho}_s(x, z)\gamma(dz)$$

for all $B_1, B_2 \in \mathfrak{B}$. Hence (3.14) holds for γ-almost all z, and $\gamma(E \setminus E') = 0$.

Now if $z \in E'$ and $0 < r_1 < r_2 < s$ are rational, then

$$\begin{aligned}\rho_s^{r_2}(x, z) &= \int p_{s-r_2}(x, dy_1)\bar{\rho}_{r_2}(y_1, z) = \int p_{s-r_2}(x, dy_1)\rho_{r_2}^{r_1}(y_1, z) \\ &= \int p_{s-r_2}(x, dy_1)p_{r_2-r_1}(y_1, dy_2)\bar{\rho}_{r_1}(y_2, z) \\ &= \int p_{s-r_1}(x, dy_2)\bar{\rho}_{r_1}(y_2, z) = \rho_s^{r_1}(x, z),\end{aligned}$$

and we can define $\rho_s(x, z)$ for $z \in E'$ by the formula $\rho_s(x, z) = \rho_s^r(x, z)$ for any rational $r \in (0, s)$. For $z \,\overline{\in}\, E'$ we put $\rho_s(x, z) = \rho_s(x, z_0)$ where z_0 is a fixed element of E'. Obviously ρ satisfies (2.6) and (2.7).

If γ is excessive, then, for each $B \in \mathfrak{B}$, $s > r > 0$

$$\int \gamma(dx)\int_B \rho_s^r(x, z)\gamma(dz) = \int \gamma(dx)p_s(x, B) \leqslant \gamma(B)$$

hence for γ-almost all z

$$\int \gamma(dx)\rho_s(x, z) \leqslant 1. \tag{3.15}$$

Put $z \in E''$ if $z \in E'$ and (3.15) holds. Since $\gamma(E' \setminus E'') = 0$, we can replace E' with E'' in the definition of ρ, and we get a function which satisfies (2.8) as well as (2.6) and (2.7).

3.5. The investigation of excessive functions associated with a stationary transition function p can be reduced to investigating excessive measures associated with another stationary transition function $\hat{p}$.

Suppose that $\gamma \in M$ is a reference measure for p and let $\rho_t(x, y)$ be the function defined in Lemma 2.1. Then the formula

$$\hat{p}_t(x, dy) = \gamma(dy)\rho_t(y, x) \tag{3.16}$$

defines a stationary transition function. Let $\hat{T}_t$ be the operators corresponding to $\hat{p}$. For all $\varphi, \psi \in \mathfrak{B}$, $t > 0$

$$\gamma(\varphi T_t \psi) = \gamma(\psi \hat{T}_t \varphi). \tag{3.17}$$

Hence

$$\gamma(\varphi g_u^x) = \hat{g}_x^u(\varphi), \qquad \gamma(\varphi g^x) = \hat{g}_x(\varphi) \tag{3.18}$$

where g_u^x and g^x are defined by (2.14), (2.15), and

$$\begin{aligned}\hat{g}_x^u(B) &= \int_0^u \hat{p}_t(x, B)\, dt = \int_E \gamma(dy)\int_B g_u^y(x)\gamma(dx), \\ \hat{g}_x(B) &= \hat{g}_x^\infty(B)\end{aligned} \tag{3.19}$$

are the truncated Green measure and the Green measure for $\hat{p}$. Denote by $\hat{M}$ the class of excessive measures for $\hat{p}$. Put

$$\gamma^h(dy) = h(y)\gamma(dy). \tag{3.20}$$

By (3.17) $\gamma^h(\hat{T}_t\varphi) = \gamma(\varphi T_t h)$. Hence if $h \in H$, then $\gamma^h \in \hat{M}$. Now suppose that $m \in \hat{M}$. Then $(m\hat{T}_t)(B) = \int_B a_t(y)\gamma(dy)\uparrow m(B)$ as $t\downarrow 0$ where $a_t(y) = \int m(dx)p_t(x, y)$. This implies the existence of a function $h \in H$ such that $a_t(y)\uparrow h(y)$ m-a.e. as $t\downarrow 0$ (see [**6**, §§4 and 5] for details). Obviously $m = \gamma^h$. Hence the mapping $h \to \gamma^h$ defined by (3.20) is a 1-1-splitting of H onto $\hat{M}$. It is easy to see that under this mapping the sets of minimal, conservative and dissipative elements of H correspond to analogous subsets of $\hat{M}$. Hence Lemma 2.3 follows from Lemma 2.2.

3.6. Now we prove Theorem 2.3. Let h be a conservative minimal element of H. Then γ^h is a conservative minimal element of $\hat{M}$. By Theorem 2.1, if $\varphi, \psi \in L^1(\gamma^h)$ and $\gamma^h(\psi) \neq 0$, then

$$\gamma^h(\varphi)/\gamma^h(\psi) = \lim_{u\to\infty} (\hat{g}_x^u(\varphi)/\hat{g}_x^u(\psi)), \quad \gamma^h\text{-a.e.} \tag{3.21}$$

(Since Theorem 2.1 holds for generalized stationary transition funcions, we do not need an assumption that the reference measure γ is excessive.)

Let h be integrable with respect to a measure ξ. Put $\varphi(y) = \int \xi(dz)\rho_t(z, y)$. We have

$$\gamma^h(\varphi) = \xi(T_t h) = \xi(h) < \infty \tag{3.22}$$

$$\hat{g}_x^u(\varphi) = \int_t^{t+u} ds \int \xi(dz)\rho_s(z, x) = \xi(g_{t+u}^x) - \xi(g_t^x). \tag{3.23}$$

Here $\xi(g_t^x) < \infty$ γ^h-a.e. since

$$\int \xi(g_t^x)\gamma^h(dx) = \xi\left(\int_0^t T_s h\,ds\right) = t\xi(h) < \infty. \tag{3.24}$$

It follows from (3.21), (3.22), (3.23) and (3.24) that

$$\xi(h)/\gamma^h(\psi) = \lim \xi(g_u^x)/c(u), \quad \gamma^h\text{-a.e.}$$

where $c(u) = \hat{g}_x^{u-t}(\psi)$. This implies Theorem 2.3.

4. Time-dependent excessive measures and functions.

4.1. The space M of excessive measures associated with a stationary transition function is a subset of a larger space TM of time-dependent excessive measures. Put

$$p(s, x; t, B) = p_{t-s}(x, B),$$

$$T_t^s\varphi(x) = \int p(s, x; t, dy)\varphi(y),$$

$$(nT_t^s)(B) = \int n(dx)p(s, x; t, B).$$

Suppose that for each $t \in R$ a σ-finite measure n_t on $(E, \mathcal{B})$ is given and let $n_s T_t^s \uparrow n_t$ as $s\uparrow t$. Then we say that n is a *time-dependent excessive measure* and we write $n \in TM$.

An important example of time-dependent excessive measures are entrance laws. We say that an element $n \neq 0$ of TM is an *entrance law* at time s_0 $(-\infty \leq s_0 < +\infty)$ if $n_t = 0$ for $t \leq s_0$, $n_s T_t^s = n_t$ for $s_0 < s < t$.

4.2. Let a positive measurable function f^t on E be given for each $t \in R$ and let f^t be finite a.e. with respect to all measures $p(s, x; t, -)$. We say that f is a *time-dependent excessive function* and we write $f \in TH$ if $f^t < \infty$ a.s. with respect to all measures $p(s, x; t, \cdot)$ and $T_t^s f^t \uparrow f^s$ as $t \downarrow s$. An element $f \neq 0$ of TH is called an *exit law* at time u_0, $-\infty < u_0 \leq +\infty$, if $f^t = 0$ for $t \geq u_0$, $T_t^s f^t = f^s$ for $s < t < u_0$. It is easy to see that $f^s < \infty$ a.s. n_s for every $f \in TH$, $n \in TM$.

4.3. All these definitions are applicable also to nonstationary transition functions $p(s, x; t, \cdot)$. (In the nonstationary case, the state space $(E_t, \mathfrak{B}_t)$ can depend on t and $p(s, x; t, B)$ is defined for $s < t \in R$, $x \in E_s$, $B \in \mathfrak{B}_t$.) To get the definition of such functions, we replace conditions (2.1)–(2.2) by

$$p(s, x; t, E) \leq 1 \quad \text{for all } s < t \in R, x \in E, \tag{4.1}$$

$$\int_E p(s, x; t, dy) p(t, y; u, B) = p(s, x; u, B) \tag{4.2}$$

for all $s < t < u \in R$, $x \in E$, $B \in \mathfrak{B}$.

We put $p(s, x; t, B) = 0$ for $s \geq t$. Obviously $n_t(B) = p(s, x; t, B)$ is an entrance law at time s, and $f^s(x) = p(s, x; t, B)$ is an exit law at time t.

4.4. Condition 2.2.B implies that, for each $n \in TM$ and every $\varphi \in \mathfrak{B}$, $n_t(\varphi)$ is measurable in t. Indeed, for every finite set $\Lambda = \{t_1 < t_2 < \cdots < t_n\}$, the function

$$F_\Lambda(t) = 0 \quad \text{for } t \leq t_1 \text{ and } t > t_n,$$
$$F_\Lambda(t) = n_{t_k}(T_t^{t_k}\varphi) \quad \text{for } t_k < t \leq t_{k+1}$$

is measurable in t, and $F_{\Lambda_k}(t) \to n_t(\varphi)$ if Λ_k is an increasing sequence with the union everywhere dense in R. The same arguments show that if, for each t, $n_t(\varphi)$ is a measurable function of a parameter ω, then it is measurable in t, ω.

4.5. Let $c(t)$, $t \in R$, and $l(x)$, $x \in E$, be positive measurable functions and γ be a measure on E. Then

$$n_t(B) = \int_{-\infty}^{t} ds\, c(s) \int_E \gamma(dx) p(s, x; t, B) \tag{4.3}$$

is a time-dependent excessive measure, and

$$f^s(x) = \int_s^{\infty} dt\, c(t) \int_E p(s, x; t, dy) l(y) \tag{4.4}$$

is a time-dependent excessive function.

4.6. Let $p(s, x; t, B)$ be a nonstationary transition function. Put $x \in E_s^0$ if $p(s, x; t, B) = 0$ for all t, B. (For a stationary transition function, E_s^0 does not depend on s and coincides with E^0 defined in Subsection 3.2.) If $f \in TH$, then $f^s(x) = 0$ for all $x \in E_s^0$. We set $f \in TH^+$ if $f^s(x) > 0$ outside E_s^0.

For each $q \in TH^+$, the formula

$$\begin{aligned} p^q(s, x; t, dy) &= q^s(x)^{-1} p(s, x; t, dy) q^t(y) \quad \text{for } 0 < q^s(x) < \infty, \\ &= 0 \quad \text{if } q^s(x) = 0 \text{ or } q^s(x) = \infty \end{aligned} \tag{4.5}$$

defines a new transition function. A function f is a time-dependent excessive function for p if and only if

$$f_q^t(x) = f^t(x)/q^t(x) \quad \text{for } 0 < q^t(x) < \infty,$$
$$= 0 \quad \text{otherwise}$$

is a time-dependent excessive function for p^q. Analogously n is a time-dependent excessive measure for p if and only if $n_t^q(dx) = q^t(x)n_t(dx)$ is a time-dependent excessive measure for p^q.

5. Markov processes.

5.1. *A stochastic process on a random time interval* is determined by the following elements:

(i) a measure: $(\Omega, \mathfrak{F}, P)$,

(ii) two measurable functions $\alpha(\omega) < \beta(\omega)$ on Ω with values in the extended real line $[-\infty, +\infty]$,

(iii) for each $t \in R$, a measurable mapping $x_t(\omega)$ of the set $\{\omega; \alpha(\omega) < t < \beta(\omega)\}$ into a measurable space $(E_t, \mathfrak{B}_t)$.

The moments α and β are called *the birth time and the death time*.

We say that *a path* ω is given if a point $\omega(t)$ of E_t is fixed for each t of an open interval $I \subset R$, and we say that a process x_t is *canonical* if Ω coincides with the space of all paths, if $x_t(\omega) = \omega(t)$, $t \in I = (\alpha(\omega), \beta(\omega))$, and if $\mathfrak{F}$ is the minimal σ-algebra in Ω which contains the sets

$$\{\alpha < t\}, \quad \{\beta > t\}, \quad \{\alpha < t, x_t \in B, \beta > t\} \tag{5.1}$$

for all $t \in R$, $B \in \mathfrak{B}$.

THEOREM 5.1. *Let p be a transition function on a standard Borel space $(E, \mathfrak{B})$ and let $n \in TM, f \in TH$. Then there exists a canonical stochastic process (x_t, P_n^f) such that*

$$P_n^f\{\alpha < t_1, x_{t_1} \in B_1, \ldots, x_{t_k} \in B_k, t_k < \beta\}$$
$$= \int_{B_1} \cdots \int_{B_k} n_{t_1}(dx_1)p(t_1, x_1, t_2, dx_2) \cdots p(t_{k-1}, x_{k-1}, t_k, dx_k)f^{t_k}(x_k) \tag{5.2}$$

for all $k = 1, 2, \ldots,$ all $t_1 < t_2 < \cdots < t_k \in R$ and $B_1, \ldots, B_k \in \mathfrak{B}$. We have

$$P_n^f(\Omega) = \langle n, f \rangle \tag{5.3}$$

where $\langle n, f \rangle$ is the supremum, over all finite sets $\Lambda = \{t_1 < \cdots < t_k\}$, of expressions

$$c_\Lambda = \sum_{i=1}^{k} n_{t_i}(f^{t_i}) - \sum_{i=2}^{k} n_{t_{i-1}}\left(T_{t_i}^{t_{i-1}} f^{t_i}\right). \tag{5.4}$$

(We put $c_\Lambda = \infty$ if $n_{t_i}(f^{t_i}) = \infty$ for some i.) Also $\langle n, f \rangle = \lim c_{\Lambda_j}$ for every increasing sequence Λ_j with the union everywhere dense in R.

Theorem 5.1 has been proved in **[6]** for the case $\langle n, f \rangle = 1$ and in **[12]** for the general case. (An even more general situation has been discussed in **[8]**.)

5.2. Evidently $\langle n, f\rangle$ is linear in n and f. We remark that $\alpha = s_0$ a.s. P_n^f if and only if n is an entrance law at time s_0, and, in this case,

$$n_t(f^t)\uparrow\langle n, f\rangle \quad \text{as } t\downarrow s_0. \tag{5.5}$$

Analogously $\beta = u_0$ a.s. P_n^f if and only if f is an exit law at time u_0. In this case,

$$n_t(f^t)\uparrow\langle n, f\rangle \quad \text{as } t\uparrow u_0. \tag{5.5a}$$

A measure P_m^f corresponding to $m_t(B) = p(s, x; t, B)$ is denoted by $P_{s,x}^f$. It follows from (5.5) and (2.5a) that $P_{s,x}^f(\Omega) = f^s(x)$.

A simple calculation shows that

$$\langle n, f\rangle = \int_R \gamma(f^s)c(s)\,ds \tag{5.6}$$

if n is defined by (4.3), and

$$\langle n, f\rangle = \int_R n_t(l)c(t)\,dt \tag{5.7}$$

if f is defined by (4.4).

5.3. With each interval I, we associate a sub-σ-algebra $\mathcal{F}(I)$ of the σ-algebra $\mathcal{F}$ generated by the sets (5.1) with $t \in I$, $B \in \mathcal{B}$. We use the following abbreviations

$$\mathcal{F}_{<t} = \mathcal{F}(-\infty, t), \qquad \mathcal{F}_{\leq t} = \mathcal{F}(-\infty, t], \qquad \mathcal{F}_{>t} = \mathcal{F}(t, +\infty),$$

$$\mathcal{F}_{\geq t} = \mathcal{F}[t, +\infty), \qquad \mathcal{F}_{<t+} = \bigcap_{u>t} \mathcal{F}_{<u}, \text{ etc.}$$

We put $A \in \mathcal{F}_\alpha$ if $\{A, \alpha < t\} \in \mathcal{F}_{<t}$ for all $t \in R$ and $A \in \mathcal{F}_\beta$ if $\{A, \beta > t\} \in \mathcal{F}_{>t}$ for all $t \in R$, and we call $\mathcal{F}_\alpha$ and $\mathcal{F}_\beta$ the *germ σ-algebras at time α and β*.

5.4. We need the following properties of the measure P_n^f.

5.4.A. If $f \in TH$, $q \in TH^+$, $n \in TM$, then for each $Y \in \mathcal{F}_{\leq t}$

$$P_n^f Y 1_{\alpha<t<\beta} = P_n^q Y f^t(x_t)/q^t(x_t).^4$$

If $A \in \mathcal{F}_{\leq t}$ and $P_n^q(A) = 0$, then $P_n^f(A) = 0$. The same is true for $A \in \mathcal{F}_\alpha$.

5.4.B. Let $f \in TH^+$. If $X \in \mathcal{F}_{\leq s}$, $Y \in \mathcal{F}_{>s}$, then

$$P_n^f(X 1_{\alpha<s<\beta} Y) = P_n^f\big(X f^s(x_s)^{-1} P_{s,x_s}^f Y\big).$$

5.4.C. Let $f \in TH^+$. A measure P on $(\Omega, \mathcal{F})$ coincides with one of the measures P_n^f, $n \in TM$ if and only if

$$P\{x_t \in B | \mathcal{F}_{\leq s}\} = p^f(s, x_s; t, B) \quad \text{a.s. } P \text{ on } \{\alpha < s < \beta\} \tag{5.8}$$

for all $s < t \in R$ and all $B \in \mathcal{B}$.

5.4.D. A restriction of a measure P_n^f, $n \in TM$, $f \in TH^+$, to an arbitrary set $A \in \mathcal{F}_\alpha$ is again a measure of the form $P_{\tilde{n}}^f$ for some $\tilde{n} \in TM$.

Properties 5.4.A and 5.4.B follow directly from the definition of the measures P_n^f. The necessity of (5.8) follow from 5.4.B and the sufficiency was proved in [6] (see Theorem 3.1). To prove 5.4.D it is sufficient to check that $P(d\omega) = 1_A(\omega)P_n^f(d\omega)$ satisfies (5.8). This is easy to do using 5.4.B.

---

[4] We omit the factor $1_{\alpha<t<\beta}$ since $f^t(x_t)/q^t(x_t)$ is not defined outside the set $\{\alpha < t < \beta\}$.

5.5. The shift $\omega' = \theta_s \omega$ of the path ω is given by the formula $\omega'(t) = \omega(t + s)$, $\alpha - s < t < \beta - s$. We put $\theta_s Y(\omega) = Y(\theta_s \omega)$ for each function $Y(\omega)$. Obviously $\theta_s \alpha = \alpha - s$, $\theta_s \beta = \beta - s$ and the σ-algebras $\mathfrak{F}_\alpha$ and $\mathfrak{F}_\beta$ are invariant with respect to θ_s.

We put $A \in \mathfrak{F}^0_\alpha$ if $A \in \mathfrak{F}_\alpha$ and $\theta_s 1_A = 1_A$ a.s. P^f_m for all $s \in R$, $m \in M$, $f \in TH$.

Let the transition function p be stationary.

5.5.A. The formulae

$$(k_s n)_t = n_{t+s} \qquad (k_s f)^t = f^{t+s}$$

define transformations k_s of the classes TM and TH, and

$$P^f_n(\theta_s Y) = P^{k_s f}_{k_s n} Y \tag{5.9}$$

for all $Y \in \mathfrak{F}$.

5.5.B. A restriction of a measure P^f_m, $m \in M$, $f \in TH^+$, to a set $A \in \mathfrak{F}^0_\alpha$ is a measure of the form $P^f_{\tilde{m}}$ where $\tilde{m} \in M$. If m is a minimal element of M, then P^f_m is trivial on $\mathfrak{F}_\alpha$ (i.e., each $\mathfrak{F}_\alpha$-measurable function is constant P^f_m-a.e.).

The statement 5.5.A is an implication of formula (5.2), and 5.5.B follows from 5.4.D, 5.4.A and 5.5.A.

6. Three lemmas.

6.1. In this section we prove three lemmas which make possible the computation of dissipative minimal elements. In the first lemma the behaviour of the ratio of two time-dependent excessive functions along a path is studied. The second one establishes a fundamental identity involving two time-dependent excessive functions and the ratio of their integrals with respect to t. The third lemma gives an approximation of the birth time α by stationary stopping times.

6.2. We denote by TK the class of all measures P^f_n, $n \in TM$, $f \in TH$. It follows from Subsection 4.6 that the classes TK corresponding to the transition functions p and p^q are identical for each $q \in TH^+$.

Let $Y_t(\omega)$ be a positive function defined for all $\omega \in \Omega$, $\alpha(\omega) < t < \beta(\omega)$. We say that Y_{t+} is a right TK-modification of Y_t if $Y_{t+} \in \mathfrak{F}_{<t+}$ and if, for each countable everywhere dense subset Λ of R,

$$Y_{t+} = \lim_{\substack{r \downarrow t \\ s \in \Lambda}} Y_r \quad \text{for all } t \in (\alpha, \beta) \text{ a.s. } TK. \tag{6.1}$$

The left TK-modification Y_{t-} of Y_t is defined in an analogous way.

LEMMA 6.1. *There exist a right TK-modification Y_{t+} and a left TK-modification Y_{t-} of the function*

$$Y_t = f^t(x_t)/q^t(x_t) \tag{6.2}$$

for every $f \in TH$, $q \in TH^+$. For each $P \in TK$,

$$Y_{t+} = Y_t = Y_{t-} \quad a.s.\ P \ \text{ on } \{\alpha < t < \beta\} \tag{6.3}$$

for all t except at most a countable set (depending on P).

PROOF. Fix Λ and denote by $N(v_1, v_2; s, u)$ the number of upcrossings of $[v_1, v_2]$ by Y_t over the set $\Lambda \cap (s, u)$, i.e., the maximal positive integer k such that there exist $s_1 < u_1 < \cdots < s_k < u_k \in \Lambda \cap (s, u)$ with the property $Y_{s_1}, \ldots, Y_{s_k} \leq v_1$, $Y_{u_1}, \ldots, Y_{u_k} \geq v_2$. Put $A(s, u) = \{\alpha < s, \beta > u, N(v_1, v_2; s, u) = \infty$ for some $v_1 < v_2\}$. The existence of Y_{t+} and Y_{t-} and the equality (6.3) will be proved if we show that $P_n^h A(s, u) = 0$ for all $s < u$, $h \in TH$, $n \in TM$ (cf. [1, Theorem 11.2]). By 5.4.A, it suffices to check this only for $h = q$, and since P_n^q is σ-finite on the σ-algebra $\mathcal{F}_{<s}$, it is sufficient to prove that

$$\int_{A(s,u)} Z P_n^q(d\omega) = 0 \tag{6.4}$$

for each P_n^q-integrable $Z \in \mathcal{F}_{<s}$. By 5.4.A and (6.2)

$$P_n^q Z 1_{\alpha<s} X Y_t = P_n^f Z 1_{\alpha<s,t<\beta} X \quad \text{for all } s \leq t, X \in \mathcal{F}_{<t}. \tag{6.5}$$

Put $P'(d\omega) = 1_{\alpha<s} Z\, P_n^q(d\omega)$. It follows from (6.5) that $(Y_t, \mathcal{F}_{<t}, P')$ is a supermartingale on $[s, \infty)$. By Doob's inequality

$$P' N(v_1, v_2; s, u) \leq (v_2 - v_1)^{-1} P'(Y_s + v_1)$$

which means that

$$P_n^q Z\, 1_{\alpha<s} N(v_1, v_2; s, u) \leq (v_2 - v_1)^{-1}\left(P_n^f Z\, 1_{\alpha<s<\beta} + v_1 P_n^q Z\, 1_{\alpha<s}\right).$$

This implies (6.5).

6.3. Now we suppose that the transition function p is stationary. It follows from (5.9) that, if all measures P of TK vanish on $A \in \mathcal{F}$, then they vanish on all sets $\theta_s A$, $s \in R$.

Let

$$\theta_s Y_t = Y_{t+s} \quad \text{for all } s \text{ and } t \tag{6.6}$$

and let Y_{t+} be a right TK-modification of Y_t. Then both $\theta_s Y_{t+}$ and Y_{t+s+} are right TK-modifications of Y_{t+s}. Hence

$$\theta_s Y_{t+} = Y_{t+s+} \quad \text{for all } t \text{ a.s. } TK. \tag{6.7}$$

If $m \in M$, $h \in H$, then, by (5.9), $P_m^h\{Y_t \neq Y_{t+}\}$ is independent of t. It follows from (6.3) that

$$P_n^h\{Y_t \neq Y_{t+}\} = 0 \quad \text{for all } t. \tag{6.8}$$

6.4. A function $\tau(\omega)$ is called a *stopping time* if $\alpha(\omega) < \tau(\omega) \leq \beta(\omega)$ for each ω and $\{\tau < t\} \in \mathcal{F}_{<t}$ for every $t \in R$. If $\theta_s \tau = \tau - s$ a.s. TK for all $s \in R$, we say that τ is *stationary*.

We put $A \in \mathcal{F}_\tau$ if $A \in \mathcal{F}$ and $\{A, \tau < t\} \in \mathcal{F}_{<t}$ for all $t \in R$, and we put $A \in \mathcal{F}_\tau^\theta$ if, in addition, $\theta_s A = A$ a.s. TK for all s.

LEMMA 6.2. *Let $m \in M, f \in TH^+, h \in TH$ and let functions*

$$\bar{f}(x) = \int_R f^t(x)\, dt, \qquad \bar{h}(x) = \int_R h^t(x)\, dt$$

be finite m-a.e. Suppose that τ is a stationary stopping time, $X \in \mathcal{F}_\tau^\theta$ and Y_{t+} is a right TK-modification of

$$Y_t = \bar{h}(x_t)/\bar{f}(x_t). \tag{6.9}$$

Then

$$P_m^f X Y_{\tau+} = P_m^h X 1_{\tau<\beta}. \tag{6.10}$$

PROOF. Put

$$F(\delta) = P_m^f X \int_\tau^{\tau+\delta} Y_{t+}\, dt, \qquad \delta > 0,$$

$$Z_t = X 1_{\tau<0<\tau+\delta} Y_{0+} f'(x_0)/\bar{f}(x_0). \tag{6.11}$$

We have

$$F(\delta) = \int_R P_m^{\bar{f}}(\theta_t Z_t)\, dt = \int_R P_m^{\bar{f}} Z_t\, dt$$

and, by Fubini's theorem,

$$F(\delta) = P_m^{\bar{f}} \int_R Z_t\, dt = P_m^{\bar{f}} X 1_{\tau<0<\tau+\delta} Y_{0+}.$$

By (6.8), $Y_{0+} = Y_0$ a.s. $P_m^{\bar{f}}$, and by 5.4.A and (6.9)

$$F(\delta) = P_m^{\bar{h}} X 1_{\tau<0<\tau+\delta}.$$

The right side is independent of f. Hence $F(\delta)$ does not change if we replace f by h, and, by (6.11), (6.9),

$$P_m^f X \int_\tau^{\tau+\delta} Y_{t+}\, dt = P_m^h X \int_\tau^{\tau+\delta} 1_{\alpha<t<\beta}\, dt.$$

Dividing by δ and tending δ to 0, we get (6.10).

6.5. LEMMA 6.3. *If $m \in M$ is dissipative, then there exists a sequence of stationary stopping times τ_n such that*

$$\tau_n \downarrow \alpha \quad a.s.\ P_m^h \tag{6.12}$$

for all $h \in TH$.

PROOF. Fix $l \in L_+^1(m)$ such that $m(l) = 1$ and put

$$q(x) = \int_0^\infty T_t l(x)\, dt = g_x(l), \qquad q_1(x) = \int_0^\infty e^{-t} T_t l(x)\, dt,$$

$$f(x) = g_x(q_1).$$

We have $f, q \in TH^+$ and $1 - f/q = q_1/q > 0$. Denote by Y_{t+} the right modification of $Y_t = f(x_t)/q(x_t)$, and put

$$a_u = \int_\alpha^{u\wedge\beta} (1 - Y_{t+})\, dt.$$

We prove that

6.5.A. For P_m^q-almost all ω,

(i) $0 < a_{u_1} < a_{u_2}$ for all $\alpha < u_1 < u_2 < \beta$,

(ii) $a_u < \infty$ for all $\alpha < u < \beta$,

(iii) a_u is continuous in u and $a_u \downarrow 0$ as $u \downarrow \alpha$.

6.5.B. $a_u \in \mathcal{F}_{<u}$.

6.5.C. $\theta_s a_u = a_{u+s}$ for all u a.s. TK.

First, we remark that $q - f = q_1 > 0$ on $E \setminus E_0$ and $q < \infty$ m-a.e. By Lemma 6.1 there exists an at most countable set Δ such that

$$1 - Y_{t+} = 1 - Y_t = q_1(x_t)/q(x_t) > 0 \text{ a.s. } P_m^q \text{ on } \{\alpha < t < \beta\} \text{ for each } t \,\overline{\in}\, \Delta \tag{6.13}$$

which implies 7.1.A, (i). We have

$$P_m^q a_u \frac{q_1(x_u)}{q(x_u)} = \int_0^\infty F(s)\, ds$$

where

$$F(s) = P_m^q \frac{q_1(x_{u-s})}{q(x_{u-s})} \frac{q_1(x_u)}{q(x_u)} = m\left(\frac{q_1}{s} T_s q_1\right).$$

Therefore

$$P_m^q a_u \frac{q_1(x_u)}{q(x_u)} = m\left(\frac{q_1 f}{q}\right) \leqslant m(q_1) \leqslant m(l) = 1.$$

Taking into account (6.13), we have 6.5.A, (ii). The property (iii) is an obvious implication of (ii), and 6.5.B, C follow from the fact that $Y_{t+} \in \mathcal{F}_{<t+}$ and (6.7).

For each $\varepsilon > 0$, we set

$$\tau_\varepsilon = \inf\{t \colon a_t > \varepsilon\}, \quad \tau_\varepsilon = \beta \text{ if } a_t \leqslant \varepsilon \text{ for all } t.$$

This is a stationary stopping time. Indeed, $\{\tau_\varepsilon < t\} = \{a_t > \varepsilon\} \cup \{\beta < t\} \in \mathcal{F}_{<t}$ by 6.5.B, and $\theta_s \tau_\varepsilon = \tau_\varepsilon - s$ a.s. TK by 6.5.C. If $\varepsilon_n \downarrow 0$, then $\tau_{\varepsilon_n} \downarrow \alpha$ a.s. P_m^q, and (6.12) follows from 5.4.A.

7. Dissipative minimal elements.

7.1. THEOREM 7.1. *Let m be a dissipative element of M, $f \in TH^+$, $h \in TH$ and let $\langle m, f\rangle < \infty$, $\langle m, h\rangle < \infty$. Suppose that the functions*

$$\bar{f}(x) = \int_R f^t(x)\, dt, \qquad \bar{h}(x) = \int_R h^t(x)\, dt \tag{7.1}$$

are finite m-a.e. Then there exists a function $Y_{\alpha+} \in \mathcal{F}_\alpha^\theta$ such that

$$Y_{\alpha+} = \lim_{\substack{s \downarrow \alpha \\ r \in \Lambda}} \left(\bar{h}(x_r)/\bar{f}(x_r)\right) \quad a.s.\ P_m^q \tag{7.2}$$

for each countable everywhere dense subset Λ of R and each $q \in TH$. Moreover

$$P_m^f X Y_{\alpha+} = P_m^h X \quad \text{for all } X \in \mathcal{F}_\alpha^\theta \tag{7.3}$$

and, if m is minimal, then

$$Y_{\alpha+} = \langle m, h\rangle/\langle m, f\rangle \quad a.s.\ P_m^f. \tag{7.4}$$

PROOF. If τ is a stationary stopping time, then so are $\tau + u$ for all $u > 0$. By Lemma 6.2, if $s < t$, $X \in \mathcal{F}^{\theta}_{\tau+s}$, then

$$P^f_m X Y_{\tau+t+} = P^h_m X 1_{\tau+t<\beta} \leq P^h_m X 1_{\tau+s<\beta} = P^f_m X Y_{\tau+s+}$$

where Y_t is defined by (6.9). Hence

$$(Y_{\tau+t+}, \mathcal{F}^{\theta}_{\tau+t+}, P^f_m)_{t>0}$$

is a supermartingale. It follows from (6.10) that $P^f_m Y_{\tau+} \leq \langle m, h\rangle$. Let N_τ be the number of upcrossings of $[v_1, v_2]$ by Y_{t+} over (τ, ∞). Since Y_{t+} is right-continuous in t a.s. P^f_m, we have

$$P^f_m N_\tau \leq (v_2 - v_1)^{-1}(\langle m, h\rangle + v_1\langle m, f\rangle).$$

Applying this inequality to the sequence τ_n and passing to the limit, we get the inequality $P^f_m N_\alpha < \infty$. Put $\Omega' = \{N_\alpha < \infty\}$, $\Omega'' = \Omega \setminus \Omega'$. Obviously Ω' and Ω'' belong to $\mathcal{F}^{\theta}_\alpha$ and $P^f_m(\Omega'') = 0$. Put $Y_{\alpha+} = \lim_{t\downarrow\alpha} Y_{t+}$ on Ω', $Y_{\alpha+} = 0$ on Ω''. Then $Y_{\alpha+} \in \mathcal{F}^{\theta}_\alpha$ and (7.2) holds for $q = f$. By 5.4.A it holds for all $q \in TH$.

By Fatou's lemma, it follows from (6.10) that

$$P^f_m Y_{\alpha+} \leq \lim P^f_m Y_{\tau_n+} = \lim P^h_m 1_{\tau_n<\beta} = \langle m, h\rangle < \infty.$$

Hence $Y_{\alpha+} < \infty$ a.s. P^f_m and, by 5.4.A, $Y_{\alpha+} < \infty$ a.s. P^h_m as well. Thus, for each $\varepsilon > 0$, there exists c_ε such that $P^h_m\{Y_\alpha > c_\varepsilon\} \leq \varepsilon$, and, by (6.10)

$$P^f_m 1_{Y_{\alpha+}>c_\varepsilon} Y_{\tau_n+} \leq P^h_m\{Y_{\alpha+} > c_\varepsilon\} \leq \varepsilon.$$

Hence Y_{τ_n+} are uniformly integrable relative to P^f_m and (6.10) implies (7.3).

If m is minimal, then, by 5.5.B, there exists a constant C such that $Y_{\alpha+} = C$ a.s. P^f_m. We have $P^f_m Y_{\alpha+} = CP^f_m(\Omega) = C\langle m, f\rangle$ and, by (7.3), $C = \langle m, h\rangle / \langle m, f\rangle$. This proves (7.4).

7.2. We apply Theorem 7.1 to the excessive functions

$$h^s(x) = \int p(s, x; t, dy)\varphi(y) = T_{t-s}\varphi(x) \tag{7.5}$$

and

$$f^s(x) = \int_s^\infty dt\, c(t) \int p(s, x; t, dy)\psi(y) = \int_0^\infty T_u\psi(x)c(s-u)\, du \tag{7.6}$$

where $\psi > 0$, $c(t) > 0$ and $\int_R c(t)\, dt = 1$. Obviously $f \in TH^+$,

$$\langle m, f\rangle = m(\psi), \qquad \langle m, h\rangle = m(\varphi), \qquad \bar{f}(x) = g_x(\psi), \qquad \bar{h}(x) = g_x(\varphi).$$

Hence, in this case,

$$Y_{\alpha+} = \lim_{\substack{r\downarrow\alpha \\ r\in\Lambda}} (g_{x_r}(\varphi)/g_{x_r}(\psi)) \quad \text{a.s. } P^q_m \tag{7.7}$$

and we arrive at the following result.

THEOREM 7.2. *Let m be a dissipative minimal element of M. Then for all $\varphi \in L^1(m)$, $\psi \in L^1_+(m)$, $q \in TH$*

$$m(\varphi)/m(\psi) = \lim_{\substack{r\downarrow\alpha \\ r\in\Lambda}} (g_{x_r}(\varphi)/g_{x_r}(\psi)) \quad a.s.\ P^q_m. \tag{7.8}$$

(The meaning of Λ is the same as in Theorem 7.1.)

7.3. COROLLARY. *If*

$$m = \int_R \nu_s \, ds \tag{7.9}$$

where $\nu \in TM$, *then* (7.8) *holds a.s.* P_ν^q.

Indeed, the relation (7.9) implies that $m = \int_R k_s\nu \, ds$. Let Ω' and Ω'' be the sets defined in the proof of Theorem 7.1. We have

$$0 = P_m^f(\Omega'') = \int_R P_{k_s\nu}^f(\Omega'') \, ds$$

and, by (5.9),

$$P_{k_s\nu}^f(\Omega'') = P_\nu^f(\theta_s\Omega'') = P_\nu^f(\Omega'').$$

Hence $P_\nu^f(\Omega'') = 0$, and (7.8) holds a.s. P_ν^f. By 5.4.A, it holds a.s. P_ν^q.

REMARK. It has been proved in [6] that each $m \in TM$ has an integral representation $m_s(B) = \int n_s(B)\mu(dn)$ where μ is a finite measure on the space of minimal elements of TM (which are entrance laws). It has been proved also that if m is a minimal element of M and if $l > 0$, $m(l) = 1$, then for each $\varphi \in L^1(m)$ and μ-almost all n

$$n_I(\varphi)/n_I(l) \to m(\varphi) \quad \text{as } I \uparrow R$$

where $n_I(\varphi)$ means the integral of $n_s(\varphi)$ over a finite interval I. If m is a null-excessive element (i.e., if $mT_t \to 0$ as $t \to \infty$), then μ is concentrated on the entrance laws at finite times and $n_R(l) < \infty$. Hence $m(\varphi) = n_R(\varphi)/n_R(l)$, and (7.9) holds with $\nu_s = n_R(l)^{-1}n_s$. It holds also for all entrance laws $k_u\nu$. Thus (7.8) is satisfied a.s. P_ν^q for some entrance law ν at time 0. This justifies the construction described in Subsection 2.7.

7.4. To investigate dissipative elements of H we introduce a backward transition function

$$\hat{p}(s, dx; t, y) = \gamma(dx)\rho_{t-s}(x, y) \tag{7.10}$$

and we denote by $T\hat{M}$, $T\hat{H}$ the corresponding classes of time dependent excessive measures and functions. The notations $\hat{M}$, $\hat{N}$, $T\hat{M}^+$ have an analogous meaning. To each $m \in T\hat{M}, f \in T\hat{H}$ there corresponds a measure $\hat{P}_m^f$, and we set $\langle m, f\rangle' = \hat{P}_m^f(\Omega)$.

Considering reversed time direction, we get the following version of Theorem 7.1.

THEOREM 7.3. *Let m be a dissipative minimal element of $\hat{M}$ and let $f \in T\hat{H}^+$, $q \in T\hat{H}$, $\langle m, f\rangle' < \infty$, $\langle m, q\rangle' < \infty$,*

$$\bar{f}(y) = \int_R f^s(y) \, ds < \infty, \qquad \bar{q}(y) = \int_R q^s(y) \, ds < \infty \quad m\text{-a.e.}$$

Then there exists a function $Y_{\beta-} \in \mathcal{F}_\beta^0$ such that

$$Y_{\beta-} = \lim_{\substack{r \uparrow \beta \\ r \in \Lambda}} (\bar{q}(x_r)/\bar{f}(x_r)) \quad a.s.\ \hat{P}_m^f$$

for each countable everywhere dense Λ. Moreover

$$\hat{P}^f_m X\, Y_{\beta-} = \hat{P}^q_m X \quad \text{for all } X \in \mathcal{F}^0_\beta$$

and, if m is minimal, then

$$Y_{\beta-} = \langle m, q\rangle' / \langle m, f\rangle' \quad \text{a.s. } \hat{P}^f_m.$$

7.5. The following result follows immediately from Theorem 7.3.

THEOREM 7.4. *Let h be a dissipative minimal element of H. If $f \in TM$ and if h is integrable with respect to measures ξ and η, then*

$$\xi(h)/\eta(h) = \lim_{\substack{r \uparrow \beta \\ r \in \Lambda}} (\xi(g^{x_r})/\eta(g^{x_r})) \quad \text{a.s. } P^h_{\gamma f}. \tag{7.11}$$

To prove this statement, we apply Theorem 7.3 to the measure $m(dx) = \gamma^h(dx) = h(x)\gamma(dx)$ which, according to Subsection 3.5, is a minimal and dissipative element of $\hat{M}$, and to the functions

$$q^s(y) = \int \xi(dx)\rho_s(x, y), \quad f^s(y) = \int_0^\infty c(s+u)\,du \int \eta(dx)\rho_u(x, y).$$

By simple computations we get the formulae

$$\langle \gamma^h, q\rangle = \xi(h), \qquad \langle \gamma^h, f\rangle = \eta(h), \qquad \bar{q}(y) = \xi(g^y), \qquad \bar{f}(y) = \eta(g^y)$$

and we notice that the measure $\hat{P}^f_{\gamma^h}$ corresponding to $\hat{p}$, γ^h, f coincides with the measure $P^h_{\gamma f}$.

7.6. As in Subsection 7.3, we prove that, if

$$h = \int_R \varphi^t\, dt \tag{7.12}$$

where $\varphi \in TH$, then (7.11) is fulfilled a.s. $P^\varphi_{\gamma f}$. If h is null-excessive, then (7.12) holds for some exit law at time 0.

Appendix. Decomposition into minimal elements.

0.1. Let $p_t(x, B)$ be a stationary transitive function in a standard Borel space $(E, \mathcal{B})$ and let M and H be the corresponding classes of excessive measures and functions. For every $l \in \mathcal{B}$, we put $M^l = \{m : m \in M, m(l) = 1\}$ and we denote by $\mathcal{B}(M^l)$ the σ-algebra in M^l generated by the sets $\{m : m \in M^l, m(B) < u\}$, $B \in \mathcal{B}$, $u \in R$. For every measure ν on $\mathcal{B}$, we put $H^\nu = \{h : h \in H, \nu(h) = 1\}$ and we denote by $\mathcal{B}(H^\nu)$ the σ-algebra in H^ν generated by the sets $\{h : h \in H^\nu, \xi(h) < u\}$, ξ is a measure on $\mathcal{B}$, $u \in R$. Our objective is to prove the following two results.

THEOREM 0.1. *Let Γ be the set of all minimal elements of M which belong to M^l. Suppose that $l \in \mathcal{B}$ is strictly positive. Then $\Gamma \in \mathcal{B}(M^l)$ and, for each $m \in M^l$, there exists one and only one probability measure μ on $\mathcal{B}(M^l)$ concentrated on Γ such that*

$$m(B) = \int_\Gamma n(B)\mu(dn) \quad \text{for all } B \in \mathcal{B}. \tag{0.1}$$

THEOREM 0.2. *Let Γ be the set of all minimal elements of H belonging to H^ν. Let ν be a σ-finite measure with the property*:

0.1.A. *There exists an excessive reference measure γ and a strictly positive function l such that*

$$\gamma(lh) \leqslant \nu(h) \quad \text{for all } h \in H. \tag{0.2}$$

Then $\Gamma \in \mathfrak{B}(H^\nu)$ and, for every $h \in H^\nu$, there exists one and only one probability measure μ on $\mathfrak{B}(H^\nu)$ concentrated on Γ such that

$$h(x) = \int_\Gamma f(x)\mu(df) \quad \text{for all } x \in B. \tag{0.3}$$

0.2. *Comments to Theorem* 0.2.

(a) If γ is a reference measure and if $h \in H$, then $h < \infty$ a.s. γ. If $h \neq 0$, then $\gamma(h) > 0$ and $\gamma(lh) = 1$ for some strictly positive function l. The measure $\nu(dx) = l(x)\gamma(dx)$ obviously satisfies 0.1.A, and $h \in H^\nu$. Hence each $h \in H$ can be decomposed into minimal elements. (The analogous statement for $m \in M$ is obvious.)

(b) Let γ satisfy condition 2.2.C and let g^y be the corresponding Green function defined by (2.14). Suppose that the function $q(y) = \nu(g^y)$ is finite γ-a.e. and strictly positive. Then the measure ν has property 0.1.A. In fact,

$$\gamma_1(B) = \int_E \nu(dx)g_x(B) = \int_B q(y)\gamma(dy)$$

is an excessive reference measure and $\gamma_1(lh) \leqslant \nu(h)$ for all $h \in H$ where

$$l(y) = q(y)^{-1}\int_0^\infty dt\, e^{-t}\int_E \nu(dx)p_t(x, y)$$

is strictly positive.

(c) In case of Brownian motion remark (b) is applicable to Lebesgue measure γ and any measure ν concentrated at one point.

(d) If γ is a reference measure, the σ-algebra $\mathfrak{B}(H^\nu)$ is generated by the sets $\{h : h \in H^\nu, \gamma(lh) < u\}$, $l \in \mathfrak{B}$, $u \in R$. This follows from the relation

$$\xi(h) = \lim_{t\downarrow 0} \gamma(l_t h) \quad \text{where } l_t(y) = \int \xi(dx)p_t(x, y).$$

(e) The function $f(x)$, $f \in H$, $x \in E$ is $\mathfrak{B}(H^\nu) \times \mathfrak{B}$-measurable. In fact, it is easy to see that $\int r(x, y)\gamma(dy)f(y)$ is $\mathfrak{B}(H^\nu) \times \mathfrak{B}$-measurable if $r(x, y)$ is $\mathfrak{B} \times \mathfrak{B}$-measurable, and measurability of $f(x)$ follows from the formula $f(x) = \lim_{t\downarrow 0} \int p_t(x, y)\gamma(dy)f(y)$.

(f) By remark (e) and the Fubini theorem, it follows from (0.3) that $\xi(h) = \int_\Gamma \xi(f)\mu(df)$ for each measure ξ.

0.3. Theorem 0.2 follows easily from Theorem 0.1. First, if $\nu(dx) = l(x)\gamma(dx)$ where γ is an excessive reference measure and $l > 0$, then the mapping (3.20) establishes a 1-1 correspondence between H^ν and $\hat{M}^l$ (we use the notations of Subsection 3.5). Obviously this mapping is measurable. By 0.2.d the inverse mapping is also measurable, and the representation (0.1) of the measure $m = \gamma^h \in \hat{M}^l$ is equivalent to the representation (0.3) of $h \in H^\nu$.

Now let ν satisfy condition 0.1.A and let $\nu_1(dx) = l(x)\gamma(dx)$. We note that $0 < \nu_1(h) < 1$ for all $h \in H^\nu$. Put $H_0^{\nu_1} = \{h : h \in H^{\nu_1}, \nu(h) < \infty\}$. Formulae $F_1(h) = h/\nu_1(h)$ and $F(h) = h/\nu(h)$ determine inverse measurable mappings F_1 of H^ν onto $H_0^{\nu_1}$ and F_2 of $H_0^{\nu_1}$ onto H^ν. Evidently they preserve minimal elements. Therefore to prove Theorem 0.2 for the measure ν, it is sufficient to check that each element $h \in H_0^{\nu_1}$ has a unique integral representation through minimal elements of H which lie in $H_0^{\nu_1}$. Since Theorem 0.2 has been proved for the measure ν_1, there corresponds to every $h \in H^{\nu_1}$ a unique probability measure μ on $\mathfrak{B}(H^{\nu_1})$ concentrated on the set Γ_1 of minimal elements such that

$$h(x) = \int_{\Gamma_1} f(x)\mu(df).$$

By 0.2.f this implies an equality

$$\nu(h) = \int_{\Gamma_1} \nu(f)\mu(df)$$

and since $\nu(h) < \infty$, the measure μ is concentrated on $\Gamma_1 \cap H_0^{\nu_1}$.

0.4. The rest of the Appendix is devoted to proving the following statement.

THEOREM 0.3. *Let $l > 0$ and let*

$$f^s(x) = \int_0^\infty T_u l(x) c(s-u)\, ds \tag{0.4}$$

where $c(t) > 0$ and $\int_R c(t)\, dt = 1$. Let $\mathfrak{F}_\alpha^\beta$ be the σ-algebra defined in Subsection 5.5. There exists an M^l-valued function n^ω on Ω with the properties:

0.4.A. *$n^\omega(B)$ is $\mathfrak{F}$-measurable for every $B \in \mathfrak{B}$.*

0.4.B. *$P_m^f\{Z|\mathfrak{F}_\alpha^\beta\} = P_{n^\omega}^f Z$ a.s. P_m^f for all $Z \in \mathfrak{F}$, $m \in M^l$.*

Properties 0.4.A, B mean that $\mathfrak{F}_\alpha^\beta$ is an H-sufficient σ-algebra for the class K of probability measures P_m^f, $m \in M^l$, and Theorem 0.1 follows from Theorem 0.3 because of the general relation between H-sufficient statistics and minimal elements established in [7] (see Theorem 3.1).

0.5. In each standard Borel space $(E, \mathfrak{B})$ there exists a support system W i.e., a countable family of positive bounded functions with the properties:

0.5.A. If μ_n is a sequence of probability measures on $\mathfrak{B}$ and if $\lim \mu_n(\varphi) = q(\varphi)$ exists for every $\varphi \in W$, then there is a probability measure μ such that $\mu(\varphi) = q(\varphi)$ for all $\varphi \in W$.

0.5.B. If a class $\mathfrak{S}$ of positive functions contains W and is closed under addition, multiplication by positive constants and if $\mathfrak{S}$ contains together with each increasing sequence its limit, then $\mathfrak{S}$ contains all functions $\varphi \in \mathfrak{B}$.

We put $x \in E'$ if the limit

$$\lim_{u\to\infty} (g_x^u(\varphi l)/g_x^u(l)) \tag{0.5}$$

exists for all $\varphi \in W$. By 0.5.A for each $x \in E'$ there exists a probability measure μ_x such that the limit (0.5) coincides with $\mu_x(\varphi)$. We put $\varphi' = \varphi l^{-1}$, $n_x(dy) = \mu_x(dy)l^{-1}(y)$. Obviously $n_x(l) = 1$ and

$$\lim_{u\to\infty} (g_x^u(\varphi)/g_x^u(l)) = n_x(\varphi) \quad \text{if } x \in E', \varphi' \in W. \tag{0.6}$$

Let Λ_0 be the set of all rational numbers. We put $\omega \in \Omega'$ if $x_r \in E'$ for all $r \in \Lambda_0$ and if

$$\lim_{\substack{r \downarrow \alpha \\ r \in \Lambda_0}} n_{x_r}(\varphi)$$

exists for all $\varphi^l \in W$. Again by 0.5.A, for every $\omega \in \Omega'$, there exists a measure $n^\omega(\varphi)$ such that $n^\omega(l) = 1$ and

$$n^\omega(\varphi) = \lim_{\substack{r \downarrow \alpha \\ r \in \Lambda_0}} n_{x_r}(\varphi) \quad \text{for all } \omega \in \Omega', \varphi^l \in W. \tag{0.7}$$

We shall see that E' and Ω' are not empty (except the case where M^l is empty, in which case our theorem is trivial). Let

$$E'' = E \setminus E', \qquad \Omega'' = \Omega \setminus \Omega'.$$

Fix arbitrary points $x' \in E'$ and $\omega' \in \Omega'$ and put $n_x = n_{x'}$ for $x \in E''$, $n^\omega = n^{\omega'}$ for $\omega \in \Omega''$.

The function $n^\omega(\varphi)$ is $\mathcal{F}$-measurable if $\varphi^l \in M$. By 0.5.B, the same is true for all $\varphi \in \mathcal{B}$. Therefore n^ω satisfies the condition 0.4.A. Theorem 0.3 will be proved if we show that, for each $m \in M^l$,

$$P_m^f\{n^\omega \,\overline{\in}\, M\} = 0 \tag{0.8}$$

and 0.4.B is fulfilled. We check this separately for dissipative and conservative m.

0.6. Suppose that $m \in M^l$ is dissipative. Obviously (0.6) holds with

$$n_x(\varphi) = g_x(\varphi)/g_x(l), \tag{0.9}$$

if $g_x(l) < \infty$. Hence $E'' \subset \{x: g_x(l) = \infty\}$ and $m(E'') = 0$.

Let $\varphi^l \in W$. Comparing (0.7), (0.8) and (7.7) and applying Theorem 7.1, we conclude that $P_m^f(\Omega'') = 0$, that the function $n^\omega(\varphi)$ is $\mathcal{F}_\alpha^\beta$-measurable and that

$$P_m^f X n^\omega(\varphi) = P_m^h X \tag{0.10}$$

with $h^s(x) = T_{t-s}\varphi(x)$ for all $X \in \mathcal{F}_\alpha^\beta$, $t \in R$. By 5.4.B and 5.4.A, for $s < t$,

$$\begin{aligned} P_m^f\left[X 1_{\alpha<s} f^t(x_t)^{-1}\varphi(x_t)\right] &= P_m^f\left[X f^s(x_s)^{-1} P_{s,x_s}^f\left(\varphi(x_t) f^t(x_t)^{-1}\right)\right] \\ &= P_m^f\left[X f^s(x_s)^{-1} h^s(x_s)\right] = P_m^h(X 1_{\alpha<s<\beta}). \end{aligned}$$

Since $\beta = t$ a.s. P_m^h, we get by setting $s \uparrow t$

$$P_m^f\left[X f^t(x_t)^{-1}\varphi(x_t)\right] = P_m^h X. \tag{0.11}$$

It follows from (0.10) and (0.11) that

$$P_m^f f^t(x_t)^{-1}\varphi(x_t) X = P_m^f X n^\omega(\varphi). \tag{0.12}$$

Established for $\varphi^l \in W$, equality (0.12) can be extended for all $\varphi \in \mathcal{B}$ using 0.5.B.

Suppose that

$$\varphi(x) = P_{t,x}^f Z, \qquad Z \in \mathcal{F}_{>t}. \tag{0.13}$$

Since $X 1_{\alpha<t} \in \mathcal{F}_{<t}$, formula (0.12) and 5.4.B imply that

$$P_m^f X n^\omega(\varphi) = P_m^f X 1_{\alpha<t<\beta} Z. \tag{0.14}$$

We apply this formula to $t = -u < 0$ and $Z = \psi(x_0)/f^0(x_0)$ and we get

$$P_m^f X n^\omega(T_u\psi) = P_m^f X 1_{\alpha < -u} \frac{\psi(x_0)}{f^0(x_0)}. \tag{0.15}$$

On the other hand, it follows from (0.12) that

$$P_m^f X n^\omega(\psi) = P_m^f X \psi(x_0)/f^0(x_0). \tag{0.16}$$

Comparing (0.15) and (0.16) we have

$$n^\omega(T_u\psi) \leqslant n^\omega(\psi) \quad \text{a.s. } P_m^f \tag{0.17}$$

and

$$n^\omega(T_{1/k}\psi) \to n^\omega(\psi) \quad \text{a.s. } P_m^f. \tag{0.18}$$

The set $C = \{(u, \omega): n^\omega(T_u\psi) > n^\omega(\psi)$ for some $\psi' \in W\}$ is $\mathfrak{B}_R \times \mathfrak{F}$-measurable and all its u-sections have P_m^f measure zero. By Fubini's theorem, there exists a set $\Omega_0 \in \mathfrak{F}$ such that $P_m^f(\Omega \setminus \Omega_0) = 0$ and $(u, \omega) \overline{\in} C$ for almost all u and all $\omega \in \Omega_0$. Using 0.5.B and the semigroup property of T_t, we prove that $n^\omega(T_u\psi) \leqslant n^\omega(\psi)$ for all $\omega \in \Omega_0$ and all $u \in R$. Because of (0.18), there exists a set $\Omega_1 \subset \Omega_0$ such that $P_m^f(\Omega \setminus \Omega_1) = 0$ and $n^\omega(T_{1/k}\psi) \to n^\omega(\psi)$ for all $\omega \in \Omega_1$ and all $\psi' \in W$. Applying 0.5.A and 0.5.B, it is easy to prove that $n^\omega \in M$ for all $\omega \in \Omega_1$.

To prove 0.4.B, we establish that, for each $Z \in \mathfrak{F}$,

(i) $P_{n^\omega}^f Z$ coincides P_m^f-a.e. with an $\mathfrak{F}_\alpha^\theta$-measurable function,

(ii) $P_m^f XZ = P_m^f(X P_{n^\omega}^f Z)$ for all $X \in \mathfrak{F}_\alpha^\theta$.

It is sufficient to check this for $Z = 1_{\alpha < t} f(x_{t_1}, \ldots, x_{t_n}) 1_{u < \beta}$ where $t < t_1 < \cdots < t_n < u$ and f is a measurable function on E^n. It follows from (5.2) that in this case $P_{n^\omega}^f Z = n^\omega(\varphi)$ where φ is given by (0.13). Both conditions (i) and (ii) are satisfied (the second one follows from (0.14)).

0.7. Now let m be a conservative element of M^l. Then by Theorem 3.2, $m(E'') = 0$ and (0.6) holds with

$$n_x(\varphi) = m^l\left(\frac{\varphi}{l} \middle| \mathfrak{B}_m^l\right), \quad m\text{-a.e.} \tag{0.19}$$

The following two lemmas establish relations between the σ-algebra $\mathfrak{B}_m^T$ and $\mathfrak{F}_\alpha^\theta$.

Lemma 0.1. *If F is a bounded $\mathfrak{B}_m^T$-measurable function, then there exists a $\mathfrak{F}_\alpha$-measurable function Y_F such that, for each countable everywhere dense set Λ,*

$$Y_F = \lim_{\substack{r \downarrow -\infty \\ r \in \Lambda}} F(x_r). \tag{0.20}$$

For each $F \in \mathfrak{B}_m^T$ and each $\varphi \in \mathfrak{B}$, $t \in R$

$$P_m^f Y_F n^\omega(\varphi) = P_m^f Y_F \varphi(x_t) f^t(x_t)^{-1}. \tag{0.21}$$

Proof. Let Z and φ be as in (0.13). By 5.4.B and (5.2), we have for $s < t$, $\psi \in \mathfrak{B}$,

$$P_m^f F(x_s)\psi(x_t) Z = P_m^f F(x_s)\psi(x_t)\varphi(x_t) f^t(x_t)^{-1} = m[FT_{t-s}(\psi\varphi)]. \tag{0.22}$$

Analogously

$$P_m^f F(x_t)\psi(x_t) Z = m(F\psi\varphi). \tag{0.23}$$

Since $F \in \mathfrak{B}_m^T$, the terms on the right side of (0.22) and (0.23) coincide. This implies the relation

$$P_m^f\{F(x_s)|\mathfrak{F}_{>t}\} = F(x_t) \quad \text{a.s. } P_m^f \text{ on } \{\beta > t\}. \tag{0.24}$$

Hence, for each u, $(F(x_t)1_{\beta>u}, \mathfrak{F}_{>t}, P_m^f)$ is a martingale on $(-\infty, u)$ and the existence of Y_F follows from the theorem on convergence of a bounded martingale. By (0.19) and (3.8)

$$\int m(dx)F(x)n_x(\varphi)T_u l(x) = \int m(dx)n_x(\varphi)l(x)F(x) = m(F\varphi) = m(FT_{t-r}\varphi).$$

Therefore, by (0.4)

$$P_m^f F(x_r)n_{x_r}(\varphi) = \int m(dx)F(x)n_x(\varphi)f^r(x) = m(FT_{t-r}\varphi)$$
$$= P_m^f F(x_r)\varphi(x_t)f^t(x_t)^{-1}.$$

Letting $r \to -\infty$, we get (0.21), first, for $\varphi^l \in W$ and then, using 0.5.B, for all $\varphi \in \mathfrak{B}$.

LEMMA 0.2. *Every bounded function $Y \in \mathfrak{F}_\alpha^\theta$ coincides P_m^f-a.e. with Y_F for some $F \in \mathfrak{B}_m^T$.*

PROOF. We choose a function $F \in \mathfrak{B}$ such that $P_m^f\{Y|x_0\} = F(x_0)$ a.s. P_m^f. For all $t > 0$, $\psi \in \mathfrak{B}$, we have

$$m(FT_t\psi) = P_m^f F(x_0)\psi(x_t)f^t(x_t)^{-1} = P_m^f\{Y\psi(x_t)f^t(x_t)^{-1}\}. \tag{0.25}$$

Since $Y1_{\alpha<t} \in \mathfrak{F}_{<t}$ and $\theta_t Y = Y$ a.s. P_m^f, it follows from 5.5.A and 5.4.B that the right side of (0.25) does not depend on t. Hence $F \in \mathfrak{B}_m^T$.

Now, by 5.4.B, $P_m^f Y\psi(x_0)Z = P_m^f F(x_0)\psi(x_0)Z$ for all $\psi \in \mathfrak{B}$, $Z \in \mathfrak{F}_{>0}$. Hence $P_m^f\{Y|\mathfrak{F}_{>0}\} = F(x_0)$ a.s. P_m^f and, by (0.24)

$$P_m^f\{Y|\mathfrak{F}_{>t}\} = P_m^f\{F(x_0)|\mathfrak{F}_{>t}\} = F(x_t) \quad \text{a.s. } P_m^f.$$

Letting $t \to -\infty$, we see that $Y = P_m^f\{Y|\mathfrak{F}_\alpha\} = Y_F$ a.s. P_m^f.

0.8. Suppose that φ is given by formula (0.13). It follows from (0.21) and 5.4.B that

$$P_m^f Yn^\omega(\varphi) = P_m^f Y1_{\alpha<t<\beta}Z \tag{0.26}$$

for $Y = Y_F$. By Lemma 0.2, (0.26) holds for all $Y \in \mathfrak{F}_\alpha^\theta$. Formula (0.26) coincides with (0.14) and we establish (0.8) and 0.4.B in the same way as in Subsection 0.6.

REFERENCES

1. J. L. Doob, *Stochastic processes*, Wiley, New York, 1953.

2. ———, *Discrete potential theory and boundaries*, J. Math. and Mech. **8** (1959), 433–458.

3. E. B. Dynkin, *Exit space of a Markov process*, Uspehi Mat. Nauk **24** (148) (1969), 89–152. (English translation: Russian Math. Surveys **24** (1969), 89–157.)

4. ———, *Excessive measures and entrance laws for a Markov process*, Mat. Sb. **84** (126) (1971), 218–253. (English translation: Math. USSR-Sb. **13** (1971), 203–246.)

5. ———, *Initial and final behaviour of trajectories of Markov processes*, Uspehi Mat. Nauk **26** (160) (1971), 153–172. (English translation: Russian Math. Surveys **26** (1971), 165–185.)

6. ———, *Integral representation of excessive measures and excessive functions*, Uspehi Mat. Nauk **27** (163) (1972), 43–80. (English translation: Russian Math. Surveys **27** (1972), 43–84.)

7. ______, *Sufficient statistics and extreme points*, Ann. Probability **6** (1978), 705–730.
8. ______, *On duality for Markov processes*, Stochastic Analysis, A. Friedman and M. Pinsky, (eds.), Academic Press, New York, 1978, pp. 63–77.
9. G. A. Hunt, *Markov chains and Martin boundaries*, Illinois J. Math. **4** (1960), 313–340.
10. ______, *Transformation of Markov processes*, Proc. Internat. Congress Math., Stockholm, 1962, pp. 531–535.
11. H. Kunita and T. Watanabe, *Markov processes and Martin boundaries*, Part 1, Illinois J. Math. **9** (1965), 485–526.
12. S. E. Kuznecov, *Construction of Markov processes with random birth and death times*, Teor. Verojatnost. i Primenen. **18** (1973), 596–601. (English translation: Theor. Probability Appl.)
13. R. S. Martin, *Minimal positive harmonic functions*, Trans. Amer. Math. Soc. **49** (1941), 137–172.
14. J. Neveu, *Bases mathématiques du calcul des probabilités*, Masson, Paris, 1964.

Department of Mathematics, Cornell University, Ithaca, New York 14853